Fish and Fishery Products Analysis

Saleena Mathew • Maya Raman
Manjusha Kalarikkathara Parameswaran
Dhanya Pulikkottil Rajan

Fish and Fishery Products Analysis

A Theoretical and Practical Perspective

Springer

Saleena Mathew
School of Industrial Fisheries
Cochin University of Science and Technology
Kochi, Kerala, India

Manjusha Kalarikkathara Parameswaran
School of Industrial Fisheries
Cochin University of Science and Technology
Kochi, Kerala, India

Maya Raman
Department of Food Science and Technology, School of Ocean Science and Technology
Kerala University of Fisheries and Ocean Studies
Kochi, Kerala, India

Dhanya Pulikkottil Rajan
Department of Aquaculture
M.E.S Asmabi College
Kochi, Kerala, India

ISBN 978-981-32-9573-5 ISBN 978-981-32-9574-2 (eBook)
https://doi.org/10.1007/978-981-32-9574-2

This Springer imprint is published by the registered company Springer Nature Singapore Pte Ltd.
The registered company address is: 152 Beach Road, #21-01/04 Gateway East, Singapore 189721, Singapore

Contents

About the Authors

Saleena Mathew has got over 35 years of research experience of which 30 years of teaching at postgraduate level. She was Director, School of Industrial Fisheries, Cochin University of Science and Technology, for one term of 3 years from 2005 to 2008 and retired from the University as Professor in 2016. For the last 27 years of service, she was engaged in teaching and research in the field of fish processing technology and fish biochemistry at the School of Industrial Fisheries, Cochin University of Science and Technology, and catered trained personnel for the seafood industry. She is a Recognized Guide of Cochin University of Science and Technology for the last 25 years and served as Expert Member and Chairperson of various scientific and academic bodies of Cochin University of Science and Technology and as Expert Member for various programs of other universities and research institutes. She was appointed as Associate Member of The World Academy of Sciences (TWAS), Italy, in 1998, and has visited Argentina, Italy, and other foreign countries as part of her research work. She has coordinated and conducted various symposia, workshops, and training programs at the School in the field of fish processing, quality assurance, and food safety. She has got more than 50 research publications in the specialized areas of biochemistry, fish preservation, and bioactive substances from fish and fish discards.

Maya Raman is presently working as Associate Professor (Meat and Fish Processing), Kerala University of Fisheries and Ocean Studies, Cochin. She graduated in Botany and received her M.Sc. (Industrial Fisheries) and Ph.D. degrees from the School of Industrial Fisheries, Cochin University of Science and Technology, India, in 2006. Her doctoral research was in fish biochemistry and fish processing technology. She also worked as Postdoc at Applied Nutrition and Food Chemistry, Centre for Chemistry and Chemical Engineering, Lund University, with the scholarship from Erasmus Mundus External Cooperation Windows Lot 15 Program. She also worked as Women Scientist at Indian Institute of Technology Madras with the Women Scientist Scheme (WOS-A) funded by the Department of Science and Technology, India. She was selected as Young Student Speaker IFCON 2003 by the Association of Food Scientists and Technologists, India. She also worked as Faculty at the Department of Food Science and Technology, MACFAST, as Certified Technologist (Export Inspection Authority, India) at Koluthara Exports Limited, and as Research Officer at NGIL, India. She has 17 peer-reviewed publications to her credit and has published a book and 8 book chapters. She has also presented 14 papers at national and international symposia/conferences. Her area of expertise is fish biochemistry (collagen), fish processing technology, nutrition and health (cancer), nanotechnology, bioactive marine compounds, dietary fibers (cereal, vegetable, marine algae) and colon cancer (in vitro and in vivo), molecular modelling and dynamics, gut simulation techniques, drug encapsulation, hydrogels, antimicrobial biodegradable food wraps, etc. She is also Life Member of the AFST(I), SOFT(I), and SASNET, and member of ACS. She also qualified the National Eligibility Test (NET) conducted by the ASRB in 2001. A patent on hydrogel from cyclic beta-glucan for cosmetics/food applications has also been filed in 2016.

Manjusha Kalarikkathara Parameswaran graduated in Zoology at the St. Teresa's College Ernakulam and attained her M.Sc. in Biochemistry from School of Biosciences, Mahatma Gandhi University, in 2002. She qualified the joint UGC-CSIR National Eligibility Test (NET) in 2002 and was awarded Junior Research Fellowship (2004–2009). She has also qualified the NET (Plant Biochemistry) conducted by the ASRB-ICAR in 2018. She was awarded Ph.D. in Marine Sciences (Biochemistry) for the studies pertaining to the bioactive potential of glycosaminoglycan's in cephalopods from the School of Industrial Fisheries, Cochin University of Science and Technology (CUSAT), India, in 2011. She has worked as Faculty in Biochemistry at Mar Athanasius College, Kothamangalam; St. Mary's College for Women, Thiruvalla; School of Aquatic Food Products and Technology, Kerala University of Fisheries and Ocean Studies (KUFOS); and the Department of Marine Biology, Microbiology and Biochemistry, School of Marine Sciences, CUSAT. She is a Lifetime Member of the Society of Marine Biologists (SOMB) and Annual Member of the Society of Fishery Technologists India (SOFT-I). She has published three papers and presented six articles in various national and international conferences. Her research interests are marine bioactives and biopolymer hydrogels.

Dhanya Pulikkottil Rajan is presently working as Assistant Professor at MES Asmabi College, Thrissur. She also worked as Faculty at School of Industrial Fisheries, Cochin University of Science and Technology (CUSAT) and St. Albert's College, Ernakulam. She is a postgraduate in Industrial Fisheries from CUSAT and holds doctoral degree in Marine Sciences. Her doctoral study investigated the effects of spice oleoresins in microbial decontamination and their potential application in quality stabilization of tuna (*Euthynnus affinis*) during storage. She has qualified the National Eligibility Test under the Indian Council of Agricultural Research in 2013 and is a Lifetime Member of the Society of Fisheries Technologists (India). She has 5 years of teaching and 4 years of research experience in the field of fish processing. She had worked as Certified Technologist at CAP Seafoods, Kerala, before pursuing research. Her research interests include biopreservation, enhancement of safety and sensory qualities, value addition, and microbial spoilage of seafood products by psychrotrophic and histamine-forming bacteria. She has presented/published 14 research papers in various international and national journals/conferences and published 1 book chapter.

Abbreviations

°C	Degree Celsius
μg	Microgram
μl	Microliter
μm	Micrometer
^{1}H-NMR	Proton Nuclear magnetic resonance
AAS	Atomic absorption spectrometer
ACN	Acetonitrile
AFB	Aflatoxin
AGE	Agarose gel electrophoresis
AMPs	Antimicrobial peptides
APHA	American Public Health Association
a_w	Water activity
AWWA	American Water Works Association
BHC	Benzene hexachloride
BOD	Biological oxygen demand
BoNT	Botulinum toxin
BSA	Bovine serum albumin
CBB	Coomassie brilliant blue
CE	Capillary electrophoresis
CHNS-O	Carbon-hydrogen-nitrogen-sulfur-oxygen
CLLE	Continuous liquid/liquid extraction
COD	Chemical oxygen demand
CP	Cold plasma
CS	Chondroitin sulfate
CTX	Ciguatoxin
DCM	Dichloromethane
DDD	Dichlorodiphenyldichloroethane
DDE	Dichlorodiphenyldichloroethylene
DDT	Dichlorodiphenyltrichloroethane
DGGE	Denaturing gradient gel electrophoresis
DO	Dissolved oxygen
DS	Double strength
DS	Dermatan sulfate

DSC	Differential scanning calorimeter
EC	Enzyme Commission
ECM	Extracellular matrix
EDTA	Ethylenediaminetetraacetic acid
ELISA	Enzyme-linked immunosorbent assay
ETP	Effluent treatment plant
FBD	Fluidized bed drying
FBS	Fetal bovine serum
FFA	Free fatty acid
FID	Flame ionization detector
FTIR	Fourier transform infrared
FTU	Formazin Turbidity Unit
GAG	Glycosaminoglycan
Gal	Galactose
GalN	Galactosamine
GalNAc	N-acetylgalactosamine
GC	Gas chromatography
GlcA	Glucuronic acid
GlcN	Glucosamine
GlcNAc	N-acetylglucosamine
GM Counter	Geiger Muller counter
H_2SO_4	Sulfuric acid
HA	Hyaluronic acid
HACCP	Hazard Analysis Critical Control Point
HCB	Hexachlorobenzene
HCH	Hexachlorocyclohexane
HHP	High hydrostatic pressure processing
HMWM	High-molecular-weight markers
HPD	Heat pump drying
HPLC	High-performance liquid chromatography
HPTLC	High-performance thin layer chromatography
HRMS	High-resolution mass spectrometry
i.d.	Internal diameter
ICP	Inductively coupled plasma spectrophotometer
ICP-AES	Inductively Coupled Plasma - Atomic Emission Spectroscopy
IdoA	Iduronic acid
IF	Infrared
IQF	Individually quick frozen
ISO	International Organization for Standardization
JTU	Jackson Turbidity Unit
KS	Keratan sulfate
MPN	Most probable number
MT	Metallothioneins
MTX	Maitotoxin
NMR	Nuclear magnetic resonance

NTU	Nephelometric Turbidity Unit
PCB	Polychlorinated biphenyl
PEF	Pulsed electric field processing
PITC	Phenylisothiocyanate
PL	Pulsed light
ppm	Parts per million
RF	Radio frequency
rpm	Revolutions per minute
SDS	Sodium dodecyl sulfate
SEM	Scanning electron microscope
SFE	Solid funnel extraction
SS	Single strength
TDS	Total dissolved solids
TEMED	Tetramethylethylenediamine
TOC	Total organic carbon
TSS	Total suspended solids
TTX	Tetrodotoxin
UV	Ultraviolet
WHO	World Health Organization
WPCF	Water Pollution Control Facilities
YTX	Yessotoxin

Assessment of Nutritional Quality of Fish

1

1.1 Nutritional Quality of Fish

1.1.1 Water: The Universal Solvent

Water is the most profusely found substance in living systems, making up 70% or more of the weight of most organisms. Water is an elemental and often predominant constituent in many foods. It supports chemical reactions and also reactant in hydrolytic processes. In the food preservation point of view, the removal or binding of water hinders many reactions and therefore inhibits the growth of microorganisms and improves the shelf lives of foods. Water also contributes considerably to the texture of food because it interacts with other constituents, viz., proteins, polysaccharides, lipids, and salts.

Water forms hydrogen bonds with itself and with solutes. Hydrogen bonds between water molecules provide the cohesive forces that make water a liquid at room temperature. Water has a higher melting point, boiling point, and heat of vaporization than most other common solvents. Attractive forces between adjacent water molecules give the liquid state of water great internal cohesion. Polar biomolecules dissolve readily in water because they can substitute water-water interactions with more energetically favorable water-solute interactions. The water molecule and its ionization products, H and OH, influence the structure, self-assembly, and properties of all cellular components, including proteins, nucleic acids, and lipids. The attractive forces between water molecules and the slight tendency of water to ionize are of crucial importance to the structure and function of biomolecules.

1.1.1.1 Chemistry of Water Molecule

Each hydrogen atom of a water molecule shares an electron pair with the central oxygen atom. The H-O-H bond angle is 104.5°, i.e., 5° less than the bond angle of a perfect tetrahedron (109.5°). The nonbonding orbitals of the oxygen atom slightly compress the orbitals shared by hydrogen. The oxygen nucleus attracts electrons more strongly than does the hydrogen nucleus (i.e., the proton), and hence oxygen

S. Mathew et al., *Fish and Fishery Products Analysis*,
https://doi.org/10.1007/978-981-32-9574-2_1

is more electronegative. The electrons are more often in the vicinity of the oxygen atom than of the hydrogen; therefore the sharing of electrons between H and O is unequal. This unequal electron sharing creates two electric dipoles in the water molecule, one along each of the H–O bonds (Fig. 1.1).

Each hydrogen bears a partial positive charge (δ+), and the oxygen atom bears a partial negative charge (δ−), thus making a charge equal to the sum of the two partial positives. The resulting electrostatic attraction between the oxygen atom of one water molecule and the hydrogen of another water molecule constitutes a hydrogen bond (Lehninger et al. 2000).

In liquid water each molecule is hydrogen bonded to approximately 3.4 other water molecules. In ice, on the other hand, each molecule is hydrogen bonded to four other molecules to yield a regular lattice structure. Breaking a sufficient proportion of hydrogen bonds to destabilize the crystal lattice of ice requires much thermal energy, which accounts for the relatively high melting point of water (Fig. 1.2).

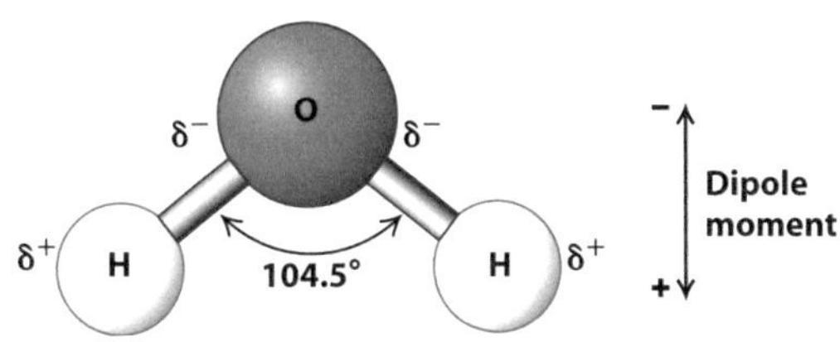

Fig. 1.1 Structure of water molecule. (https://alevelbiology.co.uk/notes/water-structure-properties/)

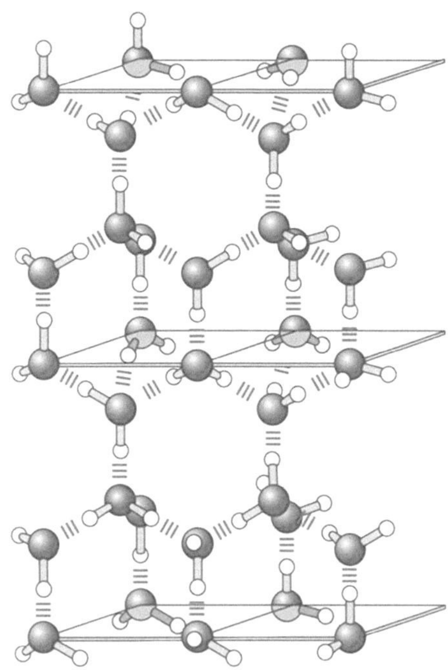

Fig. 1.2 Hydrogen bonding in ice – a regular crystal lattice. (Lehninger et al. 2000)

1.1.1.2 Bound Water and Free Water

The state of the water in food can be free or bound. The water in a sample appears to be "bound" by forces to varying degrees. This is a continuum of energy states, rather than a static "boundness." Water activity is sometimes defined as "free" or "available water" in a system. Water activity instruments measure the amount of free (sometimes referred to as unbound or active) water present in the sample. A portion of the total water content present in a product is strongly bound to specific sites on the chemicals that comprise the product. These sites may include the hydroxyl groups of polysaccharides, the carbonyl and amino groups of proteins, and other polar sites. Water is held by hydrogen bonds, ion-dipole bonds, and other strong chemical bonds. Some water is bound less tightly but is still not available (as a solvent for water-soluble food components). Many preservation processes attempt to eliminate spoilage by lowering the availability of water to microorganisms. Reducing the amount of free or unbound water also minimizes other undesirable chemical changes that occur during storage. The processes used to reduce the amount of free water in consumer products include techniques like concentration, dehydration, and freeze-drying. Freezing is another common approach to controlling spoilage. Water in frozen foods is in the form of ice crystals and therefore unavailable to microorganisms for reactions with food components. The ability of fish muscle to retain moisture is key to many meat quality parameters held in high regard by the industry and consumers. Improper handling, processing, and freezing of food stuffs leads to considerable "drip loss" upon thawing the product. Drip loss is the loss in nutrients, flavor, and tissue juice after thawing of frozen foods or ingredients leading to unfavorable appearance, texture, nutritional value, and attractiveness, thereby compromising the quality of fresh meat.

The storage life of the raw materials and also processed food products basically does not depend on the absolute value of the water content. The storage life also depends on the water content available for the microorganisms. Microbes cannot utilize the chemically bound water and that of frozen from food. Foods with the same water content differ in their perishability; hence, water content solely could not be responsible for food spoilage (Damodaran et al. 2007). Because water is present in varying degrees of free and bound states, analytical methods that attempt to measure total moisture in a sample do not always imply, thence making the concept of "water activity" relevant.

1.1.1.3 Water Activity (a_w)

Water activity is derived from fundamental principles of thermodynamics and physical chemistry. Water activity or a_w was developed to account for the intensity with which water associates with various nonaqueous constituents and solids. Simply stated, it is a measure of the energy status of the water in a system. It is defined as the vapor pressure of a liquid divided by that of pure water at the same temperature; therefore, pure distilled water has a water activity of exactly one.

Water Activity as a Thermodynamic Principle

In the equilibrium state, $\mu = \mu_o + RT\ \ln(f/f_o)$, where μ (J mol^{-1}) is the chemical potential of the system, i.e., thermodynamic activity or energy per mole of substance; μ_o is the chemical potential of the pure material at the temperature T (°K); R is the gas constant (8.314 J mol^{-1} K^{-1}); f is the fugacity or the escaping tendency of a substance; and fo is escaping tendency of pure material (Van der Berg and Bruin 1981). The activity of a species is defined as $a = f/f_o$. When dealing with water, a subscript is designated for the substance, $a_w = f/f_o$.

a_w is activity of water or the escaping tendency of water in system divided by the escaping tendency of pure water with no radius of curvature. For practical purposes, under most conditions in which foods are found, the fugacity is closely approximated by the vapor pressure (f ~ P); hence, $a_w = f/f_o \sim P/P_0$, where P is the vapor pressure of water in the substance and P_0 is the vapor pressure of pure water at the same temperature (Damodaran et al. 2007).

Equilibrium is obtained in a system when μ is the same everywhere in the system. Equilibrium between the liquid and the vapor phases implies that μ is the same in both phases. It is this fact that allows the measurement of the vapor phase to determine the water activity of the sample.

Relative humidity of air is defined as the ratio of the vapor pressure of air to its saturation vapor pressure. When vapor and temperature equilibrium are obtained, the water activity of the sample is equal to the relative humidity of air surrounding the sample in a sealed measurement chamber. Multiplication of water activity by 100 gives the equilibrium relative humidity (ERH) in percent.

$$a_w = \frac{P}{P_0} = \frac{ERH}{100}$$

Therefore, equilibrium relative humidity (ERH) = $a_w \times 100\%$, where P = partial vapor pressure of food moisture at temperature T, P_0 = saturation vapor pressure of pure water at T, and ERH = equilibrium relative humidity at T.

The value of water activity(a_w) represents the ratio of the available water content within the whole water content of food. Water activity is a measure of the energy status of the water in a system. There are several factors that control water activity in a system. Colligative effects of dissolved species (e.g., salt or sugar) interact with water through dipole-dipole, ionic, and hydrogen bonds. Capillary effect arises where the vapor pressure of water above a curved liquid meniscus is less than that of pure water because of changes in the hydrogen bonding between water molecules. Surface interactions in which water interacts directly with chemical groups on undissolved ingredients (e.g., starches and proteins) through dipole-dipole forces, ionic bonds (H_3O^+ or OH^-), van der Waals forces (hydrophobic bonds), and hydrogen bonds. It is a combination of these three factors in a food product that reduces the energy of the water and thus reduces the relative humidity as compared to pure water. These factors can be grouped under two broad categories, osmotic and matric effects. Due to varying degrees of osmotic and matric interactions, water activity describes the continuum of energy states of the water in a system.

Water activity is temperature dependent. Temperature changes water activity due to changes in water binding, dissociation of water, solubility of solutes in water, or the state of the matrix. The effect of temperature on the water activity of a food is product specific. Some products increase water activity with increasing temperature, and others decrease a_w with increasing temperature, while most high moisture foods have negligible change with temperature. One can therefore not predict even the direction of the change of water activity with temperature, since it depends on how temperature affects the factors that control water activity in the food.

Water activity is an important consideration for food product design and food safety.

Food Product Design

Food designers use water activity to formulate products that are shelf stable. If a product is kept below a certain water activity, then mold growth or bacterial spoilage is inhibited that results in a longer shelf life. Measuring water activity values can also help to limit moisture migration within a food product, made with ingredients having different water contents. Overall, water activity monitoring helps to limit or slow certain undesirable reactions in food products such as protein denaturation, fat oxidation, vitamin degradation, starch gelatinization and retrogradation, nonenzymatic browning, and enzymatic reactions.

Controlling Nonenzymatic Reactions

Foods containing proteins and carbohydrates, for example, are prone to nonenzymatic browning reactions, called Maillard reactions. The likelihood of Maillard reaction browning in a product increases as the water activity increases, reaching a maximum at water activities in the range of 0.6–0.7. In some cases, further increases in water activity can hinder the Maillard reactions. Hence, for some food samples, measuring and controlling water activity is a good way to control Maillard browning problems.

Slowing Down Enzymatic Reactions

Enzyme and protein stability are influenced significantly by water activity due to their relatively fragile nature. Most enzymes and proteins must maintain conformation to remain active. Maintaining critical water activity levels to prevent or entice conformational changes is important to food quality. Most enzymatic reactions are slowed down at water activities below 0.8. But some of these reactions occur even at very low water activity values. This type of spoilage can result in formation of highly objectionable flavors and odors. Of course, for products that are thermally treated during processing, enzymatic spoilage is usually not a primary concern.

Food Safety

Water activity (a_w) is one of the most critical factors in determining quality and safety of the goods. Water activity affects the shelf life, safety, texture, flavor, and smell of foods. While temperature, pH, and several other factors can influence if and how fast organisms will grow in a product, water activity may be the most important factor in controlling spoilage. For many years, researchers tried to correlate bacterial growth potential with moisture content and found that the values were not universal but specific to each food product. It was W J Scott in 1953 who first established that it is water activity not water content that correlated with bacterial growth. It is firmly established that growth of bacteria is inhibited at specific water activity values. Most bacteria, for example, do not grow at water activities below 0.91, and most molds cease to grow at water activities below 0.80. By measuring water activity, it is possible to predict which microorganisms will and will not be potential sources of spoilage. In addition to influencing microbial spoilage, water activity can play a significant role in determining the activity of enzymes and vitamins in foods and can have a major impact in their color, taste, and aroma.

Water activity is used as a critical control point, for Hazard Analysis and Critical Control Points (HACCP) programs. Samples of the food product are periodically taken from the production course and tested to ensure water activity values are within a specified range for food quality and safety. Measurements can be made in as little as 5 min and are made regularly in most major food production facilities. Water activity (a_w) can also significantly impact the stability, potency, and consistency of pharmaceuticals and cosmetics. The US Food and Drug Administration (FDA) regulations for intermediate moisture foods are based on water activity (a_w) values.

1.1.2 Fish Proteins

Protein is the second major component in muscle tissues of fish and is generally present in the range of 16–18% (on wet weight basis). The term protein is derived from the Greek word proteos, meaning first. They are complex organic nitrogenous substances formed from amino acids. Proteins occupy a central position in the architecture and functioning of living matter. The proteins in fish muscle tissue can be divided into sarcoplasmic, myofibrillar, and connective tissue proteins based on their solubility in salt solutions.

1.1.2.1 Sarcoplasmic Proteins

Sarcoplasmic proteins or myogens are water-soluble proteins and constitute between 20 and 30% of total proteins. These proteins are globular and exhibit solubility in water and diluted salt solutions and are constituted mainly by enzymes involved in cellular metabolism. The sarcoplasmic proteins of fish muscle are similar to that of higher vertebrates and include enzymes, myoglobin, albumins, antifreeze proteins, etc. They do not directly participate in the structural organization of muscle, and their functions are related to the metabolic activities within the cells. The electrophoretic pattern of the sarcoplasmic proteins can be used as a finger print for fish authentication. The sarcoplasmic proteins can significantly contribute to the textural quality of gels made from myofibrillar proteins. Functional properties such as emulsifying and foaming property render them potential to be used as ingredients in the food industry.

Enzymes Major seafood enzymes that have influence in the quality deterioration and applications in food industry are myosin ATPases, nucleotide-degrading enzymes, phospholipases, lipases, transglutaminases, TMAO-degrading enzymes, digestive proteinases, trypsin isozymes, polyphenol oxidases, chitinolytic enzymes, ligninolytic enzymes, and lipoxygenases. These enzymes and their products find applications in seafood quality testing, flavor, taste, texture manipulation, etc. Transglutaminases have an important role in *surimi* production. Some bacterial enzymes play specific roles in the degradation of amino acids and amines in seafoods, subsequently producing toxic end products that can be used as indicators/diagnostic tools for food safety evaluation (Muzaddadi et al. 2016).

Myosin ATPase is an important factor determining the onset of spoilage in freshly caught fish. Its activity is associated with pre-rigor change and the onset of rigor mortis. Deterioration of fish muscle during processing and storage is greatly influenced by the denaturation of myosin. Fish myosins are unstable and rapidly form aggregates with associated decrease of ATPase activity. Loss of ATPase activity is employed as a potential index of fish quality.

Nucleotide-degrading enzymes, hypoxanthine phosphoribosyl transferase, and xanthine oxidase are principle enzymes in the ATP degradation pathway (Fig. 1.9) that are used as indices of spoilage.

Lipases, triacylglycerol acylhydrolases (EC 3.1.1.3), promote the hydrolysis of tri-, di-, and monoglycerides to glycerol and fatty acids, in the presence of excess water, while in water-limiting conditions, they promote ester synthesis. They often express other activities, namely, phospholipase or other esterase types of activity, all of which have acknowledged industrial relevance. Lipases are of particular interest for the isolation of oil and fats from seafood by-products as well as in the preparation of ω-3-polyunsaturated fatty acids (ω-PUFAs) and enriched marine oils.

Transglutaminases (TGAs), protein-glutamine γ-glutamyltransferase (EC 2.3.2.13), promote acyl transfer reactions. These involve the γ-carboxyamide group of a peptide-bound glutamine residue as acyl donors and several primary amines as acyl acceptors, for example, the ε-amino group of lysine. Concomitantly, intra- and intermolecular covalent bonds are formed, namely, ε-(γ-glutamyl)lysine, resulting in the cross-linking of peptides and proteins resulting in polymerization. Given their cross-linking ability, MTGases are used for the modification/improvement of the functional and mechanical properties of fish and seafood products and therefore are used as binding ingredients for the restructuration of raw meats and in the production of surimi. Also, TGAs are used in formulations of fish meat mince, modification of finfish texture, processing of shark fin, formation of collagen and gelatin bonds, and minimization of drip after thawing.

Proteases are widely used in fish and seafood processing. Digestive proteinases and trypsin isozymes are used for deskinning, for descaling, and for the production of fish protein hydrolysate.

Trimethylamine oxide is found in most marine fish, enabling osmoregulation and other functions. TMAO is degraded into trimethylamine (TMA), dimethylamine (DMA), and formaldehyde, by the activity of TMAO oxidoreductase and demethylase. The endogenous activity of these enzymes is important in determining the quality deterioration of frozen-stored fish.

Polyphenol oxidase is the enzyme responsible for discoloration, melanosis, and black spot in crustaceans, as a result of enzymatic browning during spoilage. The enzyme acts on tyrosine, giving rise to diphenol or dihydroxyphenylalanine (DOPA), and finally to O-benzoquinones, which gives rise to melanins or brown-colored products. Melanosis indicates spoilage.

Lipoxygenases catalyze the incorporation of oxygen in the 5, 12, or 15 positions of various eicosanoic acids to produce hydroperoxides. The enzyme is also responsible for the formation of C6, C8, and C9 carbonyls and alcohols (that impart pleasant green, seaweed and melon like aromas in fresh fish) from long-chain PUFAs.

Other enzymes in fish and seafood, viz., proteolytic enzymes, chitinolytic enzymes, and ligninolytic enzymes, are used for the production of caviar; the recovery of chitin, collagen, flavor molecules, minerals, and pigments from seafood by-products; the removal of unwanted odors; and the improvement of shelf life and color retention.

Pigments Two classes of pigments that are commonly studied in fish are carotenoids (usually yellows, orange, and red) and melanin (browns, blacks, and grays), which are deposited in the integument. Other pigments include carotenoproteins, blue purple indigoids, brown melanins and melanoproteins, flavins, purines and pterins, polycyclic aromatic compounds in cephalopods, and tetrapyrroles such as biliverdin. Important pigments associated with fish meat are oxymyoglobin and oxyhemoglobin. Myoglobin imparts red color to the meat. Hemoglobin is absent in crustaceans, mollusks, and Antarctic fish. Shellfish contain copper-containing proteins called hemocyanin as the respiratory pigments. The characteristic blackening produced in canned shellfish, in presence of H_2S due to reaction with copper in hemocyanin, is an indicator of spoilage.

Parvalbumins Parvalbumins are a group of calcium-binding proteins. They are heat-stable, water-soluble, low-molecular-weight (typically 9–11 kDa) acidic proteins that are structurally related to calmodulin and troponin C. They are members of the calcium-binding EF-hand protein family characterized by a conserved protein structure. The major fish allergens are the parvalbumins (Kuehn et al. 2014).

Antifreeze Glycoproteins Antifreeze glycoproteins (AFGPs) constitute the major fraction of protein in the blood serum of Antarctic fishes. These compounds allow the fish to survive in subzero ice-laden polar oceans by kinetically depressing the temperature at which ice grows in a noncolligative manner. AFGP consists of a varying number of repeating units of (Ala-Ala-Thr)n, with minor sequence variations, and the disaccharide beta-D-galactosyl-(1-- > 3)-alpha-N-acetyl-D-galactosamine joined as a glycoside to the hydroxyl oxygen of the threonine residues.

1.1.2.2 Myofibrillar Proteins/Structural Proteins

Protein fractions soluble in solutions of higher ionic strength (>0.5 M) are known as myofibrillar proteins. Myofibrillar proteins are the most abundant proteins in fish muscle, constituting 65–75% of the total muscle proteins (compared to 40% in mammals). Among myofibrillar proteins are included contractile proteins, such as myosin and actin; regulatory proteins, such as tropomyosin and troponin; and other minor proteins. These fish proteins impact primarily on the sensory attributes relating to the texture of products. Fish fillet is composed of parallel muscle fibers arranged in blocks known as myotomes (Fig. 1.3). The myotomes are separated by collagenous sheaths and form the flakes of the cooked fillet. The texture of fish is determined to a large extent by the myofibrillar proteins within the muscle fibers. The handling of fish, throughout the supply chain, impacts on the state of fish proteins and thereby influences the final textural quality.

Myosin is the major myofibrillar protein in fish and constitutes about 50–60% of the myofibrillar fraction. It is particularly present in the thick muscle filaments. Fish myosin is similar to mammalian myosin and consists of two heavy chains (200 KDa) covalently associated with two pairs of myosin light chains (30 KDa). The myosin

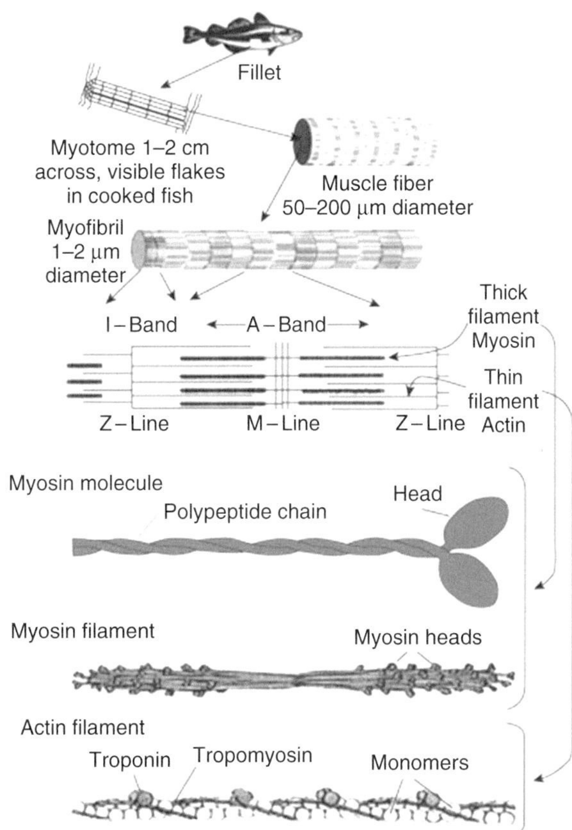

Fig. 1.3 Structure of fish muscle tissue. (Source: http://www.ucl.ac.uk/~sjjgsca/MuscleSarcomere.gif)

chain is a long structure (2 nm × 160 nm) with globular heads having ATPase activity and ATP bound to structure conferring it a dual function of enzyme catalysis and contractile protein.

Actin constitutes about 20% of the total amount of myofibrillar proteins in fish muscle and can be easily extracted. It is particularly present in the thin muscle filaments. It exists in two forms F actin (fibrous actin) and G actin (globular actin). F actin is a double helical structure, whereas G actin is the monomeric form of the protein. G actin can spontaneously polymerize to form F actin, in the presence of AT and Mg^{2+} ions. Actin spontaneously forms actomyosin complex with myosin in the solution, and, therefore, actomyosin is the main form of salt-soluble fish muscle proteins.

Tropomyosin and troponin regulate muscle contraction. The molecular weight of tropomyosin is 68 kDa, and it has two subunit chains (α and β). Tropomyosin is the most heat-stable muscle protein and remarkably resistant to acid or alkali. Troponin is a necessary protein for tropomyosin to act as a relaxation factor during muscle contraction. Troponin exists in several isoforms in fish and shellfish muscle. Troponin has three subunits C, T, and I and confers ca^{2+} sensitivity to actomyosin ATPase. TnC is the calcium-binding protein, TnI inhibits actin from binding to myosin heads, and TnT binds tropomyosin. In comminuted fish products, the long thin filaments of actin, along with troponin and tropomyosin, are connected with numerous myosin molecules and participate in gel formation.

Connectin is a well-known scaffold protein that strengthens the integrity of fish myofibrillar protein by providing inter- and intrafibrillar cross bridges. It contributes to loss of muscle elasticity, postmortem. Paramyosin is another characteristic myofibrillar component present in the striated smooth muscle of invertebrates. It is present in significant quantity in the adductor muscles of clam, in the foot muscles of abalone, and in the muscles of scallop, squid, and oyster and has significant effect on the rheological properties of gels prepared from invertebrate meat.

Water solubility of myofibrillar protein varies depending on the temperature, pH, and ionic strength. Extreme pH and high temperature cause protein denaturation resulting in low solubility. Generally, seafood myofibrillar proteins are less thermally stable than the proteins isolated from terrestrial animals. The pH and ionic strength affects thermal stability of seafood myofibrillar proteins and, hence, heat-induced denaturation. Myofibrillar proteins isolated from cold-water species are typically less thermally stable than warm water species.

1.1.2.3 Connective Tissue/Stroma Proteins

Connective tissue or stroma proteins are the fraction that is insoluble in solutions of higher ionic strength. They can be solubilized only by subjecting to conditions of extreme temperature and pH. They constitute about 2–3% of the total protein in fish. Elasmobranchs contain nearly 10% and contributes to its firm texture. This low content (3%) of connective tissue protein makes fish a source of easily digestible protein, compared with the 17% connective tissue in mammals.

Collagen is the major protein of the connective tissues (Sikorski and Borderias 1994). These tissues are composed of a highly hydrated, amorphous ground substance in which several types of cells, as well as the fibrous proteins collagen and elastin, are embedded. The cells fulfill many metabolic functions, including the synthesis of collagen and elastin, and they are shed to the extracellular spaces. The ground substance is composed mainly of water, of proteoglycans which are composed of a protein core with covalently bound glycosaminoglycans with sulfate and carboxylate groups, and of glycoproteins. The components of the ground substance interact with collagen and affect the collagen structures. The fibrous proteins form fibrous, filamentous, or network structures, depending on the kind of tissue. The mechanical properties of the connective tissues depend on the size, orientation, and cross-linking of the collagen fibrils and of elastic fibers, as well as on the proportion of all components of the tissue, including the mineral deposits. Connective tissue proteins contribute to the meat texture of fish. The species with firmer raw meat texture contains higher collagen content than the species with softer texture.

Collagen is comparatively poor in essential amino acids and therefore has little nutritive value. But, it has significant role in conferring tensile strength and integrity to the muscles and significantly affects the rheological properties of meat. The degree of cross-linking as a result of the hydroxylated proline and lysine residues provides mechanical strength in collagen. Gaping of the fish fillets is the result of the changes in the properties of collagen. "Gaping" is a phenomenon in which slits or holes appear between the muscle blocks, so that in a completely denatured state, the fillet falls to pieces and cannot be processed or marketed.

Elastin is one of the major fibrous constituents of the extracellular matrix of fish, birds, animals, and human. Elastin gives elasticity to tissues, while collagen gives tissue mechanical strength. Elastin is rich in elastic tissues, such as aorta, skin, lung, ligament, and so on. Elastin is synthesized as a soluble protein of approximately 70 kDa, referred to as "tropoelastin," by fibroblasts, smooth muscle cells, vein endothelial cells, chondrocytes, etc. In addition to providing elasticity to tissues, elastin also plays a significant role in tissue development (Love and Robertson 2007) (Fig. 1.4).

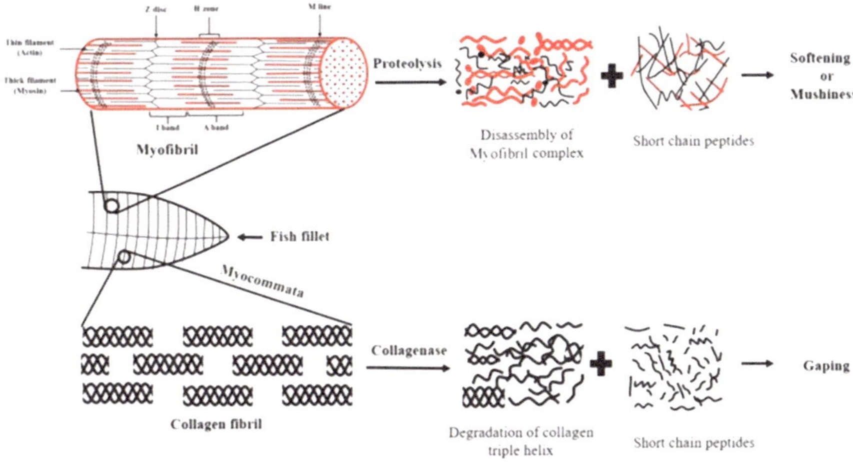

Fig. 1.4 Proteolytic breakdown of fish muscle during handling, processing and storage. (Source: Singh and Benjakul 2018)

1.1.3 Fish Lipids

The lipid fraction in fish/seafood is the component showing the greatest seasonal variation, in inverse resonance with the moisture content. Often, the variation within a certain species displays a characteristic seasonal curve with a minimum, around the time of spawning.

1.1.3.1 Fish Oils

Lipids are hydrophobic or amphiphilic small molecules that are chemically diverse and possess certain biological function. The term *lipid* is sometimes used as a synonym for fats; fats are usually regarded as triacylglycerols, which is a subgroup of lipids. Crude fat is heterogeneous material, constituting a mixture of triacylglycerols, phospholipids, fatty acids, sterols, waxes, and pigments. Total lipid is usually referred to as content of crude lipids measured gravimetrically, as an estimate for energy content and nutritional values in biological material has been criticized because of the content of non-fat and nondigestible substances. The US Nutrition Labeling and Education Act of 1990 (NLEA) has defined total fat as the sum of all fatty acids obtained from a total lipid extract expressed as triacylglycerols.

Generally lipids are classified into simple lipids, complex lipids, and derived lipids. Derived lipids arise from simple and complex lipids. Triacylglycerols, sterol esters, and wax esters are included in simple lipids, as they yield the most two types of primary products (fatty acids and glycerol), per mole on hydrolysis. The group of complex lipids yields three or more primary hydrolysis products per mole on hydrolysis which includes phospholipids, glyceroglycolipids, ether lipids, and sphingolipids. Derived lipids arise from simple and complex lipids and include fatty acids, glycerol, cholesterol, and other sterols, fat-soluble vitamins, etc.

Fish oils are triacylglycerols that are liquid at room temperature but generally solidify below 15 °C to 10 °C. The phospholipids make up the integral structure of the unit membranes in the cells, whereas the triglycerides are lipids used for storage of energy in fat depots. There are two types of fish oil: fish liver oil (generally derived from the liver of demersal fish, viz., cod, halibut, or shark) and fish body oil (generally derived from the flesh of pelagic fish, viz., herring, sardine, or anchovy). Both fish body oil and fish liver oil are rich sources of omega-3 long-chain polyunsaturated fatty acids (LCPUFAs): eicosapentaenoic acid (EPA), docosapentaenoic acid (DPA), and docosahexaenoic acid (DHA). Fish liver oil contains vitamin A (750–1200 mcg/daily dose) and vitamin D (2.5–10 mcg/daily dose) in addition to vitamin E. Recommended amounts of DHA and EPA in the form of dietary fish or fish oil supplements reduces the risk of heart attack, lowers triglycerides, prevents thrombosis, reduces abnormal heart rhythms and prevents stroke in people with known cardiovascular diseases. DHA and EPA also help in the inhibition of atherosclerosis, inflammation and lowers immune response and prevents stroke in people with known cardiovascular diseases, all of which are thought to result from a reduction in inflammatory and thrombotic prostaglandins, leukotrienes, and inflammatory cytokines.

1.1.3.2 Fatty Acids

The component fatty acids present as esters in triglycerides, phospholipids, and sterols show wide variation and diversity in fish lipids. Fatty acids may be saturated or unsaturated (mono and poly), with polyunsaturated fatty acids conferring great nutritional significance to fish oils. Myristic, palmitic, and stearic acids are the important saturated fatty acids present in fish. Among the monounsaturated group, palmitoleic and oleic acids are the important fatty acids, and the polyunsaturated group (PUFA) include arachidonic acid, eicosapentaenoic acid (EPA), and docosahexaenoic acid (DHA). There are two related families of essential fatty acids (EFAs) consisting of ω-3 and ω-6 polyunsaturated fatty acids (PUFA). Omega-3 fatty acids are long-chain PUFA commonly having 18, 20, or 22 carbon atoms in chain length with the first of the 3–6 double bonds adjacent to the third carbon atom when counting from the methyl carbon end of the fatty acid molecule. The fish-based omega-3 PUFA consists mainly of EPA (20 carbon atoms, 5 double bonds – 20:5n−3) and DHA (22 carbon atoms, 6 double bonds – 22:6n−3). Much smaller amounts of ω-6 PUFAs, docosapentaenoic acid (DPA, 22:5n−6) and arachidonic acid (20:4n−6), are found in fish and fish oils although they represent up to 5% of the fatty acids in certain marine mammal sources such as seal oils (Fig. 1.5).

Fatty acids in each of the families are interconvertible usually through alternate use of desaturases and elongases. The extent to which a given species at a given life stage can convert one ω-3 fatty acid to another or one ω-6 fatty acid to another determines the degree of essentiality of the fatty acid for that species at that life

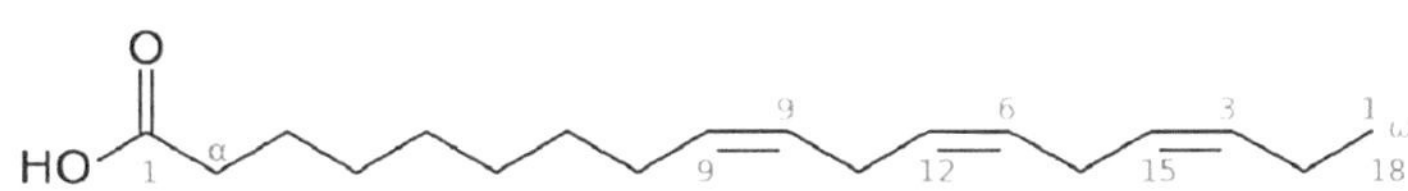

a. α-linolenic acid (18:3, n-3; ALA)

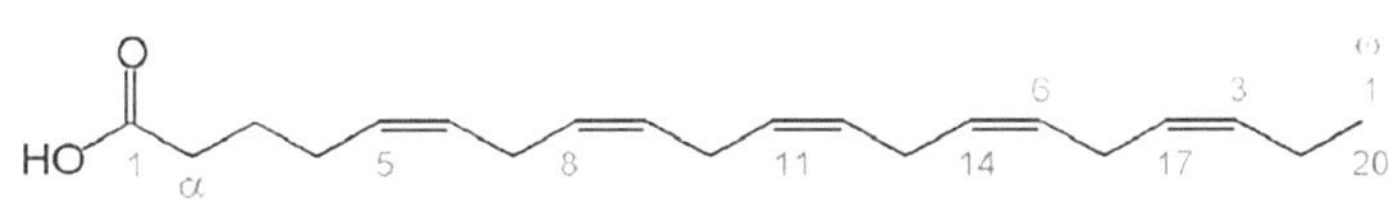

b. Eicosapentaenoic acid (20:5, n-3; EPA)

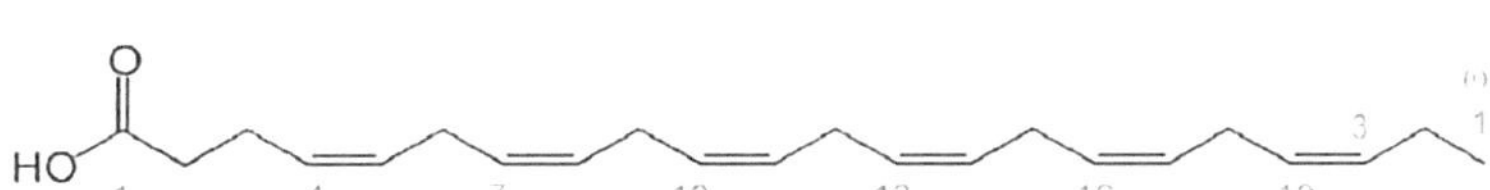

c. Docosahexaenoic acid (22:6, n-3; DHA)

Fig. 1.5 Structure of omega-3 fatty acids. (https://www.researchgate.net/figure/Chemical-structures-of-ALA-EPA-and-DHA_fig1_289536833)

Table 1.1 Omega 6/ Omega 3 fatty acid ratio in fish oils

Fish oil sample	ω-6/ω-3 ratio
Cod liver oil	1/7
Catfish oil	1/3
Mackerel oil	1/8
Salmon	1/2
Bream	1/2
Phytophagous 1	1/3
Phytophagous 2	1/3

Source: https://www.researchgate.net/publication/288545789

Cod liver and mackerel have the best value for the ω-6/ω-3 ratio

stage. There are pathways of PUFA synthesis which do not require desaturation and elongation of saturated fatty acids. Polyketide synthase pathway conducts the same reactions in an abbreviated sequence and has been speculated to be a significant contributor to PUFA in marine fish. Fatty acids have also been broadly used as biomarkers in trophic transfer studies in aquatic food webs (Iverson 2009) (Table 1.1).

1.1.3.3 Sterols and Steroids

Sterols are a large group of naturally occurring compounds having 1,2-cyclopentanophenanthrene skeleton that occur as minor components of fish lipids. Sterols are responsible for regulating membrane fluidity and permeability. They probably play an important role in the adaptation of membranes to temperature, whereas steryl esters (esters of cholesterol and fatty acids) are located intracellularly and are mainly a storage form of sterols. Cholesterol is a ubiquitous sterol present in all animal tissues. Cholesterol content in fish meat, shellfish, and mollusks varies widely as a result of species affiliation, sex, season, environmental conditions, quantity, and quality of food and breeding conditions. The cholesterol content varied from 70 mg/100 g in badejo fish to 107.6 mg/100 g of fish in white grouper (Scherr et al. 2015). Crab, lobster, and shrimp contain significant amount of cholesterol (6.2–27 mg%), whereas scallops and clams have lower concentrations (23.4–30.1 mg%).The major sterols in the mollusks are brassicasterol (12.6–45.6 mg/100 g) and 24-methylenecholesterol (16.7–41.9 mg/100 g), with the highest concentrations in oysters (Phillips et al. 2012). The prominent non-cholesterol sterols in mollusks are 22-dehydrocholesterol, isofucosterol, clionasterol, campesterol, and 24-norcholesta-5,22-diene-3β-ol. A diet with too much cholesterol raises the cholesterol level in the blood, leading to atherosclerosis and, in turn, to the inherent risk of a heart attack. The maximum recommended human intake of cholesterol is 300 mg/day.

1.1.3.4 Quality Implications

Two types of changes take place in lipids during processing and preservation of fish are hydrolytic rancidity and oxidative rancidity. Lipid hydrolysis in the tissue is mainly due to the action of lipases. As fish spoils, enzymes split the oil into its component fatty acids. Lipid hydrolysis results in release of free fatty acids and can cause protein denaturation. Denatured protein loses its characteristic properties, and this leads to loss of quality, especially the texture.

Oxidative rancidity, one of the major causes of quality deterioration in foods, is caused by the oxidative deterioration of lipids by atmospheric oxygen. Lipids oxidize through a complex series of reactions giving rise to a myriad of nonvolatile and volatile toxic compounds that are responsible for off-flavors even at concentrations in the parts-per-billion range. Oxidative rancidity in fish lipids is a complex problem affecting the quality of processed fish. Due to the high degree of unsaturation of the fatty acids present in fish lipids, its susceptibility to oxidation is extremely high. Contact with oxygen or other oxidizing agents will initiate the process of oxidation and is accelerated at elevated temperatures and presence of catalysts like copper or iron (Fig. 1.6).

The oxidation reaction is a chain reaction, which progresses through the propagation of free radicals. The process is characterized by an induction period, during which the oxidation is slow, followed by an accelerating rate of oxygen absorption with concurrent development of hydroperoxides, which are the primary products of oxidation. The hydroperoxides undergo decomposition to various products like aldehydes, ketones, alcohols, carboxylic acids, etc. some of which are volatile and some nonvolatile. The number and nature of these decomposition products depend on the position of the double bond being oxidized and the conditions under which the hydroperoxides are decomposed. Oxidation of highly unsaturated fatty acids will lead to formation of polymerized products and render the fish or oil totally unacceptable. The unsaturated fatty acid oxidation product, oxylipins, are cytotoxic and have also been implicated in toxicity towards mammalian cells causing hemolysis and ichthyotoxicity. The toxicity of free fatty acids raises some interesting physical chemical queries because of their low solubility in water.

Ideally, free fatty acid content should be below 2%, and there should be little oxidation. Oxidation produces peroxides, aldehydes, and ketones. The degree of oxidation can be determined by methods which measure these, as TOTOX value based on peroxide and anisidine values, which should ideally be below 20.

1.1.4 Nonprotein Nitrogenous Compounds

Nonprotein nitrogen (NPN) compounds are found mostly in the fiber sarcoplasm and include free amino acids, peptides, amines, amine oxides, guanido compounds, nucleotides, urea, and ammonia/quaternary ammonium molecules. NPN compounds

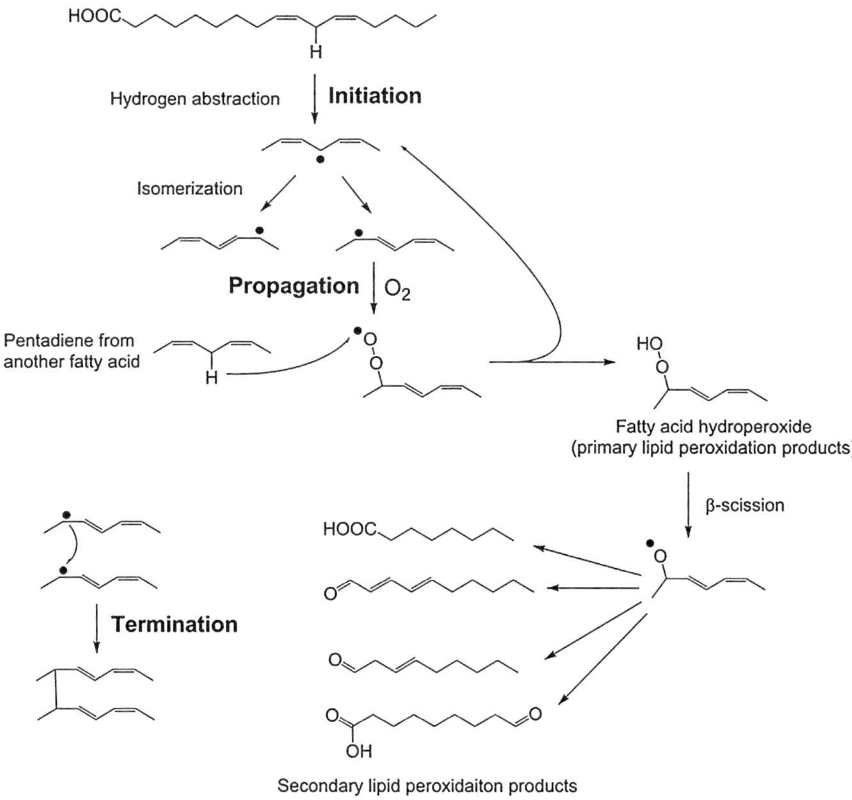

Fig. 1.6 Schematic representation of lipid peroxidation. (https://www.researchgate.net/figure/A-simplified-scheme-of-lipid-peroxidation_fig1_312051167)

account for a relatively high percentage of the total nitrogen in the muscles of some aquatic animals, 10–20% in teleosts, about 20% in crustaceans and mollusks, and 30–40% (and in special cases up to 50%) in elasmobranchs. NPN compounds in land animals usually represent no more than 10% of the total nitrogen content.

1.1.4.1 Free Amino Acids and Peptides

Free amino acids (FAAs) are among the most important fractions of nonprotein nitrogen in crustaceans and fish. Nonprotein nitrogenous compounds comprise between 0.5% and 1% of total weight of the fish muscle (Spinelli and Dassow 1982) and are involved in osmoregulation in fish. Crustaceans have higher contents of free amino acids than finfish. Free amino acids constitute about 630 mg/ 100 g light muscle in mackerel (*Scomber scombrus*), 350–420 mg/100 g in herring (*Clupea harengus*), and 310–370 mg/100 g in capelin (*Mallotus villosus*). The relative importance of the different amino acids varies with species. Taurine, alanine, glycine, and imidazole-containing amino acids seem to dominate in most fish.

The importance of amino acids in fish has been well established from the perspective of nutrition, fish meal production, fish flavors, and bacterial and chemical spoilage.

Unique nonprotein amino acids found in seafood are taurine (a sulfonic acid), sarcosine (N-methylglycine), α-aminobutyric acid, β-alanine, 3-methylhistidine, etc. Free amino acids found in relatively high amounts in fish are glycine, taurine, alanine, and lysine, with varying degree of the constituent amino acids among different species. Migratory fishes such as mackerel and tuna contain high amounts of histamine, and white-fleshed fishes are high in taurine content. Histidine is the most prominent FAA in milkfish and accounts for about 80% of the total FAA, whereas tilapia contains high amounts of taurine and glycine and relatively small amount of histidine. Red snapper contains higher levels of lysine than histidine (Gunlu and Gunlu 2014). Histidine levels are higher in the tissues of Scombridae family, which on decarboxylation gives rise to histamine. Red snapper contains higher levels of lysine than histidine (Gunlu and Gunlu 2014). Quantitatively, glycine and alanine are found to be the two most important amino acids measured in the muscle of rainbow trout (*Salmo gairdnerii* Richardson), with specific roles in osmotic pressure regulation.

Peptides

Fish can be used not only for nutritional purposes but also as a source of unique peptides with a broad spectrum of biological activities. Fish peptides have a wide range of pharmacological activities, including antihypertensive, immunomodulatory, antithrombotic, antioxidant, antimicrobial, antiviral, antitumor, and other effects. Biologically active peptides can either be absorbed through the gut wall, thus coming intact into the blood system and exerting certain physiological effects, or can exert some local effects directly in the gastrointestinal tract. A detailed description of biologically active fish peptides is given in Chap. 7.

Quality Implications

The free amino acids are enzymatically decarboxylated to the respective biogenic amines and are considered as indicators of fish quality and safety (refer to Chap. 2). Biogenic amines are defined as aliphatic, alicyclic, and heterocyclic organic bases of low molecular weight. They are not only biosynthesized in animal and vegetable cells but also produced by the decarboxylation activity of bacterial enzymes. Biogenic amines have been studied extensively because of their involvement in foodborne diseases. The most important biogenic amines occurring in food are histamine, putrescine, cadaverine, spermidine, spermine, tyramine, and agmatine. Seafood quality is affected by handling gutting, processing, and storage temperature. For the seafood industry, the monitoring of indicator compounds such as biogenic amines, associated with seafood safety, is as important as sensory evaluation. Scombroid poisoning is a foodborne intoxication caused by the consumption of scombroid fish containing hazardous levels of histamine in the muscle tissue. The group of fish most frequently implicated in scombroid poisoning is the family Scombridae, which includes the tuna, mackerel, and bonito. Histamine is the main compound responsible for this intoxication, and its toxicity is increased by other

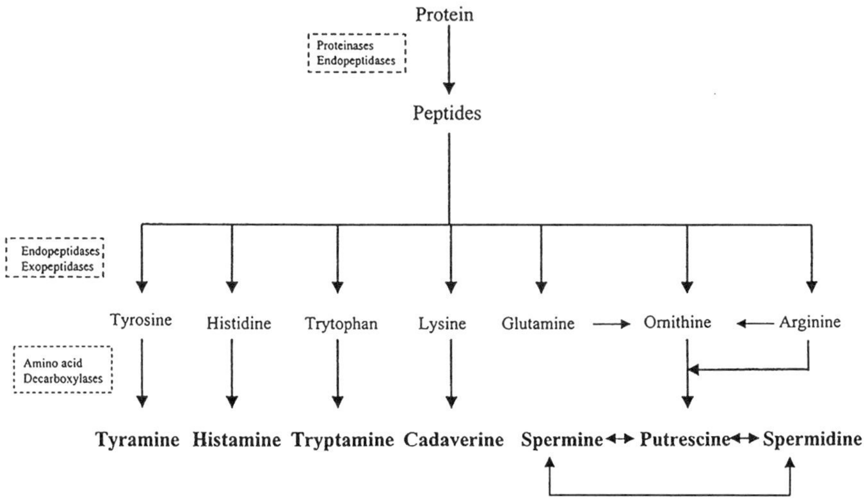

Fig. 1.7 Biogenic amines – decarboxylation products of amino acids. (https://www.researchgate.net/figure/FIGURA-1-Descarboxilacao-Fonte-RUIZ-CAPILLAS-JIMENEZ-COLMENERO-2004_fig1_266395501/download)

biogenic amines such as putrescine and cadaverine present in fish. Putrescine is the decarboxylation product of amino acid ornithine, and cadaverine arises from the decarboxylation of lysine.

The biological pathway for the formation of nonvolatile amines putrescine, histamine, cadaverine, spermidine, agmatine, and spermine from amino acids is shown in Fig. 1.7.

1.1.4.2 Nucleotides and Degradation Products

The purine derivative, ATP, constitutes 90% of the nucleotides in fish muscle. Postmortem biochemical changes associated with the catabolism of ATP occurs at different rates in different fish species. Other nucleotides in fish muscle are nicotinamide adenine nucleotides (NAD/NADH), nicotinamide adenine nucleotide phosphates (NADP/NADPH), and other coenzyme nucleotides (Fig. 1.8).

Quality Implications

ATP is the second abundant muscle metabolite after amino acids and provides the energy for muscle contraction. It is a high-energy molecule that enables to keep actin and myosin filaments separated in the muscle. In live fish, muscle ATP that is hydrolyzed to ADP gets constantly resynthesized by the mitochondrial chain. But after the death of the organism, ATP continues to be hydrolyzed to ADP by contracting muscles, but its resynthesis is blocked due to respiratory arrest. Even though ATP is resynthesized by less efficient pathways (anaerobic glycolysis, creatine kinase, and myoadenylate kinase), ATP concentration drops leading to gradual accumulation of ADP. The ADP is gradually dephosphorylated to AMP and

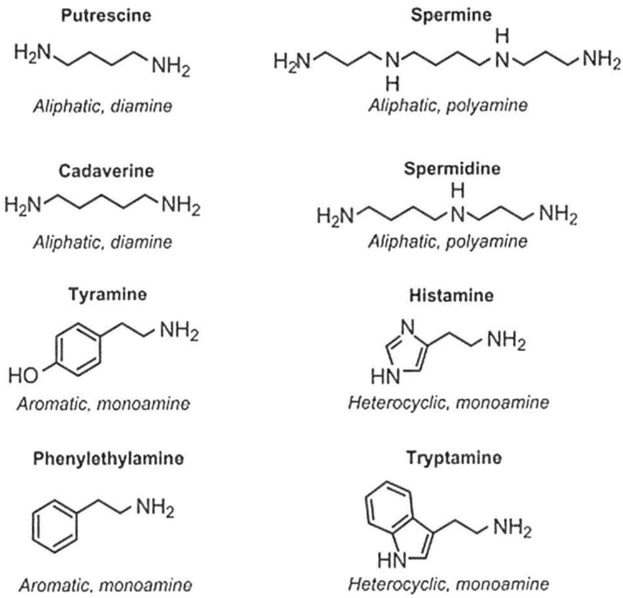

Fig. 1.8 Structure of biogenic amines. (https://www.sciencedirect.com/science/article/pii/B9780128098684000121)

deaminated to IMP by muscle AMP deaminase. IMP is predominant nucleotide in in-rigor muscle. Postmortem degradation of IMP to inosine and hypoxanthine results from both autolytic and bacterial enzymes. The rate of IMP degradation varies considerably between species and depends on handling and storage conditions. The content of ATP degradation metabolites (K value) has been widely used as biological indicators of fish freshness (refer to Chap. 2) (Fig. 1.9).

1.1.4.3 Guanido Compounds

This group of NPN compounds includes phosphagens, creatine, and creatinine phosphate. Creatine is quantitatively the main component of the NPN fraction. This molecule plays an important role in fish muscle metabolism in its phosphorylated form; it is absent in crustaceans and mollusks.

Creatine phosphate serves as a cellular storage species of high-energy phosphate that is readily available for the formation of ATP from ADP by the action of creatine kinase (EC 2.7.3.2). Creatine phosphate is a high-energy phosphate (AG′ -10 kcal mol-1 for hydrolysis) by virtue of being a phosphoramidate. The reversibility of the reaction of creatine kinase allows creatine phosphate to be formed whenever the concentration of ATP is high, and it makes creatine phosphate an efficient reserve of high-energy phosphate to generate ATP under conditions of high-energy demand. The action of creatine kinase requires a divalent metal ion, either Mg^{2+} or Mn^{2+} in complex with ADP, with Mg^{2+} being the more prevalent activating metal ion in cells (Fig. 1.10).

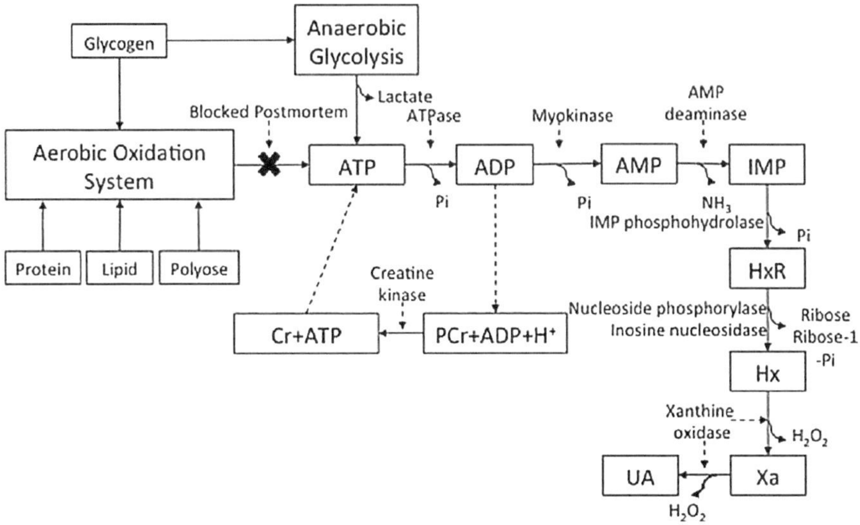

Fig. 1.9 Postmortem ATP metabolism in fish muscle. (Source: Hong et al. 2017). *ATP* adenosine triphosphate, *ADP* adenosine diphosphate, *AMP* adenosine monophosphate, *IMP* inosine monophosphate, *HxR* inosine, *Hx* hypoxanthine, *Xa* xanthine, *UA* uric acid, *Cr* creatine, *PCr* phosphocreatine.https://www.researchgate.net/figure/Post-mortem-ATP-metabolism-in-fish-muscle-modified-from-Huss-1995_fig1_279301560

Fig. 1.10 Creatine phosphorylation. (Source: https://www1.udel.edu/chem/C465/senior/fall00/Performance1/phosphocreatine.htm.html)

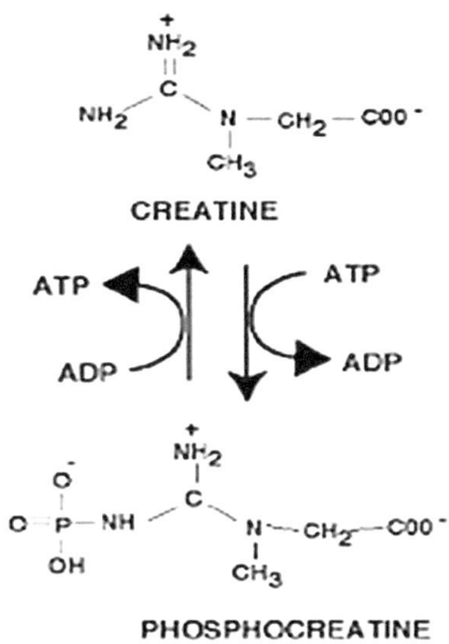

1.1.4.4 Nonnutritive Components

Amines and Amine Oxides Most marine fishes contain trimethylamine N-oxide (TMAO), a colorless, odorless, and flavorless compound that is degraded to trimethylamine, which gives a "fishy" odor and causes consumer rejection. This compound is not present in land animals and freshwater species (except for Nile perch and tilapia from Lake Victoria). TMAO participates in osmoregulation of marine species, and it stabilizes elasmobranch protein structure in the presence of large amounts of urea. Other methylamine compounds, viz., sarcosine and betaine, also participate in osmoregulation and protein structure stabilization. The amount of TMAO in the muscle tissue depends on the species, season, fishing ground, etc. In general, the highest amount is found in elasmobranchs and squid (75–250 mg N/100 g), and cod has comparatively less amount (60–120 mg N/100 g), while flatfish and pelagic fish have the least.

Urea and Ammonia Ammonia and urea are the two major forms of nitrogen that are produced and excreted to the surrounding environment in most fishes. Ammonia exists as NH^{4+} at physiological pH and plasma ionic strength. The majority of ammonia produced by typical fish species results from amino acid catabolism via specific deaminases or by transdeamination. Embryonic and larval fishes produce a substantial ammonia load as a by-product of an amino acid-based metabolism. Urea is found in the tissues of almost all teleost fish. The decomposition of urea to ammonia is catalyzed by bacterial urease. The high urea content (2.0% and 2.5%) that characterizes the blood, body fluids, and tissues of the Elasmobranchii owes its origin to the relative impermeability of the gills and integument and to the fact that the urea is actively conserved by the elasmobranch kidney. Urea retention enables the Elasmobranchii to maintain a considerably greater rate of urine formation (water excretion) than is observed in the marine teleosts.

Quality Implications

TMAO reductase catalyzes the conversion of TMAO to trimethylamine (TMA) and is found in several fish species (in the red muscle of scombroid fishes and in the white and red muscle of gadoids) and in certain microorganisms (Enterobacteriaceae, *Shewanella putrefaciens*). TMA is believed to be generated by the action of spoilage bacterial. Total volatile basic amine (TVB) is one of the most widely used measurements of seafood quality. It is a general term which includes the measurement of trimethylamine (produced by spoilage bacteria), dimethylamine (produced by autolytic enzymes during frozen storage), ammonia (produced by the deamination of amino acids and nucleotide catabolites), and other volatile basic nitrogenous compounds associated with seafood spoilage. Although TVB analyses are relatively simple to perform, they generally reflect only later stages of advanced spoilage and are generally considered unreliable for the measurement of spoilage, since it does not correlate with bacterial numbers. This phenomenon is now thought to be due to the presence of small numbers of "specific spoilage" bacteria which do not always represent a large proportion of the total bacterial flora but which are capable of producing large amounts of spoilage-related compounds such as TMA. The chief advantages of TMA analysis over the enumeration of bacterial numbers are that TMA determinations can be performed far more quickly and often reflect more accurately the degree of spoilage (as judged organoleptically) than do bacterial counts.

TMAO dimethylase (TMAO-ase) converts TMAO into equimolar quantities of dimethylamine (DMA) and formaldehyde (FA). DMA is produced along with FA in frozen storage with the accompanying FA-induced toughening of the proteins. The amount of protein denaturation is roughly proportional to the amount of FA/DMA produced, but it is most common to monitor the quality of frozen-stored fish by measuring DMA rather than FA. Much of the FA becomes bound to the tissue and is thus not extractable and cannot be measured quantitatively.

1.1.5 Vitamins, Minerals, and Trace Elements

Vitamins are defined as complex organic substances required in small amount for the maintenance and growth of living organisms. Their deficiency may lead to certain specific diseases or symptoms which can be cured by the administration of that specific vitamin only. In 1906, the British Biochemist Sir Frederick Hopkins demonstrated that foods contain certain accessory factors in addition to proteins, carbohydrates, fats, minerals, and water. Credited as the discoverer of vitamins, the Polish American Biochemist *Casimir Funk* (1884–1967) identified that the anti-beriberi substance in unpolished rice was an amine which is a type of nitrogen-containing compound. He coined the term "vitamine" a combination word from vita and amine, meaning amine of life and considered that amines are vital for life. Later it was found that not all vitamins are "nitrogen containing" or "amines," and in 1920, Drummond proposed to drop the final "e" from the word vitamines and consequently termed them as vitamins.

Early in 1915, McCollum and Davis classified the vitamins based on their solubility in water or oil (fats). Fat-soluble vitamins – vitamins A, D, E, and K – dissolve in fat before they are absorbed in the bloodstream to carry out their functions. Excesses of these vitamins are stored in the liver and adipose tissue. Because they are stored, they are not needed every day in the diet. Fat-soluble vitamins perform a multitude of different biochemical functions. By contrast, water-soluble vitamins – the B-complex group and vitamin C – dissolve in water and are not stored. Excess amounts of these vitamins are eliminated in urine; hence we require a continuous supply of them in our diets. Generally, water-soluble vitamins play a catalytic role in many metabolic reactions of the cells and act as coenzymes or part of coenzymes and enzyme systems.

1.1.5.1 Fat-Soluble Vitamins

Vitamins A, D, E, and K structurally resemble partially cyclized isoprenoid polymers and are soluble mainly in lipids or oils and thus called fat-soluble vitamins.

Absorption and transportation of these vitamins in the body are mainly associated with lipids in the intestine and stored in the liver and adipose tissue. They are eliminated slowly from the body owing to their lipophilic character. High intake of fat-soluble vitamins may result in their accumulation in the body known as hypervitaminosis.

Vitamin A

Retinol, or vitamin A_1, is a necessary nutrient in all higher animals. It plays an important role in vision, in the maintenance of epithelial cell layers, in spermatogenesis, and in fetal development. Fish liver oils, viz., tuna, shark, and cod liver oil, are important sources of vitamin A_1. Vitamin A_2 (3, 4-dehydroretinol) is the predominant form in freshwater fishes. Beta-carotene, the precursor molecule of vitamin A, is found naturally in many fruits, vegetables, and other plant sources. Beta-carotene, alpha-carotene, and beta-cryptoxanthin are the most common carotenoids, with beta-carotene being the most easily converted into retinol. Retinol must be obtained from the diet, as either from organic compounds that are in animal products, such as retinyl esters (e.g., fish oils and liver provide retinyl palmitate), or from compounds in plants, most notably β-carotene. Retinol and its derivatives are collectively known as retinoids, i.e., retinol, retinal, and retinoic acid.

Chemically, retinol is an isoprenoid; isoprenoids are a group of compounds that includes vitamins E and K and cholesterol (indeed, the precursor of vitamin D), which are synthesized from isoprene units. Retinol is a hydrocarbon molecule with a single hydroxyl group at one end. This hydroxyl group can be oxidized to form an aldehyde group (yielding retinal) or to a carboxylic acid group (yielding retinoic acid). Retinal and retinoic acid are the biologically active forms of retinol. In addition to these different oxidation states, retinol occurs as a number of different cis and trans isomers. Retinol esters serve directly as the substrate for the formation of the visual chromophore 11-*cis*-retinal, which binds with the protein-opsin, thereby forming the visual pigment, rhodopsin. The other active form of vitamin A is retinoic acid (DeLuca et al. 2000). Retinoic acid exerts its physiological functions via its interactions with a family of nuclear receptors called the retinoic acid receptors (RARs). RARs are members of a superfamily of nuclear receptors that include steroid hormone and thyroid hormone receptors. RARs are involved in regulating the transcription of many genes, including genes involved in the cell growth and cell differentiation of developing organisms (Fig. 1.11).

Vitamin D

Vitamin D, the sunshine vitamin, is particularly known for its antirachitic activity (treatment of rickets). As early as 1870, people knew there was something in cod liver oil that prevented rickets, a disease resulting in soft, deformed bones as a result of calcium deficiency.Vitamin D is commonly referred to as the sunshine vitamin, because the sterols present in the skin can be converted to vitamin D with the help of UV rays emitted from the sun. Other potential sources include fish liver oil, fish products, eggs (in D-supplemented hens), liver, milk (fortified), etc. Vitamin D is one of the most important biological regulators of calcium metabolism, through

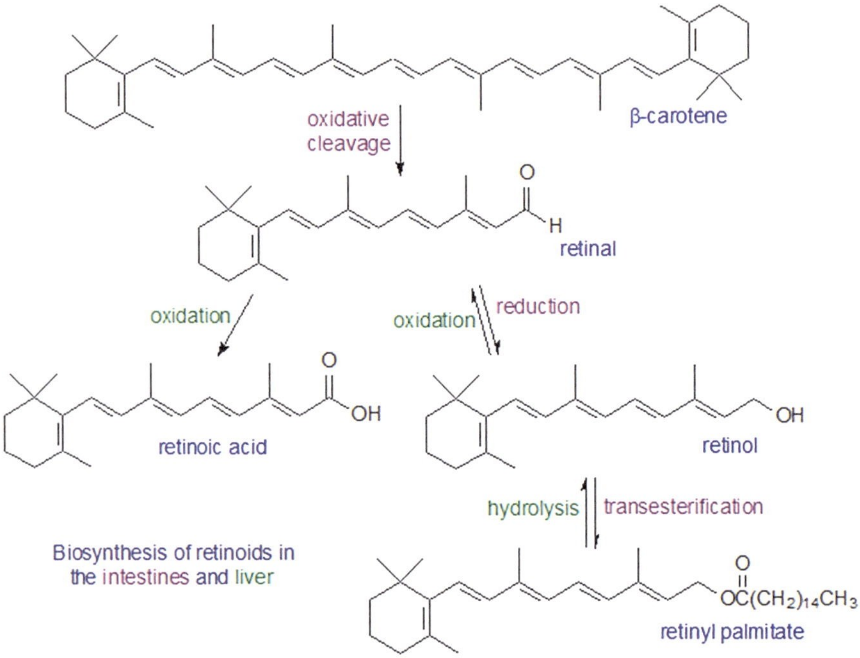

Fig. 1.11 The conversion of β-carotene to vitamin A. It involves two steps catalyzed by iron containing deoxygenase enzyme present in intestinal mucosa and alcohol dehydrogenase (Retinol dehydrogenase). (Source: http://www.lipidhome.co.uk/lipids/simple/retinoids/index.htm)

stimulating the absorption of calcium from food across the intestine and participating in the incorporation of the absorbed calcium in to the skeleton.

At least five different forms of vitamin D have been discovered. Vitamin D_1 is a molecular compound containing lumisterol and calciferol in 1:1 ratio; vitamin D_2 is ergocalciferol, derived from ergosterol of plants; vitamin D_3 is cholecalciferol and obtained from 7-dehydrocholesterol of animals; vitamin D_4 is obtained from 22-dihyroergocalciferol; and vitamin D_5 is generally referred to as sitocalciferol. The nutritionally important forms of vitamin D are vitamin D_2 (ergocalciferol) which is found in plants and vitamin D_3 (cholecalciferol) synthesized in animal body from provitamin D (7-dehydrocholesterol). Liver oils of fatty fishes are excellent sources of vitamin D_3.

Vitamins D_2 and D_3 are biologically inactive, whereas the metabolite 1,25-dihydroxycholecalciferol or calcitriol is biologically active (DeLuca 2008). Calcitriol exerts its action as a steroid hormone in the intestinal mucosal cells. 1,25-Dihydroxycholecalciferol (calcitriol) regulates the synthesis of calcium-binding proteins (CaBP) by the intestinal mucosal cells, thereby enhancing the uptake of calcium present in the diet. Thus, along with parathyroid hormone and calcitonin, vitamin D maintains the homeostasis of calcium and phosphorus in the tissues and body fluids. The critical effect of vitamin D on the bone is to provide the

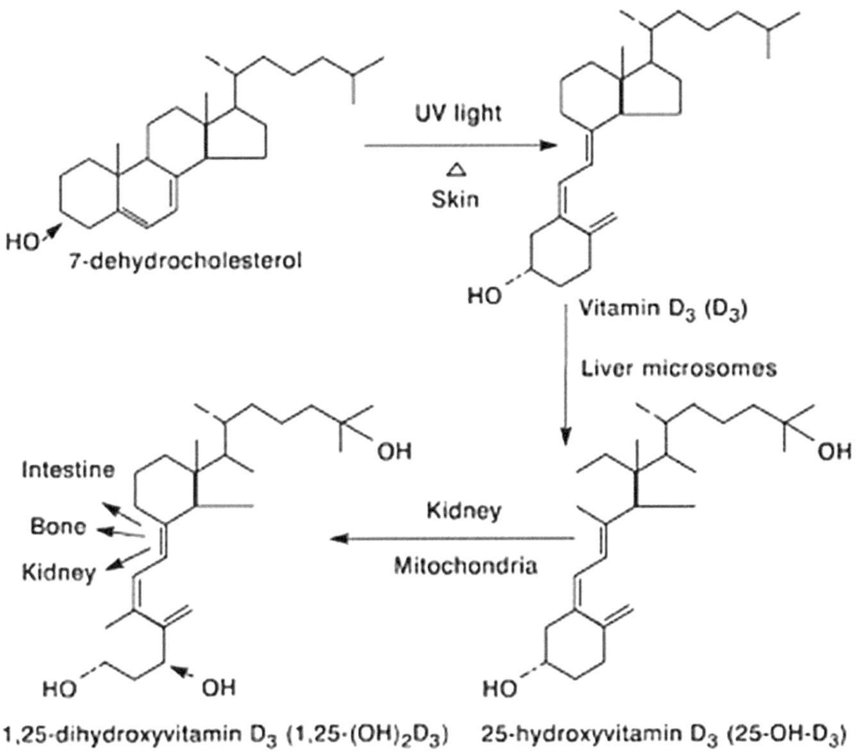

Fig. 1.12 Biosynthesis of calcitriol (1,25-dihydroxy vitamin D_3) from vitamin D_3 (cholecalciferol). (Source: https://www.dsm.com/markets/anh/en_US/Compendium/companion_animals/vitamin_D.html)

proper balance of calcium and phosphorus to support mineralization. Calcitriol, the biologically active form of vitamin D_3, is involved in mineral metabolism and bone growth and prevents rickets, a disease resulting in soft, deformed bones as a result of calcium deficiency (Fig. 1.12).

Vitamin E

Natural vitamin E is the mixture of two classes of compounds, tocopherols (tokos means child birth; phero indicates to bear; ol implies alcohol) and tocotrienols. They are amphipathic and lipid-soluble compounds that are easily oxidized when subjected to heat, light, and alkaline conditions (Eitenmiller and Lee 2004). Both the tocols (tocopherols) and tocotrienols consist of a hydroquinone nucleus and an isoprenoid side chain. The four forms of tocopherols and the tocotrienols are alpha (α), beta (β), gamma (γ), and delta (δ) forms (Cohen et al. 1981). Alpha-tocopherol has the highest biopotency among the eight naturally occurring derivatives of vitamin E and is present in relatively larger proportions in fish liver oil, apart from fruits, fortified cereals, legumes, seeds and seed oils, olives, green leafy vegetables,

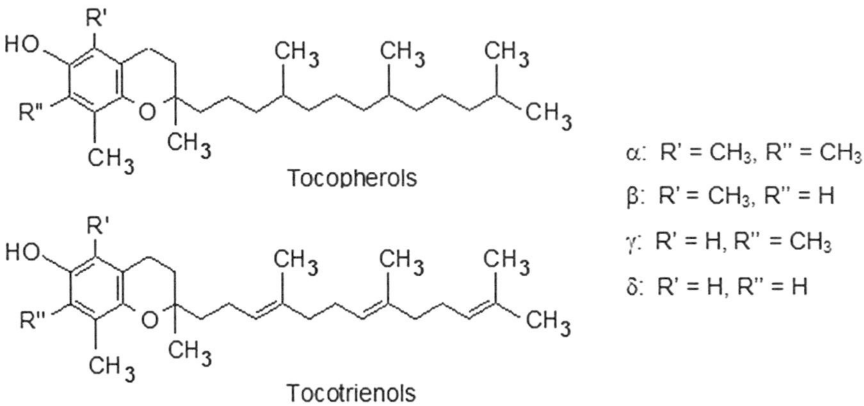

Fig. 1.13 Chemical structure of tocopherols and tocotrienols. (Source: http://lipidlibrary.aocs.org/Analysis/content.cfm)

liver, eggs, rice bran, almond oil, wheat germ oil, etc. Being an antioxidant, vitamin E plays an important role to protect the body cells from the damage caused by free radicals which are highly reactive and destructive compounds formed due to oxidative deterioration (metabolism) of polyunsaturated fats (Pesillo et al. 2004; Yu and Paetau-Robinson 2006) (Fig. 1.13).

Vitamin K

Vitamin K refers to a group of fat-soluble quinone compounds that have characteristic antihemorrhagic effects. The natural forms of vitamin K include vitamin K_1 known as phylloquinone and vitamin K_2 called menaquinone. The simplest form of vitamin K is the synthetic menadione (K_3), which has no side chain. The Nobel Prize in Physiology or Medicine (1943) was divided equally between Henrik Carl Peter Dam (for the discovery of vitamin K_1 from alfalfa leaves) and Edward Adelbert Doisy (for the discovery of the chemical nature of vitamin K_2 from putrefied fish meal, synthesized by the intestinal bacterial flora).

Vitamin k is the only fat-soluble vitamin with a coenzyme function. Coagulation time of blood is increased when vitamin K is deficient because the vitamin is required for the synthesis of prothrombin (factor II). Plasma clotting factors VII (proconvertin), IX (Christmas factor), and X (Stuart-Prower factor) also depend on vitamin K for their synthesis. These four blood-clotting proteins are synthesized in the liver in inactive precursor forms and then converted to biologically active proteins by the action of vitamin K (Suttie and Jackson 1977). Vitamin K-dependent reactions are present in most tissues and not just blood and that a reasonably large number of proteins are subjected to this posttranslational carboxylation of specific glutamate residues to gamma-carboxyglutamate residues (Vermeer 1986). Vitamin K is involved in the regulation of multiple enzymes involved in sphingolipid metabolism within the myelin-rich regions in the brain (Denisava and Booth 2005).

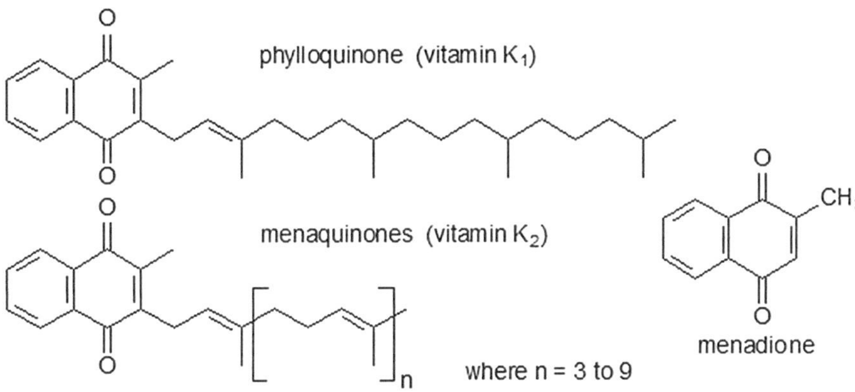

Fig. 1.14 Chemical structure of vitamin K_1 (phylloquinone), K_2 (menaquinone), and K_3 (menadione). (Source: http://www.lipidhome.co.uk/lipids/simple/isoprene/index.htm)

Vitamin K-dependent γ-carboxylation is essential to several bone-related proteins, including osteocalcin (Gundberg et al. 2012) (Fig. 1.14).

1.1.5.2 Water-Soluble Vitamins

Fish is not an important reliable dietary source of the water soluble – vitamin C – but they are important in the general metabolism of the organism. In general, fish meat is a good source of the B vitamins.

Vitamin B Complex

Most of the B vitamins have been recognized as coenzymes (substances that participate with enzymes in accelerating the interconversion of chemical compounds), and they all appear to be essential in facilitating the metabolic processes of all forms of animal life. The complex includes thiamin (vitamin B_1), riboflavin (vitamin B_2), pantothenic acid (vitamin B_3), niacin (vitamin B_5), biotin (vitamin B_7), folic acid (vitamin B_9), pyridoxine (vitamin B_6), and cobalamin (vitamin B_{12}). The compounds carnitine, choline, lipoic acid, inositol, and para-aminobenzoic acid perform analogous functions and are included among the B vitamins (Fig. 1.15).

The B vitamins participate in biochemical reactions in the intermediate energy metabolism, and high concentrations of these vitamins are found in metabolically active fish tissues. Cold-water carnivorous fish may not have a regular intestinal microflora that supplies additional vitamins. The exogenous supply of B vitamins is therefore the natural content of the live feed, feed ingredients, and additional enrichment or supplementation of the diet through vitamin premixes. While vitamin C deficiency causes drastic changes, lack of B vitamins is less obvious to reveal, especially at larvae and juvenile stages of farmed fish (Chatterjee et al. 2017) (Tables 1.2 and 1.3).

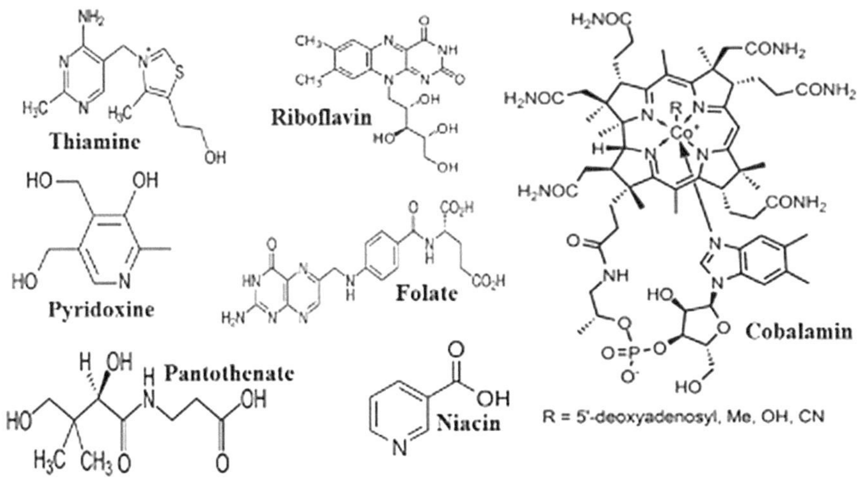

Fig. 1.15 Chemical structure of B-complex vitamins. (https://www.researchgate.net/figure/Structure-of-vitamin-B-complex-in-green-vegetables_fig1_317132087)

1.1.5.3 Minerals and Trace Elements

Minerals are inorganic compounds. They remain back after incineration of food of animal and plant origin. They can be sorted according to the magnitude of their intake requirements for human beings. The main elements occurring in foods are Na, K, Ca, Mg, Cl, P, and S. They are essential for human in amounts more than 50 mg/day. Iron, I, Mo, Co, F, Zn, Se, Cu, Mn, Cr, and Ni are assigned to the group of trace elements. Some shellfish, such as clams and oysters, are a good source of iodine, iron, zinc, magnesium, copper, and other trace minerals. Most fish contain moderate to small amounts of these minerals. They are essential in concentrations less than 50 mg/day. Most of the trace elements has important role as a constituent of metalloproteinases. The requirement of ultra-trace elements such as Al, As, Ba, Bi, B, Br, Cd, Cs, Ge, Hg, Li, Pb, Rb, Sb, Si, Sm, Sn, Sr, Tl, Ti, and W has also been proven based on some animal experiments, accounting to approximately 72 trace elements required for the proper functioning of the human body.

The knowledge of the mineral content and their bioavailability in foods is important in order to estimate whether food intake provides for the micro and macroelement needs of the organisms. Minerals have widespread biological role. They can be present both in inorganic form in solutions (e.g., electrolytes constituents) or building materials (e.g., in bones) and essential part of organic macromolecules (e.g., enzyme prosthetic group). The roles of major and minor minerals and the recommended daily intake (RDI) values are summarized in Tables 1.4 and 1.5, respectively.

Minerals exert also an effect on the food quality, e.g., they can activate or inhibit enzyme catalyzed reactions or nonenzymatic reactions in food and therefore they can have an impact on organoleptic properties (color and flavor). They initiate lipid

Table 1.2 Dietary source and functions of B-complex vitamins

Vitamin		Dietary source	Functions
B_1	Thiamin ($C_{12}H_{17}N_4OS$)	Liver, whole grains, peas, meat, legumes, enriched grain products	Functional part of coenzyme TPP in pyruvate and α-ketoglutarate dehydrogenases, transketolase; poorly defined function in nerve conduction
B_2	Riboflavin ($C_{17}H_{20}N_4O_6$)	Liver, dark green vegetables, milk products, oysters, enriched grain products, eggs	Functional part of coenzymes FAD, FMN in oxidation-reduction reactions
B_3	Niacin ($C_6H_5O_2N$)	Liver, fish, peanuts, meat, enriched grain products, eggs, synthesized from tryptophan	Functional part of coenzymes NAD, $NADP^+$ in oxidation-reduction reactions
B_5 (PP)	Pantothenic acid ($C_9H_{17}NO_5$)	Liver, kidney, legumes, meat, enriched grain products, eggs yolk, synthesized by intestinal flora	Functional part of coenzyme CoA (universal acyl carrier in Krebs cycle, fatty and other carboxylic acid metabolisms) and phosphopantetheine (acyl carrier protein in fatty acid synthesis)
B_6	Pyridoxine, Pyridoxal, Pyridoxamine (C_5H_5N derivatives)	Whole grains, dark green vegetables, peas, fish, meat, legumes, enriched grain products	Functional part of coenzyme PLP in transamination and decarboxylation of amino acids and glycogen phosphorylase
B_7 (H)	Biotin ($C_{10}H_{16}O_3N_2S$)	Liver, kidney, legumes, meat, green vegetables, eggs yolk, synthesized by intestinal flora	Coenzyme (biocytin) in carboxylation reactions in gluconeogenesis and fatty acid synthesis
B_9	Folic acid ($C_{19}H_{19}O_6N_7$)	Liver, kidney, legumes, fish, green vegetables, peas, meat, enriched grain products, citrus fruits	Functional part of coenzyme THFA in transfer of one-carbon fragments
B_{12}	Cobalamin ($C_{63}H_{88}O_{14}N_{14}$ P co)	Found only in animal foods -liver, kidney, fish, meat, eggs yolk, milk products, oysters, shellfish	Functional part of coenzymes adenosylcobalamin (methylmalonyl-CoA mutase) and methylcobalamin (methionine synthase) in transfer of one-carbon fragments and metabolism of folic acid

peroxidation processes and therefore facilitate the formation of off-flavors in the products.

When these elements are not adequately provided to the body, mainly by dietary intake, the individual may suffer from mineral deficiency diseases, for example, anemia, osteoporosis, goiter, stunted growth, and genetic disorders.

Table 1.3 Vitamin content in fish

Fish	A (IU/g)	D (IU/g)	B_1(thiamine) (μ/g)	B_2 (riboflavin) (μ/g)	Niacin (μ/g)	Pantothenic acid (μ/g)	B_6 (μ/g)
Cod fillet	0–50	0	0.7	0.8	20	1.7	1.7
Herring fillet	20–400	300–1000	0.4	3.0	40	10	4.5
Cod liver oil	200–10.000	20–300	–	3.4	15	4.3	–

Source: FAO

Table 1.4 Major elements in fish/seafood

Sodium	Sodium is important in muscle contraction and nerve transmission. Sodium ions are the main regulators of extracellular fluid and volume. It maintains the osmotic pressure of the extracellular fluid and also can be an enzyme activator. The Recommended Daily Intake (RDI) value is 2400 mg. An excessive intake can lead to hypertension that can be avoided with non-salty diet or using diet salt. The intake of chloride is correlated with that of sodium. Chloride is a counterion for hydrogen ions in gastric juice.
Potassium	It is the most abundant cation in the intracellular fluid regulating osmotic pressure within the cell. Some foods have low levels of potassium (e.g., white bread); if their consumption is predominant, potassium deficiency may occur. Potassium and sodium work together in muscle contraction nerve impulse transmission. The Recommended Daily Intake (RDI) value is 3500 mg.
Magnesium	Magnesium has several roles including supporting the functioning of the immune system and assists in preventing dental decay by retaining the calcium in tooth enamel. It has an important role in the synthesis of proteins, fat, nucleic acids, glucose metabolism, as well as membrane transport system of cells. Magnesium also plays an important role in muscle contraction and cell integrity. Magnesium influences the activity of several enzymes and also can be an enzyme constituent. The RDI value of magnesium is 350 mg.
Calcium	Calcium is responsible for strong bones and teeth and accounts for 90% of the calcium in the body. Calcium present in circulating body fluids functions in the transmission of nerve impulses, contractions of muscles, blood clotting, secretion of hormones, and activation of some enzymes. The prerequisite for the adequate absorption of calcium is an adequate intake of vitamin D. The main sources of calcium are milk and milk products, meat, fish, seafood, some fruits, and vegetables. The Recommended Daily Intake (RDI) value is 1000 mg.
Phosphorus	Phosphorus is an essential structural component of cell membranes and nucleic acids and also involved in several biological processes, including bone mineralization, energy production, cell signaling through phosphorylation reactions, and regulation of acid-base homeostasis. The Recommended Dietary Allowance (RDA) of phosphorus for healthy adults is 700 mg/day.

Table 1.5 Minor elements in fish/seafood

Iron	Iron is the central metal in the hemoglobin molecule for oxygen transport in the blood and also occupies a major portion in the myoglobin of muscles. It is also present in several enzymes, and the Recommended Daily Intake (RDI) value is 15 mg.
Copper	Copper assists in the formation of hemoglobin, helping to prevent anemia and also acts as a constituent of several enzymes. Copper together with iron can enhance unwanted reactions during food processing. Moreover the copper(II)-ions are taste bearing. The Recommended Daily Intake (RDI) value of copper is 2 mg.
Selenium	Adequate supply of selenium is contributed to the sufficient function of antioxidant activity of the body (e.g., glutathione peroxidase). Selenium has several roles such as regulating the thyroid hormone as well as assisting in deactivation of heavy metals. The Recommended Daily Intake (RDI) value of selenium is 35 μg.
Chromium	The function of chromium is related to stabilizing blood sugar levels with respect to insulin required for release of energy from glucose. Chromium has an important role in the glucose metabolism; its deficiency can cause a decrease in glucose tolerance and therefore enhanced risk of cardiovascular diseases. The Recommended Daily Intake (RDI) value is 120 μg.
Zinc	Zinc plays a very important role in the production, storage, and use of insulin, the hormone that lowers high blood glucose levels. This transition metal is an essential cofactor in insulin-degrading enzyme and several key Zn^{2+} finger transcription factors that are required for β cell development and insulin gene expression. It is crucial to the immune system, enhances the senses of smell and taste, involves in the metabolism of macronutrients, promotes tissue growth and cell reproduction, and also acts as an antioxidant, protecting cells from damage, and it is associated with hundreds of enzymes. Oysters are the richest source of zinc. The Recommended Daily Intake (RDI) value of zinc is 15 mg.
Manganese	Manganese is one of the cofactors in a number of enzymes as in the case of molybdenum. It is relatively nontoxic even in higher amounts. The Recommended Daily Intake (RDI) value of manganese is 5 mg.
Molybdenum	Molybdenum is present in bacterial nitrate reductase that is involved in meat curing and pickling processes. The Recommended Daily Intake (RDI) value of Mo is 75 μg.
Nickel	Nickel affects optimal growth, healthy skin, and bone structure. It is involved in iron metabolism and plays a role in the creation of red blood cells. It is necessary in the metabolism of sugars, fats, hormones, and cell membranes. Nickel has been shown to enhance insulin activity. The Recommended Dietary Allowance (RDA) of nickel for healthy adults is 170 μg/day.
Fluorine	Fluorine in the form of fluoride inhibits the growth of microorganisms responsible for the development of dental caries. Important constituent in the material of teeth (fluorapatite) and therefore retards the solubilization of tooth enamel in an acidic pH. Fluoridated water is the primary source of dietary fluorine intake. The Recommended Daily Intake (RDI) value of fluorine is 4 mg.

(continued)

Table 1.5 (continued)

Iodine	Iodine is an essential component of the thyroid hormones thyroxine (T4) and triiodothyronine (T3). Thyroid hormones regulate many important biochemical reactions, including protein synthesis and enzymatic activity, and are critical determinants of metabolic activity. The occurrence of goiter has been shown to correlate with iodine deficiency. Most fruits and vegetables contain very low levels of this element. The most abundant sources are seaweeds, seafoods, milk, and eggs. The Recommended Daily Intake (RDI) value of iodine is 150 μg.
Cobalt	Cobalt is very important for the health of the thyroid gland as it is a necessary cofactor for making the thyroid hormone, thyroxine. Cobalt is also important for healthy red blood cells. It helps to create properly formed and sufficient amounts of hemoglobin, the oxygen carrying pigment in the body. The best food sources of cobalt are clams, fish, and green leafy vegetables. The average adult intake of cobalt is 5–8 μg per day.

1.1.6 Carbohydrates

Carbohydrates are one of three basic macronutrients needed to sustain life (the other two are proteins and fats). Carbohydrates make up three-fourths of the biomass of plants but are present only in small quantities in the animal body as glycogen, sugars, and their derivatives. Glycogen is often referred to as animal starch because it is not present in plants. Derived monosaccharides such as the sugar acids, amino sugars, and the deoxy sugars are constituents of all living organisms. Carbohydrate also occurs as part of the chemical constituents of nucleotides and forms the source of ribose liberated as a consequence of the postmortem autolytic changes.

The contents of carbohydrate including dietary fiber in shellfish tissue are low, compared to protein and fat. Carbohydrate varies from 1.3% in cooked lobster meat to 2% to 3% in oyster and the green mussel. Glycogen contents of 1.0–1.2% are seen during winter, in red shrimp, pink shrimp, and Norway lobster. Pacific oyster has a glycogen content of 6.5 ± 3.0%, of dry weight, during winter. Asian hard clam has a maximum of 7.9% carbohydrates (Karnjanapratum et al. 2013). The mussels (*Mytilus* spp.) contain mytilan, a noncovalently linked complex of 95% polysaccharide and 5% protein, and another polysaccharide, a (1–4)-D-glucan (Venugopal and Gopakumar 2017).

1.1.6.1 Carbohydrates of Nutritional Significance

In a nutritional point of view, fish has only negligible amount of carbohydrates, compared to protein and fat. But they are a very important factor affecting the quality of fish during processing. The major carbohydrates present in the muscle of fish/seafood are glycogen. The glycogen content also contributes to the organoleptic properties of edible bivalve meat.

Glycogen is a water-soluble polysaccharide, composed by polymerized glucose molecules joined through glycosidic bonds, forming a network-like structure. It represents the main form of energy storage in the animal bodies along with body fat. In particular bivalves present higher percentage of glycogen (1–7%) compared to fat, and its contents are influenced by internal factors, such as growth and sexual maturation, and external factors, such as food availability and other environmental factors (Yurimoto 2015) (Fig. 1.16).

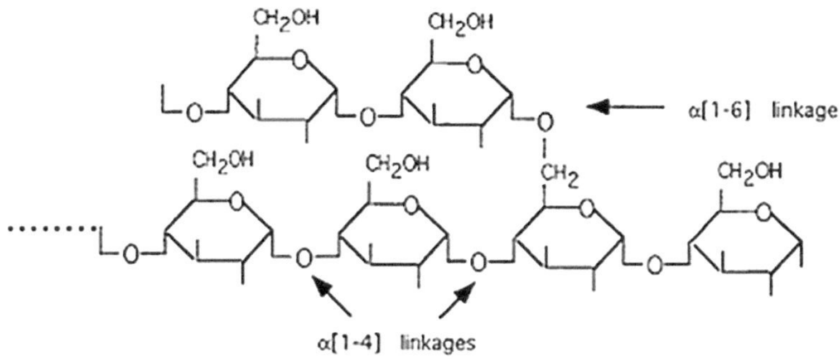

Fig. 1.16 Structure of glycogen

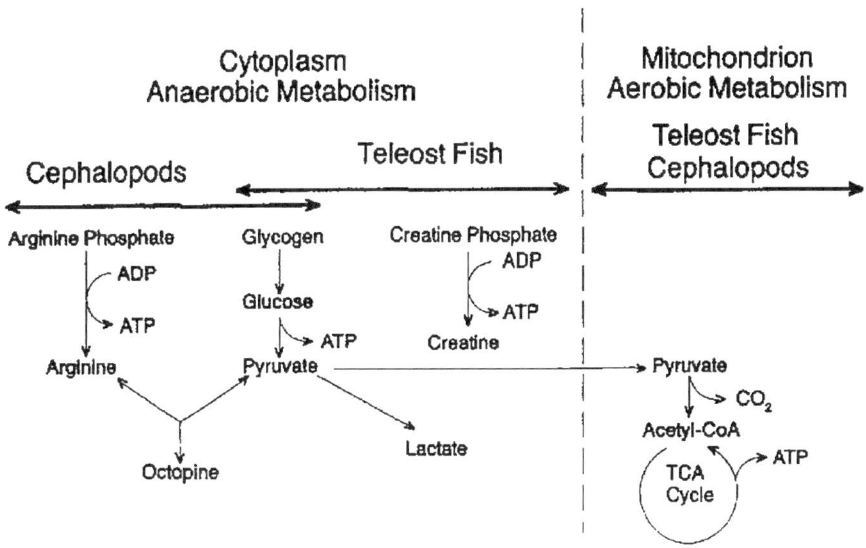

Fig. 1.17 Aerobic and anaerobic degradation of glycogen in fish muscle. (www.fao.org/3/v7180e/V7180E0a.htm)

Quality Implications

Glycogen content declines rapidly after death of the organism, particularly with the stress associated during capture. Thus it varies depending on the physiological state of the organism, and it can be used to evaluate physiological condition in different species. Glycogen generally decomposes into lactic acid and small amounts of glucose, sugar phosphates, and pyruvic acid. The accumulation of lactic acid can change the pH of muscle and significantly affects the quality during processing. Instead of lactic acid, the decomposition of glycogen produces a mixture of alanine, succinic acid, and octopine in some species of mollusks (Fig. 1.17).

1.1.6.2 Carbohydrates of Non-nutritional Significance

Chitin is the second most abundant biopolymer, which can be found both in plant and animal sources. Naturally, chitin is found as ordered crystalline microfibrils, which associate with other materials, such as proteins, lipids, polysaccharides, calcium carbonate, and pigments. Plant chitin is highly crystalline with some disordered or paracrystalline regions that arise from defects. Chitin and cellulose are structurally analogous, where in the former the hydroxyl group of glucose moiety is replaced by the acetamide group. However, degree of acetylation in the polysaccharide molecule depends on the source and chitin isolation process.

Chitin is structurally composed of β(1,4)-linked units of N-acetylglucosamine (GlcNAc), which can be degraded by chitinases (EC 3.2.14) to generate GlcNAc and chitooligosaccharides. It exists as two different crystalline forms, α-chitin and β-chitin, while a third form, γ-chitin, is a combination of the α and β forms. α-Chitin chains are the most abundant form that are arranged in an antiparallel fashion, are very stable, and are present in insect cuticles, shells of crabs, lobsters, shrimp, and fungal cell walls. β-Chitin occurs in diatoms, the pens of squid, the chaetae of annelids, and the tubes secreted by tubeworms of the Siboglinidae family (Annelida). It is observed that length of the yeast chitin is only with 100 GlcNAc residues, whereas the crab chitin contains 5000–8000 GlcNAc residues. The distinct molecular packing patterns of the chitin polymorphs confer different physicochemical properties to their corresponding crystalline structures. Chitin and its derivatives are used in wide range of applications (refer to Chap. 7) (Fig. 1.18).

1.1.6.3 Nutraceutical Carbohydrates

Glycosaminoglycans (GAGs) are a fascinating group of molecules that are attracting increased interest among scientists, owing to their several applications in the biomedical, veterinary, pharmaceutical, and cosmetic field, and are therefore regarded as nutraceuticals. Glycosaminoglycans (GAGs), formerly named as mucopolysaccharides, are important components of the extracellular matrix of connective tissues and also on the cell surface of many cell types and in intracellular granules. They occur as long chains of linear acidic polysaccharides which are generally associated with matrix proteins (core proteins) via covalent bond to form

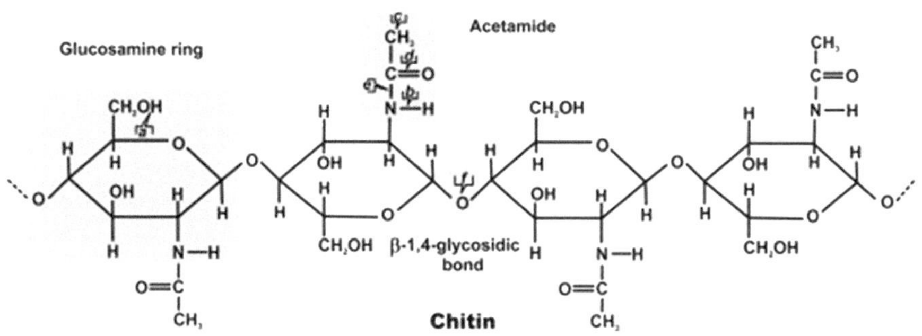

Fig. 1.18 Structure of chitin – oligosaccharide

Table 1.6 Structural and functional properties of GalGs (the galactosaminoglycan – GAGs)

Synonyms/ etiology	Properties	Structure of the disaccharide repeating unit	Functional significance
Chondroitin sulfate			
Chondroitin sulfate -O, A, C, D, E, F, H, K. "Chondrous" referring to cartilage	Anionic, sulfated heteropolysaccharide abundant in cartilage. Regulates growth factors, cytokines, chemokines, lipoproteins	D-glucuronic acid (GlcA) and sulfated N-acetyl-D-galactosamine (GalNAc) linked together via alternating β-1,4 and 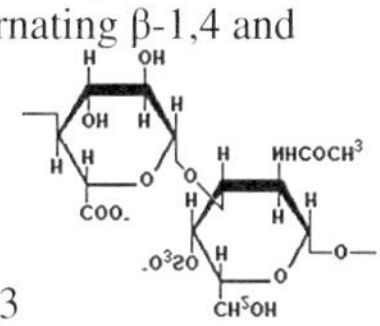 β-1,3	Treatment of osteoarthritis, biomaterial scaffold
Dermatan sulfate			
Chondroitin sulfate B	Dermatan sulfate derives its name from the dermis, its primary source. Role in coagulation, cardiovascular disease, carcinogenesis, infection, wound repair, fibrosis, and stimulate cell growth	N-Acetylgalactosamine (GalNAc) and l-iduronic acid (l-IdoA) joined by β-1, 4 and β-1, 3 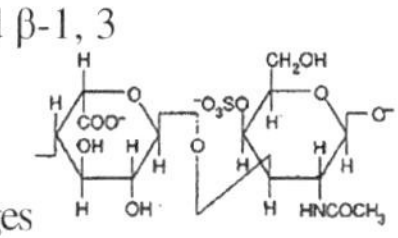 linkages	Important biomaterial scaffold for wound and skin tissue repair

proteoglycans (PGs) and thereby involved in a wide range of biological processes. These heteropolysaccharides are unbranched polysaccharide chains, generally composed of repeating disaccharide units that are highly negatively charged, due to the presence of sulfate or carboxyl groups on most of their sugar residues. They are called glycosaminoglycans because one of the two sugar residues in the repeating disaccharide is always an amino sugar (N-acetylglucosamine or N-acetylgalactosamine), which in most cases is sulfated. Depending upon the hexosamine unit present, GAGs can be classified mainly into galactosaminoglycans (GalGs) and glucosaminoglycans (GlcGs). Chondroitin sulfates A and C and dermatan sulfate, also termed chondroitin sulfate B (CS-B), are the galactosaminoglycans (GalGs). The glucosaminoglycans include hyaluronic acid, heparin, heparan sulfate, and keratan sulfate. The structure, properties, and functional significance of GAGs are given in Tables 1.6 and 1.7.

Other potential uses of GAGs are as viscosupplements, antiviral, anti-infective, and anti-inflammatory agents. Novel documented areas of interest in particular are the suitability of hyaluronan and chondroitin sulfate for tissue engineering and regenerative medicine. Both these polymers are gaining popularity as a biomaterial scaffold which can significantly be improved with cross-linking of collagen, gelatin, chitosan, etc. producing a hydrogel, for wound healing applications. Nutraceutical carbohydrates present in seaweeds are discussed in Sect. 7.4.

Table 1.7 Structural and functional properties of GlcGs (the glucosaminoglycan – GAGs)

Synonyms/etiology	Properties	Structure of the disaccharide repeating unit	Functional significance
Hyaluronic acid			
Biological glue, hyaluronate, hyaluronan	Mw: 1000 Da to 10,000,000 Da anionic, non-sulfated GAG, abundant in synovial fluid of human and other vertebrates.	D-glucuronic acid and D-N-acetylglucosamine, linked together via alternating β-1,4 and β-1,3 glycosidic 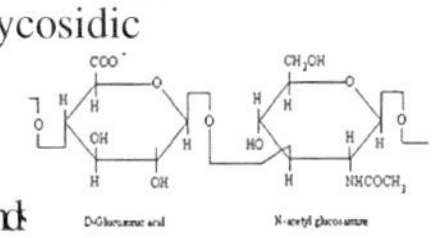 bond	Functions as a biological lubricant in joints. It is a biodegradable, biocompatible, non-immunogenic biopolymer
Heparin			
"Heparin" is derived from *hepar* (Greek for "liver"), originally isolated from canine liver cells	Mw: 3 kDa to 50 kDa. Heavily sulfated with the highest negative charge density of any known biological molecule, hence a naturally occurring anticoagulant	2-O-sulfated iduronic acid (IdoA-2S) and 6-O-sulfated, N-sulfated glucosamine (GlcNS-6S), with varying degree of 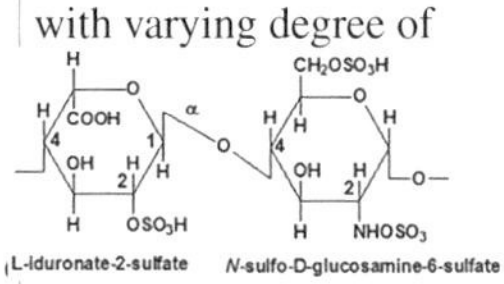	Heparin is coated in medical devices – stents, vacutainers, renal dialysis machines, and blood oxygenators used in heart-lung machines
Heparan sulfate			
HS	Very closely related in structure to heparin, with lesser degree of sulfation. Mw 5–70 kDa	**Glucuronic acid (GlcA) linked to *N*-acetylglucosamine (GlcNAc) via** β (1→4) 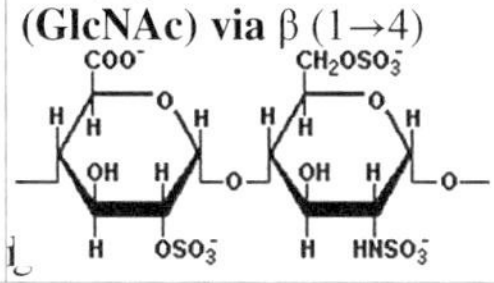	Wound repair, contribute to pathophysiology
Keratan sulfate			
Keratosulfate – KSI, KSII, and KSIII Word derived from keratinized tissues	Main function of keratan sulfate is the maintenance of corneal transparency	**Galactose and N-acetylglucosamine** via alternating β-1,4 and β-1,3 glycosidic bonds **(3Gal -β1–4-** 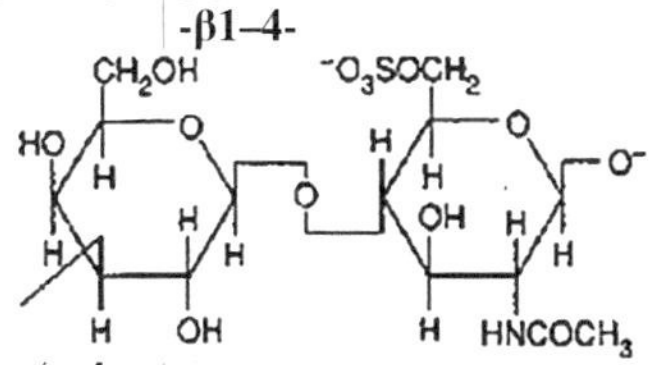	The important property of KS is its anti-adhesive character

1.1.7 Flavor Bearing Constituents

Flavor is an overall sensation during food consumption. It is the interaction of three elements: taste, odor, and textural feeling. The compounds that are responsible for taste are usually nonvolatile at room temperature and therefore interact only with taste receptors. Molecules that are responsible for odors are called aroma substances, and they are volatile. Some compound can provide both sensations. The most important substances defining the characteristic aroma of a given food are called key odorants. Components that can cause a faulty odor or taste or both are called off-flavor.

It is well known that the flavor of fish and shellfish principally originates from the extractive components. Taurine, glycine, alanine, glutamic acid, arginine, AMP, and glycine betaine are major constituents in shellfish which were considered as taste-active components in abalone, snow crab, scallop, and short-necked clam. Sarower et al. (2012) found that glutamate, glycine, alanine, arginine, proline, valine, methionine, phenylalanine, tyrosine, inosine 5′-monophosphate (IMP), adenosine 5′-monophosphate (AMP), guanosine 5′-monophosphate (GMP), trimethylamine, trimethylamine oxide (TMAO), glycine betaine, lactate, and succinate as important contributors to the taste of raw and processed fishery products. Sweet, salty, bitter, sour, and umami are the basic tastes, defined by these taste-active components. Sweet taste is imparted by glycine, alanine, and TMAO while bitter taste by arginine and other hydrophobic amino acids. Glutamate has a role in sour taste and contributes to umami taste through synergetic effects in coexistence of IMP, GMP, and AMP. The large amount of alanine or glutamate suppresses the sweetness effect of glycine through antagonistic effect.

Nonprotein nitrogenous compounds have been widely used as indicators for evaluation of the taste of fish and shellfish. Some peptides and FAAs are very important taste substances; many of the latter – such as alanine, glutamic acid, and glycine – are responsible for flavor and taste in foods. Alanine, glycine, and glutamic acid have the "umami" taste typical of crustaceans (Yamanaka and Shimada 1996).

Fish and seafood reveal a diverse group of desirable flavors and a range of deteriorative flavors. These aromas and flavors are generally classified into six different types (Lindsay 1990), namely:

- Extremely fresh fish- and seafood-like flavors
- Oxidized, stale, and stored flavors
- Spoiled or putrid flavors
- Species-related characterizing flavors
- Derived or processing flavors
- Environmentally derived flavors

Very Fresh Fish- and Seafood-Like Flavors This flavor is highly sought by consumers. Very fresh fish aromas and flavors are characterized by mild, green, and planty notes which are easily recognized and readily associated with fresh fish that are contributed by 6-, 8-, and 9-carbon aldehydes, ketones, and alcohols derived from long-chain polyunsaturated fatty acids via specific lipoxygenase activity. C8 compounds such as 1-octen-3-ol, 1-octen-3-one, 1-cis-5-octadiene-3-ol, and 1-cis-5-octadien-3-one occur in most types of fish and seafood and contribute heavy plant like odors and metallic-like flavors. Dimethyl sulfide is related with cooked oyster flavors, and dimethyl-β-propiothetin acts as the thermally labile precursor for this sulfur compound.

Oxidized, Stale, and Stored Flavors Loss of flavors in refrigerated fish is caused by degradation of fresh fish compounds to less flavorful derivatives and the concomitant production of odorous microbial metabolites. Microbial contributions are often supplemented with simultaneous production of lipid oxidation products and a reasonably predictable range of flavors (from fresh flat, sweet to stale putrid in progression). During this progression, both potent fishiness caused by trimethylamine formation and fishiness caused by the 2,4,7-decatrienals from lipid oxidation are prone to develop and also contribute strongly to negative consumer acceptance.

Spoiled or Putrid Flavors Under aerobic conditions of refrigerated storage, putrid aromas and flavors caused by volatile sulfur compounds render the product completely unfit for consumption. Under carbon dioxide-modified atmospheres, spoilage in refrigerated fresh fish is altered so that noxious sulfur compounds, such as methanethiol, dimethyl disulfide, and dimethyl trisulfide, are not formed.

Species-Related Flavors Depending on the nature of the fish, mollusk, or crustacean, very fresh flavors are often accompanied by easily recognized species-related flavors as in salmon, shrimp, clams, or crabs; and these highly valued by consumers.

Derived or Processing Flavors These include fish and seafood flavors derived through processing such as flavors of canned tuna-like and the flavors that develop by the addition of condiments as in fish, oyster, and clam sauces.

Environmentally Derived Flavors Generally, environmentally derived flavors in fish and shellfish contribute negatively to fish and seafood flavors. These compounds can cause significant detrimental flavors in aquaculture and wild-captured fish. Environment plays an important role in contributing desirable flavors. Musty or muddy off-flavors caused by geosmin and methylisoborneol, produced by bacteria and algae, have been studied extensively in relation to fish flavors.

Table 1.8 Principal nutritional constituents (%) in fish. (Source: FAO)

Constituent	Fish fillet (g%)
Protein	16–21
Lipid	0.2–25
Carbohydrate	<0.5
Ash	1.2–1.5
Water	66–81

1.2 Proximate Composition/Nutritional Quality Evaluation

The percentage composition of the four major constituents of fish, viz., water, protein, lipid, and ash (minerals), is referred to as proximate composition. The chemical composition of fish varies greatly from one species and one individual to another depending on age, sex, environment, and season. The variation in the chemical composition of fish is closely related to feed intake, migratory swimming, and sexual changes in connection with spawning. The general composition of water, lipid, protein, and ash in fish is shown in Table 1.8.

1.2.1 Analysis of Moisture (Robert 2014)

Moisture content may be determined by numerous techniques. These may be divided into two major categories, primary and secondary measurement.

Primary moisture techniques involve direct chemical determination of water content, usually by extracting the moisture from the product. All primary methods are destructive and time-consuming. Primary methods are performed offline but are usually very accurate. Small sample size may not adequately represent bulk product. The most common primary method is weight loss, in which a sample is weighed and dried (in an ordinary, microwave or vacuum oven) until no further weight loss and then reweighed. Other methods include the use of volumetric moisture meter, based on Karl Fischer titration. The accuracy of all offline primary methods is dependent upon laboratory instrument accuracy and the skill of lab personnel. Since offline methods require taking a product sample from the process, the sampling method must provide consistent product samples for testing.

Moisture content is normally expressed as a percentage by weight of either total product (wet basis) or dry product (dry basis).

$$\text{Moisture content (wet basis)} = \frac{100 \times (\text{Wet weight} - \text{Dry weight})}{\text{Wet weight}}$$

$$\text{Moisture content (dry basis)} = \frac{100 \times (\text{Wet weight} - \text{Dry weight})}{\text{Dry weight}}$$

From the above equations, wet basis moisture content cannot exceed 100%. Dry basis moisture may exceed 100% and is a nonlinear function.

Secondary moisture techniques measure a property of the variable (moisture) rather than the variable directly. Secondary measurement principles must be calibrated against a primary reference technique and have the advantage of continuous or rapid sampling measurement without sample destruction. Among online moisture measurement techniques, dielectric measurement and near infrared reflectance are the most reliable and accurate methods of moisture monitoring (refer to Sect. 5.1).

1.2.2 Analysis of Protein (AOAC 2005)

1.2.2.1 Microkjeldahl Method

In the Kjeldahl procedure (1883), proteins and other organic food components in a sample are digested with sulfuric acid in the presence of catalysts. The total organic nitrogen is converted to ammonium sulfate, and finally the digest is neutralized with alkali and distilled into a boric acid solution. The borate anions formed are titrated with standardized acid, which is converted to nitrogen in the sample. The result of the analysis represents the crude protein content of the food since nitrogen also comes from nonprotein components. This approach is based on two assumptions: that dietary carbohydrates and fats do not contain/contain only negligible amounts of nitrogen and that nearly all of the nitrogen in the diet is present as amino acids in proteins. It should be noted that the Kjeldahl method would also estimate the nitrogen content in any ammonia and ammonium sulfate.

The steps include:

(i) Digestion with sulfuric acid, with the addition of a suitable catalyst ($CuSO_4.5H_2O:K_2SO_4$; 1:8) to complete the oxidation and conversion of nitrogen to ammonium sulfate
(ii) Neutralization of the diluted digest, followed by distillation into a known volume of standard (2% boric acid)
(iii) Titration of the distillate against a suitable acid (N/50 H_2SO_4)

Nitrogen content is then multiplied by a factor to arrive at protein content. On the basis of early determinations, the average nitrogen (N) content of fish proteins is found to be about 16%, which directs to use of the calculation $N \times 6.25$ ($1/0.16 = 6.25$) to convert nitrogen content into protein content.

1.2.2.2 Colorimetric Methods

The water-soluble protein estimation and be done by colorimetric methods using Biuret (sensitive for mg quantities), Lowry's method (sensitive for μg quantities), or Bradford's (sensitive for μg quantities, even in presence of other interfering chemicals).

(a) *Biuret Method* (Lubran 1978)

Principle This is the most common method for the estimation of compounds having a peptide bond. Compounds containing two or more peptide bonds react with alkaline copper sulfate to give a violet- or pink-colored product. This reaction is due to the formation of coordination complex of cupric ions with unshared pairs of peptide nitrogen and oxygen of water.

This reaction is also answered by compounds containing peptide bonds. The –CONH- bond in urea gives positive test since it gets converted to biuret during the reaction. This reaction is also given by compounds such as glutathione, carnosine (β-alanyl-histidine), and anserine (β-alanyl-methylhistidine).

Materials

Biuret reagent: Dissolve 3.0 g $CuSO_4$ and 9.0 g sodium potassium tartrate in 500 ml of 0.2 N NaOH. Add 5.0 g KI (optional). Make up the volume to a liter using 0.2 N NaOH.

Standard BSA: Dissolve Std. BSA in minimum volume of 0.2 N NaOH without frothing. The solution is finally made up to 100 ml in a standard flask.

Procedure

Pipette out standard BSA solution into different test tubes and make up the volume to 3.0 ml using distilled water. Pipette 3.0 ml each of the made up unknown samples into tubes labeled as test. 3 ml of deionized double-distilled water is used as blank. Add 3.0 ml of Biuret reagent to all the tubes, including blank, mix well, and keep undisturbed for 10 min at room temperature. Absorbance is determined at 540 nm after setting zero using the blank.

(b) *Lowry's Method* (Lowry et al. 1951)

Principle This is the most common and versatile method for the determination of proteins because of its high sensitivity and reliability to detect quantities as low as 20 μg protein in solution. The –CO-NH- (peptide) bonds in polypeptide chain react with copper sulfate in an alkaline medium to give a blue-colored complex. In addition the tyrosine and tryptophan residues of proteins cause reduction of the phosphomolybdate and phosphotungstate components of the Folin-Ciocalteu reagent to give bluish products which contribute towards enhancing the sensitivity of this method.

However it is important to remember that several compounds like EDTA, Tris, carbohydrates, NH_4^+, Mg^{++} ions, thiol reagents, phenol, etc. interfere with color development, and it should be ensured that such substances are not present in samples.

Materials

Alkaline Copper reagent. Solution A, 2% sodium carbonate in 0.1 N NaOH; solution B, 0.5% Cu SO_4 in 1% Rochelle salt solution (Na-K tartrate). Mix 200 ml of solution A + 4.0 ml of solution B immediately before use.

Standard Bovine serum albumin (BSA): Conc. µg/ml.

Folin-Ciocalteu Phenol reagent: Diluted the reagent appropriately so that it is 1 N in respect of its acid content. (Diluted 1:1 volume using distilled water)

Procedure: Pipette out 0.1–1.0 ml of standard BSA solution into test tubes and make up the volume to 1.0 ml in each tube. Pipette 1.0 ml each of unknown samples into tubes labeled test. One ml of deionized double-distilled water is as used blank. Add 5.0 ml of alkaline copper reagent to all the tubes, mix well, and kept undisturbed for 10 min at room temperature. Add 0.5 ml of Folin's phenol reagent to all tubes and mix well. Absorbance is determined at 670 nm.

(c) *Bradford Method* (Bradford 1976)

Principle The protein in solution can be measured quantitatively by different methods. The methods described by Bradford uses a different concept – the protein's capacity to bind to a dye, quantitatively. The assay is based on the ability of proteins to bind to Coomassie Brilliant Blue and form a complex whose extinction coefficient is much greater than that of free dye.

Materials

Dissolve 100 mg of Coomassie Brilliant Blue G250 in 50 ml of 95% ethanol. Add 100 ml of 85% phosphoric acid, and make up to 600 ml with distilled water. Filter the solution and add 100 ml of glycerol, and then make up to 1000 ml. The solution can be used after 24 h. Standard BSA (µg/ml).

Procedure

Prepare various concentrations of standard protein solutions from the stock solution (0.2, 0.4, 0.6, 0.8, and 1.0 ml) into series of test tubes and make up the volume to 1 ml. Pipette out 0.2 ml of the sample in two other test tubes and make up the volume to 1 ml. A tube with 1 ml of water serves as blank. Add 5.0 ml of Coomassie Brilliant Blue to each tube, and mix by vortex or inversion. Wait for 10–30 min and read each of the standards and each of the samples at 595 nm. Plot the absorbance of the standards verses their concentration. Plot graph of optical density versus concentration. From graph find amount of protein in unknown sample.

1.2.3 Analysis of Crude Lipid (David and Wayne 2014)

Lipids are soluble in ether, chloroform, or other organic solvents but are sparingly soluble in water. Therefore, water insolubility is the essential analytical property used as the basis for the separation of lipids from proteins, water, and carbohydrates in foods. Liquid triacylglycerols at room temperature are referred to as oils, whereas solid triacylglycerols at room temperature are referred to as fats. Quantitative and qualitative analysis of lipids in foods is important for accurate nutritional labeling.

Semicontinuous Extraction Method Solvent extraction is the most common methods of lipid extraction from fish and seafood. Ideal solvents for fat extraction should have a high solvent power for lipids and minimal solvent power for other macromolecules. They should evaporate readily with no residue, have relatively low boiling point, and should be nonflammable and nontoxic in either states (liquid/vapor). The ideal solvent should penetrate sample particles readily, be in single component form to avoid fractionation, and be inexpensive and nonhygroscopic. Fat-soluble organic solvents can be used for the extraction of lipids from moisture-free samples. Ethyl ether and petroleum ether are regularly used for extraction. Ethyl ether has a boiling point of 34.6 °C and is a better solvent for fat than petroleum ether. It is generally expensive compared to other solvents, has a greater danger of explosion and fire hazards, is hygroscopic, and forms peroxides. Petroleum ether is the low boiling point fraction of petroleum and is composed mainly of pentane and hexane. It has a boiling point of 35–38 °C and is more hydrophobic than ethyl ether. It is selective for more hydrophobic lipids, cheaper, less hygroscopic, and less flammable than ethyl ether. The semi-discontinuous fat extraction method can be performed using a Soxhlet apparatus as mentioned in Chap. 5.

For continuous solvent extraction, solvent from a boiling flask flows incessantly over the sample held in a ceramic thimble. Fat content is calculated by the weight loss of the sample or by weight gain of the beaker. These methods give faster and efficient extraction results than semicontinuous extraction methods. However, they may cause channeling which results in incomplete extraction. The Goldfish and the Wiley and Underwriters tests are examples of continuous lipid extraction methods.

$$\%\text{Fat content}(\text{dry basis}) = \frac{\text{g weight of sample}}{\text{g weight of dried sample}} \times 100$$

1.2.4 Analysis of Ash Content (Maurice 2014)

Ash content signifies the total mineral content in food, as a part of proximate analysis for nutritional evaluation. It refers to the inorganic residue retained after ignition or complete oxidation of organic matter in a foodstuff. It is the initial step for specific elemental analysis. Because certain foods (fish/seafood) are high in particular minerals, determination of ash content becomes vital. A basic knowledge of the characteristics of various ashing procedures and types of equipment is essential to ensure reliable results. Two major types of ashing used are dry ashing and wet ashing (done by conventional/microwave unit). Dry ashing is used primarily for proximate composition, whereas wet ashing (oxidation) is used for the analysis of certain minerals. The advantages of conventional dry ashing are that it is safe and requires no complicated instrumentation other than muffle furnace (refer to Chap. 5) and the crucibles used for ashing. This method requires no added reagents or blank deduction. Usually several samples can be handled at once, and the resultant ash can be used for other analyses (specific elements, acid-insoluble ash, and water-soluble and insoluble ash). The only drawback of the method is the duration of time taken for analyses (12–18 h or overnight) and the expensive equipment. There will be a loss of the volatile elements and interactions between mineral components and crucibles. Volatile elements at risk of being lost include As, B, Cd, Cr, Cu, Fe, Pb, Hg, Ni, P, V, and Zn (Maurice 2014). The amount and composition of ash content in a sample are also depended upon the nature of the ignited sample and on the method of ashing.

Crucible selection is critical in ashing (refer to Chap. 5). All crucibles should be ignited and cleaned thoroughly before use, and it should be marked suitably for identification. Identification marks with an ordinary marking pen will disappear during ashing at high temperature. An iron nail dissolved in concentrated HCl forms brown stain and is suitable for marking the crucible. Crucibles also may be etched with a diamond point and marked with a 0.5 M solution of $FeCl_3$, in 20% HCl. Laboratory inks scribed with a steel pin are also available commercially. Utmost care must be taken when placing samples in and out of the furnace, and the crucibles should always be held with metal tongs.

Dry Ashing Dry ashing refers to the use of a muffle furnace capable of maintaining temperatures of 500–600 °C. Water and volatiles are vaporized, and organic substances are burned in the presence of oxygen in air to CO_2 and oxides of N_2. Most minerals are converted to oxides, sulfates, phosphates, chlorides, and silicates. Partially volatilizing elements such as Fe, Se, Pb, and Hg may require wet ashing procedure before analysis.

Wet Ashing Wet ashing is used for oxidizing organic substances by using acids/oxidizing agents or their combinations. Minerals are solubilized without volatilization in the wet ashing and are often preferred for specific elemental analysis. Wet ashing requires a hood (a special hood if perchloric acid is used), corrosive reagents, and constant operator attention.

Moisture free samples are used for ashing; and in the case of high-fat fish meats, fat is extracted before ashing. The ash content can be expressed on wet weight or dry weight basis.

The ash content is calculated as follows:

$$\%\text{Ash content}\left(\text{dry basis}\right)=\frac{\left(\text{Weight after ashing}-\text{Tare weight of crucible}\right)}{\text{Original sample weight}\times\text{dry matter coefficient}}\times 100$$

$$\text{Where Dry matter coefficient} = \frac{\%\text{solids}}{100}$$

$$\%\text{Ash content}(\text{wet basis}) = \frac{(\text{Weight after ashing} - \text{Tare weight of crucible})}{\text{Original sample weight}} \times 100$$

1.2.5 Analysis of Carbohydrates

Extraction of Glycogen from Fish Liver with Trichloroacetic Acid (TCA)
A 100 mg sample of the liver is weighed and finely ground with 20 ml of 5% TCA, using a mortar and pestle or homogenizer. The precipitate of proteins is filtered off continuously and finally made up to 100 ml in a standard flask. The clear filtrate submitted to analysis of glycogen. If the iodine reagent is used for the determination of glycogen in the extracts, the concentration of trichloroacetic acid (TCA) should not exceed the limits 4.9–5.1% (w/v). This can be checked by titration with 0.1 N alkali and the solution adjusted if necessary by adding water or a concentrated solution of TCA.

Analysis of tissue glycogen can also be done by dissolution of entire tissue in hot alkali and further precipitation of glycogen from the neutralized extract using alcohol (95%). The precipitate is filtered and vacuum dried, and the yield is estimated gravimetrically. The glycogen content can also be estimated by acid hydrolysis of the precipitate and colorimetric determination of the glucose formed, using anthrone method or phenol-sufuric acid method.

1.2.5.1 Estimation of Glycogen Using Iodine Reagent

Iodine Reagent 16.5 ml of Lugol's solution, prepared by dissolving 1 g of iodine and 2 g of KI in 20 ml of water, is added to 990 ml of an aqueous solution, containing 25% (w/v) of KCl.

Procedure

In a colorimeter tube (1.2 cm. diameter), 2 ml of a liver TCA extract is added to 3 ml of iodine reagent. After mixing, the optical density is read in a spectrophotometer at 650 nm against a blank, obtained by adding 2 ml of 5% TCA to 3 ml. of reagent in the same way. The amount of glycogen is estimated from a calibration curve, prepared using glycogen of the same origin as the sample to be estimated.

The color intensity of the glycogen-iodine complex decreases with increase in the concentration of TCA in the extract, but within the limits of 4.9–5.1% TCA, variations in optical density due to different acidity can be neglected.

1.2.5.2 Estimation of Total Free Sugars by Anthrone Method (Hedge and Hofreiter 1962)

Reagents

Anthrone reagent: 0.2% anthrone in 95% $H_2 SO_4$. This reagent deteriorates rapidly, and hence it should be freshly prepared before each determination.

Glucose standard: Stock standard (1% solution of glucose). Dissolved 1.0 g dextrose in 100 ml water. Working standard: 1.0 ml of the above 1% glucose solution was diluted to 100 ml with distilled water. [Conc: 100 μg/ml].

Principle All carbohydrates get hydrolyzed into constituent monosaccharides in presence of Conc. H_2SO_4. Further the monosaccharides get dehydrated to produce furfurals or their derivatives that combine with anthrone to form bluish-green-colored complexes. Color intensity produced is a function of the amount of carbohydrate present in the test sample. A standard curve is plotted to determine the concentration range within Beer-Lambert's law.

Procedure Pipette 0.1–1.0 ml of diluted glucose standard solution as well as samples into test tubes, and make up the volume to 1.0 ml in each tube using deionized double-distilled water (Milli-Q water). One ml of deionized double-distilled water is taken as blank. Place the tubes in an ice bath, and add 5.0 ml of anthrone reagent to each tube. Mix thoroughly, and keep the tubes in a boiling water bath for 15 min. Cool in dark at room temperature for 30 min. Determine the optical density at 620 nm.

1.3 Amino Acid Composition Analysis by HPLC

Quantitative determination of the amino acids can be performed using a reverse phase high-performance liquid chromatography (RP-HPLC) after precolumn derivatization using PITC (phenyl isothiocyanate) according to Shang and Wang (1996). All the standard amino acids except tryptophan can be analyzed using this method.

Materials

Pico-Tag Eluent A: Pico-Tag Eluent A is prepared by dissolving 19.0 g sodium acetate trihydrate in 1 l of deionized Milli-Q water. Add 0.5 ml triethylamine (TEA) to the above solution and titrate to pH 6.4 using glacial acetic acid. To the resulting solution, add 63.8 ml of acetonitrile and filter using sterile membrane filters (cellulose nitrate, 47 mm Dia, 0.45 μm pore size, WCN type, Whatman Ltd., Maidstone, England).

Pico-Tag Eluent B: Pico-Tag Eluent B is prepared by mixing HPLC-grade acetonitrile and deionized Milli-Q water in the ratio 6:4 by volume.

Sample diluent solution: The diluent solution is prepared as follows. 710 mg of Na_2HPO_4 is added to 1 l of water and titrate to pH 7.4 with 10% phosphoric acid (H_3PO_4). The resulting solution is mixed with HPLC-grade acetonitrile, so that acetonitrile is 5% by volume.

Sample redrying solution: The redrying solution consists of HPLC-grade methanol, Millipore water, and triethylamine (TEA) in the ratio 1:1:0.5 by volume.

Sample derivatization solution: The derivatization reagent consists of HPLC-grade methanol, triethylamine, Millipore water, and phenyl isothiocyanate (PITC) in the ratio 7:1:1:1 by volume.

1.3.1 Sample Preparation for Amino Acid Analysis

Samples (100–200 mg tissue) are hydrolyzed using 10.0 ml of 6 M HCl containing 2% phenol at 110 °C for 24 h in heat-sealed tubes after flushing with nitrogen and finally filtered using Whatman No. 1. The hydrolyzed sample solutions thus obtained are then subjected to derivatization using PITC (phenyl isothiocyanate). Briefly, 5 μl each of the above-filtered sample solutions are transferred to a 1.0 ml sample tube and dried under vacuum to remove all traces of HCl. The samples in the tubes are treated with 20 μl of redrying solution (reagent 1.4.1 (d)), by gentle vortexing and dried under vacuum. Subsequently, the amino acids are derivatized by addition of 10 μl of derivatization solution (reagent 1.4.1(e)), to the above dried samples by gentle vortexing for 20 min at room temperature. The sample derivatization solution is further removed by drying under vacuum for 30 min. After derivatization, 1 ml of sample diluent solution (1.4.1(c)) is added and mixed by vortexing for a few seconds and filtered using 0.45 μm Nylon filters (Whatman, Maidstone, England). For chromatographic analysis, 10 μL each of the above samples is injected into the HPLC.

1.3.2 Apparatus

HPLC system equipped with a binary pump, solvent delivery system, column oven, a dual λ absorbance detector, manual/automated injector and a 5 μm Pico-Tag Reversed-Phase column (3.9 mm i.d. × 150 mm length), and a suitable data analysis software can be used for the analysis of amino acids.

1.3.3 Chromatographic Conditions

Chromatographic separation makes use of continuous gradient elution with Pico-Tag Eluent A and Pico-Tag Eluent B for hydrolysate amino acid analysis. The gradient starts at 100% Eluent A and is decreased to 54% and finally increases to 100% in 14.5 min. The total separation time is less than 14 min, and the gradient is

Table 1.9 HPLC gradient profile for the analysis of amino acids

Sl. No.	Flow (mL min^{-1})	Time (min)	% Eluent A	% Eluent B	% C	Curve
1	1.00	0.01	100.0	0.0	0.0	6
2	1.00	2.00	100.0	0.0	0.0	5
3	1.00	12.00	54.0	46.0	0.0	5
4	1.00	12.50	0.0	100.0	0.0	6
5	1.00	13.50	0.0	100.0	0.0	6
6	1.50	14.00	0.0	100.0	0.0	6
7	1.50	14.50	100.0	0.0	0.0	6
8	1.50	22.00	100.0	0.0	0.0	6
9	1.00	22.50	100.0	0.0	0.0	6

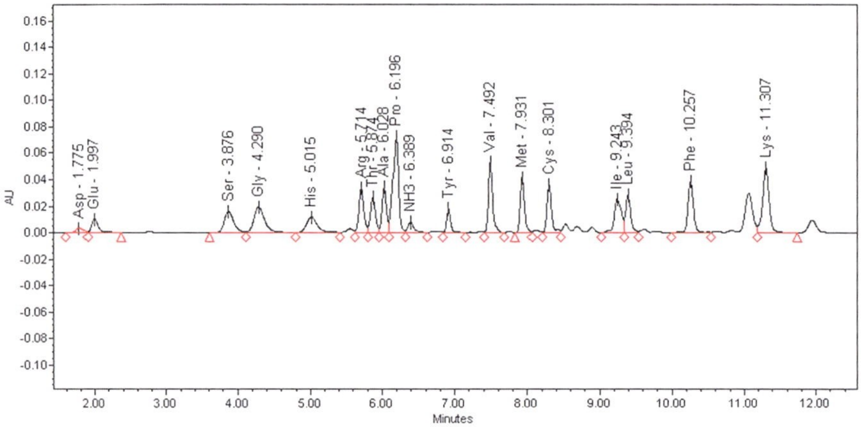

Fig. 1.19 Amino acid profile of standard amino acids obtained by reverse phase HPLC

further run for 22.5 min to ensure full separation. Detection can be monitored at 254 nm. The concentrations of standard amino acids in the range 1.25 µmoles/ml (for cysteine) and 2.5 µmoles/ml for all other standard amino acids may be used for quantification. Quantification of the peak areas of individual amino acids is compared with standard amino acids, and concentration is calculated after injecting 20 µl amino acid solutions. HPLC gradient profile developed for the separation of amino acids is given in Table 1.9 (Fig. 1.19).

1.4 Analysis of Fatty Acids

Marine organisms (fish, seafood, algae) are the main natural sources of essential fatty acids in human diet (mainly EPA and DHA). Fish oil is considered to have the highest amounts of ω-3 PUFA.

1.4.1 Analysis of Fatty Acids Using Gas Chromatography-Mass Spectrometry (GC-MS) (AOAC 2000)

The nutritional value of a seafood sample is dependent on its fatty acid profile, and hence the estimation the fatty acid profile of a sample, based on gas chromatography (GC), is an important and routine analytical method. The oil samples are initially extracted according to the Soxhlet protocol (Chap. 5) or accomplished by a solvent extraction using chloroform and methanol preferably in an inert atmosphere of nitrogen. Sample derivatization prior to analysis increases the volatility of lipid components, thus providing a quick and better resolution of the component fatty acids. An important derivatization procedure involves the conversion of fatty acid components into their corresponding esters. The analytes are derivatized into free fatty acids (FA) by a base-catalyzed reaction and esterified to form fatty acid methyl esters (FAMEs) in an acid-catalyzed reaction using the Lewis acid boron trifluoride in methanol. Preparation of fatty acid methyl esters (FAME) consists essentially of conversion of one ester to another (i.e., transesterification); by cleavage of an ester bond via methanol, the reaction is referred to as methanolysis or transmethylation. When the fatty acid methyl esters are formed as products of reaction between fatty acids and methanol, the process is termed methylation. Both methylation and transmethylation are reversible reactions, normally accomplished in the presence of an acid catalyst (BF_3). HCl is considered as a mild and useful derivatization reagent, because it produces almost quantitative yields with an intermediate methylating power, and low transmethylating capacity, usually requiring reaction times of about 30 min. The analytes are injected on a gas chromatography (GC) system for separation and are subsequently detected by a flame ionization detector (FID). Analyte identity can be further verified by an additional injection on a GC where the analytes are separated and identified by a mass spectrometric detector (MSD).

Extraction and Transesterification

Samples are stored at or below −60 °C prior to extraction in order to minimize oxidation and degradation of samples. Samples are subsequently thawed at room temperature, and approximately ~0.01 mg sample is weighted out using an analytical balance. The samples are then extracted with 65% chloroform and 35% methanol at 2000 psi pressure and 100 °C for 10 min. Approximately 40 ml of a 0.7% NaCl saline solution is added to the crude extract in order to remove methanol and water-soluble components of the extract. The aqueous layer is siphoned off, and the organic phase is dried by passing it through a sintered glass funnel containing several grams of anhydrous sodium sulfate, followed by thorough rinsing of the funnel with chloroform. The pooled dried extract is concentrated to approximately 1.0 ml under a stream of nitrogen using a bath temperature of 35 °C. The remaining solvent is evaporated by impinging with nitrogen from the extraction concentrate. Two milligrams of the recovered lipids are reconstituted in 1 ml of 0.5 M methanolic KOH and subjected to alkali hydrolysis at 80 °C for 1 h (or using 0.5 M NaoH at 100 °C for 5 min.). This is followed by transesterification using 1 ml of fresh 12% (v/v) BF_3 in methanol, at 100 °C for 30 min. After transesterification, 2 ml deionized H_2O and

Table 1.10 Chromatography conditions for fatty acid (FAME) analysis by GC

Parameter	Conditions
Column	Capillary column 0.2 μm film thickness, 0.25 i.d × 100 m length
Carrier gas	Helium
Injection solutions	CH_2Cl_2 of HPLC purity grade
Oven temperature	Oven initial temperature 140 °C to final temperature 240 °C, 4 °C/min
Sample injection volume	1 μL
Split rate	100:1
MS detector	Triple axis detector

1 ml hexane are added to the sample to quench the reaction. The recovered organic phase is pooled and spiked with the methyl ester of an internal standard (henicosanoic acid methyl ester, C21:0) to a final concentration of 25.0 μg/ml.

GC-Flame Ionization Detector Analysis

The samples can be analyzed on a GC-FID using a 60 m, capillary column with I.D. 0.25 mm and 0.25 μm film thickness. Helium or nitrogen is used as the carrier gas with a constant flow of 1 ml/min. The split ratio is 20:1. The injection temperature is set at 300 °C, and temperature range increase from 125 to 240 °C at 3°/min ensures separation, and analysis of fatty acids within a total runtime is 40 min, with data acquisition from 8 to 40 min.

GC-MS Analysis

Nearly 10% of samples are measured with GC-MS for peak confirmation. The same measurement conditions as for GC-FID can be employed or as given in Table 1.10. The mass range is 50–400 m/z, scan time 0.5 s/scan, and emission current of 10uA, with an acquisition time from 8 to 40 min.

1.4.2 ^{1}H-NMR Spectroscopy of Fatty Acids (Knothe and Kenar 2004)

The ^{1}H-NMR spectra of fish oils can be recorded on an NMR spectrometer (refer to Chap. 5). ^{1}H-NMR spectra can be employed as a reliable technique for the analyses of saturated and unsaturated fatty acids. The major advantage of this technique is that the sample preparation gets simply reduced to the dilution of 20 μL of fish oil in 80 μL of deuterochloroform ($CDCl_3$).

Standard operating conditions at 9.4 Tesla, corresponding to the resonance frequency of 399.95 MHz for the ^{1}H nucleus, gives accurate idea of the contents of fatty acids in a given sample. The chemical shifts are reported in ppm, using the TMS as internal standard. Typical parameters for ^{1}H-NMR spectra are 45° pulse, 2.05 s acquisition times, 6.4 KHz spectral window, 32 scans, and 26 K data points.

Another advantage of NMR is that the average acquisition time of the ^{1}H-NMR spectra is approximately 2 min and it can be applied directly on triglycerides, without any sample preparation.

1.5 Analysis of Phospholipids

Phospholipids are a group of complex lipids. They are essential building blocks of cell membranes in plants, animals, and microorganisms. They provide molecular strength for the construction and stability of membranes and other cellular constituents. Besides the structural significance, their dynamic roles in the highly active process like methylation provide an important mechanism for the biological signal transduction through membranes. Phospholipids also serve a source of arachidonic acid and other polyunsaturated fatty acids and are metabolized into biologically active eicosanoid and lipoids.

1.5.1 Direct Estimation of Phospholipids (Colorimetric Method)

This colorimetric method is based on the formation of a complex between phospholipids and ammonium ferrothiocyanate (Stewart 1980).

Reagents

Dissolve 27 g of ferric chloride ($FeCl_3$ $6H_2O$) and 30 g of ammonium thiocyanate (NH4SCN) in 1 l water (this solution is stable for months at room temperature), standard solution of phosphatidylcholine in chloroform (1 mg/ml).

Procedure

Evaporate completely aliquots of phospholipid solutions in glass tubes.

Dissolve the phospholipid residue in 2 ml chloroform and add 1 ml of thiocyanate reagent. Vortex 1 min and centrifuge at low speed; remove the red lower layer (chloroform) with a Pasteur pipette.

Read the absorbance of this solution at 488 nm and compare with known amounts of a standard phospholipid solution (range: 10–100 μg per tube).

All phospholipids do not give the same response, but this method is useful for measuring rapidly phospholipids in mixtures before further analyses.

1.5.2 Phospholipid Determination by Phosphorous Assay

Phospholipids in lipid extracts or in chromatographic fractions are satisfactorily estimated by phosphorus determination through an acidic digestion. The released inorganic phosphate is reacted with ammonium molybdate, and the complex formed gives a strong blue color. Two methods are discussed below based on the sensitivity of reaction.

1.5.2.1 Phosphorus Assay for Amounts Higher than 1–2 μg (25–50 μg Phospholipids) (Rouser et al. 1970)

Materials

Electrically heating block set at 180 °C. Boiling water bath, glass tubes (to be centrifuged) washed before use in water acidified with conc. nitric acid (about 1% final) and dried in an oven at 100 °C.

Conc. perchloric acid (70%), ammonium molybdate solution (2.5 g in 100 ml water), ascorbic acid solution (10 g in 100 ml water). Prepare freshly (do not keep more than 1 week at 4 °C). Standard solution of KH2PO4: dilute ten times a stock solution of KH2PO4 (439 mg per liter water, i.e., 100 μg P/ml).

Procedure Lipid samples are transferred into clean glass tubes, and the solvent is completely evaporated. Add 0.65 ml perchloric acid, and place the tubes in the heated block for about 30 min or until the yellow color has disappeared. Silica gel spots may be digested in the same way. When cool, add to the tubes 3.3 ml water, 0.5 ml of molybdate solution, and then 0.5 ml of ascorbic acid solution. Agitate on a vortex after each addition. The tubes are placed in a boiling water bath for 5 min. The absorbance of cool samples (including the standards) is read at 800 nm. Standards (1–5 μg P/tube) are diluted in 3.3 ml water and 0.65 ml perchloric acid. Digestion is not necessary before adding reagents.

Classically, 5 μg P gives an absorbance of 0.9.

The amount of phospholipids is calculated directly on a molar basis from the amount of P and on a weight basis after multiplying the amount of P by 25.

1.5.2.2 Phosphorus Assay for Amounts Lower than 1 μg (25 μg Phospholipids)

This is an improved method developed for the determination of lipid phosphorus in the sub-μg range (Zhou and Arthur 1992).

This method is based on the formation of a complex between phosphomolybdenum and malachite green and its resultant shift in absorption maximum. It is convenient for the determination of phospholipid distribution after TLC. In the presence of silica gel (scraped from TLC plates), the samples are resuspended in a suitable solvent and are centrifuged at low speeds to remove silica.

Materials

Solution A: 0.4% malachite green (w/v) in water. Prepare by vigorously stirring with a magnetic stirrer for 30 min.

Solution B: 4.2% ammonium molybdate (w/v) in 5 M HCl. Prepare by vigorously stirring with a magnetic stirrer for 30 min.

Solution C: 1.5% Tween 20 (w/v) in water. Prepare by vigorously stirring with a magnetic stirrer for 30 min.

Solution D: mix 1 vol. solution B with 3 vol. solution A. Prepare by vigorously stirring with a magnetic stirrer for 30 min. Filter through Whatman paper to obtain a yellow clear solution. This stock solution is stable 2–3 weeks in the dark.

Working solution: before use, mix 1 ml of solution C with 32 ml of solution D. Prepare by vigorously stirring with a magnetic stirrer for 30 min.

Standard solution of KH2PO4: dilute 20 times a stock solution of KH2PO4 (439 mg per liter water, i.e., 100 μg P/ml).

Procedure Lipid samples are transferred into clean glass tubes, and the solvent is completely evaporated. Add 0.2 ml perchloric acid, and place the tubes in the heated block for 30–60 min or until the yellow color has disappeared.

When cool, add to the tubes 0.2 ml water and 2 ml of working solution. Agitate on a vortex, and measure the absorbance after 20 min at 660 nm.

Standards (0–0.2 ml) are diluted to 0.2 ml before adding 0.2 ml perchloric acid.

Classically, 0.6 μg P gives an absorbance of 1.

Phospholipid Samples with Silica Gel (Scraped Spots of TLC)

Scraped spots are transferred into clean glass tubes; add 0.2 ml water and 0.4 ml perchloric acid. Heat at 180 °C. When cool, centrifuge at low speed, and collect 0.2 ml of the supernatant for phosphorus determination. Multiply the measured value by 2 to get the total amount present in the tube.

1.5.3 HPLC of Phospholipids (Rehman et al. 2017)

The high-performance liquid chromatography (HPLC) technique has emerged to be the most potent tool of phospholipids analysis, with ultraviolet (UV) coupled detection having greater sensitivity over refractive index (Table 1.11).

Table 1.11 Chromatography conditions for phospholipid separation by HPLC

Parameter	Conditions
Column	250 mm L × 4.6 mm ID, 5 μm silica, 80 Å pore
Isocratic solvent system	Acetonitrile-methanol-phosphoric acid (100:10:1.8, v/v)
Sample injection solvent	n-hexane and 2-propenol (3:1, v/v)
Sample injection volume	10–20 μL
Flow rate	1.5 ml/min, 10 min separation time
Detection	203 nm, UV absorption
Elution order	PS, PE, LPS, PI, LPE, PC, PG, SM, LPC, PA
Resolution time	9.5 min

PS phosphatidylserine, *PE* phosphatidylethanolamine, *LPS* lysophosphatidylserine, *PI* phosphatidylinositol, *LPE* lysophosphatidylethanolamine, *PC* phosphatidylcholine, *PG* phosphatidylglycerol, *SM* sphingomyelin, *LPC* lysophosphatidylcholine, *PA* phosphatidic acid

1.6 Analysis of Vitamins

Vitamin assays generally involve the extraction of a vitamin from its biological matrix by one or several treatment methods such as solvents, acid, alkali, enzymes, and heat. The extraction procedures are specific for each vitamin and are designed to stabilize the vitamin. The water-soluble vitamins B_1 and B_2 are extracted by boiling or autoclaving in acid and an additional treatment with enzymes. Niacin is extracted by autoclaving in acid for noncereal products or using alkali in the case of cereal products. Folate is generally subjected to enzyme extraction with α-amylase, protease, and γ-glutamyl hydrolase (conjugase). Typical extraction procedures for fat-soluble vitamins (A, D, or E) include organic solvent extraction, saponification, and re-extraction with organic solvents. Analysis of fat-soluble vitamins may require saponification, generally either overnight at room temperature or by refluxing at 70 °C. For unstable vitamins, antioxidants are routinely added to inhibit oxidation. High-performance liquid chromatography provides a robust chromatographic method for the separation and analysis of fat-soluble and water-soluble vitamins.

1.6.1 Analysis of Water-Soluble Vitamins

The separation of eight different water-soluble B vitamins in samples can be achieved using HPLC with photodiode array (PDA) detection. Due to the diverse absorptivity range among these vitamins, multiple analytical wavelengths are used to provide the best overall response. Method conditions and performance data, including linearity and repeatability, are given in Table 1.12.

1.6.2 Analysis of Fat-Soluble Vitamins

Analysis of fat-soluble vitamins in fish can be performed using a simple and rapid detection technique by the method of Stancheva and Dobreva (2013) (Table 1.13).

Table 1.12 Chromatography conditions for the separation of B-complex vitamins by HPLC

Parameter	Conditions
Column	C_{18}, 27 μm, 3.0 × 100 mm
Gradient solvent system	Mobile phase A: 5 mM ammonium formate (NH_4 formate) buffer; pH 4.8–4.9
	Mobile phase B: acetonitrile (ACN)
Oven temperature	40 °C
Sample injection volume	3 μL
Flow rate	0.6 mL/min. (max. pressure: ~2300 psi/~153 bar)
Detection	PDA; wavelengths used for calibration/quantitation
	214 nm: for B_3, B_3', B_6, B_7, B_9, and B_{12}
	267 nm: for vitamins B_1 and B_2
Sampling rate	10 pts./s
Analysis time	10 min.; 6-min injection delay time between injections

Table 1.13 HPLC gradient profile for the analysis of B-complex vitamins

Sl. No.	Flow (mL min^{-1})	Time (min)	% Eluent A	% Eluent B	%C	Curve
1		0.01	100.0	0.0	0.0	
2	0.60	2.00	100.0	0.0	0.0	6
3	0.60	8.00	70.0	30.0	0.0	6
4	0.60	10.0	70.0	30.0	0.0	6
5	0.60	10.01	100.0	0.0	0.0	6

Table 1.14 Chromatography conditions for the fat-soluble vitamins by HPLC

Parameter	Conditions
Column	ODS_2 Hypersil™ 250 × 4.6 mm, 5 μm
Isocratic solvent system	Mobile phase: MeOH:H_2O (97:3)
Sample injection volume	20 μL
Flow rate	1.0 mL/min.
Detection	UV: retinol, λ_{max} = 325 nm; cholecalciferol λ_{max} = 265 nm
	Fluorescence: α-tocopherol λ_{ex} = 288 nm and λ_{em} = 332 nm

Sample Preparation

An aliquot of the homogenized sample (1000 g) is weighed into a glass tube with a screw cap. Two milliliters (2 mL) of 1% methanolic solution of L-ascorbic acid and 5 mL of 1 M methanolic solution of potassium hydroxide are added. The samples of edible fish tissue are prepared and subjected to saponification at 80 °C for 20 min.

The fat-soluble vitamins from the hydrolysate are extracted with *n*-hexane (2 mL) by vortexing for 1 min. The combined extracts should be evaporated to dryness under a gentle stream of nitrogen. The dry residue of the fat-soluble vitamin fraction is reconstituted with 200 μL methanol, and 20 μL aliquot is used for injecting into the HPLC column (Table 1.14).

1.7 Analysis of Minerals

Seafood is a reasonably rich source of minerals such as calcium, phosphorus, potassium, and selenium, and hence the accurate and precise elemental analysis of food materials is critical. Some shellfish, such as clams and oysters, are a good source of iron, zinc, magnesium, copper, iodine, and other trace minerals. Most fish contain moderate to small amounts of these minerals. Certain trace elements are toxic, including arsenic, mercury, and lead, and their consumption has been associated with serious health problems. Exposure to harmful trace elements occurs most commonly through ingestion of food that has been contaminated by bioaccumulation and during production/processing.

The methods best suited for elemental analysis include flame atomic absorption spectrometry (FAAS), graphite furnace atomic absorption spectrometry (GFAAS), inductively coupled plasma atomic emission spectrometry (ICP/AES), and, more recently, the inductively coupled plasma mass spectrometry (ICP/MS). FAAS and ICP/AES offer similar detection limits up to ng/ml, whereas GFAAS and ICP/MS can provide sub-ng/ml levels. ICP/AES and ICP/MS offer the advantage of providing simultaneous multielement measurements, making them well suited for the analysis of large numbers of elements in food samples. If one or a few elements are to be determined, less expensive atomic absorption spectrometry (AAS) methods might be more suitable. The coupling of chromatographic separation and atomic spectroscopic detection has proven useful for elemental composition analysis.

1.7.1 Atomic Absorption Spectrophotometry (Mary 2012; Paul et al. 2014)

Principle

Atomic absorption spectroscopy (AAS) is an important technique for the analysis of specific elements in foodstuffs, as it is a robust technique that provides excellent sensitivity, accuracy, and precision. A liquefied sample is aspirated, aerosolized, and mixed with combustible gases such as acetylene and air or acetylene and nitrous oxide and burned in a flame to release the individual atoms. On absorbing UV light at specific wavelengths, the ground-state metal atoms in the sample are transitioned to higher state, thus reducing its intensity. The instrument measures the change in intensity, and the intensity is converted into an absorbance related to the sample concentration by a computer-based software. Combining flame, furnace, and vapor techniques, AA facilitates the analysis of a large number of trace elements across a wide analytical range, from parts per million (ppm) down to subparts per billion (ppb) making it an ideal choice for the analysis of trace elements in foodstuffs. Advances in automation through online sample dilution and automated standard preparation simplify routine tasks and offer increased productivity and sample throughput. The general instrumentation of an AAS is given in Chap. 5 (Table 1.15).

Table 1.15 Elemental composition analysis by AAS - detection wavelengths and flame gases used

Element	Wavelength	Flame – gases
Copper	324.8	Air-acetylene
Lead	217.0	Air-acetylene
Zinc	213.9	Air-acetylene
Cadmium	228.8	Air-acetylene
Iron	248.3	Air-acetylene
Aluminum	309.3	Air-acetylene/nitrous oxide-ac
Nickel	232.0	Air-acetylene

Procedure

Ions or atoms in a sample must undergo desolvation and vaporization in a high-temperature source such as a flame atomization or a graphite furnace.

At first sample solution is aspirated into the spray chamber through the capillary tube. The liquid sample is aerosolized and mixed with combustible gases such as acetylene and air or acetylene and nitrous oxide in the spray chamber and burned in a flame, and the individual atoms of the sample are released to form a cloud inside the flame. The sample solution is sprayed ("nebulized") continuously into the flame (similarly to the flame photometer). Sample solutions are usually aspirated with the gas flow into a nebulizing/mixing chamber to form small droplets before entering the flame. The nebulizer produces an aerosol. The basic operation involves drawing up the sample at high velocity through a capillary tube to a fine jet, which forms a liquid aerosol or mist before injection into the flame. The high-speed gas flow creates a turbulence, which breaks up the solution as it passes out of the capillary, to produce an aerosol. Nebulization efficiency depends upon solvent properties.

The graphite furnace has several advantages over a flame atomization. It is a much more efficient atomizer than a flame, and it can directly accept very small absolute quantities of sample. It also provides a reducing environment for easily oxidized elements. Samples are placed directly in the graphite furnace, and the furnace is electrically heated in several steps to dry the sample, ash organic matter, and vaporize the analyte atoms. The measurements can be done at ppb level (part per billion, ppb = 10^{-3} ppm, i.e., $\mu g\ kg^{-1}$ or $\mu g\ dm^{-3}$ in case of dilute solutions). The drying, combustion, vaporization, and atomization of sample happen in a heated graphite tube that is placed in the way of light. This "graphite furnace" is protected against oxidation by an inert gas (e.g., argon).

The hydride generation technique or vapor system is an accessory for measuring the hydride forming elements such as arsenic, selenium, and mercury with better analytical sensitivity than flame atomization. In this method the analytes are reduced to volatile hydrides using sodium borohydride and are carried in a carrier gas stream to a heated measurement cell for measurement.

When the atoms of the element are released, to bring it from ground state into an excited state, UV light is passed through hollow cathode lamp. On absorbing UV light at specific wavelengths, the ground-state metal atoms get transitioned to higher electronic state. The region of the spectrum to be measured is selected by a monochromator. The specific spectral line falls on the photomultiplier, the detector, and the output is amplified and sent to a readout device meter, digital or analog or through a computer data processing system. The data is processed through software. Absorption of a selected wavelength is measured by the change in light intensity striking the photomultiplier detector and is directly related to the amount of element in the sample. The waste is drained out finally, through a drain pipe.

Sample Preparation

There are at least five methods of sample preparation used by the analyst, viz., wet ashing or acid digestion, fusion, pressure dissolution, dry ashing, and microwave digestion. The most useful sample preparation methods for a wide range of analytes and sample matrices are based on dry ashing or wet ashing of the sample.

An acid digestion procedure may be used for sample preparation of many elements in fish and seafood tissue including K, Na, Zn, Cu, Cr, Cd, Fe, Ni, and Pb. A weighed sample is placed in a digestion vessel, acid is added, and the mixture is heated for several hours. The samples are digested with HNO_3 and $HClO_4$ or HNO_3 and H_2SO_4 depending on the technique and heating vessel used. After the digestion, the samples are diluted to a specific volume and analyzed directly or chelated and extracted into an organic solvent if the element of interest is present in low concentration. The main advantage of wet digestion is that it eliminates elemental loss by volatilization because the digestion takes place at a low temperature. The main disadvantages of a wet digestion procedure are that it is subject to reagent contamination and requires operator attention.

Wet Ashing

Weigh about 5 g of sample (dry weight) into a digestion tube. Add 5 ml of HNO_3 and then 5 ml of H_2SO_4 to the sample. Allow the reaction to proceed. When the reaction slows, place the tubes in a hot-block digestion apparatus and heat at a low temperature (60 °C) for 30 min. Remove the tubes from the hot block, allow to cool, add 10 ml of HNO_3, return tubes to digestion rack, and heat slowly to 120 °C. Increase the temperature to 150 °C. Remove the tubes when the samples go black, allow to cool, and then add 1 ml of H_2O_2. A vigorous reaction may occur. Return the tubes to the block. Repeat the H_2O_2 additions until the samples are clear. Remove the tubes and make up to 50 ml with deionized water.

Most elements can be determined directly. However, to determine Pb and Cd, solvent extraction is used to concentrate these elements. Take 40 ml of the digest and make up to 100 ml; add 5 ml of ammonium pyrrolidine dithiocarbamate (APDC) and 5 ml of MIBK (methyl isobutyl ketone). Shake vigorously for 5 min. Determine Pb and Cd in the MIBK phase. Standards should be treated in the same way as samples. It is important that the standards contain the same amount of acid as the samples, especially H_2SO_4, as it will have a viscosity effect that will suppress sensitivity.

Dry Ashing

Dry ashing is a method that can be used for the determination of several elements in fish and seafood samples including Pb, Cd, Cu, Zn, Cr, Mn, Co, Na, and K. It has been reported that the major drawback to dry ashing is loss of metal due to volatilization. However, if the temperature in the muffle furnace is held at 450–500 °C, loss from volatilization is minimal. The dry ashing method is less time-consuming than wet digestion methods. When levels of Pb and Cd are too low to be determined directly, solvent extraction can be used to concentrate these elements.

Weigh out 25 g of wet tissue (5 g dry weight) into crucible, dry samples at 135 °C for 2 h, and obtain a dry weight. Transfer to a cool muffle furnace, and slowly raise temperature to 450–500 °C. Ash overnight. Remove samples, let cool to room temperature, cautiously add 2 ml HNO_3, and swirl. Evaporate carefully just to dryness on warm hot plate or steam bath. Transfer to cooled furnace, slowly raise temp to 450–500 °C, and hold at this temperature for 1 h. Remove crucible and cool. Repeat HNO_3 treatment, if necessary, to obtain clean practically carbon-free ash. Add 10 ml of 1 N HCl and dissolve ash by heating cautiously on a hot plate. Transfer to a 25 ml volumetric flask and add HCl as necessary. Cool and dilute to volume.

Microwave Digestion

The microwave digestion technique is quicker and safer. The microwave system functions by using a combination of acid, temperature, and pressure. The technique is also useful for the measurements of volatile metals. The major advantage of this technique is the analysis time, which is in minutes, rather than hours. It has been successfully used for acid dissolution of the bone, biological samples, steels, environmental samples, botanical matrices, ores, rocks, metals from sediments, etc. In the case of fish sample, acid digestion and microwave digestion are most useful.

General Instrumental Procedure

1. Light the hollow cathode lamp or electrode discharge lamp and D2 lamp if such background correction is used. Set the lamp current to the value specified by the manufacturer.
2. Position the monochromator at a selected wavelength.
3. Carefully balance the intensity of the hollow cathode lamp and the D2 lamp if such background correction required.
4. Align the burner head to assure that the center of the light beam passes over the burner slot.
5. Light the flame, and regulate the flow of fuel and oxidant to produce an oxidizing flame (lean blue).
6. Aspirate calibration blank and establish a zero point.
7. Aspirate standard solutions (use at least three standard solutions, having a concentration of 1000 ppm), and construct a calibration curve. A quality control standard should also be analyzed to verify the calibration.
8. Aspirate distilled water after each standard and sample.
9. Aspirate unknown samples. Samples that are found to have concentration higher than the highest standard should be diluted and reanalyzed.

Calculation

The concentration of metal is detected in mg/liter or ppm.

$$\text{The Metal}(\text{mg\%}) = \frac{\text{Conc.of standard metal in ppm or mg / L} \times \text{Made up volume}}{\text{Sample weight}}$$

Nutrient-rich foods like fish and seafood contain a good proportion of proteins, phospholipids, essential fatty acids, vitamins, and minerals and are few in calories. They also are low in sugar, sodium, starches, and bad cholesterol. The nutritional value of food defines what a food is made of and its impact on the body. In this context it is vital to understand the nutritional value of food fishes because of the effect on the body as it relates to fat, cholesterol, mineral, carbohydrate, and protein intake.

References

AOAC. (2000). *Association of Official Analytical Chemist, Official Methods of Analysis* (16th ed.). Gaithersburg: AOAC International.

AOAC. (2005). *Association of Official Analytical Chemist, Official Methods of Analysis* (18th ed.). Gaithersburg: AOAC International.

Bradford, M. M. (1976). A rapid and sensitive for the quantitation of microgram quantitites of protein utilizing the principle of protein-dye binding. *Analytical Biochemistry, 72*, 248–254.

Chatterjee, N. S., Ashok, K. K., Ajeeshkumar, K. K., Remya, K. R., Vishnu, K. V., Anandan, R., Mathew, S., & Ravishankar, C. N. (2017). Screening natural content of water-soluble B vitamins in fish: Enzymatic extraction, HILIC separation, and tandem mass spectrometric determination. *Journal of AOAC International, 100*(3), 579–585. https://doi.org/10.5740/jaoacint.17-0056.

Cohen, N., Scott, C. G., Neukon, C., Lopresti, R. L., Weber, G., & Saucy, G. (1981). Total synthesis of all 8 stereoisomers of alpha-tocopheryl acetate. Analysis of their diastereomeric and enantiomeric purity by gas chromatography. *Helvetica Chimica Acta, 64*, 1158.

Damodaran, S., Parkin, K. L., & Fennema, O. R. (2007). Water and ice (chapter 2). In O. R. Fennema (Ed.), *Fennema's food chemistry* (4th ed.). Boca Raton: CRC.

David, B. M., & Wayne, C. E. (2014). Ash analysis, chapter. In S. S. Nielsen (Ed.), *Food analysis laboratory manual* (Food science texts series) (4th ed., p. 8). Boston, MA: Springer.

DeLuca, H. F. (2008). Evolution of our understanding of vitamin D. *Nutrition Reviews, 66*(2), 573–587.

DeLuca, F., Uyeda, J. A., Veronica, M., Mancilla, E. E., Yanovski, J. A., Barnes, K. M., Zile, M. H., & Bacon, J. (2000). Retinoic acid is a potent regulator of growth plate chondrogenesis. *Endocrinology, 141*, 346–353.

Denisava, N. A., & Booth, S. L. (2005). Vitamin K and sphingolipid metabolism: Evidence to date. *Nutrition Reviews, 63*(4), 111–121.

Eitenmiller, R. R., & Lee, J. (2004). *Vitamin E, food chemistry, composition, and analysis*. New York: Marcel Dekker Inc. 540 pp.

Gundberg, C. M., Lian, J. B., & Booth, S. L. (2012). Vitamin K-dependent carboxylation of osteocalcin: Friend or foe? *Advances in Nutrition, 3*(2), 149–157.

Gunlu, A., & Gunlu, N. (2014). Taste activity value, free amino acid content and proximate composition of mountain trout (Salmo trutta macrostigma Dumeril, 1858) muscles. *Iranian Journal of Fisheries Sciences, 13*(1), 58–72.

Hedge, J. E., & Hofreiter, B. T. (1962). In R. L. Whistler & J. N. Be Miller (Eds.), *Carbohydrate chemistry* (Vol. 17). New York: Academic.

Hong, H., Regenstein, J. M., & Luo, Y. (2017). The importance of ATP-related compounds for the freshness and flavor of post-mortem fish and shellfish muscle: A review. *Critical Reviews in Food Science and Nutrition, 57*(9), 1787–1798.

Iverson, S. J. (2009). Chapter 12: Tracing aquatic food webs using fatty acids: From qualitative indicators to quantitative determination. In M. T. Arts, M. T. Brett, & M. J. Kainz (Eds.), *Lipids in aquatic ecosystems* (pp. 281–307). Dordrecht: Springer.

Karnjanapratum, S., Benjakul, S., Kishimura, H., & Tsai, Y. (2013). Chemical compositions and nutritional value of Asian hard clam (Meretrix lusoria) from the coast of Andaman Sea. *Food chemistry., 141*, 4138–4145. https://doi.org/10.1016/j.foodchem.2013.07.001.

Knothe, G., & Kenar, J. A. (2004). Determination of the fatty acid profile by ^{1}H-NMR spectroscopy. *European Journal of Lipid Science and Technology, 106*, 88–96. https://doi.org/10.1002/ejlt.200300880.

Kuehn, A., Swoboda, I., Arumugam, K., Hilger, C., & Hentges, F. (2014). Fish allergens at a glance: Variable allergenicity of parvalbumins, the major fish allergens. *Frontiers in Immunology, 5*, 179. https://doi.org/10.3389/fimmu.2014.00179.

Lehninger, A. L., Nelson, D. L., & Cox, M. M. (2000). *Lehninger principles of biochemistry*. New York: Worth Publishers.

Lindsay, R. C. (1990). Fish flavours. *Food Reviews International, 6*(4), 437–455.

Love, R. M., & Robertson, I. (2007). The connective tissues of fish. *International Journal of Food Science & Technology, 3*, 215–221. https://doi.org/10.1111/j.1365-2621.1968.tb01459.x.

Lowry, O. H., Rosebrough, N. J., Farr, A. L., & Randall, R. J. (1951). Protein measurement with the folin phenol reagent. *Journal of Biological Chemistry, 193*, 265–275.

Lubran, M. M. (1978). The measurement of total serum proteins by the biuret method. *Annals of Clinical and Laboratory Science, 8*(2), 106–110.

Mary, M. (2012). Nutritional metals in foods by AAS. In M. A. Farrukh (Ed.), *Atomic absorption spectroscopy*. ISBN: 978-953-307-817-5. InTech. Available from: http://www.intechopen.com/books/atomic-absorption-spectroscopy/nutritional-metals-in-foods-by-aas

Maurice, R. M. (2014). Ash analysis, chapter. In S. S. Nielsen (Ed.), *Food analysis laboratory manual* (Food science texts series) (4th ed., p. 7). Boston, MA: Springer.

Muzaddadi, A. U., Devatkal, S., & Oberoi, H. (2016). *Seafood Enzymes and Their Application in Food Processing*. https://doi.org/10.1016/B978-0-12-802392-1.00009-5.

Paul, B. N., Chanda, S., Das, S., Singh, P., Pandey, B. K., & Giri, S. S. (2014). Mineral assay in atomic absorption spectroscopy. *The Beats of Natural Sciences, 4*, 1), 1–1),17.

Pesillo, S. A., Freeman, L. M., & Rush, J. E. (2004). Assessment of lipid peroxidation and serum vitamin E concentration in dogs with immune-mediated hemolytic anemia. *American Journal of Veterinary Research, 65*(12), 1621–1624.

Phillips, K. M., Ruggio, D. M., Jacob Exler, J., & Patterson, K. Y. (2012). Sterol composition of shellfish species commonly consumed in the United States. *Food & Nutrition Research, 56*. https://doi.org/10.3402/fnr.v56i0.18931.

Rehman, S.-U., Rehman, S., Welter, D., Wildenauer, D., & Ackenheil, M. (2017). Simple and rapid separation and determination of phospholipids by HPLC-UV system. *Annals of Pharmacology and Pharmaceutics, 2*(12), 1063.

Robert, L. B. (2014). Moisture and total solids analysis. In S. S. Nielsen (Ed.), *Food analysis laboratory manual* (Food science texts series) (4th ed.). Boston, MA: Springer.

Rouser, G., Fkeischer, S., & Yamamoto, A. (1970). Two dimensional then layer chromatographic separation of polar lipids and determination of phospholipids by phosphorus analysis of spots. *Lipids, 5*, 494–496.

Sarower, G. M., Hasanuzzaman, A. F., Biswas, B., & Abe, H. (2012). Taste producing components in fish and fisheries products: A review. *International Journal of Food and Fermentation Technology., 2*, 113–121.

Scherr, C., Gagliardi, A. C., Miname, M. H., & Santos, R. D. (2015). Fatty acid and cholesterol concentrations in usually consumed fish in Brazil. *Arquivosbrasileiros de cardiologia, 104*(2), 152–158. https://doi.org/10.5935/abc.20140176.

Shang, S. F., & Wang, H. (1996). Sensitive determination of amino acids in kelp by reverse phase high performance liquid chromatographywith pre column derivatization using phenylisothiocyanate. *Chromatographia, 43*, 309–312.

Sikorski, Z. E., & Borderias, J. A. (1994). Collagen in the muscles and skin of marine animals. In Z. E. Sikorski, B. S. Pan, & F. Shahidi (Eds.), *Seafood proteins*. New York: Springer.

Singh, A., & Benjakul, S. (2018). Proteolysis and its control using protease inhibitors in fish and fish products: A review. *Comprehensive Reviews in Food Science and Food Safety, 17*. https://doi.org/10.1111/1541-4337.12337.

Spinelli, J., & Dassow, J. A. (1982). Fish proteins: Their modification and potential uses in the food industry. In R. E. Martin, G. J. Flick, C. E. Hebard, & D. R. Ward (Eds.), *Chemistry & biochemistry of marine food products* (pp. 13–37). Westport: AVI Publishing Company.

Stancheva, M., & Dobreva, D. A. (2013). Bulgarian marine and freshwater fishes as a source of fat-soluble vitamins for a healthy human diet. *Foods (Basel, Switzerland), 2*(3), 332–337. https://doi.org/10.3390/foods2030332.

Stewart, J. C. (1980). Colorimetric determination of phospholipids with ammoniumferrothiocyanate. *Analytical Biochemistry, 104*(1), 10–14.

Suttie, J. W., & Jackson, C. M. (1977). Prothrombin structure, activation, and biosynthesis. *Physiological Reviews, 57*, 1–70.

Van der Berg, C., & Bruin, S. (1981). Water activity and its estimation in food systems. In L. B. Rockland & G. F. Stewarts (Eds.), *Theorical aspects in water activity: Influence on food quality* (pp. 12–45). New York: Academic.

Venugopal, V., & Gopakumar, K. (2017). Shellfish: Nutritive value, health benefits, and consumer safety. *Comprehensive Reviews in Food Science and Food Safety, 16*(6), 1219–1242. https://doi.org/10.1111/1541-4337.12312.

Vermeer, C. (1986). Comparison between hepatic and nonhepatic vitamin K-dependent carboxylase. *Haemostasis, 16*, 239–245. www.fao.org

Yamanaka, H., & Shimada, R. (1996). Post-mortem biochemical changes in the muscle of Japanese spiny lobster during storage. *Fisheries Science, 62*, 821–824.

Yu, S., & Paetau-Robinson, I. (2006). Dietary supplements of vitamins E and C and beta-carotene reduce oxidative stress in cats with renal insufficiency. *Veterinary Research Communications, 30*, 403–413.

Yurimoto, T. (2015). Seasonal changes in glycogen contents in various tissues of the edible bivalves, pen shell *Atrina lischkeana*, ark shell *Scapharca kagoshimensis*, and manila clam *Ruditapes philippinarum* in West Japan. *Journal of Marine Biology, 2015*, Article ID 593032, 5p. https://doi.org/10.1155/2015/593032.

Zhou, X., & Arthur, G. (1992). Improved procedures for the determination of lipid phosphorus by malachite green. *Journal of Lipid Research, 33*, 1233–1236.

Fish and Fishery Products: Quality Indices

2

2.1 Processing of Various Fish and Seafood

Even though the nutritive value and the unique taste of seafood (fish, shell fish, mollusks, and echinoderms) have gained immense popularity in all continents, the susceptibility of these to postmortem spoilage and rapid deterioration, and hence, the shortened shelf life, is of great concern. Species, handling, and storage conditions affect the nutritional and microbial qualities of seafood. Further, two most important concerns with respect to marketing and distribution of seafood are their high perishability and poor hygienic quality. To overcome these issues, various preservation methods are being adopted where time-temperature association is the key determinant (Boziaris 2013; Mohan et al. 2008; Borda et al. 2018). Feeding habits, primary microbiological load, season, geographical origin, and handling conditions influence the shelf life of seafood. Therefore, it is decisive to have a science-based comprehension of the growth and activity of spoilage microorganisms in seafood, in order to reduce the losses and preserve the food. Hence, the chief aim of fish preservation is to delay, reduce, or inhibit the enzymatic, chemical, and microbial spoilage by managing the storage temperature, water activity, and pH or by using chemical preservatives. The preservation of seafood can be attained by conventional methods (thermal) such as chilling, drying, smoking, refrigeration or freezing, and canning or by employing novel techniques (nonthermal) such as pulsed light, high hydrostatic pressure, pulsed electric fields, ultraviolet light, high-intensity ultrasound, oscillating magnetic fields, ionizing radiation, etc. Currently, consumers are in continuous demand for minimally processed food with extended shelf life and safety assurance.

2.1.1 High- and Low-Thermal Preservation Techniques

High- and low-thermal preservation can be done either for short and long duration.

S. Mathew et al., *Fish and Fishery Products Analysis*,
https://doi.org/10.1007/978-981-32-9574-2_2

2.1.1.1 Chilling and Freezing

Chilling (preservation for short duration) is the first and simplest method to preserve and process fish (Hall 2012). This is obtained by covering the fish with layers of ice. For chilling, the ratio of fish to ice is generally 1:1 or 1:2. Ice alone is not effective for long preservation as flavor and certain nutritional contents leach out along with the melting water. Ice is effective only for short-term preservation. Chilling preserves seafood by decreasing the rate of undesirable biochemical and chemical reactions and by hindering the growth and spoilage activity of microorganisms. Autolytic enzymatic activities are checked by lowering the temperature. Various methods are adopted for chilling such as conventional flake ice, slurry ice, refrigerated seawater, ozonized refrigerated water, and dry ice (solid carbon dioxide). The storage life of chilled fish is limited to 2–18 days depending on the species.

Ice works in two ways:

1. It reduces the growth rate of bacteria by dropping the temperature of the fish.
2. It washes the bacteria and slime away as it melts.

Freshness and keeping quality of chilled fish could be improved by the amalgamation of chemical preservative treatment and innovative packaging techniques such as modified atmospheric packaging, etc. Recently, active packaging is employed to improve chilling of fish (Boziaris 2013; Mohan et al. 2008).

Freezing is another fish processing technique used from earlier years to preserve fish at low temperature. It modifies the structural and chemical properties of muscle and increases the free fatty acid content (FFA) and lipid oxidation products. The main objective of freezing is to minimize microbial load, reduce enzymatic activity, and preserve the flavor and the nutritional properties. The ice crystal formation during freezing is a critical point; larger the crystals, higher the risk of texture damage and membrane disruption, which results in increased oxidation. Hence, formation of small ice crystals during freezing is preferred to minimize the thaw-loss. Small uniform ice crystals are formed during faster and homogeneous freezing. Elastic nature of cells withstands the textural damage to certain extent; however, it disrupts during rapid freezing such as cryogenic freezing, which facilitates the access of oxygen into the cells/tissues. Thawing and refreezing lead to bigger crystal formation.

Some recently developed fast-freezing techniques such as high-pressure freezing or pressure shift freezing, cryogenic freezing, liquid immersion freezing techniques and plate freezers, and air blast freezers facilitate freezing of seafood by small crystal formation. In plate freezer, the product is in direct contact with the plates or cooled surface; these are ideal for regular-shaped fishery products (Hall 2012). These freezers function generally at a temperature of −40 °C. Air blast freezers are the simplest and very effective systems for products in various shapes. The products are placed in a room or moved through a tunnel where cold air flows at variable speed. When the products are moving on a belt, the air flow could be applied horizontally and/or vertically in the direction of conveyor. The freezing speed depends on belt speed, product load, and airflow characteristics. Air blast freezers operate at between −20 and − 40 °C. In immersion freezing, the packed products are either dipped into or sprayed with a liquid freezing medium such as brine, ethylene glycol, or propylene glycol. The refrigerant comes in direct contact with the entire product surface; hence, this can operate at higher temperatures (between −6 and − 20 °C) compared with plate and air blast freezers, with higher efficiency. The cryogenic freezing is similar to immersion freezing, but the liquids used are liquid nitrogen or carbon dioxide. In high-pressure or pressure shift freezing, the products are exposed to elevated pressure and cooled under pressure to a temperature just above the melting temperature of ice. Pressure is then fast released resulting in super cooling, which results in instant and homogeneous ice crystal formation. This concept is specifically useful for freezing of large-sized foods. The pressure variation may cause alteration in the myosin fraction resulting in the more stable texture properties. Other freezing techniques reported include the injection or dipping of antifreeze proteins.

2.1.1.2 Drying and Curing

The term curing refers to the addition of salt, sugar, nitrites, nitrates, seasonings or spices, and phosphates to preserve food, which aid in the development of characteristic color and flavor in the products. Salt has been the most widely used ingredient for curing of seafood. It creates low water activity (a_w) environment in the food, thus forcing the microorganisms to face prolonged lag phase and eventually to enter the death phase (Boziaris 2013). There are different kinds of salt used in fish curing including table salt, sea salt, salt mined from earth, etc. Two chief techniques of fish curing are wet salting and dry salting.

Wet Curing/Salting The principle is to retain the fish in brine (1:4, salt/water) for longer period in watertight containers such as tin, drum, canoe, barrel, etc. The water used could be freshwater or seawater. If the salt is coarse, it has to be ground or pounded prior to dissolving. The small fish can be salted as whole and large fish must be cleaned scaled, beheaded, and gutted. The slashes are made for easy penetration of salt. Very large fish are cut into thin fillets before curing. A plank or matting is laid over it and weighed with rocks in order to keep fish entirely soaked in the brine and kept for a long time in dark or in shady place. The remaining brine can be reused (three times), but each time, water and salt must be added. However, fresh brine is always the best.

Dry Curing/Salting In drying curing method, the fish is salted (1:2, salt/fish) in layers (layers of fish separated by layers of salt), and the exudates including slime and brine can be drained. It is usually done in an old canoe or on mats, leaves, boxes, etc. The fish are kept whole. The excess salt can always be removed by soaking it in freshwater before use.

Drying Natural or artificial drying techniques are employed for drying the seafood. Tiny and lean fish can be sun dried (by spreading on mats or platforms or suspended) or dried using different mechanical driers (solar, tunnel, kiln, etc.). The dry cured fish (after 3 days of curing) may be stored to longer duration by subjecting it to drying. Temperature, relative humidity, air speed, surface area, and volume of fish are the key factors that need to be controlled to ensure optimum drying conditions.

2.1.1.3 Smoking

Smoking of fish is another method of fish preservation which combines preservation value of the smoke (burning wood contains a large number of compounds specially phenols that kill microorganisms), drying (smoke generates heat that will dry products), and cooking (fish flesh is cooked as the temperature generated is very high, which in turn destroys enzymes and kills bacteria). In some cases, later two phenomena play a major role in longer storage life. The burning of wood or sawdust creates smoke during incomplete combustion, and this varies with source of fuel or the ventilation of fire (Boziaris 2013).

2.1.1.4 Canning

Nicolas Appert, a confectioner, brewer, distiller, and wholesale caterer, developed the canning technology in 1809. This method of preservation averts spoilage by killing microorganisms through heat. During canning, spoilage microorganisms in the food are eliminated, and the entries of others are restricted. The process involves pretreatment of fish, preparation of can, filling and closure of can, heating (exhausting), sterilization (to kill microorganisms), cooling, cleaning, and storage. The canning technique mainly considers the killing/elimination of *Clostridium botulinum*. These are found in protein rich food with pH 6–7 and are non-acidic. There are other heat-resistant bacteria like *Clostridium sporogenes* that can be eliminated at a temperature of 5–6 times more than *Clostridium botulinum*. It needs a temperature of 120 °C for 4 min or at 115 °C for 10 min. Canning includes following process.

Filling Clean empty cans are packed carefully manually or mechanically, with no air pockets that cannot be removed by exhausting. Too tight packing is avoided. A head space is left on the top of can to accommodate the gases liberated while processing and to prevent can deformation. Fatty fish are canned with salt, while non-fatty fish are canned with additives and texture improvers. Brine enhances the flavor of the fish. Monosodium glutamate (MSG, 1.6 ppm), vegetable oil, and olive oil are other ingredients used for can fill.

Exhausting The air/gas is removed from the can before sealing. Exhausting minimizes the strain on the can through expansion of air during heating. Removal of oxygen is essential to avoid internal corrosion creating of vacuum upon cooling. It also checks the growth of microorganisms.

Sealing It provides an air tight seal between cover and body of the container so that spoilage organisms cannot enter from outside while sterilization/processing.

Processing Complete air removal is possible in steam processing. The container with the contents inside is heated in a retort at a temperature sufficient enough to kill the potential spoilage organisms (vegetative bacteria, 80 °C/30 min; spore formers, 110 °C/30 min; destroying spores, 121 °C/3 min). Retort system could be batch or continuous; open boilers are used currently. The pressure processing could be either by using steam or water. Retorts could be horizontal or vertical and rotating and are fed by steam. The retort has an inlet at the top through which steam enters. The weight of the steam (entered and the incoming) both drives the air out from the bottom without mixing. Air pockets if present in a retort may give rise to uneven processing and lead to under processing while interfering between pressure and temperature. The processing time and temperature vary with the types of pack, size of cans, etc. By adopting the above heating process, the majority of the spoilage agents or bacteria are killed. The elimination of thermophilic spoilage bacteria can be ensured by subjecting it to rapid cooling immediately after processing. The can should be cooled to a temperature of 36–42 °C. Chlorinated water of 5 ppm can be used for cooling purpose. It is advised to store for 3 months before final quality control.

Cooled containers are immediately dried. Heated bed drier is found to be very efficient, which rolls cans over a surface heated by steam to 127–130 °C covered by an absorbent cloth. Contaminants on the surface are thus rapidly eliminated. Commercial sterility means the absence of microorganisms capable of growing in the food at normal non-refrigerated conditions at which the food is likely to be held during manufacture, distribution, and storage (Codex Alimentarius Commission (WHO/FAO) CAC/RCP 40–1993). Commercial sterility (incubation of cans at 30 °C for 15 days and 55 °C for 7 days) will demonstrate the complete sterility of the product. Improper sterilization/leakage may lead to spoilage of cans (swell, flipper, springer, leakage, overfilled can, flat souring, rust, damage, sulfiting, etc.).

The various drawbacks of seafood preservation (Hall 2012):

1. Chilling brings about denaturation of flesh. This is because of ice crystals formed during chilling and causing mechanical damage to the muscles. Cell walls burst, structure gets deformed, and the flesh loses much of flavor and taste. The flesh also becomes dehydrated and losses texture.
2. If proper hygienic measures are not taken during the processes like washing, guttation, evisceration, etc., more harm would be done to the preserved material, owing to increase in the bacteria population.
3. Incomplete or poor preservation leads to decarboxylation of histidine of fish flesh into histamine. The latter some other related substances, collectively called saurine, are common causes of food poisoning.

4. Drying reduces weight, nutritive value, and the digestibility of the flesh.
5. Excess salting allows growth of salt tolerant bacteria, causing pink eye spoilage of fish flesh.
6. Salting combined with smoking results in loss of protein, about 1–5% due to salting and 8–30% due to smoking.
7. Smoking also accelerates rancidity of fat and so reduces digestibility of fat products.
8. Canning leads to much loss of vitamin B1, pantothenic acid, vitamin C, and pteroylglutamic acid.

2.1.2 Novel Nonthermal Processing and Preservation Techniques

The thermal food processing techniques are classic and ensure microbiological safety of foods; however, it leads to unwanted changes in the food such as sensory attributes or low nutritional value. The enhanced consumers' interest in high-quality foods with higher nutritive value and fresh-like sensory attributes has led to the development of a number of nonthermal food processing technologies as alterative to conventionally heat treatments. These techniques also provide safety for food and environment (Borda et al. 2018).

2.1.2.1 High-Pressure Processing (High Hydrostatic Pressure)

High hydrostatic pressure (HHP) or high-pressure processing (HPP) has earned more consideration in the recent years. It has been employed as it improves the nutritional and functional properties and quality and safety of a broad range of food with imperceptible damage on their dietary and sensory distinctiveness.

During HHP processing, food is exposed to high pressure (200–1000 MPa) for a period of time using water (a most apt pressure-transmitting fluid). HHP is based on the isostatic principle, in which food products are consistently and immediately compressed from all directions. When pressure is removed, food retains its original shape (Borda et al. 2018). This hydrostatic compression is competent of inactivating microorganisms. Heating can be pooled with HHP for improved results. Process temperatures can vary from sub-zero to temperatures beyond the boiling point of water. Pressurized vessel, system controls, an intensifier to produce higher pressures, a low-pressure pump, temperature regulator, and product holding system comprise the equipment. The pressurized vessels vary in diameter and have different operating pressures. The common designs include a monolithic chamber separately forged, a multiwall chamber shrunk-fit from a series of concentric tubes, and a wire-winded stainless steel core tube. The former is typically designed for working pressures of 400–600 MPa for small-scale vessels with inner diameters below 15 cm. A wire-wound vessel or pre-stressed multilayered (multiwall) vessel is employed for higher pressures and longer dimensions. A larger capacity (525 L) monolithic chamber is used to reduce microbial load in seafoods, particularly oysters. Quick closing and opening systems that permit quick product stacking and

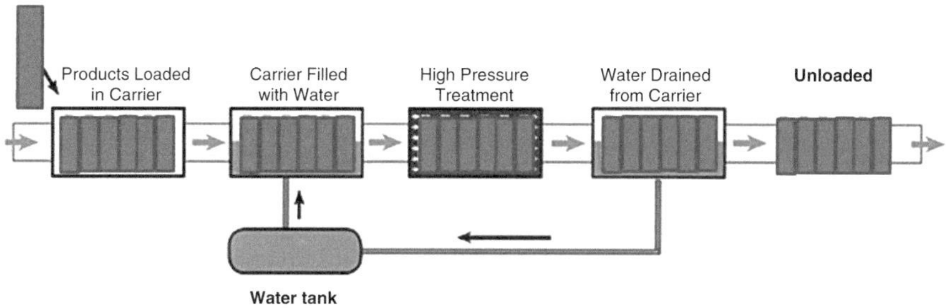

Fig. 2.1 Schematic diagram of high-pressure processing (HPP)

un-stacking are required for HPP. Breech-type, threaded closures or pin closures are used for small-diameter or lower-pressure vessels. A siphoning heating/cooling medium in pressure vessel is used to attain constant temperature in a large or more modern system.

Mode of action of HPP in inactivating microorganisms involves collapse of non-covalent bonds in macromolecules, biochemical effects, disturbance and increased permeability of cell membranes, morphological alterations, and modulation of the genetic mechanisms of cells. Different microbial groups and strains have different ranges of barotolerances. Different pressure conditions for different durations decrease the microbial load differently and increase the shelf life, with no significant changes in organoleptic and chemical characteristics. HPP destabilize functional protein molecules such as enzymes by ionic and hydrophobic – hydrophobic interactions – which may affect the organoleptic parameters. It can accelerate lipid oxidation in seafood during storage possibly by releasing inorganic transition metal ions from their respective compounds. It also gives a cooked appearance to fishery products (˃200Mpa). It also induces the formation of formaldehyde, which stimulate protein cross-linking and increase the hardness (Fig. 2.1).

2.1.2.2 Pulsed Electric Field Processing

Pulsed electric field (PEF) processing is a new nonthermal preservation technique that kills the microorganisms and extends the shelf life, preserves the nutritional quality, and improves the sensory attributes. A source of high voltage, treatment chamber, switch, and capacitor bank makes up the PEF processing system and engages a quick discharge of energy within a short duration. Uninterrupted (multiple electrodes) microsecond high-voltage pulses (10–60 kV) are involved in PEF processing (Gómez et al. 2019). The use of high-voltage pulses induces electroporation in microbial cell membranes and initiates loss of barrier function, intracellular content leakage, and loss of vitality. The product is subjected to the high-voltage pulses, into which liquid and semi-viscous liquid materials are pumped, while solid materials are transported through the treatment chamber. Treatment duration is less than 1 s. To guarantee sufficient treatment, pulses are applied at repetition rates of up to 1000/s. PEF treatment intensity is affected by specific energy input, field strength, product composition, and temperature. PEF can be applied at slightly above, below, or at ambient temperature in the form of bipolar, an exponentially decaying, oscillatory pulse, and square wave.

Based on electroporation, several applications of high-intensity pulsed electric fields (HIPEF) in different food matrixes have been investigated. PEF technology has been productively applied for the pasteurization of fish soups, tomato juice, and liquid eggs. Targeted microorganisms and their growth and physiological states are key factors influencing PEF efficacy. The inherent parameters of the microorganism such as shape, size, growth state, or species conclude its susceptibility to PEF. The Gram-negative vegetative cells are less resistant to PEF than Gram-positive bacteria. PEF sensitivity is higher in yeasts than bacteria. PEF treatment of 50 kV/cm for 50 pulses at 25 °C in salt solution showed 5 log and 3.42 log reductions of *Bacillus cereus* and *Bacillus subtilis* spores, respectively. PEF treatment (1.36 kV/cm for 40 pulses at room temperature) for longer duration may produce gaping and less water loss at 2.0 kV/cm and 90 pulses/min, in fish samples (frozen cod loins, Iceland cyprine, cod fresh fillets, common whelk, frozen haddock loins, and pollock fillets). High initial cost and the lack of any effects on enzymes are the chief technical drawbacks of PEF. The application of PEF processing is limited to food products with low electrical conductivity and no air bubbles. PEF is unsuitable for solid food products that cannot be pumped.

2.1.2.3 Ohmic Processing

Ohmic heating is a thermal-based process involving electric field strength (50–60 Hz), residence time and the electric conductivity of a food product, temperature generated, and ohmic heating efficacy (Ahmed et al. 2016). Ohmic heating is based on capacitive dielectric, radiative dielectric, inductive, and radiative magnetic heating. It is analogous to microwave heating but with diverse frequencies. The success of ohmic heating depends on the rate of heat generation in the system, the electrical conductivity of the food, electrical field strength, residence time, and the method by which the food flows through the system. The food behaves as an electrical resistor and is heated by the diffusion of electrical energy. Electrical resistance heating permits particles and liquids to heat at the same rate and allows the rapid heating of mixtures with high solid fractions.

The electrodes are in contact with the food to pass electricity of varying voltage and current combinations. The most critical factors are coldest spot and the shortest residence times spent in the heating and holding system. If a particle of low conductivity is encompassed by a high-conductivity environment, this particle will thermally lag the fluid. If isolated, low-conductivity particles enter the system, and the risk of under processing prevails. A distribution-free method was the most appropriate method to determine reliably the characteristic fastest particle of the system. The ohmic processing unit consists of a power supply (generator) to produce the electricity and electrodes. The electrode gap can oscillate depending on the size of the system, but by changing this distance, the electric field strength can be varied. Currently, ohmic heaters are specifically engineered to interface with continuous, as opposed to batch, process lines. Two possible geometries can be considered, the transverse field and collinear field modes: with the transverse mode, the applied electric field and current flux are at right angles to the mass flow, and with the collinear mode, they are parallel.

Surimi is a stabilized fish myofibrillar protein and is a major ingredient for imitation products. The microbiological quality of these seafood products is very high. Ohmic heating experiments showed that Pacific whiting surimi paste and mince were stabilized in the 20–70 °C range at frequencies from 55 to 200 kHz. Ohmic heating is also applicable in thawing frozen foods by placing it between two electrodes and subjecting it to alternating current. The advantages of this process include no generation of water and wastewater during thawing, relatively uniform due to volume heating, and it is easily controlled. The major disadvantage of ohmic heating is the flow of current through the thawed portion of shrimp block, which may give it a cooked appearance. This is called runaway heating or formation of hot spots. The other advantages of ohmic heating include better product quality, less cooking time, lower capital cost, better energy efficiency, and environment friendliness. It involves lower capital cost compared to other electro-heating methods. It is continuous and penetrates deep into the food material. Ohmic heaters are easy to clean due to reduced product fouling on the food contact surface. The major advantages asserted for ohmic heating technology are processing without heat transfer surfaces, uniform heating of liquids, and carrier fluids at similar rates, thus making it possible to use as high-temperature short-time (HTST) technique. The advantage of ohmic heating is that it uniformly heats unlike microwave heating. It avoids the degradation of thermosensitive compounds through overheating and reduces the fouling of treated food surfaces during processing. It improves the food quality and saves cost and energy to processors, produces safe high-quality food, and validates any commercial process by experimentally demonstrating its application. The shelf life of ohmically processed foods is comparable to that of canned and sterile, aseptically processed products (Fig. 2.2).

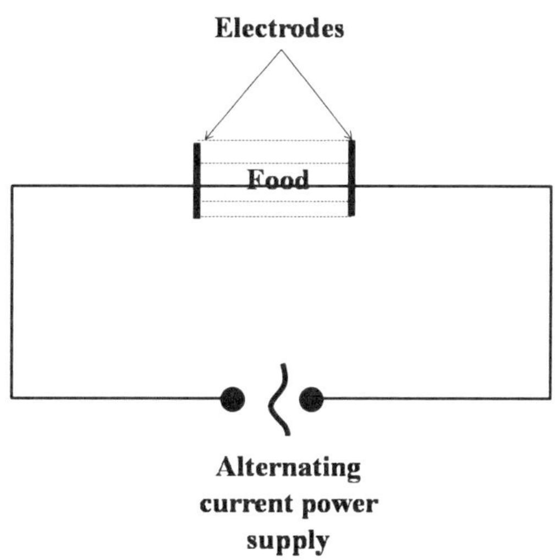

Fig. 2.2 Schematic diagram of ohmic heating

2.1.2.4 Ultraviolet Processing

Recently, ultraviolet (UV) processing has been confirmed to be a potential method for lengthening the shelf life of seafood. UV-A with a wavelength of 320–400 nm, UV-B with a wavelength of 280–320 nm, and UV-C with a wavelength of 200–280 nm are the three regions in the UV spectrum. UV-C light technology application is restricted to liquids and free-flowing foods. UV treatment is a surface treatment because of its low depth penetration ability (Koutchma 2008). The germicidal property of UV-C is accountable for four log reductions or more in viruses, yeasts, and bacteria at 1000 J/m^2 dose rate. When a microbial cell is subjected to UV rays, several alterations happen in the cellular environment. The lethality of this technique depends on treatment time, target species, and range of wavelengths adopted for the irradiation of the cells. The UV rays interact with microbial nucleic acid resulting in microbial inactivation and formation of photoproducts such as pyrimidine dimers amid two neighboring pyrimidine bases (thymine and cytosine) on same strand of DNA. The rate of pyrimidine adduct formation is lower than pyrimidine dimers. At high doses, cross-links between DNA and proteins are formed; and at even higher doses, DNA strand breaks. Apart from the wavelength, the concentration of hydrogen ions, the relative humidity, the treatment temperature, and the microbial load also influence the efficacy of UV rays. UV-C in combination with MAP enhance the shelf life of fish by reducing microbial growth, lower growth rate and number of colonies in the stationary phase, and decrease biogenic amines, putrescine content, total ammonia, and total volatile base nitrogen (TVB-N).

UV lamps are typically made of distinct quartz glass that permits 70–90% of UV rays to pierce. Mercury vapor lamps contain mercury sealed within the glass tube and an inert gas carrier. Mercury ionizes and vaporizes and produces UV radiation when an electric arc is formed. Accelerated senescence and surface discoloration may occur in seafood. UV radiation induces lipid oxidation in treated seafoods as hydrogen peroxide, superoxide radicals, and lipid radicals are indirectly formed by UV light. Peroxide formation diminishes the pigments and the fat-soluble vitamins. Protein, aromatic amino acids, and enzymes are denatured by UV radiation that modulates the seafood composition.

2.1.2.5 Microwave Processing

The microwave heating is used for various fish processing methods such as drying, pasteurization, sterilization, thawing, tempering, baking, etc. Electromagnetic waves of varying frequency 915 MHz to 2450 MHz between 10 and 200 kW are used in this technique. The equipment contains a power unit (power supply and circulating matching circuit), vacuum tube, hybrid system, microwave or radiofrequency applicatory, and a control and automation.

Microwave heating is caused by the capability of the food products to absorb microwave energy and transfer it into heat by dipolar and ionic mechanisms. In an oscillating electric field, the permanently polarized dipolar water molecules try to realign in the direction of the electric field; and at high-frequency electric field, this realignment occurs at a million times per second, causing internal friction of molecules and volumetric heating of the material. Studies showed that microwave cooking practices of skipjack tuna retained omega-3 fatty acids. Microwave blanching retained the color and inactivated the enzymes. Microwave drying is used to eliminate moisture from fish and fishery products. The drying happens at a faster rate while maintaining the quality of the product. In microwave drying, due to volumetric heating, the vapors are formed inside, and an internal pressure gradient is developed that forces the water outside. Thus, shrinkage of food materials is prohibited in microwave drying. One of the shortcomings of microwave drying is that extreme temperature is the scorching along the edges of the food and the production of off-flavors. Microwave collectively with other drying techniques such as air-drying/infrared/vacuum drying/freeze-drying gives improved drying characteristics in comparison to these alone (Chandrasekaran et al. 2013).

2.1.2.6 Radio-Frequency Heating and Infrared Radiation

Electromagnetic radiation is conveyed through the space at an enormous velocity (speed of light). The heat generation in material subjected to this could be due to vibrational (infrared) or rotational movement (radiofrequency and microwave) of molecules. Application of electromagnetic radiation in food processing is gaining popularity due to its specific advantages such as quicker and efficient heat transfer, low processing cost, uniform heating and better organoleptic, and nutritional characteristics in the processed material. In electromagnetic processing, the waves penetrate the material causing fast heat and mass transfer, leading to reduced processing time and enhanced product quality (Jiao et al. 2018).

Radiofrequency (RF) heating refers to the use of electromagnetic waves of certain frequencies to generate heat in the material. It could be executed in one of the three frequencies (13.56 MHz ± 6.68 kHz, 27.12 MHz ± 160 kHz, and 40.68 MHz ± 20 kHz). The principle includes heat generation due to the resistance to the flow of high-frequency electric current in dielectric materials when a high-frequency electric field is applied across them. In other cases, the heat is generated from the resonant absorption mechanism whereby the frequency of applied electric field is such that the dipoles within the material reorient incessantly to reversing electric field. The amount of heat generated is determined by the frequency, the square of the applied voltage, dimensions of the product, and the dielectric loss factor of the material. The advantages include increased heating and processing speed, better product quality and yield due to uniform heating, less floor space, instant on/off and temperature alteration, and higher energy saving and "self-limiting" property, which controls the consumption of RF energy according to the workload.

Infrared spectrum occupies the region between visible and microwave radiation and has wavelength in the range of 0.78 μm to 1000 μm. Infrared radiation is categorized into near-infrared (NIR), mid-infrared (MIR), and far-infrared (FIR) radiation. NIR is appropriate for HTST processes and FIR for surface heating. Hybrid drying involving MIR and hot air is reported to be apposite for drying of high moisture materials. The advantages of IR heating include its adaptability for any heating applications, energy efficiency of properly, and efficient heat transfer to the food material; accordingly, reducing the processing time and energy cost may be maintained at ambient temperature, uniform heating, less damage, retention of nutrients and organoleptic properties, and easy housing of IR heating with convective, conductive, and microwave heating making it ideal for hybrid processes. The limitations of IR heating include surface heating, thus making it inappropriate for heating material requiring deep heat penetration and high heat insulation. Moreover, the radiations travel in straight line; hence, the product shape is also important (Krishnamurthy et al. 2008).

Although IR and RF waves have wide applications in nonfood industry, its practice in food industry is restricted. In the recent past, hybrid processes have gained recognition in food processing operations where electromagnetic waves are used together with other techniques such as drying, baking, roasting, blanching, pasteurization, and sterilization. The IR spectrum is used for selective heating and effective microbial inactivation. The RF radiation finds application in precooking, blanching, sterilization, tempering, thawing, and baking. Dielectric thawing (fish, meat and meat products), defrosting (meat and fish at 35 MHz frequency radio waves), and pasteurization (small cured hams) also have been observed. During radio-frequency thawing fishery products pass through oven/heater and are subjected to a direct or volumetric heating process in the form of a radio-frequency (RF) energy source. The RF heating process depends on the ionic conductivity of the material being heated. Temperature rise within the product is relatively uniform and depends on the size and composition of the product. At the lower frequencies of RF, penetration of the RF energy into foods is much greater.

2.1.2.7 High-Intensity Pulsed Light Technology

Pulsed light (PL) is nonthermal technology for decontamination of food surfaces and food packages, comprising of short-time high-peak pulses of broad-spectrum white light. The term light is used as wavelength range from 180 to 1100 nm (UV 180 to 400 nm, roughly subdivided into UV-A, 315 to 400 nm; UV-B, 280 to 315 nm; UV-C, 180 to 280 nm), visible light (400 to 700 nm), and infrared rays (IR 700 nm to 1100 nm). It is used for rapid inactivation of microorganisms on food surfaces, equipment, and food packaging materials. The UV light causes thymine dimerization in the DNA chain preventing replication, ultimately leading to cell death. This technique has been approved by the FDA in 1996 under the code 21CFR179. Pulsed light can be used to inactivate microorganisms (green technology) on the surface of food packaging materials; thereby, it would lead to a reduced use of preservatives or chemical sterilizing agents (Sun 2014). Studies reported that this can reduce the microbial load of *Listeria monocytogenes*, *Salmonella enteritidis*, *Pseudomonas aeruginosa*, *Bacillus cereus* and *Staphylococcus aureus*, and *Escherichia coli*. These also reduce the psychrotrophs and coliform population on the surface of fillets.

2.1.2.8 Irradiation

Subjecting food/materials to ionizing radiations such as electron beams, X-rays, and gamma rays is called irradiation. During irradiation molecules absorb energy, and chemical bonds are cleaved to give free radicals or highly unstable and reactive

Fig. 2.3 Radura symbol indicates the product has been processed using irradiation

ions. High-energy isotope sources such as cobalt-60 are used commercially to generate γ-rays, and X-rays and high-energy electrons can be produced from machine sources (Sun 2014).

Machine sources are electron accelerators that consist of an evacuated tube, and electrons are supplied by heated cathode, in which high-voltage electrostatic field excites the electron. Through amplifying the cascade of collisions, these electrons can ionize other atoms. The absorbed dose, the overall economics of the process, food density, and food thickness establish the effectiveness of the application. The formation of reactive species induced by irradiation plays a leading role in killing microbes. DNA, proteins, or cell membranes are interrupted by reactive species such as free radicals that are released from water molecules by high-energy electrons. The principal oxygen accessibility and the dose affect the microbial inactivation. Reduction in water activity prevents bacterial inactivation; hence, reduced accessibility of reactive water molecules by freezing increases the microbial resistance to irradiation. The sensitivity of microorganisms to ionizing radiation is in the ascending order of viruses $<$ bacteria $<$ parasites. The shelf life of foods can be extended by radurization (0.4–10 kGy, in which viable spoilage microorganisms, molds, and yeasts are destroyed), radicidation (0.1–8 kGy, where viable non-spore-forming foodborne pathogens are reduced), or radappertization (10 to 50 kGy), in which both spores and vegetative bacteria are completely destroyed. Irradiation is harmful to humans. However, being in low doses in seafood pretreatment, these are safe for consumption. At high doses, some amino acids are cleaved, which alter the flavor and aroma of foods. Irradiation accelerates autooxidation of lipids, thereby producing hydroperoxides and off-flavors (Fig. 2.3).

2.1.2.9 Cold Plasma Processing

Cold atmospheric plasma (CP) is a novel nonthermal process, where electricity and carrier gases (oxygen, nitrogen, air, and argon) are the key elements. Because of the charged particles, free radicals, photons, chemical reactive species, and ultraviolet radiations are produced; hence, this technique can be used for sanitizing foods and packaging containers. The antimicrobial property of the technique aids in extending the shelf life of the product and ensures food safety (Sun 2014). It does not alter the nutritional profile and sensory attributes of food products. It is a chemical-free, contactless, and waterless method that can prevent and reduce microbial growth. Mixture of ions, electrons, and free radical species in gaseous plasmas are accountable for microbial destruction. The gas becomes ionized when in contact with electrons and form new active species when passed through plasma. The equipment efficiency depends on the numerous CP jet devices installed for applying plasma.

The strong oxidizing agents such as ozone and atomic oxygen, UV radiations, photons, chemical reactive species, and charged particles contribute to antimicrobial effect. Agricultural products (mango, lettuce, almond, apple, and melon), real food systems (cheese and cooked meat), and egg surface have been decontaminated by CP in the food industry. A very limited attempt has been made at the use of plasma technology to enhance the quality and safety of seafoods. The application of CP in reducing microbial growth has been studied in smoked salmon, cold smoked salmon, and dry fish. Treatment with CP can induce lipid oxidation and formation of short-chain fatty acids, aldehydes, hydroxyl acids, and keto acids, which cause off-flavors and off-odors during storage. Undesirable textural properties, acidity, and discoloration were also reported in processed food. High installation cost is a major drawback.

2.1.2.10 Ozonification

The interaction of molecular diatomic oxygen (O_2) with an oxygen atom gives rise to ozone (O_3), which is a highly unstable triatomic oxygen molecule; ozone, which is employed as an oxidizing agent in water treatment and in the food industry, is a powerful sanitizer (Zhao et al. 2018). Recently, ozonification has gained interest in seafood preservation due to its ability to reduce spoilage organisms and improve the quality and safety of the product. The principle involves production of ozone in a ventilation of electrical discharge of voltage in the presence of unadulterated oxygen, whereby the oxygen molecules split into free oxygen atoms (O) and oxygen molecules (O_2) that can collide with these atoms. This leads to the formation of ozone molecules (O_3). Immense energy is necessary to cleave the O-O bond; and the sources to excite oxygen could be chemi-nuclear sources, electrolytic processes, corona discharge (CD), and ultraviolet light (wavelengths <200 nm). Commercially, ozone is produced with the CD method.

Bacteria, vegetative cells, and spores are drastically affected by ozonification. Alteration of glycolipids, glycoprotein, amino acids (tryptophan), and sulfhydryl groups results in bacterial cell disruption and eventually death. Chemical composition, pH, additives, temperature, bacteria load, and ozone contact time and surface area affect the effectiveness of the treatment. Although ozonification can prolong the shelf life of seafood, it may induce oxidation which may lead to changes in sensory parameters. It may also cause protein oxidation, thereby leading to changes in protein functionality and poor-quality products. There is a high risk of post-contamination of water as ozone only lowers the initial microbial load but has less effect on prevention of contamination after treatment. Other processing technologies such as high hydrostatic pressurization, ionizing radiation, cold plasma, ultraviolet light, and pulsed electric fields could be employed together with ozone treatment for microbial inactivation and product safety.

2.1.2.11 Microwave Vacuum Drying

Microwave vacuum drying is sparely used in seafood industry. During vacuum drying, high-energy water molecules are distributed to the surface and evaporate due to low pressure. Due to the vacuum condition, oxidation is prevented, and color, texture, and flavor of dried products are maintained. Microwave drying is widely used to dry food due to the high drying rate and uniform drying (Sun 2014; Zhao et al. 2018). In microwave drying, microwaves (MW) penetrate into dielectric materials and generate internal heat, which establishes a vapor pressure and gently pushes the moisture to the surface. Because of this moisture pumping effect, the moisture is forced to the surface, and case hardening does not occur, enabling improved drying rates and product quality. However, due to its high initial capital investment and low-energy efficiency, it has not been very popular. The advantages of microwave drying include enhanced diffusion of heat and mass, development of internal moisture gradients which enhance drying rates, increased drying rates without increased surface temperatures, and better product quality. It also improves the quality of food materials by enhancing the drying rate during falling rate period (final stage of drying).

During conventional drying shrinkage of the samples and low surface moisture content are observed. During microwave drying, vapors are generated in the interior of food materials due to volumetric heating, which are forced to move outside due to the pressure gradient. In order to improve the drying efficiency and quality of dried product, microwave drying can be combined with other drying methods. Microwave vacuum drying is a combination of microwave drying and vacuum drying. The rate of microwave vacuum drying is much higher than the rate of vacuum drying. A polypropylene container can be used as a vacuum chamber which is placed inside microwave cavity, and the vacuum chamber can be connected to vacuum pump to create low pressure. It is used prominently in drying of fruits and vegetables. Central Institute of Fisheries Technology (CIFT) has developed an innovative project-microwave vacuum oven, first of its kind in the country, for drying fish that will ensure a super quality of the product, have longer shelf life, and are suitable for exploring export markets.

2.1.2.12 Hybrid Drying Technologies

Heat Pump Drying (HPD) HPD technology is expensive and is used to enhance the quality of foods where low-temperature drying and well-controlled drying conditions are required. The food products such as meat, seafood that are heat sensitive and usually subjected to freeze-drying can utilize this technology. The advantages of the heat pump dryer include higher-energy efficiency with improved heat recovery to meet specific production requirements, wide range of drying conditions usually 20–100 °C (with auxiliary heating), and relative humidity 15–80% (with a humidification system). The restrictions of the HPD are non-environmental friendliness as CFC is used, regular maintenance of compressor, refrigerant filters, etc., charging of refrigerant, increased capital costs, limited drying temperature, and need for process control and design. With proper control, it is also possible for HPD to produce freeze-drying conditions at atmospheric pressure.

Fluidized Bed Drying (FBD) It is usually used for the drying of granular solids in the food (50–2000 mm), ceramic, pharmaceutical, and agriculture industries. The advantages of FBD include high drying rates due to exceptional gas-particle contact leading to high heat and mass transfer, smaller flow area, high-thermal efficiency, lower capital, and maintenance costs compared to conventional rotary dryers. The limitations may include high-power consumption, high potential of attrition, granulation or agglomeration of sample, low flexibility, and potential of defluidization.

Superheated Steam Drying (ISSD) SSD is a nonpolluting and safe drying method with low-energy consumption. The principle behind this drying mechanism is based on using superheated steam for drying incorporating a vapor recompression cycle to recover heat. The entire system comprises a heat treatment chamber, a compressor, a heat exchanger for heat recovery, and a blower system. The drying medium is superheated steam that performs drying in a closed-cycle picking up moisture from the product in the heat treatment chamber and condensing the evaporated water in a heat exchanger. The advantages of the technique include improved drying efficiency, environmental friendliness and oxidation-free products, and control over the drying operations. The application of hot steam aids in destroying all stages of insects, molds, and microorganisms found in foodstuffs.

Infrared Drying (IRD) This technique reduces heating time as the IR energy is transferred from the heating element to the product surface without heating the surrounding air. The advantages of IRD include high heat transfer rates, easy to direct the heat source to drying surface, quick response, and low cost.

Radio-Frequency Drying (RFD) In RFD, RF waves generate heat volumetrically within the wet material by the combined mechanisms of dipole rotation and conduction effects which speed up the drying process. Food materials that are difficult to dry with convection drying are good candidates for RF-assisted drying. Radiofrequency heats all parts of the product mass simultaneously and evaporates the water in situ at relatively low temperatures usually not exceeding 80–82 °C. Since water moves through the product in the form of a gas rather than by capillary action, migration of solids is avoided. RF drying improves the color of products and avoids cracking and uneven shrinkage. Simultaneous external and internal drying considerably reduces the drying time to reach the desired moisture content.

2.1.2.13 Hurdle Technology/Chemical and Biochemical Hurdles

Hurdle technology is used in the preservation of meat and fruits and vegetables. Hurdles are physical or chemical parameters that can be attuned to ensure the microbial stability and safety of foods. The physical parameters include the processing and storage temperatures, water activity, pH, and redox potential that could be controlled to inhibit or inactivate the microorganisms. Salt, lactic acid bacteria, antimicrobial agents (chitosan), and pulsed light technology are some of the hurdle technologies used in seafood processing. Hurdle drying technology also follows the same principle and could be of different types.

Acids (and salts) are supplemented to food to control pH and thwart oxidation or to serve as preservatives. The utilization of organic acids in foods is most regular conventional methods of chemical preservation. These enter into the living cell and dissociate or aid in proton uptake by the cell. The microorganism attempts to maintain its internal pH, extruding the organic acids or protons, thereby using all its energy. Furthermore, sorbic, propionic, and benzoic acids interfere with the membrane proteins or change the membrane fluidity. Other ways include inactivation of metabolic enzymes, protein synthesis systems, or the genetic material (Eklund 1989; Pundhir and Murtaza 2015). Some of the chemical and biochemical hurdles are as follows.

Acetic Acid It is an extensively used preservative in food in low concentrations. Its flavor and taste are acceptable in canned products (fruit and vegetables), mayonnaise, salad dressings, mustard, ketchup, and marinades for meat, poultry, and fish. The fermentation of lactic acid bacteria as in pickles, sauerkraut, and fermented milk products leads to acetic acid formation in these food. Compared to other organic acids, it has a high pK (4.75), thus enabling it as an appropriate agent for low acid foods. Acetic acid inhibits bacteria than yeasts and molds, even though it is effectual against specific bread molds. It interacts with the cell membrane and neutralizes the electrochemical potential, restrains amino acid uptake in membrane vesicles, and denatures proteins inside the cell. Due to its unpleasant taste and odor, it is often used along with lactate or citrate. It is used as a part of a hurdle concept; the effective acetate concentration can be minimized, so that the negative sensory influence is minimized. The acetic acid lowers the pH, escalating the amount of undissociated acid.

Propionic Acid This is not as extensively used in the food industry due to its pungent, disagreeable odor and, hence, is used in amalgamation with other organic acids (and their salts). It is largely used as a mold inhibitor for cheese, bread, and cakes and against rope-forming bacteria in bread dough, in low doses. The pK value is 4.87 and the upper limit for efficacy is around 5.5. Nevertheless, it also functions at higher pH but at pH 6.0 only 6.7% remains undissociated. It is produced in cheeses by fermentative action of propionic acid bacteria. These are lipophilic and soluble in the cell membrane, where they apparently act as proton ionophores. These assist in proton transport into the cells, thus compelling the microorganisms to utilize more energy and pump out protons in order to maintain their internal pH.

Lactic Acid It has a reasonable strong acid bitter taste and is used for acidification (pK 3.08), flavor augmentation, and microbial inhibition. It also finds application in sugar confectionary, dairy products, beer, wine, beverages, and bakery products. It is a most widespread organic acid formed from during lactic acid fermentation. It also manipulates the membrane potential of microorganisms and lowers the pH of the foods below the growth range. The effect of lactic acid banks on the pH of the food, buffering capacity, and the form it has been added. Lactate (2–4%) is used in meat products as a part of hurdle concept (alone or in combination with acetate or propionate). It is also used as a meat sanitizer for fresh/cut meat. It transforms the spoilage organisms and Gram-positive and Gram-negative pathogens. It also enhances the sensory freshness of meat. Recently, it has been used in combination with the nisin.

Sorbic Acid It is chiefly used in foods as salts of calcium, sodium, or potassium. Calcium sorbate is tasteless and flavorless. Sorbates are principally used as fungistats in cheese, bakery products, fruit juices, beverages, and salad dressing. In meat products these function as antibotulinal agents. These are principally effective against yeasts and molds and bacteria (catalase-positive cocci, spore-forming bacteria, and aerobes). The acid acts best below pH 6.0 and is normally futile above pH 6.5. Between pH 4.0 and 6.0, these are more effective than benzoates. The pK of sorbic acid is 4.8. These are linked with energy depletion of the living microorganism, amendment of the electrochemical membrane potential, inhibition of enzymes, and interference with the microbial genetic material.

Benzoic Acid These were the first chemicals allowed in the USA to be used as food preservatives. The pK of benzoic acid is quite low (pK 4.20); hence, its key antimicrobial effect is due to the undissociated acid as seen in high acid foods (ciders, soft drinks, and dressings). It is also used in margarine, fruit salads, sauerkraut, jams, and jellies. These act as mold and yeast inhibitors. Benzoic acid occurs naturally in cranberries, prunes, strawberries, apples, and yogurts. In some cases, it may impart "peppery" or burning flavor. Its mechanism of action against microbes is similar to other acids. It also inhibits amino acid uptake and certain enzymes (alpha-ketoglutarate, succinate dehydrogenase, 6-phosphofructo-2-kinase, and lipase).

Parabens (Esters of p-Hydroxybenzoic Acid) These are used in foods chiefly as methyl, propyl, and heptyl esters in bakery products, soft drinks, fish products, pickles, and salad dressings. The esterification of the carboxyl group of benzoic acid permits the parabens to remain undissociated (antimicrobial) in a broader pH range (pH 3–8); hence, these are effectual against a broad range of microorganisms (Gram-positive and Gram-negative bacteria, yeasts, and molds). The microbial inhibition amplifies with increasing alkyl length. Some Gram-negative bacteria are resistant to the higher parabens. These interfere with the synthesis of protein, RNA and DNA.

Citric Acid It is the most extensively used acid in the food industry. It is a water-soluble tricarboxylic acid with pK values of 3.14, 4.77, and 6.39 (for each carboxylic group). This augments the flavor of citrus-based foods (tomato juice, ice cream, sherbets, beverages, salad dressings, jams, and jellies). The antimicrobial property is because of the acidulation property of the acid. It also acts as an antioxidant because of its chelating property that catalyzes oxidation. It is also considered as a nitrite-saving agent in meat products.

Malic Acid It is a dicarboxylic acid with pK values of 3.40 and 5.11. It has a smooth, tart taste that lingers in the mouth without imparting a burst of flavor. It is highly water soluble and inhibits yeasts, molds, and bacteria by lowering of pH. It is used in beverages, hard candies, canned tomatoes, and fruit pie fillings.

Fumaric Acid It is also a dicarboxylic acid (pK values 3.03 and 4.44, for each carboxylic group). Though it has a strong acid taste, it blends and intensifies the aftertaste of certain flavors as in fruit drinks, gelatin desserts, pie fillings, biscuit dough, and wines, thus giving a positive flavor effect. It is naturally seen in rice, sugarcane, wine, plant leaves, mushrooms, and gelatin. It lowers the pH and acts as chelator, binding micronutrients essential for bacterial growth.

Tartaric and Adipic Acid Tartaric and adipic acids reduce pH in the foods. As such, these do not have any antimicrobial effects. Tartaric acid (pK 2.98) has a strong tart taste and augments grape-like flavor. It is used in fruit jams, jellies, and grape-flavored beverages. The application of adipic acid (pK 4.43) is found in evaporated milk, puddings, desserts, and flavorings.

GdL (Glucono Delta-Lactone) GdL is extensively used as acidulant in meat products (dry cured sausages) where a controlled release of acid is required. In the presence of water, it is converted back to gluconic acid, which lowers the pH, inhibits bacteria, and accelerates drying.

Long-Chain Fatty Acids and Phenolic Acids Long-chain fatty acids (oleic and lauric acid) and phenolic acids (chlorogenic, hydroxycinnamic, caffeic, p-coumaric, ferulic) have shown antimicrobial activities in food. Monolaurin (monoglyceride of lauric acid) inhibits the growth of *Listeria monocytogenes* in Italian Stracchino cheese. Oleic acid and potassium oleate diminish the number of bacteria on poultry skin. Aerobic bacteria, enterobacteriaceae, *Campylobacter*, and *Listeria monocytogenes* are most sensitive; *Escherichia coli* and *Pseudomonas aeruginosa* are more resistant, while *Enterobacter cloacae*, *Staphylococcus lentus*, and *Salmonella typhimurium* have the greatest resistance.

Plant-Derived Antimicrobials Plants are under constant attack from fungi, insects, worms, and other herbivorous animals. In order to self-protect, these have developed self-defense strategies by the synthesis of numerous antimicrobial chemical compounds (secondary metabolites) that are stored in specialized tissue/vacuoles. These antimicrobial compounds are created during defense mechanisms (phytoalexins). Catechol, carvacrol, thymol, and caffeic acid are preformed or constitutive antibiotics, while alliin (in garlic) and sinigrin (in cabbage) are inactive precursors, which are converted into active compounds (allicin and allyl isothiocyanate) by enzyme action (alliinase and myrosinase, respectively). The phytoalexins are only produced after infection of the plant, induced by elicitor molecules from the invading organism. Several vegetables especially from the family Leguminosae (chick peas, peanut, French bean, soya bean) or Solanaceae (potato, green pepper, eggplant) produce these antimicrobial agents. Oleoresins/essential oils/extracts from herbs/spices (cinnamon, clove, thyme, sage, rosemary, oregano, basil, allspice, nutmeg, coriander, lemongrass, cumin, and bay laurel) also exhibit antimicrobial effects. This antimicrobial activity is usually a broad-spectrum against Gram-positive and Gram-negative bacteria and fungi. The antimicrobial effect of essential oils is chiefly attributed to phenolic compounds, terpenes, aldehydes, alcohols, esters, oxides, and ketones. These interfere with the cytoplasmic membrane, resulting in cell wall permeability and leakage of the cell. Other mechanism includes inhibition of enzyme systems or enzyme denaturation. These also exert antioxidative effect and impart flavor.

Chitin/Chitosan Chitin is one of the most profusely found natural biopolymer, consisting of N-acetyl-D-glucosamine units linked by β-(1,4) bonds, while chitosan is a deacetylated form of chitin, obtained by strong alkali treatment. The degree of deacetylation varies between 60% and 95%, which leaves free amino group in chitosan and allows it to act as a weak base in acidic solutions. The degree of deacetylation and depolymerization manipulates the antimicrobial effect of chitosan. Chitin can also be degraded using enzymes such as chitinase, lysozyme, etc. resulting in oligomers or monomers. Chitosan is soluble in weak organic acids such as lactic acid or acetic acid, while chitin is soluble in dimethylformamide or hexafluoroisopropanol. Chitosan has amino group in C-2 position, which exerts positive charge at pH values below its pK of 6.3. This encourages it to interact with the negatively charged microbial cell membrane, leading to the leakage of intercellular constituents. Other mechanisms include chelating trace metal ions, binding water, and inhibition of various enzymes. The size of the molecule (the degree of polymerization), solubility, and the net charge (degree of deacetylation, pH) are important parameters in defining the antimicrobial effects of chitosan. It inhibits Gram-negative (*Enterobacter* and *Pseudomonas*) and Gram-positive (*Bacillus*, *Listeria monocytogenes*, *Brochothrix*, and *Lactobacillus*) bacteria. Chitosan is an appropriate preservative for fruit and vegetables and marine foods but less appropriate for meat and meat products.

Antimicrobial Enzymes Antimicrobial enzymes are found widespread in nature acting as a defense mechanism against microbial infection. These are chiefly hydrolases or oxidoreductases (lysozyme, lactoperoxidase and glucose oxidase). Lysozymes are natural constituents of blood, milk, eggs, and plants (figs and papaya). Hen egg-white lysozyme contains 129 amino acids with four disulfide bonds and a molecular weight of 14–15 kD; nevertheless, the size and structure vary depending on origin. The enzyme is stable at pH 3–7 and withstands temperatures up to 50 °C. The lysozyme used for food applications is primarily derived from hen egg white. Lysozyme hydrolyzes the (1–4)-glucosidic linkage of the peptidoglycan present in the bacterial cell wall. It is largely effective against Gram-positive bacteria, as they lack the outer membrane. The Gram-negative bacteria withstand lysozyme as its outer membrane contains lipopolysaccharide, phospholipid, and lipoprotein. Destabilization of outer membrane using EDTA may induce lysozyme activity on the cell wall. It leads to cell lysis and loss of cytoplasm. Lysozyme has been used as a preservative for a wide range of foods (cheese, meat, fruit, vegetables, and wine). Lactoperoxidase, secreted from mammalian glands, are found in milk, saliva, and other body fluids. Bovine milk contains approximately 20 times more lactoperoxidase activity than human milk. It catalyzes the oxidation of thiocyanate to hypothiocyanate (OSCN), which is responsible for its antimicrobial activity probably due to reaction with sulfhydryl groups on various proteins. The enzyme is fairly heat resistant and is only partially inactivated by short-time pasteurization at 74 °C. It is highly active at acidic pH with an optimum around pH 5. It shows a broad-spectrum antimicrobial effect against both Gram-positive and Gram-negative bacteria. It also helps in enhancing the shelf life of milk.

Glucose Oxidase The antimicrobial action of the glucose oxidase is due to the H_2O_2, though the effect is a short-term effect. Long-term exposure of foods to H_2O_2 may cause rancidity as a result of lipid oxidation. Glucose oxidase has not been used commercially as an antimicrobial. The O_2 depletion prevents the growth of strict aerobes.

Other Oxidoreductases Other oxidoreductases (myeloperoxidase of human origin and haloperoxidases of fungal origin) exert antimicrobial activity. Myeloperoxidase from human leucocytes is irrelevant to food industry, but the haloperoxidases could be of significance for future food applications, due to their cost-effective production.

Nisin Bacteriocins are proteins/peptides (nisin) produced by lactic acid fermentation of vegetables, meat, and dairy products resulting in increased shelf life and food safety. Addition of lactic acid bacteria as starter cultures to fermented products is a conventional food processing technique that has been approved by food authorities, but the use of bacteriocins or the bacteriocin-producing bacteria is limited.

Nisin is a polypeptide (consisting of 34 amino acids in its mature form), produced by *Lactococcus lactis*, and is a heat-stable monomer with a MW of 3.4 kD. However, it is normally found as dimer (approx. 7 kD), as it is more stable. At acidic pH, it can withstand 115 °C for 20 min and does not lose its antimicrobial activity. It is a broad-spectrum bacteriocin with antimicrobial activity against many Gram-positive bacteria and the endospores of *Clostridium* and *Bacillus* spp. It interferes with cytoplasmic membrane resulting diffusion and cell lysis. Like lysozyme, the effect of this bacteriocin may be improved with other hurdles, such as EDTA, monolaurin, lactoperoxidase, lysozyme, and lactic acid.

Lactoferrin This iron-binding glycoprotein is found in mammalian milk, saliva, tears, and other secretions, and together with lactoperoxidase and lysozyme, it contributes to the antimicrobial action of milk. Normal milk contains 2–3 g/l, and due to the enormous production of cow's milk, lactoferrin can be made available in bulk amounts at relatively lower cost. Antimicrobial activity of lactoferrin is due to its metal-binding activity and blocking effect. It is used in a commercialized product (activated lactoferrin or Activin™), where it effectively prevents bacteria from attaching to the surface of meat. Hydrolysates of lactoferrin (lactoferricins) exert 9–25 times stronger antimicrobial activity against *Listeria monocytogenes* in broth cultures. It is obtained by the enzyme action of pepsin/acid protease on glycoprotein.

Reuterin Reuterin (3-hydroxypropionaldehyde), an intermediate metabolite produced by *Lactobacillus reuteri* during anaerobic metabolism of glycerol, is water and lipid soluble and is active over a broad range of pH, exhibiting a broad range antimicrobial activity. It is resistant to heat and proteolytic enzymes. However, limited studies have been reported about its antimicrobial activities in various food systems.

Electrolyzed Water and Miscellaneous This is a newly evolved technique in Japan and is produced by passing sodium chloride solution across an electrically charged membrane resulting in two different fractions: an acidic solution (pH 3) containing hypochlorous acid and a highly reactive alkaline solution (pH 11) containing sodium hydroxide. The acidic fraction exhibits antimicrobial mechanisms by lowering pH and inducing contact with chlorine or chlorine-based compounds and a high oxidation-reduction potential (ORP 1100 mV). This newly tailor-made commercial antimicrobial product may be the future preservatives of choice, but they need more investigation in order to gain approval from authorities.

2.1.2.14 Modified Atmosphere Packaging

Modified atmosphere packaging (MAP) involves replacing the air in a packaging container by a mixture of gases or a specific gas different food from the initial air composition. Oxygen (O_2), nitrogen (N_2), and carbon dioxide (CO_2) are the main commercial gases used in MAP. The packaging container is filled with the gas mixture, sealed, and stored with no additional control during storage. Oxygen in the package is displaced by N_2, lessening oxidative rancidity and hindering any aerobic microbial growth (Sandhya 2010). N_2 prevents possible packaging collapse. O_2 retard obligate anaerobic bacterial growth, although sensitivity to oxygen differs among anaerobic microorganisms. O_2 also reduce exudation or drip in fish throughout storage. Mixtures of 30% O_2, 30% N_2, and 40% CO_2 have been recommended for fish with low-fat content, and a 40% N_2 and 60% CO_2 gas mixture for fatty fish. Temperature influences the CO_2 bacteriostatic and bactericidal effects. At low temperature, CO_2 is trapped in the packing materials; at high temperature CO_2 gains kinetic energy and escapes from the packing materials. During storage, the lack of refrigeration may permit the growth of microorganisms stored under CO_2 atmosphere. For successful use of MAP, good or hygienic quality raw material is a key factor. Temperature control and good packaging material are also essential for MAP. Initial bacterial populations, type of product, storage conditions, and concentration of CO_2 all have an influence on the bacteriostatic effect of MAP. The CO2 form carbonic acid with water and alter the cell membrane, enzyme activity, and intracellular pH. Direct alterations in the physicochemical properties of proteins are the most important mechanisms explaining the influence of CO_2 on bacterial cells. The use of antioxidants can be a means to prevent lipid oxidation (in the presence of O_2) of seafood packaged under MAP.

2.2 Biochemical Changes During Processing

Rigor Mortis Postmortem fish muscle undergoes a sequence of rigor mortis process (Hall 2012; Boziaris 2013). Immediately after death, the fish muscle is in fully relaxed condition. In this pre-rigor stage, fish muscle is very soft and pliable and texture is firm and elastic to touch. Both ATP and ADP act as plasticizers for actin and myosin, thus preventing their interaction and actomyosin cross bridge formation and keeping the muscle in state of relaxation. After some time, the fish muscle begins to stiffen, whole muscle becomes inflexible, and fish reaches in-rigor stage. Eventually, after hours or days, fish muscle gradually begins to become softer, and in this state, the fish is said to have reached post-rigor condition. Production of nitrogenous compounds as a result of degradation of complex molecules leads to an increase in post-rigor muscle pH. Several proteins are actively involved in contraction. The two major proteins, actin and myosin, combine in the presence of calcium ions to form actomyosin. ATP the biochemical energy carrier provides energy for contraction, locking the muscle called rigor mortis. This can be slowed down by chilling the fish immediately after death (Fig. 2.4).

Over the time, rigor resolves and rigidity disappears but fish muscle gets considerably deteriorated.

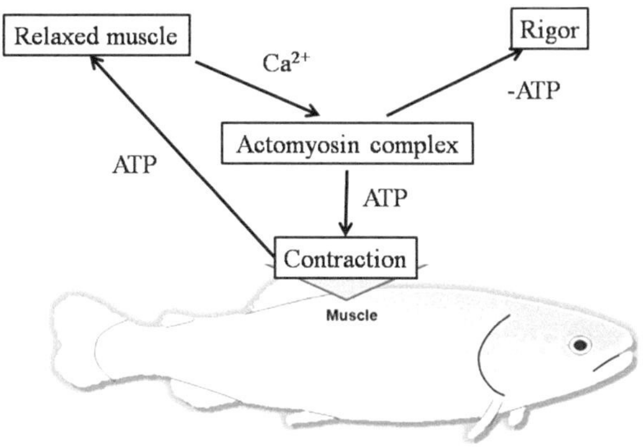

Fig. 2.4 Muscle contraction reactions

Autolysis Enzymes in the flesh and gut previously involved in metabolism now catalyze autolytic reactions (breakdown), which decomposes proteins, nucleotides, and sugars into tasteless or bitter ones. Gut enzymes attack the internal organs, turning them into a soupy mess and allowing bacteria to enter the flesh and decompose the muscle. Anaerobic bacteria produce a particularly foul type of spoilage which results in an inedible fish. A number of biochemical changes take place during postmortem, and these are outlined below.

2.2.1 Postmortem Biochemical Changes

2.2.1.1 Changes Related to Energy Metabolism

Muscle is a highly specialized tissue, and its functioning represents a classical example of conversion of chemical energy to mechanical energy in living systems. Muscle requires a large outlay of energy to rapidly operate the contractile apparatus. This energy is derived immediately from the high-energy compound ATP. Creatine phosphate (CP) then rapidly transfers its high-energy phosphate to ADP so that the ATP level may not decrease too much during activity. Resynthesis of ATP from ADP and CP is catalyzed by the enzyme creatine kinase. The enzyme adenylate kinase converts two molecules of ADP to one molecule of ATP and one molecule of AMP; this reaction also serves as a source of ATP for the muscle cell.

2.2.1.2 Nucleotide Breakdown

This involves the enzymatic breakdown of the energy carrier ATP (Fig. 2.5).

2.2.1.3 Glycogen Degradation

When the fish/crustacean are caught, they struggle immensely causing antemortem exhaustion of energy reserves chiefly glycogen and high-energy phosphates. Rigor mortis starts immediately or shortly after death if the fish is starved and the glycogen reserves are depleted. Tissue enzymes metabolize the energy reserves leading to its degradation and production of carbon dioxide, water, adenosine triphosphate, which degrade further to hypoxanthine, formaldehyde, ammonia, inorganic

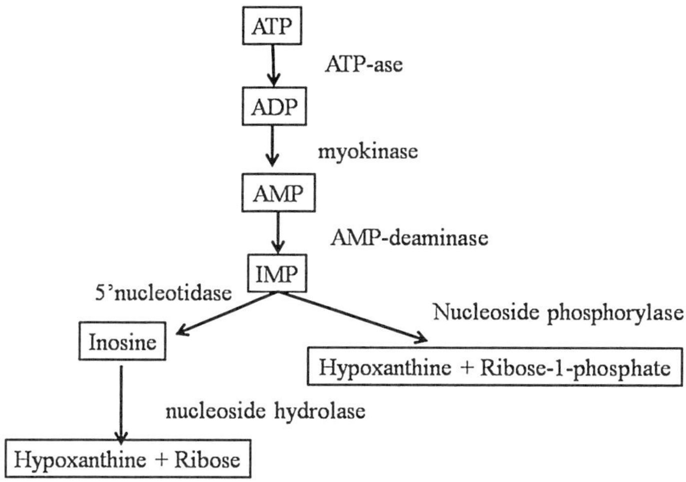

Fig. 2.5 Degradation of ATP molecules

phosphate, and ribose phosphates. The degradation of glycogen in fish follows the Embden-Meyerhof-Parnas pathway through a series of phosphorylated six-carbon and three-carbon intermediates, catalyzed by endogenous enzymes, and results in the accumulation of lactic acid, contracting the tissues and inducing rigor mortis. The production of lactic acid during contraction may lead to a temporary reduction of pH 6.6) in the living muscle cell. Obviously anaerobic glycolysis is favored only under conditions where the mitochondrial system cannot function, i.e., under conditions of limited oxygen. The lactic acid formed in fish is far less. The amount of anaerobic glycolysis that occurs in a given muscle is, therefore, dependent on many factors, including the vascular system, the muscle, the amount of myoglobin, and the number of mitochondria. The rate of glycolysis is temperature dependent and is slowed by low storage conditions. All these lead to loss of water-holding capacity of muscle resulting in gapping during high-temperature rigor.

2.2.1.4 Protein Denaturation

The conformation of proteins as it is assumed to occur in cells and tissues is termed the native state. The native structure of the protein is reversibly or irreversibly altered by denaturation. In general, this conformation is the most stable one under the particular physical and chemical conditions existing in the cells or tissues of the living organisms. Protein denaturation may be defined as any modification of the secondary, tertiary, or quaternary structure of the protein molecules. Denaturation is therefore a process by which hydrogen bonds, hydrophobic interactions, and linkages are broken and the protein is unfolded. In some cases the protein completely unfolds to random coil-type structure. This structural modification may bring about definite changes in chemical, physical, or biological properties of the proteins. The molecular parameters adopted to measure the extent of denaturation are solubility, molecular weight, electrophoretic mobility, sedimentation or diffusion constraints, absorption coefficient, and intrinsic viscosity.

During postmortem, pH reduction in fish muscle lowers the net surface charge on the muscle protein, causing it to denature and lose its water-holding capacity. This also led to muscle contraction, making it stiff and inextensible, which lowers its commercial acceptability. During frozen storage the rate of protein denaturation was found to depend on various factors such as storage temperature, state of rigor at the time of freezing, freezing rate, and time taken for thawing and refreezing. A denatured protein loses its water-holding ability which results in excessively dripping when thawed (drip loss), and fish appears white, dull, and spongy and upon chewing becomes fibrous and tasteless. Sarcoplasmic and lysosomal enzymes, hydrolases, oxidoreductases, and transferases are responsible for quality deterioration of fish after death and before bacterial spoilage. Myofibrillar proteins undergo changes during rigor mortis, resolution of rigor mortis, and long-term frozen storage. The integrity of the myofibrillar protein molecules and the fish muscle texture are affected by these alterations. A certain degree of correlation is observed between the stability of myosin and the resistance of fish muscle to the deteriorative effects of freezing, frozen storage, and dehydration. Cooked seafood no longer possesses the characteristic firm texture and shows a mushy or soft mouth-feel. For frozen seafood, these degradations are accompanied by a loss in the functional characteristics of muscle proteins, mainly solubility, water retention, gelling ability, and lipid emulsifying properties. It gets worse when proteins cross-link in the presence of formaldehyde. Such food upon cooking becomes tough, chewy, and stringy or fibrous. Repeated freezing and thawing make the situation even worse. Degradation of stroma causes microtextural loss (Raman and Mathew 2015) and may lead to honeycombing in skipjack tuna and mackerel and mushiness in freshwater prawn. The cathepsins are released into the cell juices upon physical abuse or upon freezing and thawing of postmortem muscle. The cathepsin D digests both myofibrillar protein and connective tissue protein.

Enzyme	Substrate	Changes encountered
Glycolytic	Glycogen	Lactic acid production, pH drop, loss of water-holding capacity, gaping
Autolytic enzymes	ATP, ADP, AMP, IMP	Loss of freshness, flavor, bitterness due to hypoxanthine
Cathepsin	Proteins/peptides	Tissue softening
Chymotrypsin, carboxypeptidases	Proteins/peptides	Autolysis of visceral cavity
Calpain	Myofibrillar proteins	Tissue softening and mold softening in crustaceans
Collagenases	Connective tissue	Gaping and softening
TMAO demethylase	TMAO	Formaldehyde-induced toughening

2.2.1.5 Total Volatile Base Nitrogen (TVB-N)

TVB-N is a measure of the total amount of nitrogen-containing substances which are produced during storage such as trimethylamine (TMA), which is formed from the reduction of trimethylamine oxide (TMAO). Marine fish contain a small amount of TMAO. This odorless and tasteless compound is reduced by invading bacteria to TMA, which is characterized by its “fishy” odor. TMA becomes useful as a quality index during the middle and late stages of spoilage after the bacteria have invaded the fish. Trimethylamine oxide is converted in the muscle tissue into dimethylamine (DMA) and formaldehyde by enzyme action during frozen storage. The reduction of trimethylamine oxide (TMAO), an osmoregulatory compound in many marine teleost fish, is usually due to bacterial action, but in some species an enzyme is present in the muscle tissue which is able to break down TMAO into dimethylamine (DMA) and formaldehyde (FA):

$$(CH_3)_3\,NO \rightarrow (CH_3)_2\,NH + HCHO$$

It is important to note that the amount of formaldehyde produced is equivalent to the dimethylamine formed but is of far greater commercial significance. Formaldehyde induces cross-linking of the muscle proteins making the muscle tough and readily loses its water-holding capacity. The enzyme responsible for formaldehyde-induced toughening is called TMAO-ase or TMAO demethylase and is most commonly found in the gadoid fishes (cod family). Most of the TMAO demethylase enzymes are membrane-bound and become most active when the tissue membranes are disrupted by freezing.

2.2.1.6 Oxidation of Lipids

The two major deteriorative changes occurring in fish lipids include:

- Enzymatic hydrolysis of lipids (fats) to produce free fatty acids and glycerol
- Oxidation of fatty acids leading to rancid odors and taste

These are the major issues encountered during storage (Fig. 2.6).

Oxidation of lipids is a major cause of off-flavors and textural changes of frozen seafood. The oxidation of unsaturated fatty acids involves the formation of free radicals and hydroperoxides. The study of lipid oxidation in fishery products emphasizes on the nonenzymatic or autoxidative processes. Nonenzymatic lipid oxidation is enhanced by metal ions like iron, copper, and cobalt. In vivo, the bound forms of iron like heme, nonheme iron, and chelation complexes like histidine iron may play an

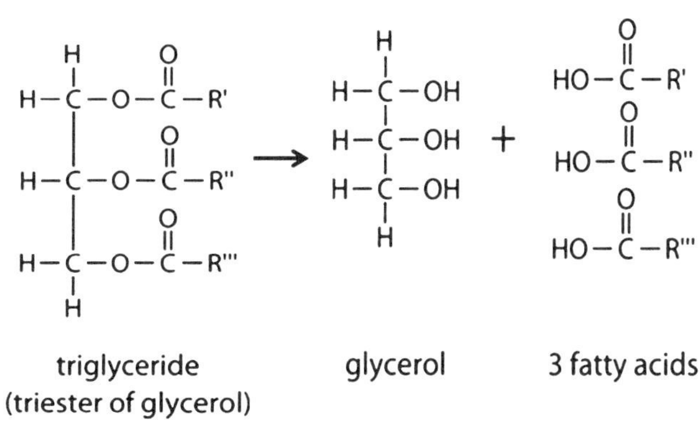

Fig. 2.6 Hydrolysis of lipids

important role in catalyzing lipid oxidation. Lipid oxidation in frozen fish tends to occur at a more rapid rate in the dark muscle and skin and is stimulated by mincing the tissue. The rate of lipid oxidation is also directly related to the fat content. Free fatty acid formed by hydrolysis can contribute to the texture toughening by influencing protein denaturation. It can also enhance lipid oxidation leading to flavor deterioration. Protein and lipids can interact by hydrophobic interactions. The free fatty acids are more prone to peroxidation than esterified fatty acids. Studies have indicated that enzyme-catalyzed reaction may also contribute to the lipid oxidation during the early stages of storage by the enzyme lipoxygenase. This may explain why fish stored in ice for a few days prior to freezing become rancid faster during frozen storage.

2.2.1.7 Other Changes

Epithelial Discoloration Following rigor mortis, the epidermal chromatophores on squid skin undergo contraction and bleeding leading to muscle discoloration and poor quality. Subcutaneous yellowing during prolonged frozen storage (due to lipid oxidation ad carbonyl-amine reactions) could also lead to epithelial discoloration.

Hemoglobin Usually, hemoglobin contributes less to the appearance of seafood than myoglobin as former is lost during handling and processing and latter is retained in the intracellular structure. The amount of residual hemoglobin in fish muscle influences the color of the meat. The green meat of raw or frozen broadbill swordfish (*Xiphias gladius*) is believed to be due to the combination of hemoglobin with hydrogen sulfide generated from the decomposition of the meat.

Hemocyanin Hemocyanin is present in shellfish and is a copper-containing protein. They combine reversely with oxygen. The blue discoloration of canned crabmeat is associated with a high content of hemocyanin.

Myoglobin Myoglobin in fish is easily oxidized to brown-colored metmyoglobin. The discoloration of tuna during frozen storage is associated with the formation of metmyoglobin, depending on temperature and location. Greening discoloration associated with cooking tunas arises from the formation of a sulfhydryl adduct of myoglobin in the presence of oxidizing agent, TMAO.

Carotenoids The carotenoids contribute to the attractive yellow, orange, and red color of several important fish and shellfish products. These may be easily oxidized by lipogenase-like enzyme.

Melanosis Melanosis (melanin formation) or "blackspot" in shrimp during postmortem storage is caused by phenol oxidase and has economic implications.

2.3 Value Addition

Large quantities of fish/shellfish are discarded due to several reasons such as uneconomic handling, processing/preservation, or transportation. Through improvement in infrastructure facilities, the quality of the landings can be upgraded for direct human consumption. The up-gradation of underutilized species may be achieved by use of improved handling and processing techniques or developing different products such as value-added products using the species in glut.

Value-added fish products may be:

(i) Mince or mince-based products
(ii) Battered and breaded or coated products
(iii) Surimi-based products

Fish mince is the fish flesh and finds application in processing several "convenience foods." Advantage of mince-based product is that it conceals identity of the original fish from which it is made and consumers may not hesitate to accept mince or mince-based products even though the original fish would have been unacceptable as whole fish. A batter can be defined as a liquid mixture containing water, flour, starch, and seasoning into which food products are dipped prior to cooking. The breading is normally a bread-based crumb or other coating like small potato chips or puffed grain such as rice also are popular. Surimi is a Japanese term for mechanically deboned fish mince from white fleshed fish that has been washed, refined, and mixed with cryoprotectants for better frozen shelf life. Because of its high gel strength, surimi is used as an intermediate in processing several value-added products with stimulated texture, flavor, and appearance (Venugopal et al. 1995; Hall 2012; Mahapatra et al. 2013).

2.3.1 Mince-Based Products

2.3.1.1 Surimi and Surimi-Based Products

Products	Ingredients and method
Chikuwa	Surimi ground with other ingredients around a brass tube or a piece of bamboo and then moved inside the cover as boiling proceeds. Hollow cylindrical shape, white inside with a brown surface
Satsuma-age	Dried fried fish cake. Surimi is made into compact paste and solidified through frying. The paste contains fish, salt, sugar, and other spices and is molded into several shapes
Hampen	Surimi is ground to fine paste along with Japanese yarn and other ingredients. HAMPEN is different from Kamaboko (steamed) and Chikuwa (boiled). It is processed by boiling. It has a soft spongy texture resulting from incorporation of air during rapid kneading

Products	Ingredients and method
Fish ham	Salt, sugar, chemical flavorings, spices, and a smoke flavoring are added to the surimi and cured for a day or two in a cold atmosphere. The cured mass is mixed with additional quality of surimi, starch, and 50% protein and grounded to a fine paste. The paste is sealed in a PVC casing and is heated in water at 85–90 °C for 40–60 min and frozen after cooling. If the product is intended for marketing fresh, it will be steamed for 4 min at 120 °C
Kamaboko	It is made from white fish, filleted and pounced into paste, and mixed with salt, sugar, egg whites, fish sauce, and Japanese sake. It contains no artificial colorings or seasonings and preservatives. It is shaped as Kamaboko and steamed, baked, fried, or poached. It has smooth chewy texture and delicious flavor
Fish sauce	An amber-colored liquid extracted from the fermentation of fish with sea salt. It is used as condiment in various cuisines. In addition to being added to dishes during the cooking process, fish sauce is also used as a base for a dipping condiment. Fish sauce, and its derivatives, imparts an umami flavor to food due to their glutamate content
Fish meal	It is a commercial product made from underutilized fish, bones and offal and small marine fish. It is powder or cake obtained by drying the fish or fish trimmings, often after cooking, and then grinding it. If fatty fish is used, it is first pressed to extract most of the fish oil

2.3.1.2 Battered and Breaded or Coated Products

Fish ball	These are very common in Nordic countries and very popular in Southern China, Hong Kong, and Southeast Asia. It is prepared from fresh fish and contains salt and pepper. The cleaned degutted fish meat is separated by filleting. To the meat, salt and pepper are added and minced thoroughly. This is then made into balls by rolling between the palms
Fish nuggets	It contains eggs, pepper, cheese, salt, corn flour, dry bread crumbs, chopped onion, minced garlic, and fish flakes. The first four ingredients are mixed and to it fish flakes are dipped and drained. It is then rolled in premixed remaining ingredients and fried
Fish finger	It contains minced fish meat, salt, sugar (as per need), pepper, green chilli, coriander leaves, ginger, garlic, cumin, bengal gram powder, biscuit powder, dry bread crumbs, potato, egg white, chopped beet, carrot, capsicum and onion, vinegar, and oil. The fish meat is cut into finger size and mixed with boiled potato, grounded onion, and other ingredients. It is then mixed with biscuit powder and made into round shape. It is dipped in egg white and breaded and fried in hot oil
Fish cutlet	It contains minced fish meat, potato, white pepper, cumin, coriander leaves, egg white, bread crumbles, salt, chopped onion, ginger, green chilli, bengal gram powder, and oil. The cleaned-degutted fish is boiled to remove bones. This minced fish meat is mixed with cooked potato and other ingredients. The dough is given desired shape, dipped in egg white, breaded, and fried in hot oil
Fish roll	It contains minced fish meat, salt, monosodium glutamate, chopped onion, garlic and ginger, green chilli, cumin, white pepper, starch emulsion, and bread crumbs. The cleaned-degutted fish are filleted and deboned and dipped in solution containing salt, vinegar, and ginger and garlic juice for 30 min. All the ingredients are mixed thoroughly. Spiced minced meat is kept on one side of the treated fillet and rolled, and two sides of roll are pressed inside. The rolls are then fried in hot oil

Fish samosa	It comprises of minced fish meat, salt, chopped onion, ginger-garlic paste, tomato, turmeric powder, chilli powder, curry leaves, pepper, bread crumbs, maida, groundnut, and coconut. The ginger-garlic paste, green chilli powder, and onion are fried, mixed with minced meat, and fried again. Fried groundnut, coconut, coriander leaves, and biscuit powder are mixed with this. Wheat flour, dalda, and water are mixed to prepare dough that is spread to a thin sheet. The samosa shape is given of the fine sheet of wheat dough, and meat is stuffed inside and fried in refine oil
Fish sandwich	It comprises of minced fish meat, mustard seed, methi (fenugreek), green chilli, minced garlic and ginger, chilli powder, cumin powder, turmeric powder, oil, salt, sugar, cinnamon powder, cardamom powder, butter, lemon and coriander leaf. To hot oil, ginger-garlic-green chilli paste is added, followed by dry chilli powder, turmeric powder and cumin powder, and sugar. To this, cinnamon and cardamom powder are added. These are fried and minced fish is added and fried thoroughly. Lemon juice is added to this. Bread is toasted and the paste is spread over it
Fish pakora	It comprises of minced fish meat/prawn, wheat flour, eggs, cream, corn flour, ginger, and garlic. Marinate overnight the fish with ginger and garlic paste. Roast wheat flour till brown in color and allow to cool. Beaten eggs are added to flour and mixed thoroughly. Add cream, warm water, and marinated fish. Separate into small pieces and fried in hot oil till brownish color

2.3.1.3 Extruded Fish Products

Fish sausage	*The fish sausage contains* 67–68% water, 14–15% protein, 5–6% fat, and 8–9% starch. It contains the following ingredients: minced fish meat, fat/oil, wheat starch, sugar, salt, and spices. Fish sausages contain about 2.5% salt in relation to the weight of the meat. Different species of fish can be mixed together. Adding pork fat will make a great sausage. Mix ground fish with all ingredients. Flour, cornstarch, bread crumbs, or cooked rice act as binder and help to develop a good texture. Egg white is also an excellent binder added to sausages. After mixing, these are stuffed into casings and cooked to 63 ° C internal meat temperature/baked in smoke/cooked in water at 80–85 ° C. These are stored in refrigerator until used later
Fish burger	The ingredients include fish, egg, mayonnaise, mustard paste, margarine (melted), salt, chopped parsley, dry mustard, ground pepper, garlic salt, red pepper, vegetable oil, red leaf lettuce leaves, tomato slices, red onion slices, crushed cornflakes cereal, jalapeno, and sandwich buns (split). Baked fish flakes mixed with egg, salt, and red pepper and shaped into patties and refrigerated (1 h). Potato may be used as binder. Fried patties placed between bun with lettuce, tomato, onion, and jalapeno, with sauces/pastes
Fish noodles	The ingredients include fish cubes, salt, ginger-garlic paste, chilli powder, turmeric powder, pepper powder, vegetable oil, chopped onion, grated carrot, grated cabbage, beans (chopped thin), coriander leaves, and noodles. The noodles are cooked in water containing oil, drained, and washed with cold water. The marinated fish is fried. The chopped garlic and chopped vegetables are cooked, and to this cooled noodles and fried fish are the mixed. Garnish with coriander leaves
Fish pappad/ wafers	*The ingredients include fish, cereal, black gram, spices, crushed green chilli, garlic, asafetida, salt, and preservatives. It's easy to prepare and enrich with nutrients. It remains fresh for a long time and has a lingering taste.* This is a high-protein, low-fat, delicious, crunchy taste fish product

2.3.2 Allied seafood delicacies

Fish/prawn pickle	It comprises of fish pieces (1.5 cm^2)/peeled and cleaned prawn, salt, mustard seed, green chilli, garlic, ginger, chilli powder, turmeric powder, gingelly/mustard oil, sugar/jaggery, and vinegar. Mix salt and fish in the ratio 1:1 (w/w) for 2–3 h. Fry the fish in oil and remove. All other ingredients, except turmeric and chilli powder, are gently fried in the same oil. Turmeric and chilli powder are then added followed by fried fish. Mix well, and add sufficient quantity of boiled cooled water just to cover the materials. The pickle is kept for aging for 2–3 months
Fish satay	Fish satay is popular in Southeast Asian countries. It is a marinated hard dried product. It comprises of fish (butterfly fillets), sugar, chilli powder, salt, ginger, and pepper. Fish fillets are dried (8–10 h), flattened (in sauce), marinated, drained, and roasted
Fish curries	The ingredients include vegetable oil, chopped onion, chopped ginger, chopped garlic, curry paste (chilli powder, turmeric powder, coriander powder), salt, chopped tomato, and fish fillets. Heat the oil in a deep pan, and gently fry the onion, ginger, and garlic until soft. Add the curry paste and stir-fry and then add chopped tomatoes. Add fish and gently cooked for 4–5 mins until the fish flakes
Sushi	Sushi is one of the most popular dishes among the Japanese. It is usually enjoyed on special occasions, such as a celebration. During the Edo period, "sushi" is referred to pickled fish preserved in vinegar. Nowadays sushi can be defined as a dish containing rice which has been prepared with sushi vinegar. There are many different types of sushi
	Nigiri: Small rice balls with fish, shellfish, etc. on top. There are countless varieties of nigirizushi, some of the most common ones being tuna, shrimp, eel, squid, octopus, and fried egg
	Gunkan: Small cups made of sushi rice and dried seaweed filled with seafood. The common gunkan includes sea urchin and various kinds of fish eggs
	Norimaki: Sushi rice and seafood, etc. rolled in dried seaweed sheets. There are countless varieties of sushi rolls differing in ingredients and thickness
	Temakizushi: Temakizushi (hand rolls) are cones made of nori seaweed and filled with sushi rice, seafood, and vegetables
	Oshizushi: Oshizushi is pressed sushi, in which the fish is pressed onto the sushi rice in a wooden box
	Inari: Inarizushi is a simple and inexpensive type of sushi, in which sushi rice is filled into deep fried tofu bags
	Chirashi: Chirashizushi is a dish in which seafood, mushrooms, and vegetables are spread over sushi rice
Chilli fish	The ingredients include fish piece (boneless), maida, corn flour, baking powder, soy sauce, chopped celery, pepper, salt, oil, chopped ginger, chopped garlic, chopped green chilli, tomato sauce, chilli sauce, and corn flour. The fish fingers are battered with corn flour, maida, baking powder, soy sauce, celery, pepper powder, water, and salt and deep fried in hot oil. Saute the garlic, ginger, and green chilli and add soy sauce, chilli, and tomato. Mix corn flour and cook until it becomes thick. Fried fish are added to this sauce and mixed well

Fish paturi	The ingredients include fish fillets (Basa), scraped coconut, banana leaves, mustard powder, poppy seeds, green chilli, fresh coriander chopped, turmeric powder, salt, mustard oil, and lemon juice. Marinate the fish in a separate bowl. Mix mustard powder and water and set aside. Blend the coconut, poppy seeds, green chillies, and coriander to a smooth paste, and mix it with mustard paste, and add turmeric, salt, and mustard oil. In a piece of banana leaf, the marinated fish and the above paste are mixed and tied thoroughly; and steam cooked
Fish dessert	There are different varieties of fish dessert recipes. In orange pomegranate salmon, thinly chopped small red onion, skinned salmon fillet, salt, orange, pomegranate seeds, virgin olive oil, and minced fresh dill are taken. Bake fish fillets, pomegranate seeds, orange slices, and onion slices with little oil

2.3.3 Miscellaneous

Fish protein concentrate	Fish protein concentrates (FPC) are protein concentrates from fish for consumption in foods such as biscuits, soups, breads, and stews. It is of three types: Type A: odorless and tasteless powder with maximum total fat content of 0.75% and up to 80% protein Type B: no specific limits for odor or flavor, has fishy flavor and maximum fat content of 3% Type C: normal fish meal produced under satisfactory hygienic conditions. Manufacture of FPC involves removal of most of the water, fat, and fishy tasting components using chemical solvents. Then, hydrolyzing fish protein by means of enzymes or chemicals and then concentrating the product into a paste or extract. Solvents used to make type A include ethanol/propanol/ethylene dichloride

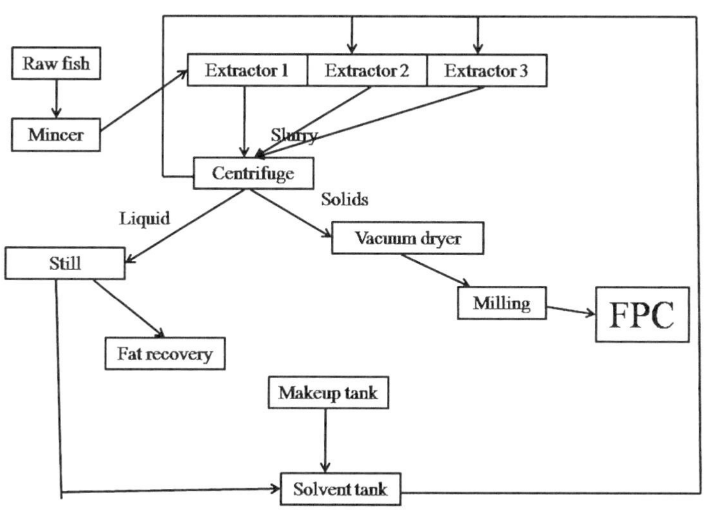

Fish soup powder	Fish soup powders are rich in protein, vitamins, fat, and minerals. Any types of fish can be used for soup powder preparation. The ingredients include cooked fish meat, salt, fat, onions, coriander, starch, milk powder, sugar, pepper, ascorbic acid, carboxymethyl cellulose, and monosodium glutamate. The cooked fish meat is blended thoroughly in water. Fried onion and other ingredients except milk powder are added to the blended fish. These are made into paste and dried at 70 °C. To the dried mix, milk powder is added and homogenized thoroughly to obtain fish soup powder
Fish hydrolysate	Fish protein hydrolysate (FPH) is an excellent amino acid product. It contains minimum 70% of hydrolyzed protein. It has excellent functional and bioactive properties. However, taste defects such as bitterness and process economics are the limiting factors for FPH applications. Fish protein hydrolysates (FPH) are a product made from fish or fish material by the method of protein hydrolyzation (breakage of proteins into peptides and amino acids) using enzymes 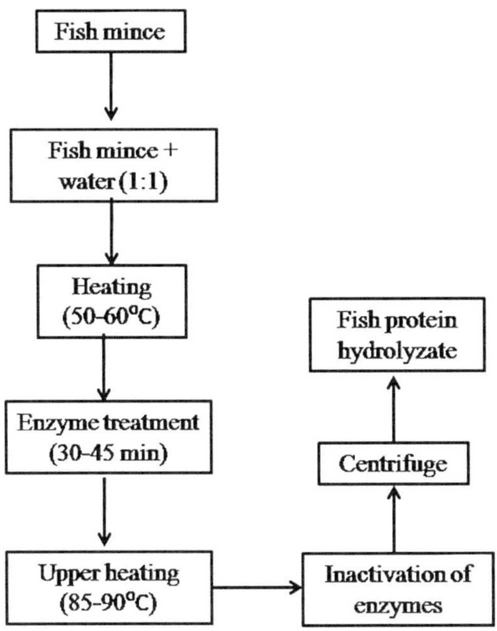
Fish paste	Fish paste includes fermentation (chemical breakdown) of fish muscle until it becomes soft creamy purée or paste. Physical breakdown of fish by pounding, grinding, pressing, mincing, blending, and/or sieving may also be employed to prepare fish/shrimp/crab paste. It is usually used as a condiment or seasoning to add flavor to food or in some cases to complement a dish. Fish paste is a thick, rich concentrate that has been cooked for a long time
Fish oil	Fish oil is taken as nutritional supplement and is rich in omega-3 fatty acids. Two of the most important omega-3 fatty acids contained in fish oil are eicosapentaenoic acid (EPA) and docosahexaenoic acid (DHA)

Masmin	Masmin is a heavily smoked, hard dried product prepared from tuna meat (flakes). The traditional method involves winding of each strip with green coconut leaves, boiling (in a mixture of seawater and freshwater (3:1)), and soaking it in the same for about 12 h. The meat is then smoked continuously using coconut husk and saw dust and sun dried. This process is repeated until dark hard sticks are obtained. The latest method includes dipping the tuna meat in saturated brine for 30 min and steaming. The steamed strips are sun dried (40% moisture). This process is repeated until the characteristic glossy appearance, brown color, and smoky flavor are developed. The smoked strips were then dried to moisture of about 10%
Imitation products	Surimi or minced meats are usually used to develop imitation seafood products, such as imitation crab (meat/sticks), abalone, shrimp, calamari, and scallop. Surimi is also used to make kosher imitation shrimp and crabmeat, using only kosher fish such as pollock. Imitation products help in utilization of low value fish proteins in producing high cost products in mass. The process includes development of surimi paste made into different shapes with added flavors and dyes to imitate the desired product
Isinglass	Isinglass is basically collagen and derived from dried swim bladders of fish belonging to the Polynemidae, Sciaenidae, and Siluridae families. It is used for clarification or fining of beer or wine and to remove haze. It is also used as glue. The hard dried fish bladder is softened by soaking in water and flattened to a low thickness, followed by drying. The treatments given should not adversely affect the quality of the collagen
Shark fin rays	These have several health benefits as documented in Chinese medical book. These are categorized into wet fin (the fin needles of the dried shark fin soft and ready for cooking) and cooked fin (that are not required for immediate use and are often re-dried or frozen). The preparation of shark fin rays includes removal of denticles by soaking in lukewarm water to 60 °C and scratching with small knife. Then cartilaginous platelet is removed and the fins are trimmed. These are bleached overnight by smoking with sulfur or treatment with 3% hydrogen peroxide for about 30 min
Beche-de-mer	The food product from dried sea cucumber (*Holothuria* spp.) is called "beche-de-mer." It is consumed as delicacy and for their medicinal properties. The processing steps include first boiling, slitting and gutting, second boiling, smoking, and finally sun drying. It is classified into four grades A, B, C, and D based on their quality
Caviar	It is a delicacy consisting of salt-cured roe of Acipenseridae family (sturgeons). The roe can be fresh or pasteurized. It is used as garnish or a spread. The roes are collected from roe bag and passing through sieve to get uniform sized roe. These are then dry salted and packed.
Ambergris	It is a natural by-produced of the sperm whales' squid based diet and is used in perfumery. It is also called floating gold or the treasure of the sea. Pure white varieties are considered as of best quality than black ambergris; other colors range from grey to brown
Fish silage	It is defined as a liquid product made from underutilized fish or wastes (offal) by enzyme degradation in the presence of an added acid. The enzymes break down fish proteins into smaller soluble units, and the acid helps to speed up their activity while preventing bacterial spoilage. The process includes treating minced fish with 85% formic acid, which cause liquefaction due to enzyme action by enzymes thriving at pH 4 or below. Introduction of slight temperature makes the reaction faster

2.4 Determination of Functional Quality of Processed Food

Food processing transforms the food specifically seafood into different forms of products that may bring changes to its physiochemical and textural properties. These functional changes may be constructive or destructive. There are several methods to assess these functional changes.

2.4.1 Determination of Color and Odor

Principle The color measurement of food products provides a standardized measure to determine the freshness and quality during processing and packaging. The color variation may be diverse for various processed food products. The color of meat may change slightly with age, and consumers associate this with freshness and quality. The color of tuna meat slightly reddens on storage, but this redness is transient. Consumers accept this as natural redness and correlate it to the excellent quality tuna. Currently, Hunter's lab colorimeter is used to measure the color of the various raw and processed food products. Mostly L-, a-, and b-type color scales (Hunter L, a, b or CIE L∗, a∗, b∗) are used for color measurement.

The system of coordinates designated as L∗, a∗, and b∗ was suggested by the International Commission on Illumination (CIE) in 1931. One color is absolutely determined by three coordinates: L∗, a∗, and b∗ or L∗, C∗, and H∗. L∗ is the vertical coordinate of a 3D system of colors, which has values from 0 (black) to 100 (for white); a∗ is the horizontal coordinate and the values range from −80 (green) to +80 (red); b∗ is the horizontal coordinate with the values ranging from −80 (blue) to +80 (yellow). The angle H∗, the color hue, is expressed in degrees (H∗ = 0 corresponds to red; H∗ = 90 to yellow; H∗ = 180 to green; H∗ = 270 to blue); C∗, the color saturation, signifies the mentioned color distance from a non-multicolored point, which corresponds to a∗ = b∗ = 0.

Procedure A HunterLab spectrophotometer with the special glass or polycarbonate covered nose cone can be used to measure the color of food products, either natural or processed. The average of several readings must be taken for meat products. Care has to be taken so that moisture does not enter into the instrument. For meat products, the illuminant/observer is set at D65/10°. C/2°. Configure the software or the instrument firmware to read using the desired color scale, illuminant, and observer. Select a meat sample that is at least 25 mm (1 in) thick for steaks, fillets, and patties or 12.5 mm (0.5 inch) thick for breaded products. If frozen, thaw and store chilled prior to measurement. The conditions of the sample should be similar to that as in display. Standardize the instrument with the glass port in place, first using the black glass to set the bottom of scale. Make sure the black glass is in solid contact with the port. Complete the standardization using the calibrated white standard. The nose cone of the instrument is placed on the sample, ensuring that the sample is in solid contact with the port. Take a single color reading. Move to another area on the sample and take a second reading. Turn the product over and take two more readings in different areas on the sample surface. Record the average color values for the sample. For tuna, if the average $a* < 16.2$, then the color is considered acceptable.

Whiteness Index (W) Assessment

The study of reflective characteristics of white object is analogous to the one of any object. Whiteness is an attribute by which an object is judged to approach the preferred white. The individual preference for whites makes optical criteria of whiteness variable from one observer to another.

It is likely that the best whiteness index will be based on an equation that locates the position of an ideal white and measures whiteness as the distance in color space from the sample to that ideal.

The UV/Vis spectroscopy and fluorescence spectroscopy could also be used to determine whiteness color index.

2.4.1.1 Determination of Odor

Principle Odor (aroma and flavor) measurement is usually by physical means (sensory evaluation). The nose detects the volatile aromas released from food. The intensity of the specific odor is recorded for a food. Humans are microsmatic and do not have keen sense of smell. About 1 lakh odors can be discriminated by human, but cannot be labeled accurately. First, odor molecules of food bind to nose receptor; signals from these receptors travel up to olfactory bulb. From there, few signals go to primary olfactory cortex and to higher-order parts of the brain. Few signals from olfactory bulb go to amygdale in the brain and hippocampus (memory). Hence, we associate some smells in our memory with a specific emotion.

Usually odors are quantified by dynamic dilution olfactometric analysis. Olfactometric analysis consists of determining the olfactory perception threshold of a gaseous sample. The olfactory perception threshold is defined as the number of dilutions at which 50% of a jury perceives the odor, while 50% do not. This does not involve the determination of the quality of the odor. By definition, the olfactory perception threshold is equivalent to 1 odor unit per cubic meter of air (1 o.u./m^3). The number of dilutions of the odor sample required to obtain 1 o.u./m^3 indicates the odor concentration of the sample in odor units per cubic meter of air [o.u./m^3].

2.4.2 Determination of Salt Content

Principle The bioelectrical measurements and a needle-type multielectrode array could be used in evaluating the salt and moisture content of structurally heterogeneous food samples (fish). Impedance measurements are performed at different frequencies to find the most appropriate range, where impedance module and phase angle are most sensitive to salt and moisture content. This is a rapid and nondestructive analytical technique, and its low cost enables its usage as a potential quality control technique for the fish industry (AOAC 2000; Nielsen 2010).

Titration Method The salt is extracted by water from the pre-weighed sample. After the precipitation of the proteins, the chloride concentration is determined by titration of an aliquot of the solution with a standardized silver nitrate solution (Mohr method).

Procedure Homogenized sample (5 g) is weighted into a 250 ml volumetric flask and vigorously shaken with approximately 100 ml water. To this 5 ml of potassium hexacyanoferrate solution (15%) and 5 ml of zinc sulfate (30%) are added and shaken well and made up to mark. After shaking, it is allowed to stand for precipitation; the flask content is filtered through a filter paper. An aliquot of the clear filtrate is transferred into an Erlenmeyer flask, and two drops of phenolphthalein are added. Sodium hydroxide (0.1 N) is added dropwise until the aliquot takes on a faint red color. The aliquot is then diluted to approximately 100 ml, and approximately 1 ml potassium chromate solution (5%) is added and titrated with silver nitrate solution (0.1 N). End point is indicated by a faint (distinct), change in color. This faint reddish-brown color should persist after brisk shaking. The blank value with no salt is also calculated. End point determination can also be made by using instruments like potentiometer or colorimeter.

Calculation

$$\text{Salt concentration}\left(\%\right) = \left(\text{V} \times \text{C} \times 58.45 \times 250 \times 100\right) / \left(\text{A} \times \text{W} \times 1000\right)$$

A = volume of aliquot (ml)
C = concentration of silver nitrate solution
V = titer volume of silver nitrate solution, corrected for blank value
W = sample weight (g)

2.4.3 Determination of Water Activity

Principle Water is a most important component in terms of food safety. Water activity (a_w) can be defined as the ratio of the partial vapor pressure of water in a food to the partial vapor pressure of pure water at the same temperature and pressure (partial pressure of water is 1). Most of the water in food is bound water (water of hydration or to surfaces of large molecule or cell structures) and is not free to support microbial growth/to participate in/support chemical or enzyme reactions. The free water supports microbial growth and participates in chemical and enzymatic reactions and spoilage process. This amount of free water is called water activity. Employing various processing techniques would reduce the water activity in a food product. It is more significant for qualitative considerations such as product stability, shelf life (microbiological and enzymatic stability, aroma retention), handling characteristics, physical properties, and chemical stability. It is a nondestructive, easy-to-use measurement in a wide range (AOAC 2000; Nielsen 2010).

The water activity (a_w) of a food is the ratio between the vapor pressure of the food (in a completely undisturbed balance with the surrounding air media) and the vapor pressure of distilled water under identical conditions. The water activity increases with temperature. The moisture condition of a product can be measured as the equilibrium relative humidity (ERH) expressed in percentage or as the water activity expressed as a decimal.

$$a_w = \frac{p}{p_0}$$

where:

P = vapor pressure of food
P_0 = vapor pressure of pure water

Procedure Water activity and moisture content can be related by a graph called a *sorption isotherm* (Fig. 1.3); measurement of both parameters and relating them (interpolation) would give the sorption isotherm characteristics (Fig. 2.7).

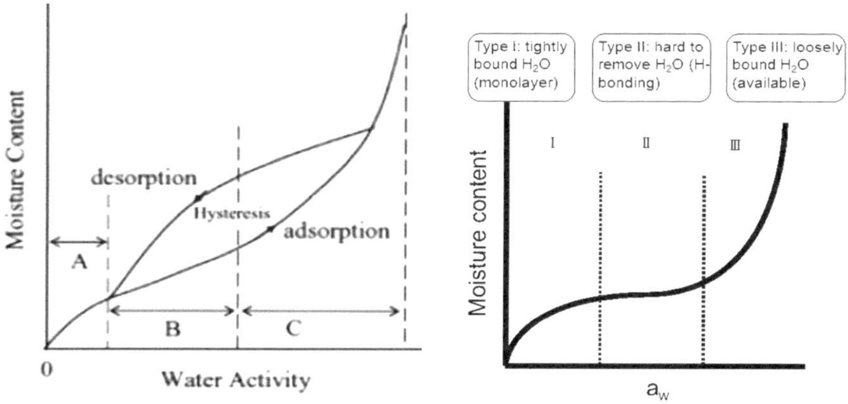

Fig. 2.7 Moisture sorption isotherm

Analysis of Moisture Content

Gravimetric Method: This involves drying a known quantity of the food product in a hot air oven until all moisture has been evaporated. By measuring the residual dry matter content, the water content can be determined. A vacuum oven can be used for heat-sensitive foods.

Karl Fisher Titration: The Karl Fisher titration is particularly adaptable to food products that show erratic results when heated or subjected to vacuum (low moisture foods such as dried fruits and vegetables, candies, chocolates, roasted coffee, oils, and fats or any low moisture foods rich in sugar or protein). The method is based on the fundamental reaction described by Bunsen in 1853 involving the reduction of iodine by SO_2 in the presence of water:

$$2H_2O + SO_2 + I_2 \rightarrow C_5H_2SO_4 + 2HI$$

This was modified to include methanol and pyridine in a four-component system to dissolve the iodine and SO_2. For each mole of water, one mole of iodine, one mole of SO_2, three moles of pyridine, and one mole of methanol are used. In the titration, if a sample containing water is titrated with a Karl Fisher reagent, iodine is consumed, while there is water in the system. When no water remains, then free iodine appears, which can be detected using two platinum electrodes immersed in the solution being titrated. As water is titrated, the voltage of the electrodes approaches virtually zero. And where there is an excess of iodine (all the water has been consumed), a current can be measured.

Analysis of Water Activity

Resistive Electrolytic Hygrometers: Resistive electrolytic hygrometers use a sensing element in the form of a liquid electrolyte held between two small glass rods by capillary force. The electrolyte changes resistance if it absorbs or loses water vapor. The resistance is directly proportional to relative air humidity and also to the water activity of the sample (once vapor-liquid equilibrium is established). The sensor does not have any physically given hysteresis and does not require regular cleaning as its surface is not the effectively sensing element. Volatiles influence the measurement performance especially those that dissociate in the electrolyte and thereby change its resistance. Such influences can easily be avoided by using chemical protection filters that absorb the volatile compound before arriving at the sensor.

Capacitance Hygrometers: Capacitance hygrometers comprise of two charged plates separated by a polymer membrane dielectric. As the membrane adsorbs water, its ability to hold charge is amplified, and the capacitance is measured. This value is approximately proportional to the water activity as determined by a sensor-specific calibration. Capacitance hygrometers are not affected by most volatile chemicals. Cleaning is not required and are less precise than dew point hygrometers ($+/- 0.015\ a_w$). They should have regular calibration checks and can be affected by residual water in the polymer membrane (hysteresis).

Dew Point Hygrometers: The temperature at which dew forms on a clean surface is directly related to the vapor pressure of the air. Dew point hygrometers work by placing a mirror over a closed sample chamber. The mirror is cooled until the dew point temperature is measured by an optical sensor. This temperature is then used to find the relative humidity of the chamber using psychrometrics charts. This method is theoretically the most precise (+/− 0.003 a_w) and often the fastest. The sensor requires cleaning if debris accumulates on the mirror.

2.4.4 Evaluation of Sensory Parameters

Principle Sensory evaluation is defined as the scientific discipline used to evoke, measure, analyze, and interpret reactions to characteristics of food as perceived through the senses of sight, smell, taste, touch, and hearing. Currently, instruments are also capable of measuring parameters included in the sensory profile (Instron, Bohlin Rheometer for measuring texture and other rheologic properties). Microscopic methods combined with image analysis are used to assess structural changes and "the artificial nose" to evaluate odor profile. Scientifically, sensory evaluation can be done in the following steps:

1. Detection of a stimulus by the human sense organs
2. Evaluation and interpretation by a mental process
3. The response of the assessor to the stimuli

The response of the same level of stimuli can vary with individuals and can contribute to a nonconclusive answer. Hence, it is very imperative to consider all these differences while selecting and training judges for sensory analysis; and they must be able to assess the stimulus both objectively and subjectively. Assessment in quality control must be objective, which can be classified into two groups: discriminative tests and descriptive tests. Discriminative testing is used to verify if a disparity exists between samples (triangle test, ranking test), while descriptive tests are used to conclude the nature and intensity of the disparity (profiling, structured scaling and quality tests). The subjective test is an affective test based on a measure of preference or acceptance (Rehbein and Oehlenschlager 2009).

Quality Assessment of Fresh Fish

Quality Index Method: Torry Research Station developed first modern and detailed method for sensory analysis of raw fish, which was further modified later. The EU scheme was used in EU countries to assess fish, where these were classified as A, B, and E; below B, the fish are discarded. A new method, the Quality Index Method (QIM) originally developed by the Tasmanian Food Research unit, is now used by the Lyngby Laboratory for fresh and frozen fish. In the Nordic countries and Europe, it is currently in use (Table 2.1).

QIM gives zero score for very fresh fish. QIM does not follow the traditionally accepted S-curve pattern for deterioration of chilled fish during storage. Based on demerit score, a linear correlation between the sensory quality and storage life on ice is obtained, which makes it possible to predict remaining storage life on ice.

Table 2.1 Quality assessment scheme used to identify the quality index demerit score

Quality parameter	Character	Score (ice/seawater)
General appearance	Skin	0 Bright, shining
		1 Bright
		2 Dull
	Bloodspot on gill cover	0 None
		1 Small, 10–30%
		2 Big, 30–50%
		3 Very big, 50–100%
	Stiffness	0 Stiff, in *rigor mortis*
		1 Elastic
		2 Firm
		3 Soft
	Belly	0 Firm
		1 Soft
		2 Belly burst
	Smell	0 Fresh, seaweed/metallic
		1 Neutral
		2 Musty/sour
		3 Stale meat/rancid
Eyes	Clarity	0 Clear
		1 Cloudy
	Shape	0 Normal
		1 Plain
		2 Sunken
Gills	Color	0 Characteristic, red
		1 Faded, discolored
	Smell	0 Fresh, seaweed/metallic
		1 Neutral
		2 Sweaty/slightly rancid
		3 Sour stink/stale, rancid
Sum of scores		(min. 0 and max. 20)

Structured Scaling: Structured scaling gives the panelist an actual scale showing several degrees of intensity. A few detailed attributes are chosen often based on work from a fully trained descriptive panel. The words must be carefully chosen and panelists must agree with the terms. Objective terms should be preferred rather than subjective terms. The standards are included at various points of the scale (hedonic scaling).

9-Point Hedonic Scale

1-Like Extremely
2-Like Very Much
3-Like Moderately
4-Like Slightly

5-Neither Like nor Dislike
6-Dislike Slightly
7-Dislike Moderately
8-Dislike Very Much
9-Dislike Extremely

Quality Assessment of Fish Products

Triangle Test: The most used discriminative test in sensory analysis is the triangle test (ISO standard 4120 1983), which indicates whether or not a detectable difference exists between two alike samples. The panelist tastes the samples from left to right. The six possible order combinations should be randomized across panelists. For samples A and B, the six possible order combinations are AAB, ABA, BAA, BBA, BAB, and ABB. The panelist is instructed to identify the odd sample and record his answer. For evaluation, the chi-square distribution ($X^2 = \Sigma\ (|O\text{-}E|)^{2/}E$, where O, observed, and E, expected) is used. To determine the number of expected correct answers, multiply the chance of choosing a correct answer by chance by the total number of panelists. In a triangle test, the probability of a correct answer by chance is 1/3. The probability of choosing an incorrect answer by chance is 2/3.

Duo-Trio Test: The panelists are given a reference sample and two coded test samples, and they have to identify which is more identical to the reference sample.

Paired Comparison Test: In the paired comparison test (UNI EN ISO 5495), the panelists (30 or more) are given simultaneously two coded samples and have to determine whether two products differ in a specified attribute, such as sweetness, crispness, redness, etc. The panelist chooses the intense in the specified sensory attribute. It is a "forced" choice and therefore the judges must give an answer in any case. It is frequently used in preference tests.

Ranking: In ranking, a number of samples are presented to the taste panel, and they arrange them in order according to the degree to which they exhibit some specified characteristics such as downward concentration of salt. It is quicker and less training is required. It is often used for preliminary screening. The method gives no individual differences among samples and is not suited for sessions where many criteria have to be judged simultaneously.

Profiling: Profiling is simple, gives a detailed description (qualitative and quantitative), and is used for the assessment of a single attribute such as, texture, flavor, and appearance. The panel members (16 judges) are handed a broad selection of reference samples and use the samples for creating a terminology that describes the product. Advanced multivariate analysis is used for statistics and makes it possible to correlate single attributes to spoilage or deteriorative changes in the food. The results can be reported in a "spider's web," and the panel uses an intensity scale normally ranging from 0 to 9. Profiling can be used for all kinds of fishery products, even for fresh fish when special attention is placed on a single attribute.

A panel used for descriptive testing shall preferably consist of minimum 8–10 persons, and test becomes statistically stronger if it is done in duplicate (Huss 1995).

2.4.5 Evaluation of Texture

Principle Food texture is a collective term of sensory experiences originated from visual, audio, and tactile stimuli. It is defined as "the sensory and functional manifestation of the structural, mechanical and surface properties of food detected through the senses of vision, hearing, touch and kinesthetics." In accordance with terminology of ISO 5492:2008, mechanical attributes are those related to the reaction of the product to stress such as hardness, cohesiveness, viscosity, elasticity, and adhesiveness; geometrical attributes are those related to the size, shape, and arrangement of the particles within a product; and surface attributes involve the sensations produced in the mouth by the moisture and/or fat in and near the surface of the products, and body attributes are those related to the sensations produced in the mouth by the moisture and/or fat in the substance of the products and the way in which these constituents are released. Since texture consists of a number of different physical sensations, it is preferable to talk about "textural properties," which infer a group of related properties rather than a single parameter as texture (Bourne 1990). Consumer concern and interest of food texture vary from one type of food to another. For solid foods, sensory experience associated with fracture and breaking could be the most relevant textural features, whereas the sensation of flow behavior could be the most critical texture-related feature for fluid foods. For semisolid or soft solid foods, different patterns of stress-strain deformation provide key information for the delicate texture variation among this type of food. The food texture can be determined instrumentally or organoleptically (sensory).

Instrumental Texture Analysis The instrumental techniques could be classified into three groups according to the type of information obtained by the test: fundamental, empirical, and imitative (Bourne 1990). Fundamental tests determine one or more physical constants to describe exactly the properties of food in terms of well-defined rheological parameters. Empirical tests usually measure parameters that are poorly defined in rheological terms but, from practical experience, are closely related to the property of interest. Imitative tests aim to reproduce the mechanical operations applied in the human evaluation of food, which are closely correlated with sensory assessment. Instruments like viscometer, rheometer, powder flow tester, and texture analyzer are used to measure texture. The latter consists of Instron machine, universal testing machine (UTMs), and texture profile analysis (TA-XT and its different series) equipped with specific probes which are used for texture analysis. Of these, later is currently used profoundly for the texture analysis of food samples. The typical parameters that can be assessed using TPA are adhesiveness, chewiness, cohesiveness, consistency, crispiness, crunchiness, elasticity, extensibility, firmness, fracturability, gel strength, gumminess, hardness, rupture strength springiness, stiffness, stringiness, texture profile analysis toughness, work to cut, work to penetrate, and work to shear (Fig. 2.8 and Table 2.2).

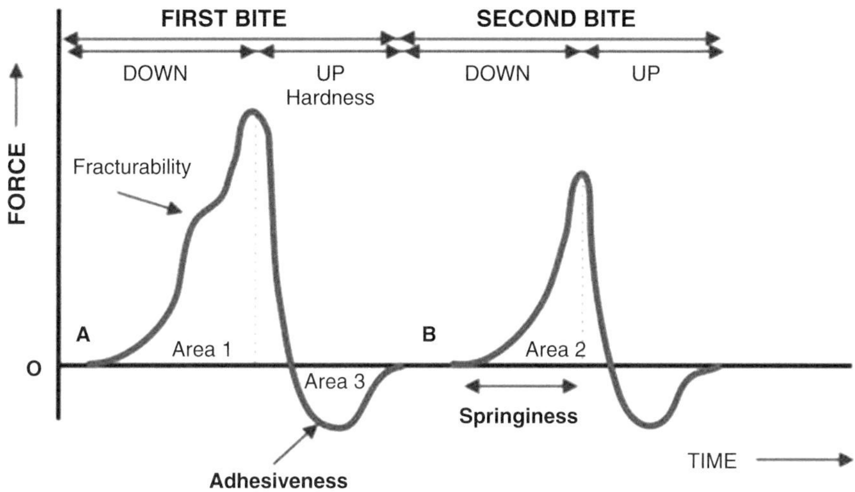

Fig. 2.8 Texture profile analysis of food

Table 2.2 Summary of instrumental techniques used to evaluate texture of food products

Parameters observed	Instrument/technique
Empirical	
Crispiness	Instron machine
Crispiness and brittleness	Universal texture analyzer
Crispiness/crunchiness	Various
Hardness	Instron machine
Breaking strength index	TA-XT2
Brittleness, breaking strength, and hardness	TA-XT2i
Hardness and total energy required to puncture	TA-XT2
Fracturability, cohesiveness, adhesiveness, hardness, chewiness	TA-XT2
Fundamental	
Shear stress	Instron machine
Shear force	TAHDi
Apparent compressive strain	
True compressive strain	
Apparent compressive stress	
True compressive stress	
Young's modulus	

The typical TPA parameters are:

Parameter		Formula
Hardness	The force required to compress a substance between the molar teeth (for solids) or between the tongue and palate (for semisolids) to a given deformation or to penetration	
Fracturability (brittleness)	The force with which a sample crumbles, cracks, or shatters; the horizontal force with which the fragments move away from the point where the vertical force is applied. Fracturability is the result of a high degree of hardness and low degree of adhesiveness	Peak Force at F1
Cohesiveness	The extent to which a material can be deformed before it ruptures	Area 2/Area 1
Springiness (elasticity)	The amount of recovery from a deforming force; the rate at which a deformed material returns to its undeformed condition after the deforming force is removed	Distance 2/Distance 1
Gumminess (semisolids)	A denseness that persists throughout mastication, the energy required to disintegrate a semisolid food to a state ready for swallowing. Gumminess is a product of a low degree of hardness and a high degree of cohesiveness	Hardness × Cohesiveness or Hardness ∗ (Area 2/Area 1)
Chewiness (solids)	The length of time or the number of chews required to masticate a solid food to a state pending for swallowing. Chewiness is a product of hardness, cohesiveness, and springiness	Hardness × Cohesiveness × Springiness or Gumminess ∗ Distance 2/Distance 1
Adhesiveness	The force required to remove material that adheres to the mouth (generally the palate) during the normal eating process	A 3
Viscosity	The force required to draw (slurp) a liquid from a spoon over the tongue	
Resilience	It indicates how well a product "fights to regain its original height." Resilience is measured on the withdrawal of the first penetration, before the waiting period is started	Area 4/Area 3

2.4.6 Determination of Electrical Properties/Redox Potential

Principle The redox potential (Oxidation Reduction Potential or ORP) is a measure of potential difference in a system or food. It is generated by a couple of reactions such as oxidation (loss of electron by a substance or electron donor or a reducing agent) and reduction (gain of electron by a substance or electron acceptor or oxidizing agent). It is measured in electrical units of millivolts designated as Eh, and unit is mV (positive or negative). The redox potential of a food is influenced by its chemical composition, specific processing treatment, and storage conditions in relation to air (vacuum packed, liquid N_2, CO_2, etc.). Fresh food of plant/animal origin are in reduced stage due to the presence of reducing substances such as sugars, ascorbic acids, and sulfhydryl group of protein. Diffusion of oxygen into fresh food would change its redox potential. Heat processing also alters the redox potential. Food stored in air can have a wide Eh range than when stored under vacuum. The presence or absence of oxygen in the environment is important in the growth of microorganisms and its metabolism (FSSAI 2015).

Redox potential of different groups of microorganisms

Aerobes	Between +500 and + 300 mV
Anaerobes	Between +100 and -250 mV
Facultative anaerobes	Between +300 and -100 mV

Materials

Dry weighed chemicals and store overnight in a desiccator before use.

1.4080 g $K_4Fe(CN)_6{\bullet}3H_20$ (potassium ferrocyanide)

1.0975 g $K_3Fe(CN)_6$ (potassium ferricyanide)

7.4557 g KCl (potassium chloride)

Dissolve these in deionized water and dilute to 1 l. Store in dark bottle under chilled condition.

Procedure The redox combination electrode consists of a platinum sensor (or other noble metal) and an Ag/AgCl reference (or calomel reference electrodes) half-cell built in a standard 12 mm electrode body. The measurements are displayed in millivolts (mV). Other redox electrodes include gold and glassy carbon electrodes. Use of the correct electrolyte filling solution is essential to proper measurement. ZoBell's solution is used as standard solution for testing redox instruments.

2.4.7 Folding Test for Surimi

Principle Surimi is a semi-processed washed fish mince protein concentrate mixed with cryoprotectants for frozen storage, which is the primary constituent of processed foods. Folding the test samples by hand and biting samples by the front teeth

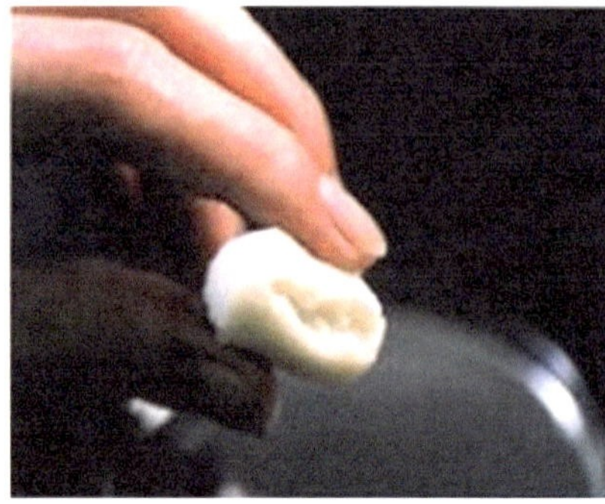

Fig. 2.9 Folding of surimi

are commonly used in Japan and South East Asia for sensory valuation of surimi (Cortez-Vega et al. 2014). This test points out about the gel-forming ability and elastic characteristics of the surimi. The high quality could be due to the washing and increased extraction of sarcoplasmic proteins (Fig. 2.9).

Procedure The folding test is conducted by folding a 5 mm thick slice of gel slowly in half and in half again while examining it for signs of structural failure (cracks). Three or more slice pieces of the same inspection sample were folded completely in half for 5 s, and changes in the shape were evaluated using five stage merit marks as shown below. The average values for three trials were calculated.

Merit	Mark property
5	No crack occurs even if folded in four
4	No crack occurs if folded in two but a crack(s) occur(s) if folded in four
3	No crack occurs if folded in two but splits if folded in four
2	Cracks if folded in two
1	Splits into two if folded in two

2.4.8 Specific Gravity of Fish Oils

Principle Specific gravity is the density of a unit object divided by the density of a reference substance. Usually, reference substance used is water, which is considered to have a density of 1 g/ml or 1 g/cm^3. The specific gravity is dimensionless. Pycnometer is used to measure the specific gravity of the oils (AOAC 2000; FSSAI 2015).

Procedure To a pre-weighed empty clean dry pycnometer (W_p), a known weight of oil is taken (up to mark). The weight of the pycnometer with oil (W_{ps}) is determined.

No air bubbles must be entrapped; hence tap gently for 10 min and apply partial vacuum. Determine the weight of distilled water in clean dry pycnometer (W_g).

Observation and Calculations

$$\text{Specific gravity} = \text{weight of oil}\left(W_{ps} - W_p\right) / \text{weight of water}\left(W_g\right)$$

2.4.9 Analysis of Starch in Breaded and Battered Products

Principle Starch is an important polysaccharide used chiefly for storage of carbohydrate in plants. It is composed of glucose molecules and is a mixture of amylose and amylopectin. It is hydrolyzed into simple sugars by dilute acids and quantified using colorimeter. *The sample is treated with 80% alcohol to remove sugars, and then starch is extracted with perchloric acid. In hot acidic medium, starch is hydrolyzed to glucose and dehydrated to hydroxymethyl furfural (anthrone method). This compound forms a green-colored product with anthrone.* The other methods to measure the amount of glucose (dextrose) include the use of glucometer or analytical method. The resistant starch cannot be measured by this method (AOAC 2000).

Procedure Homogenize 0.1–0.5 g of sample with hot 80% ethanol to remove sugars. Centrifuge and retain the residue. Wash the residue repeatedly with hot ethanol 80% till no color obtained with anthrone. Dry the residue over a water bath and add 5 ml water and 6.5 ml perchloric acid. Extract at 0 °C for 20 min. Centrifuge and collect the supernatant. Repeat this twice and pool the supernatant. Make it to 100 ml with distilled water. Pipette 0.1–0.2 ml of supernatant and make up to 1 ml with distilled water. Prepare glucose standards (0.2–1 ml) in each tube and take distilled water as blank. Add anthrone reagent to each tube and heat for 8 min in a boiling water bath. Cool rapidly and read the intensity of green to dark green color at 630 nm.

Phenol-sulfuric acid method can also be used to determine the amount of glucose. To each tube, 1 ml phenol and 5 ml 90% concentrated sulfuric acid are added and shaken well. After incubation in a water bath at 25–30 °C for 20 min, the readings are taken at 490 nm.

The sample is ground and homogenized thoroughly. To a fixed amount of sample, water, buffer (phosphate buffer, pH 6.8), and amyloglucosidase (16 h, 37 °C) are added, and the solution is placed in controlled temperature water bath for enzymatic digestion. The nonresistant starch is thus hydrolyzed to glucose. The resistant starch collected after washing with ethanol was solubilized using alkali solution and hydrolyzed as mentioned earlier. The glucose content was determined using glucometer or an enzymatic kit.

Calculations
The glucose content in the sample could be determined using standard graph. Multiply the value by a factor 0.9 to calculate starch content.

2.5 Determination of Freshness Indices of Processed Food

2.5.1 Assessment of Protein Degradation

2.5.1.1 Determination of α-Amino Nitrogen

Hydrolysis of Protein Sample A suspension of protein isolate (25 g/200 ml water, pH 8) is prepared and heated to 50 °C; and 50 ml of trypsin solution (proteolytic activity of 0.375 Anson's Unit/50 ml) is added and mixed. Enzymatic hydrolysis is performed at 50 °C, pH 8 by constant addition of a 1 M NaOH solution. To the aliquot of the hydrolysate (2.5 ml), 5 ml of a 1% sodium dodecyl sulfate (SDS) solution is added, and to inactivate the enzyme action, the mixture is heated at 90 °C for 15 min. The solution is then cautiously transferred to a volumetric flask and made up to 25 ml, and α-amino nitrogen could be determined by the following colorimetric methods (Panasiuk et al. 1998: Wang et al. 2012).

2,4,6-Trinitrobenzenesulfonic Acid (TNBS) Assay
To 0.125 ml aliquot of the hydrolyzed sample, diluted 2.5-fold, 1 ml of a phosphate buffer (pH 8.2, 0.212 M) is added along with 1 ml of a freshly prepared aqueous solution of TNB (0.1%). Tightly secured tubes are shaken in the dark at 50 °C for 1 h. The reaction is stopped by the addition of 2 ml of a 100 mM HCl solution. After 20 min, 4 ml of water is added, and after another 10 min, the absorbance read at 340 nm. The blank is prepared identically, and L-leucine is used as the standard ($0–5 \times 10^{-7}$ mol/O, 125 ml of sample).

o-Phthaldialdehyde (OPA) Assay To the aliquot, 0.25 ml of hydrolyzed samples, 10 ml of freshly prepared solution (160 mg OPA, 4 ml MeOH, 0.4 ml ethanethiol, and 200 ml 0.1 mol/l $Na_2B_40_7$) is added. After the tubes are tightly secured, these are mixed and left at room temperature for 20 min. The absorbance is measured at 340 nm. A blank is prepared identically. The standard curve with L-leucine is prepared for concentrations from 0 to 2×10^{-6} mol/0.25 ml of sample.

Ninhydrin Assay To the tubes, containing 2 ml of the test sample and 2 ml of a phosphate buffer (pH 7.0, O.O80M), 1 ml freshly prepared aqueous solution of ninhydrin (1%) is added. The tubes are loosely secured with a stopper and heated in a boiling water bath for 30 min. After cooling, the samples are carefully transferred to volumetric flasks and adjusted to a final volume of l00 ml. The absorbance is read at 570 nm. Blanks are prepared identically. The standard curve is prepared with L-leucine ($0–1 \times 10^{-5}$ mo1/2 ml of sample).

Formol titration: 5 ml of the freeze-dried protein hydrolysate (1 mg/ml) is mixed with 50 ml of deionized water. The pH is adjusted to 8.2 with NaOH (0.1 M), and to it 10 ml of 35% (w/w) formalin (pH 8.2) is added. The pH value of the solution goes down with the addition of formol. The titration using NaOH is done until the pH of the solution is 9.2. The concentration of α-amino nitrogen is estimated using the following equation:

$$C_{\alpha-\text{amino nitrogen}}(\text{mM}) = \Delta V \times \text{n} \times 10^3 / \text{V}$$

where:

ΔV (ml) = titer volume of NaOH
n (mol/l) = molar concentration of NaOH solution
V (ml) = sample volume
RP-HPLC could also be used to determine α-amino nitrogen.

RP-HPLC conditions:

Pico-Tag column	3.9 × 150 mm
Mobile phase (gradient)	pH 6.4 sodium acetate buffer, pH 6.4 (eluent A) and aqueous acetonitrile 60% (v/v) (eluent B)
	0–10 min, linear gradient 0–46%B
	10–12 min, linear gradient 46–100% B
	12–15 min, 100% B
Flow rate	1 ml/min
Column temperature	38 °C

Sample Derivatization Prior to running, the amino acids were derivatized with phenyl isothiocyanate (PITC). First a triethylamine (TEA) solution with TEA-water-ethanol (1:2:2) was added to samples and evaporated to dryness under vacuum. Then PITC solution including PITC-ethanol-TEA-water (1:7:1:1) was added to the dried samples and reacted at room temperature for 20 min.

In theory, the total molar concentration of free amino acids in pure amino acid solution (including no proline) should be equal to the α-amino nitrogen molar concentration, because each of the 20 amino acids in protein contains only 1 α-amino nitrogen except proline which contains no α-amino nitrogen and contributes only about 7% of the α-amino nitrogen. Generally, proline usually exists in fish skin, and there is very few proline in fish protein; therefore, the sum of the molar concentration of each detected amino acid could be considered equal to the total α-amino nitrogen concentration of the sample. The total α-amino nitrogen concentration (mmol/L) of the sample is obtained by summing up the molar concentration of each detected amino acid.

2.5.1.2 Analysis of Buffering Capacity of Fish Muscle

Principle Muscle buffering capacity assists in preventing acidosis caused by anaerobic metabolism in muscle, especially in white muscle. Histidine-related (imidazole) compounds, histidine, anserine (β-alanyl-π-methyl-L-histidine) and carnosine (β-alanyl-τ-methyl-L-histidine), and inorganic phosphate in white muscle act as buffers in white muscle. The inorganic phosphate maintains the underlying buffering capacity of white muscle. Very high levels of histidine-related compounds are typical for migratory pelagic marine fishes. Biotic and abiotic factors affect contents and profiles of histidine-related compounds in white muscle. Buffering capacity is defined as the mmol of base needed to change the pH of the homogenate by 1 pH unit per kg wet mass of the muscle tissue (Castellini and Somero 1981).

Procedure A weighed sample (0.5 g) of the fish muscle was homogenized with 200 volumes of salt solution (145 mM KCl, 10 mM NaCl, and 5 mM iodoacetic acid). The homogenate was adjusted to pH 6.0 and was titrated to pH 7.5 with 0.1 N NaOH solution that was added at a rate of 0.0125 ml at 10 s intervals. Changes in the pH values were measured and recorded at a specific temperature (25 °C). The buffering capacities were determined over the pH range of 6.0–7.5.

2.5.1.3 Analysis of Free Amino Acids and Biogenic Amines

Principle Amino acids (AA) are essential nutritional components of a balanced diet and occur in foods in either free form or as the building blocks of proteins. The analysis of AAs in foods is composed of a number of unit operations: the release of the AAs from the food matrix, the separation of the individual AAs, and their quantification using calibration standards. Each of these steps has their own eccentricity, such as different hydrolysis conditions are required for the optimal release of different AAs, and there are a diverse number and type of food matrices, such that most laboratories adapt methods to best suit their applications (AOAC 2000; FSSAI 2015).

Procedure HPLC (ion exchange ore reverse phase), GC-MS, LC-MS/MS, etc. are currently used for the detection of free amino acids.

HPLC The protein hydrolysis (6 M HCl under vacuum, heating, 110 °C, 24 h) is an inevitable step. Automatic protein hydrolysis system is also in practice. The ninhydrin, phenyl isothiocyanate, and o-phthaldialdehyde have been the most frequently used reagents for the pre- and post-column derivatization of AAs. In the preparation of fluorescent derivatives of free amino acids and biogenic amines, one of the most studied reagents is o-phthalaldehyde (OPA), together with different alkylthiols such as ethanethiol, tert-butylthiol, several N-acyl-L-cysteines, 2-mercaptoethanol, and 3-mercaptopropionic acid (3-MPA). The drawback of OPA is that it does not react with secondary amino acids such as proline. 9-fluorenylmethyl chloroformate (FMOC-Cl) reacts with primary amino acids and amines, also with secondary amino acids. Using both reagents, OPA (dissolve 51.2 mg of o-phthalaldehyde in 1.5 ml ethanol; 80 μl of 3-mercaptopropionic acid and 1750 mg of KSCN added; and made up to 50 ml with borate buffer (0.4 M, pH 10.4)) and FMOC-Cl (dissolve FMOC-Cl in acetonitrile to a concentration of 2.5 g/l and store at −15 °C), simultaneous quantification of primary and secondary amino acids and of biogenic amines could be done.

Standard Preparation Standard solutions are prepared from a stock solution (free amino acids and biogenic amines) by dilution with 0.1 M HCl. The average concentration of amino acids in the standards ranged from 1.1 to 11 mg/l, except for arginine (4.53–45.29 mg/l) and for proline (25.44–254.4 mg/l). Standard biogenic amine concentrations range from 0.15 to 4.20 mg/l. The concentration of norvaline (internal standard) in all standards is 5.86 mg/l.

Sample Preparation Samples are centrifuged for 10 min at 4000 rpm and then diluted (1:4 v/v) with internal standard solution in 0.1 M HCl. This solution is again filtered (0.45 μm filter) and used for injection.

Derivatization First injection: 200 μl OPA reagent is added to 200 μl borate buffer (0.4 M, pH 10.6); 40 μl of diluted sample is added and the reaction mixture allowed to stand for 2.5 min in a clean autosampler vial; then 20 μl is injected. Second injection: 15 μl of diluted sample is added to 420 μl borate buffer (0.4 M, pH 10.6); and 15 μl FMOC-CL reagent is added and the reaction mixture allowed to stand for 2 min in a clean autosampler vial, and 20 μL is injected. The second injection is made 112 min after the first.

Column	RP-18 endcapped (25 cm × 4.6 mm; 5 μm particles)
Column temperature	42 °C
Mobile phase	(A) 20 mM sodium acetate solution, with 0.018% (v/v) triethylamine, 0.3% (v/v) tetrahydrofurane, and 0.010% (v/v) of a 4% (m/v) solution of EDTA, pH 7.20 (adjusted with 0.1% acetic acid)
	(B) 20% (v/v) sodium acetate solution (100 mM, pH 6.0), 40% (v/v) of acetonitrile, 40% (v/v) of methanol and 0.018% (v/v) triethylamine
Injection volume	20 μl
Fluorescence detector	Excitation/emission wavelengths are, respectively, 340/450 nm for primary amino acids and 237/340 nm for secondary amino acids. The latter is used to enhance the sensitivity of proline detection. The change in wavelengths was made 115 min after the first injection

LC-MS/MS: Sample Preparation – Sample is diluted for analysis in 0.1% formic acid in water.

Column	0.5 × 150 mm
Flow rate	20 μl/min
Mobile phases	0.1% formic acid in water (A), 0.1% formic acid in acetonitrile (B) Gradient: 2–20% B over 5 min
Injection volume	2 μl
Mass spectrometry	Quadrupole ion trap (Q-Trap), with 50 μm i.d. electrode

GC-MS: Derivatization with propyl chloroformate in the biological samples without prior protein precipitation or solid-phase extraction of the amino acids.

2.5.1.4 Determination of Extractable Protein Nitrogen

The protein extractability is closely related to moisture (>95%) and pH (>7.5). These conditions soften the cell wall and make it easy for the extraction of protein. Several methods can be used to extract protein such as based on the solubility in water, alkali, and mild acids. The food product after adjusting the moisture and pH is pulped, filtered, coagulated, and filtered to recover the protein. The extractable protein is quantified by Kjeldahl or Dumas method (AOAC 2000).

Kjeldahl method includes digestion of sample in the presence of catalyst and concentrated sulfuric acid, which converts the protein nitrogen to ammonium sulfate; the digest containing ammonium sulfate is distilled in the presence of 40% alkali, and the released nitrogen is trapped in boric acid, which is further titrated against N/70 sulfuric acid. Factors are used to convert the nitrogen to protein.

Food products	Conversion factors
Beer	6.25
Brewing sugar and syrups	6.25
Tea	6.25
Wheat whole meal/flour	5.83
Wheat flour	5.7
Macaroni, spaghetti, wheat pastes	5.7
Wheat brain	6.31
Rice (all products)	5.95
Rye (all products)	5.83
Barley (all products)	5.83
Oats (all products)	
Groundnut	5.46
Soya bean (all products)	5.71
Almonds	5.18
Brazil nuts	5.46
Coconut	5.30
Chestnuts	5.30
Seeds (sesame, safflower, sunflower)	5.30
Other seeds	5.30

Food products	Conversion factors
Bread	5.7
Milk (all products)	6.38
Ice cream and frozen desserts	6.38
Fruit products	6.25
Meat/fish	6.25

In Dumas method, the protein content of foods can be estimated by the determination of elemental nitrogen. In this method, the nitrogen-containing constituents of the sample are combusted at high temperature in the presence of oxygen to oxides of nitrogen (NO_x) and then reduced over copper or tungsten to gaseous nitrogen which is measured by gas solid chromatograph using thermal conductivity detector (AOAC 2000).

2.5.2 Assessment of Lipid Oxidation

2.5.2.1 Determination of FFA and Acid Value

Principle FFA or acid value is determined for edible fats and oils such as butter, olive, palm or sunflower oil, or fish oil and is expressed as percentage of FFA (oleic, palmitic, or lauric acid). The acid value of an oil/fat is the quantity of base, expressed in milligrams of potassium hydroxide that is required to neutralize free fatty acids present in 1 g of sample (AOAC 2000; Lakshmanan et al. 2013).

Procedure To a known quantity of minced sample, twice the amount anhydrous sodium sulfate is added and mixed thoroughly to remove the excess water. The blend is then shaken with chloroform for 5–10 min and filtered. Take 10 ml of this extract in a clean pre-weighed petri dish, and evaporate off the chloroform on a water bath. The weight difference is noted (A). Another aliquot (10 ml) is transferred to conical flask and chloroform is evaporated off. To this, 10 ml neutral alcohol is added and titrated against 0.01 N sodium hydroxide using phenolphthalein as indicator.

Calculation

Equivalent weight of oleic acid = 280
1 ml of 1 N NaOH = 0.28 g of oleic acid in 1 L
% FFA as oleic acid = titer value of NaOH × 0.01 × 280 × 100/A
Acid value = % FFA × 2

where:

A = oil/fat weight in 10 ml chloroform

2.5.2.2 Determination of Oxidative Rancidity: Peroxide Value (PV)

Principle Rancidity happens in contact with air (oxidative rancidity) or by microorganisms (ketonic rancidity) in oil. In oxidative rancidity oxygen is taken up by the fat with the formation of peroxides. The degree of peroxide formation and the time taken for the development of rancidity differ among oils. *Peroxide value (PV) is a measure of the peroxides contained in the oil. These are determined by titration against thiosulfate in the presence of KI using starch as indicator* (AOAC 2000; Lakshmanan et al. 2013).

Procedure Minced meat is blended thoroughly with anhydrous sodium sulfate, and oil is extracted (5–10 min) using distilled chloroform. 10 ml aliquot may be used directly, or the oil/fat provided is taken in a clean dry conical flask. 20 ml glacial acetic acid and a pinch (1 g) of potassium iodide are added to it. The flask is sealed with KI and kept in the dark for 30 min. This is then titrated against N/500 sodium thiosulfate using starch as indicator. The end point is the disappearance of blue color. Another aliquot of chloroform extract (10 ml) is dried and the weight of oil/fat is calculated (A).

Calculation

$$\text{Peroxide value}\left(\textit{milliequivalent peroxide / kg sample or millimoles of peroxide / l}\right) = \text{titer value}\left(\text{test} - \text{blank}\right) \times \text{N} \times 1000 / \text{A}$$

where:

Tire value = Titer value of test minus titer value of blank
N = normality of sodium thiosulfate
A = oil/fat weight in 10 ml chloroform

2.5.2.3 Determination of TBA Number

Principle Thiobarbituric acid (TBA) method is considered to relate closely the peroxide value with sensory measurements of oxidative rancidity. It is a quantitative method for malondialdehyde and is expressed as units of malondialdehyde per unit weight of sample (AOAC 2000).

Materials
TBA reagent: Sodium citrate butter (25 ml) is added to warm thiobarbituric acid solution (0.5 g/50 ml ethanol). Use within 4 h of preparation.

Procedure

Weigh approximately 0.1–0.2 g minced meat/oil in a brown-colored flask, and to it add 15 ml alcoholic hydrochloric acid (0.6 N) and TBA reagent (6 ml). The samples are refluxed for 30 min, and to it aqueous hydrochloric acid (79 ml, 0.6 N) is added and continued refluxing for another 10 min. The contents of flask are cooled, and an aliquot (15 ml) is shaken vigorously with 15 ml petroleum ether (40–60 °C) for 30 s. The residual fat is extracted. Repeat the process and keep in dark for 16 h. The upper layer is again discarded, and optical density is determined at 532 nm against TBA reagent as blank. The TBA number is expressed as milligram malondialdehyde per kg sample or oil.

Alternate Method

Weigh 10 g of sample and mince with 50 ml distilled water. Transfer this to round-bottom flask (500 ml) and add 25 ml HCl (1:2) to adjust pH to 1.5. Collect the distillate (50 ml). To an aliquot (1 ml), 5 ml distilled water and TBA reagent (0.2 M in 90% acetic acid) are added; and these are heated in a water bath for 30 min. The red color developed is measured spectrophotometrically at 538 nm.

Calculation

$$\text{TBA Number} = \text{Sample OD} \times \text{dilution factor} \times 7.8$$

where:

7.8 = correction factor taking into consideration the % recovery on distillation is 68%.

2.5.2.4 Determination of Refractive Index

The principle and procedure are mentioned in detail in Chap. 3.

2.5.2.5 Determination of Total Volatile and Nonvolatile Acids

Principle Organic acids are formed in fermented foods as a result of hydrolysis, biochemical metabolism, and microbial activity. Quantitative determination of organic acids is of great significance in fermented foods due to technical, nutritional, sensorial, and microbial reasons. The volatile and nonvolatile organic acids in food include acetic acid, lactic acid, propionic acid, and butyric acid, generated mostly during processing and storage. The most often reported compounds in fermented food products include esters (ethyl 2-methyl butyrate), acids (acetic acid, 2/3-methyl butanoic acid), pyrazines, and phenolic compounds. However, the volatile component profiles of fermented food products vary with the microflora and processing conditions (fermentation, drying, brining, or aging) (AOAC 2000; Shukla et al. 2010).

Procedure

Chromatographic Method

Standard solutions: All nonvolatile organic acids (oxalic acid, citric acid, lactic acid, succinic acid, malonic acid, and glutaric acid) are prepared at 0.5% concentration in distilled water. Standard mix solution of volatile organic acids (acetic acid, propionic acid, and butyric acid) is prepared at 0.1% concentration in distilled water. 3-Methyl butanoic acid is dissolved separately in ethanol at the same concentration.

Nonvolatile organic acids: 100 g of sample is refluxed with 150 ml of 86.5% ethanol for 1 h at 65 °C and filtered. This is repeated twice with same volume of 65% ethanol for 1 h. The filtrate collected is centrifuged and the upper ethanol layer is vacuum-evaporated. 360 ml of methanol (90%) is added (thrice) to this, to desalt the samples and filtered. The filtrate is again vacuum-evaporated and the residue dissolved in 100 ml distilled water. Extracted components are purified by ion exchange column chromatography (amberlite IR-120 and amberlite IRA-400, as cation and anion exchangers, respectively). A 20 ml sample is eluted with distilled water and sugar-free components are collected. The remaining sample components, absorbed on amberlite IR-400, are eluted with 150 ml 2 N $(NH_4)_2CO_3$ at slow flow rate. Recovered eluent is vacuum-evaporated, dried (P_2O_5) overnight, and the nonvolatile organic acids dissolved in 15–20 ml of diethyl ether and derivatized (esterified). 10 μl of this is injected into HPLC and chromatographs are analyzed.

Volatile organic acids: For the extraction of volatile organic acids, 10 g of sample is homogenized with 20 ml of triple distilled pure water and centrifuged. Subsequently, to 0.7 ml of the filtrate, 0.37 μl of 2% H_2SO_4 is added, so as to make the final concentration of H_2SO_4 as 0.1% and filtered using 0.45 μm syringe filter. Finally, 2 μl solution and standards are injected into GC separately.

HPLC	
Column	RSpak KC-811 column (8.0 × 300 mm)
Column temperature	40 °C
Mobile phase	0.1% H_3PO_4/distilled water
Flow rate	1 ml/min
Detector	Refractive index detector
Gas chromatography	
Column	10% PEG6000 in glass column (1.2 mm × 1.5 m)
Oven temperature	170 °C
Injector temperature	200 °C
Carrier gas	Helium
Flow rate	0.9 ml/min

Distillation Method

Principle: In this procedure, it is assumed that all the low molecular weight fatty acids up to octanoic acid have significant vapor pressures at 100 °C. At pH 6.5 to 7.5, organic acids exist in ionic form and cannot be distilled. By the addition of a strong nonvolatile acid, such as sulfuric (pH 1.0), the organic acids are converted to the unionized form and can be distilled. Formic acid and water form a Class II mixture, and water is the major component of the vapor phase when dilute solutions are distilled. Acetic acid and water form a Class I mixture, and the vapor phase is predominantly water vapor. Propionic, butyric, and other fatty acids in water form Class III mixtures. The distillate contains the acid and water in a constant ratio until all the acid is distilled.

Procedure: The distilled acids are measured quantitatively by titration with a standardized NaOH solution to the phenolphthalein end point. Approximately 70% of the volatile acids is distilled from samples treated by the recommended procedure. This matter is taken into account in calculations of the volatile acids present.

2.5.3 Assessment of Carbohydrate Catabolites

2.5.3.1 Determination of Lactic Acid/Metabolites

Principle Lactic acid is a catabolic product of carbohydrate metabolism. It has several health benefits and also has a significant role in nutritional and sensory properties of food. It can be determined by several methods (spectrophotometrically, HPLC, capillary electrophoresis, tandem mass spectrometry method). Mostly lactic acid is determined by the estimation of acetaldehyde/formaldehyde (AOAC 2000; Jeong et al. 2015).

Procedure

Spectrophotometric Estimation of Lactic Acid

Stock solution of lactic acid is prepared and diluted to obtain working standards (0–50 μl). To 2 ml of working standards, 0.2% iron (III) chloride is added and stirred. The absorbance is measured as 390 nm. Blank contains 2 ml 0.2% iron (III) chloride. Calibration curve is plotted and the concentration of lactic acid in test solution is determined.

HPLC

Column	Reverse phase, Tracer Excel 120ODSA (250 × 4.6 mm, 5 μm particle size)
Column temperature	40 °C
Flow rate	10 ml/min

Mobile phase	Isocratic, solvent A: 3×10^{-2}mol/l H_3PO_4 in water, solvent B: acetonitrile (88:12 v/v)
Injected volume	20 μl
Detector	UV/Vis, 200 nm

Estimation of Acetaldehyde and Formaldehyde

Sample Preparation: Solid/semisolid samples (1 g) are mixed with 9.950 ml of 30% NaCl solution and an internal standard [50 μL AA 1,2-$^{13}C_2$ (10 μg/l, w/v)] and then sonicated at ambient temperature for 30 min. The sample is centrifuged (3000 rpm, 4 °C, 10 min) and vortexed (30 s). The supernatant (5 ml) is transferred to 10 ml headspace vials with screw caps. Liquid samples are mixed with 4.475 ml of 30% NaCl solution, and an internal standard [25 μl AA 1,2-$^{13}C_2$ (10 μg/l, w/v)] is sonicated at ambient temperature for 30 min. Then 100 mg potassium hydrogen phthalate (KHP) and 50 μl O-(2,3,4,5,6-Pentafluorobenzyl)hydroxylamine hydrochloride (PFBHA, 10 mg/ml) are added to 5 ml sample before vortexing for 30 s and derivatizing at 45 °C for 40 min. Solid-phase micro-extraction (SPME) fiber coated with 65 μm polydimethylsiloxane/divinylbenzene (PDMS/DVB) is used to adsorb volatile compounds in the headspace at 45 °C for 15 min. The derivatized samples are analyzed using GC-MS.

Column	HP-5MS, 30 m × 0.25 mm ID × 0.25 μm film thickness
Carrier gas	Helium
Flow rate	0.8 ml/min
Splitless mode GC injector (SPME)	220 °C, 5 min
Oven temperature	50 °C-2 min, raised to 120 °C @ 30 °C/min, held at 200 °C-10 min
Front inlet temperature	220 °C
Detector transfer temperature	250 °C
Detector	Mass selective detector, 70 eV electron ionization (m/z acetaldehyde, formaldehyde, internal standard: 209, 195. 211; m/z 181 for qualifier)

2.5.4 Assessment of Nonprotein Nitrogen (NPN)

2.5.4.1 Determination of TVB-N and TMA-N

Principle After death, the volatile compounds produced in the muscle are TMA and ammonia, which escape from the alkaline solution as vapor and dissolve in acid solutions, neutralizing some of the acid. The excess acid is then titrated with alkali to determine the amount, which has not been neutralized. From this volatile bases are calculated. The principle of both assays is essentially the same. The homogenized muscle is extracted with trichloroacetic acid (TCA), which contains water-soluble components such as TMA (trimethylamine nitrogen) and ammonia (majorly TVB-N, total volatile base nitrogen) (AOAC 2000; Lakshmanan et al. 2013).

Procedure Sample (20 g) is homogenized and extracted with 20% TCA (20 ml) and filtered. Repeat the extraction with 1%TCA and pool the filtrate and make up to 100 ml. Conway units are used for estimation. Petroleum jelly is used to prevent any leakage. Assay is performed in triplicates. 1 ml of filtrate is poured on the outer ring of Conway unit and spread uniformly. 1 ml saturated potassium carbonate is added to this and gently stirred for thorough mixing (in the case of TMA-N, 0.5 ml neutralized formalin is added prior to potassium carbonate). 1 ml of standard acid is poured in the center chamber and covered with the plate (for 1 h or overnight at 37 °C). The unreacted acid (A) is titrated with 0.01 N NaOH using Tashiros indicator. Blank is also repeated in similar way (B).

Calculations

TVB-N or TMA-N % = (A-B) × 0.14 × extract volume × 100/sample volume (1 ml) × sample weight
1 ml of N/100acid = 0.14 mg nitrogen

2.5.5 Assessment of Nucleotides and Nucleotide Catabolite

Principle Dietary nucleotides aid in building the DNA and RNA in our body. These also play a critical role in metabolism and energy (ATP). Food such as meat, fruits, seeds, and vegetables are natural sources of nucleotides. Each nucleotide consists of a sugar molecule, a nitrogen base, and a phosphate group. The apolar products of nucleotide breakdown may include adenosine and its catabolites such as inosine, hypoxanthine, and uric acid (AOAC 2000; Lakshmanan et al. 2013).

The degradation of ATP catabolites proceeds in the same manner with most fish, but the speed of each individual reaction greatly varies from one species to another, and a formula was postulated by Japanese for fish freshness based on these autolytic changes (Fig. 2.10).

$$K\% = \frac{[\mathrm{Ino}]+[\mathrm{Hx}]}{[\mathrm{ATP}]+[\mathrm{ADP}]+[\mathrm{AMP}]+[\mathrm{IMP}]+[\mathrm{Ino}]+[\mathrm{Hx}]} \times 100$$

where [ATP], [ADP], [AMP], [IMP], [Ino], and [Hx] represent the relative concentrations of these compounds in fish muscle measured at various times during chilled storage.

The K or "freshness" index gives a relative freshness rating based primarily on the autolytic changes which take place during postmortem storage of the muscle. Thus, the higher the K value, the lower the freshness level. But some fish species such as Atlantic cod reach a maximum K value well in advance of the shelf life as determined by trained judges, and K is therefore not considered reliable as a freshness index for all marine finfish. Also, the degradation of nucleotide catabolites is

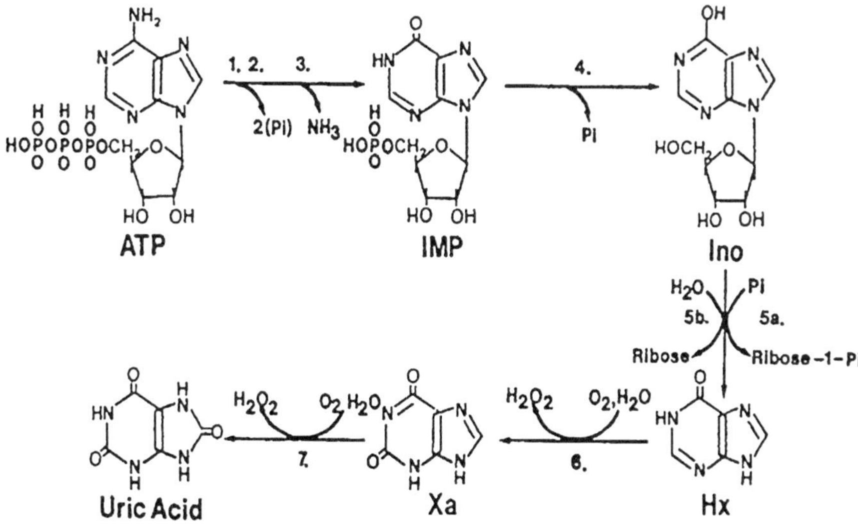

Fig. 2.10 Degradation of ATP

only coincidental with perceived changes in freshness and not necessarily related to the cause of freshness deterioration since only Hx is considered to have a direct effect on the perceived bitter off-flavor of spoiled fish. It is now widely accepted that IMP is responsible for the desirable fresh fish flavor which is only present in top quality seafood. None of the nucleotide catabolites are considered to be related to the perceived changes in texture during the autolytic process except of course ATP whose loss is associated with *rigor mortis*.

Procedure (K value) 5 g of sample is homogenized with 25 ml of 0.6 M perchloric acid at 0 °C for 1 min, centrifuged (3000 g, 10 min, 0–2 °C) and 10 ml supernatant is immediately neutralized to pH 6.5–6.8 with 1 M potassium hydroxide. Allowing to stand at 0 °C for 30 min, the precipitated potassium perchlorate is removed by filtration (Whatman filter paper No. 1), and the filtrate is diluted to 20 ml or stored (−20 °C) until analyzed using HPLC. Standard solutions of 0.166 mM are injected for calibration and quantification.

Column	Silica based, reverse phase, C18, 300 mm × 3.9 mm ID, 5 μm
Column temperature	30 °C
Mobile phase	Solvent A: 0.04 M potassium dihydrogen orthophosphate Solvent B: 0.06 M dipotassium hydrogen orthophosphate
Injection volume	20 μl
Flow rate	2 ml/min
Detector	UV detector, 254 nm

Freshness Assay kit and fresh tester strips also could be used to evaluate the freshness of the food products (or nucleotide degradation).

Filter Paper Electrophoresis ATP and its breakdown products in the sample are extracted from homogenized samples using 1 N HCl and centrifuged (8000 g, 15 min). The paper disc (6 mm diameter) containing 10 μl supernatant is subjected to paper electrophoresis immersed in pyridine acetic acid buffer solution (acetic acid/pyridine/water = 1:10:289, pH 3.7). ATP and its breakdown products were separated and analyzed under UV light at 254 nm. These were digitized and calculated by image processing software.

2.6 Miscellaneous

2.6.1 Determination of Na+/K+

Principle Sodium is the major cation of the extracellular fluid. Daily diet contains 130–280 mmol (8–15 g) sodium chloride. Potassium is the major cation found intracellularly. The average cell has 140 mM K^+ inside but only about 10 mM Na^+. The human body requires about 50–150 mmol potassium/day. Flame photometry is the simplest and traditional method to estimate sodium and potassium in biological samples. The principle relies on the fact that an alkali metal salt drawn into a non-luminous flame will ionize, absorb energy from the flame, and then emit light of a characteristic wavelength, as the excited atoms decay to the unexcited ground state. The intensity of emission is proportional to the concentration of the element in the solution.

Procedure Standards of different dilutions of sodium and potassium chlorides are prepared and the intensity of emission plotted against concentration. From this calibration curve, the concentrations of the test samples are evaluated.

2.6.2 Determination of Formaldehyde

Principle Formaldehydes are naturally formed in seafood during postmortem when trimethylamine oxide (TMAO) is converted to dimethylamine and formaldehyde. It may also be formed during the aging and deterioration of fish flesh. Besides natural formation of formaldehyde in seafood, other biochemical reactions also contribute to its formation such as oxidation of lipids by microbial action. This will lead to physical damage or production of biogenic amines or other unpleasant compounds in fish. Consumers are presently aware of the application of formaldehyde and its side effects (AOAC 2000; Kleeberg and Klinger 1982).

Materials

Nash reagent: 300 g ammonium acetate is dissolved in water, and 6 ml of acetic acid and 2 ml of acetyl acetone are added and made up to 1 l.

Procedure

Sample preparation: The samples are homogenized with 60 ml of 6% trichloroacetic acid and filtered (Whatman No.1), and pH is adjusted to 6–7 using 1 M potassium hydroxide (KOH)/hydrochloric acid (HCl). Then, 5 ml of sample solution is taken in a 50 ml of volumetric flask. Then, the sample is frozen (−20 °C, 1 h). During analysis, the sample is mixed with 2 ml of Nash reagent (indicator). Samples are then heated in the water bath at 60 °C for 30 min and absorbance is measured.

Standard preparation: Formaldehyde stock solution (6.2%) is prepared and diluted to prepare working stock of different concentrations (0–6 ppm). To these, Nash reagent is added, and respective absorbance at 415 nm is recorded, and calibrated curve is plotted. From this curve, the concentration of formaldehyde in test samples is evaluated.

The rapid detection kit developed by CIFT, India, is also used for the detection and determination of formaldehyde in fish samples.

2.6.3 Determination of Ammonia

Principle Ammonia represents a major proportion of the volatile amines produced by enzymatic processes of spoilage of protein rich food (Pranaitytė et al. 2007).

Procedure 25 g samples homogenized with 75 ml water and adjusted pH 5.2 using 0.01 M HCl. The solution is heated to 70 °C (with stirring) and immediately cooled to room temperature. The precipitate is removed by filtration and is stored at −30 °C until analysis, using capillary electrophoresis. The capillary is conditioned with the filtered (0.45 μm) running buffer for 15 min, and between each run it is again flushed with buffer for 3 min. The temperature of the capillary is set to 25 °C. Standards or samples (1:10 with water containing the internal standard) are injected hydrodynamically (2 s, 50 mbar) with a subsequent buffer plug (3 s, 50 mbar). In between, the capillary is twice dipped into water in order to avoid sample carryover into the buffer. A voltage of 30 kV is applied with the positive polarity at the sample injection end. The separated compounds are detected in the indirect mode by measuring the difference in absorbance at 310 and 215 nm. The stop time is set to 10 min.

2.7 Microbiological Parameters (Fernandes 2009)

2.7.1 Evaluation of Total Plate Count

Total plate count can be determined by spread plating method or pour plate method. For spread plating method, agar plates are poured and dried in advance. Melt one flask of TGA in a water bath, and cool one flask of the medium to 45 °C; pour it into petri dish (about 15–18 ml each). Allow to set. Dry the surface of the medium in a laminar flow chamber for 45 min.

Sampling Aseptically cut 10 g of skin with muscle of the fish sample into a sample dish. Macerate and homogenize with 90 ml diluent (normal saline) in a sterile mortar (10^{-1} dilution). Pipette 1 ml of the supernatant to 9 ml diluent and mix well (10^{-2} dilution). From the 10^{-2} dilution, pipette 1 ml of the supernatant to 9 ml diluent and mix well (10^{-3} dilution). Prepare 10^{-4} and 10^{-5} dilutions in the above manner.

Spread Plating Arrange the preset TGA plates in three rows in duplicates and label appropriately. Inoculate 0.5 ml each of the 10^{-3}, 10^{-4}, and 10^{-5} dilutions on the surface of the respective plates. It is spread on the surface of each plate using a sterile bent glass rod. Between spreading operations of each plate, the rod was sterilized by dipping in alcohol and flaming. After 30 min, incubate the plates at 37 °C for 48 h. After incubation, the colonies develop in each plate were counted using a Quebec colony counter. The colony counts of duplicate plates agreed within 10% limit, and the counts between decimal dilutions agreed decimally. TPC/g sample was calculated using the relation:

$$\mathrm{TPC} / \mathrm{g\,sample} = \text{average count} \times \text{dilution factor} \times 10 / \mathrm{w},$$

where w = weight of the sample

Only those plate counts which fell between 30 and 300 were statistically acceptable.

2.7.2 Isolation, Identification, and Characterization of Seafood Bacteria

Identification of the Bacteria

Gram's Staining and Microscopy: Only young cultures are used for staining. Usually 16–24 h cultures are considered young. A speck of young culture was emulsified with a drop of sterile water in the middle of a sterile glass slide and spread uniformly. Fix by passing the slide 3–4 times through the blue flame of Bunsen burner. Flood the smear with Gram's crystal violet stain for 1 min. Wash with water. Flood with gram's iodine for 1 min and again wash with water. Destain dropwise with addition of alcohol until washings were free from violet color. Wash once again and counterstain with safranin for 1 min. Dry in air and observe the cells under microscope (100X objective). The color, shape, size, and arrangement of the cells were noted. Purity of the culture was also checked. In case of the cells being Gram-positive rods, it was further examined for spore formation.

Penicillin Sensitivity: Prepare preset antibiotic agar plates and dry the surface at 56 °C for 45 min. Cool to room temperature. Divide each plate into four quarters by drawing lines on the bottom so that one plate could be used for four cultures. A little of the culture was smeared over about 4 cm^2 area in each quarter. A filter paper disk impregnated with penicillin (each disk contains 2.5 IU penicillin) was placed on the surface of each smear. Plates are incubated without inverting for 18–24 h. Examine for clear zones of inhibition around the discs. Cultures showing clear zones of inhibition are sensitive to 2.5 IU penicillin.

Cytochrome Oxidase Test: Smear a little of the young culture on the test paper (already impregnated with Kovac's cytochrome oxidase reagent). Development of a blue color in a few seconds indicated a positive test.

H&L glucose O/F reaction test (Hugh / Leifson's oxidative verses fermentative reaction test using glucose as substrate): Using a platinum wire, a little of the culture was stab-inoculated into the H&L glucose O/F medium, in such a way that at least 2-cm-long column of the medium at the bottom of the tube remained uninoculated. Incubate for 18–24 h and observe changes. A color change into yellow indicated acid production from glucose. A deepening of red color of the medium indicated an increase in pH to alkaline level. Growth of bacteria along the line of inoculation and a yellow color throughout the medium indicated fermentative reaction (fermentative with acid but no gas). If gas bubbles were seen trapped in the medium, reaction was fermentative with gas production (fermentative with gas; FAG). If yellow color appeared only at the top part of the medium, the reaction was oxidative. Sometimes, a deep pink color developed near the top surface, indicating the change of medium to alkaline (alkaline top). In both cases, the reaction was nonfermentative (NF).

Motility of Bacteria: A small drop of distilled water was placed in the middle of a cover slip; a speck of young culture from agar slant was emulsified with it. A cavity slide was taken; the margin of the cavity was smeared with a little paraffin wax. The slide was inverted on the cover slip in such a way that the cover slip got attached to the slide and on turning upside down, the culture drop hanged into the cavity. Observe under high-power objective (45X).

Catalase Test: Place a speck of young culture on a clean glass slide, and flood with two drops of 30% H_2O_2. Evolution of gas from the culture indicated positive test for catalase.

Spore Formation: Take a dust-free, dirt-free, and oil-free microscopic glass slide. A speck of young culture was emulsified with a drop of sterile water in the middle of the glass slide and spread uniformly. Dry in air. Add a drop of malachite green stain. Fix by passing the slide 3–4 times through the blue flame of Bunsen burner. Care was taken not to char the smear. Vegetative cells were then decolorized with water and stained pink with safranin counterstain.

2.7.3 Detection and Identification of Fecal Streptococci

Estimation of Total and Fecal Streptococci For enumeration of streptococci, most probable number (MPN) technique should be used (Surendran et al. 2006). Three tubes containing 10 ml double strength Azide Dextrose broth should be inoculated with 10 ml of water samples, while two other sets, each of three tubes containing 5 ml single strength broth, should be inoculated with 1 ml and 0.1 ml water sample, respectively. Well-mixed tubes should be incubated at 35 ± 0.5 °C for 24–48 h. Turbidity should be noted and interpreted from MPN table. Five ml of single strength Azide Dextrose broth should be inoculated with 1 ml, shaken well, and incubated at 44.5 °C for 48 h. The tubes that showed growth should be considered positive for fecal streptococci. MPN should be computed. For further confirmation KF streptococcal agar should be streaked with turbid culture and incubated at 35 ± 0.5 °C for 48h. Pink to dark red colonies should be examined, purified, and further identified by different biochemical tests. The biochemical test for streptococci is summarized in table below. For the antibacterial susceptibility test, turbid suspension of pure culture of streptococci is swabbed uniformly on the nutrient agar plate and tested with four antibiotics such as penicillin, methicillin, tetracycline, and erythromycin. The plates are incubated at 37 °C for 24 h and zone of inhibition are measured.

2.7.4 Detection and Identification of *Salmonella typhimurium*

Transfer 25 g of sample to 225 ml of lactose broth and incubate at 37 °C for 24 ± 2 h. Transfer 1 ml each from above to 9 ml selenite cysteine broth and 9 ml tetrathionate broth, and incubate both the tubes at 37 °C for 24 ± 2 h. Streak a 3 mm loopful from the incubated selenite cysteine broth on Hektoen Enteric Agar (HEA), Bismuth Sulfite Agar (BSA), and Xylose Lysine Deoxycholate Agar (XLD). Also repeat the streaking from the incubated tetrathionate broth to HEA, BSA, and XLD agar. Incubate the plates at 37 °C for 24 ± 2 h. Examine the plates for salmonella colonies. Select the suspicious colonies from each selective agar, and inoculate into Triple Sugar Iron (TSI) agar, and incubate at 37 °C for 24 ± 2 h. Typical *Salmonella* cultures produced an alkaline slant (red) and an acidic butt (yellow) with the production of H_2S (blackening of the agar) and gas.

2.7.5 Detection and Identification of *Vibrio cholerae* and *Vibrio parahaemolyticus*

Vibrio cholerae Transfer 25 g of sample to 225 ml alkaline peptone water and incubate for 6 h at 37 °C. Streak one loopful from the surface growth of the above to TCBS Agar plates. Also transfer 1 ml to 9 ml alkaline peptone water, and incubate for 18 h at 37 °C, and then streak one loopful from second enrichment to another TCBS plate for 18–24 h at 37 °C. Transfer the suspected colonies (yellow flat and smooth colonies with opaque centers and transparent peripheries, 2–3 mm diameter) from both TCBS plates to Kligler Iron Agar (KIA) slants, and incubate at 37 °C for 18 h. *Vibrio cholerae* gave an acidic butt (yellow) and alkaline (pink) slant with no gas and no hydrogen sulfide (no black color).

Vibrio parahaemolyticus synthesizes three major antigens, a thermostable somatic O antigen, a thermolabile capsular K antigen, and a flagellar H antigen. They show a definite sodium requirement and need at least 0.5% salt to initiate growth. Growth in water activity (a_w) of up to 0.937 has been noted. *Vibrio parahaemolyticus* is a relatively fragile organism. It is sensitive to drying and heat. Despite the general agreement that the organism exhibits poor cold resistance, it is reported to survive in iced fish/shrimp. Human illness that frequently occurs is a mild and self-limited gastroenteritis with duration of 2–3 days. Seafood are the main vehicle of this type of infection and mainly happen in places where people eat raw seafood. Hence, this type of gastroenteritis due to *Vibrio parahaemolyticus* is restricted to countries where raw seafood is a delicacy.

Detection and Identification All media for *Vibrio parahaemolyticus* contains 3% NaCl, unless otherwise specified. 25 g of the sample is taken in 225 ml alkaline peptone water with 3% salt (APW + N_3); macerate in a sterile blender or stomacher blender, and transfer to a sterile 500 ml flask, and incubate at 36 ± 1 °C for 16–18 h. Streak loopful of the APW culture on preset TCBS agar with 3% salt (TCBS +N_3) plates, and incubate at 36 ± 1 °C for 18–24 h. Examine the colonies on TCBS agar. *V. parahaemolyticus* forms around 3–5 mm green or blue-green colonies with green or blue center. Pick typical colonies onto NA slants with 3% salt (NA + N_3), and incubate at 36 ± 1 °C for 24 h. Using the culture on NA slant, the following tests are done.

Kanagawa reaction (if human blood is available): Prepare plates of Wagatsuma agar (NaCl 7%); spot or stab inoculates from overnight broth culture. Incubate at 36 ± 1 °C for 18–24 h. Check for hemolysis. A clear zone (beta hemolysis) indicates a positive reaction. Discoloration (alpha hemolysis) and no change (gamma hemolysis) are ignored.

2.7.6 Detection and Identification of *E. coli*

Sampling was done as described in 2.4.1. From the 10^{-1} dilution, pipette 0.5 ml on the surface of previously dried Tergitol-7 Agar plates. After 24 hours, yellow, circular, flat, non-mucoid colonies appeared. Biochemical tests are given in the table below.

2.7.7 Detection and Identification of *Staphylococcus aureus*

Sampling was done as described in 2.7.1.10^{-1} dilution and was inoculated on Baird Parker Agar plate. Black colonies with a clear zone are the characteristic of staphylococcus colonies on this agar. Coagulase test was performed to confirm *Staphylococcus aureus*.

2.7.8 Detection and Identification of *Listeria monocytogenes*

Primary Selective Enrichment 25 g of the sample is macerated with 225 ml of the UVM-I broth in a 500 ml flask and incubated at 28 ± 2 °C for 18–24 h.

Selective Enrichment 1 ml of the primary selective enrichment culture is inoculated to 10 ml of Frazer's broth and incubated at 36 ± 1 °C for 18–24 h. A black- or dark-colored growth indicates hydrolysis of the aesculin, indicating possible presence of *Listeria* and proceeds further. (In case there is no dark or black color developed, re-incubate for another 24 hr. If there is no darkening or blackening of the media even then, *Listeria* is not present, need not proceed further.)

Selective Plating Loopful of the culture from the Frazer's broth is streaked on to preset *Listeria* selective agar (LSA) and the plates incubated at 36 ± 1 °C for 48 h. Examine the plates after 24 hrs for typical grey dark small (2 mm dia.) colonies, occasionally with sunken center and a dark halo around. Re-incubate if necessary for another 24 h. Instead of LSA, we can use PALCAM for selective isolation. Pour the plates as given in the media section. Streak loopful from the Frazer's broth and incubate at 36 ± 1 °C; examine for typical colonies. Incubation in a carbon dioxide incubator, with 5% CO_2, would give grey-green colonies with sunken dark center and a dark halo around.

Isolation and Identification All incubations are done at 36 ± 1 °C, except for motility. Pick typical colonies with a dark/black halo around as presumptive *Listeria* colonies, and streak on to TSA slants or BHIA slants. Incubate at 36 ± 1 °C. *Listeria* produces only colorless (translucent) and thin growth. To ensure purity of the isolate, purify the culture by streak dilution on pre-dried TSA plates. *Listeria* colonies will appear light bluish-green on examination under Henry's illumination. Proceed further for biochemical identification.

Test for Umbrella Motility Since the *Listeria* shows active motility, umbrella motility is analyzed. Stab the presumptive *Listeria* culture into the umbrella motility medium, and incubate at 25–30 °C for 36–48 h, and examine for umbrella roof in the top part of the media.

Test for Hemolysis This is yet another confirmative test for *Listeria.* Inoculate the presumptively positive cultures on surface of pre-dried blood agar plate by streaking, and incubate at 36 ± 1 °C for 18–24 h. Examine for clear zone of hemolysis. Even though *Listeria monocytogenes* is beta-hemolytic (complete degradation of hemoglobin), the zone of hemolysis is very small. So, examine for hemolysis beneath the colony. Alternately, stabs inoculate a thick blood agar plate, so that hemolysis will be visible around the line of slab. If beta-hemolysis is negative, then the culture is *Listeria innocua. L. seeligeri* also produces a small zone of hemolysis, but *L. ivanovii* produces a wider (3–5 mm) and clearer zone.

CAMP Test A thin layer of sheep blood agar (5% v/v in nutrient agar) is poured on the surface of a nutrient agar base plate and dried before use. The test is carried out by streaking *Staphylococcus aureus* (NCTC 1803) or *Rhodococcus equi* (NCTC 1621) across the sheep blood agar plate. Test strains are streaked at right angles to the *S. aureus* or *R. equi* so that the cultures are at their closest about 1–2 mm apart. Plates are incubated at 37 °C overnight. Positive results are obtained when an enhanced zone of hemolysis between the two cultures occurs.

The biochemical tests of the bacteria are summarized in Table 2.3.

2.7.9 Detection and Identification of *Klebsiella*

Gram-negative rods fermenting adonitol or inositol in 48 h were accepted as possible *Klebsiella* strains (Surendran et al. 2006; Fernandes 2009). Using a short series of biochemical tests already in use in many routine laboratories, it is possible to identify strains of the genus *Klebsiella* and to differentiate *K. aerogenes* from other types. Marked production of slime is not peculiar to the group, and colonial appearance alone is not a satisfactory basis for identification (Table 2.4).

2.7.10 Detection and Identification of *Shigella*

Recent phylogenetic studies indicate that the *Shigella* are not a unique genus; rather they are an immunologic and biochemically distinct lineage of *E.coli.* However, for medical purposes and to maintain historical consistency, the *Shigella* has not been merged with *E. coli.* The genus *Shigella* is comprised of four species: *Shigella dysenteriae* (also referred to Group A), *Shigella flexneri* (also referred to Group B),

Table 2.3 Biochemical test for *E.coli*, *S. typhimurium*, *V. cholera*, *Streptococci*, and *V. parahaemolyticus*

Biochemical test	*E.coli*	*S. typhimurium*	*V. cholerae*	*Streptococci*	*V. parahaemolyticus*	*L. monocytogenes*
Catalase	+	+	+	–	+	+
Indole	+	–	+	–	+	NA
Methyl red	+	+	NA	NA	NA	+
Urease	–	–	–	–	–	–
Voges-Proskauer	–	–	–	NA	–	+
Citrate	–	+	+	NA	+	NA
H_2S	+	+(weak)	NA	NA	NA	NA
Motility	+	+	+	–	+	NA
Oxidase	–	–	+	NA	+	–
Gas production	–	–	–	NA	–	NA
Glucose	+	+	–	+	–	+
Lactose	+	–	–	NA	–	NA
Maltose	+	+	NA	+	NA	NA
Sucrose	+	–	+	+	–	NA
TSI	A/Alk	Alk/A	Alk/A	NA	Alk/A	NA
O-Nitrophenyl-β-D-galactopyranoside (for β-galactosidase)	+	NA	+	NA	+	NA
Sorbitol	+	NA	NA	NA	NA	NA
Lysine decarboxylase	+	+	+	NA	+	NA
Ornithine decarboxylase	NA	–	+	NA	+	NA
Dulcitol	NA	–	NA	NA	NA	NA
Rhamnose	NA	–	NA	NA	NA	NA
Salicin	NA	–	NA	NA	NA	NA
Arginine dihydrolase	NA	NA	–	–	–	NA

(continued)

Table 2.3 (continued)

Biochemical test	*E.coli*	*S. typhimurium*	*V. cholerae*	*Streptococci*	*V. parahaemolyticus*	*L. monocytogenes*
Growth (0% NaCl)	NA	NA	+	NA	–	NA
Growth (6% NaCl)	NA	NA	NA	+	+	NA
TCBS growth	NA	NA	Good	NA	Good	NA
Colony color on TCBS	NA	NA	Yellow	NA	Green	NA
PYR	NA	NA	NA	+	NA	NA
CAMP	NA	NA	NA	–	NA	+
Bacitracin	NA	NA	NA	Sensitive	NA	NA
10% Bile	NA	NA	NA	Insoluble	NA	NA
PNF	NA	NA	NA	+	NA	NA
Blood agar	NA	NA	NA	+	NA	NA
Crystal violet blood agar	NA	NA	NA	+	NA	NA
Tryptose soy broth	NA	NA	NA	Pink/red colonies	NA	NA
Methylene blue	NA	NA	NA	+	NA	NA
Aesculin	NA	NA	NA	+	NA	+
myo-inositol fermentation	NA	NA	–	NA	–	NA
Nitrate reduction	NA	NA	+	NA	+	–
Wagatsuma agar (Kanagawa reaction)	NA	NA	NA	NA	+, with clearance of blood	NA

Table 2.4 Biochemical tests for *Klebsiella* spp.

Biochemical tests	*K. aerogenes*	*K. edwardsii-edwardsii*	*K. pneumoniae*	*K. edwardsii-atlantae*	*K. rhinoscleromatis*
Fimbriae	+	_	+	_	_
Glucose	+	_	+	+	_
Lactose (acid)	+	(+)	+	(+)	_
Methyl red	_	D	+	+	+
Voges-Proskauer	+	+	_	D	_
Citrate	+	D	+	+	_
Urease	+	+	+	+	_
Gluconate	+	+	D	D	_
Malonate	+	D	+	_	+
Lysine decarboxylase	+	+	+	+	_
44 °C test	D	_	_	_	_
Dulcitol(acid)	D	_	+	_	_
KCN	+	+	–	+	+

(+) = delayed, D = some strains -, some +

Shigella boydii (also referred to Group C), and *Shigella sonnei* (also referred to Group D). Included among these organisms is the etiologic agent of epidemic dysentery, *S. dysenteriae* serotype 1. With the exception of *S. sonnei*, each species may be further divided into serotypes on the basis of reactivity with hyperimmune serum (*S. dysenteriae*, 15 serotypes; *S. flexneri*, 6 serotypes and 2 variants; and *S. boydii*, 20 serotypes). *Shigella* group and serotype are identified by slide agglutination using antisera. Both polyvalent and monovalent *Shigella* antisera are available. Polyvalent antisera are used to determine group and contain antibodies to multiple serotypes of *Shigella*. Monovalent antisera is used to determine serotype. A final, confirmatory, identification of *Shigella* spp. is based on immunologic and biochemical tests (Table 2.5).

Shigella spp. will typically be lactose negative following overnight incubation (18–24 h). However, it is not uncommon for *Shigella* spp. to ferment lactose following prolonged incubation (5–7 days).

Differentiation of *Shigella* strains from *E. coli* is one of the problems faced by a microbiological laboratory and may reflect the fact that *E. coli* and all four *Shigella* species are very closely on the basis of the DNA relationship. The tests indicated below are suggested for the differentiation (Fernandes 2009) (Table 2.6).

2.7.11 Detection and Identification of *Campylobacter*

There is no simple standard method for the routine isolation of *Campylobacter species*. *Campylobacter* spp. are sensitive to environmental conditions such as dehydration and oxygen and both freezing and high temperatures, which can reduce their

Table 2.5 Biochemical tests for *Shigella* spp.

Biochemical tests	*Shigella* spp.
Hydrogen sulfide	–
Glucose (gas)	–
Glucose (acid)	+
Motility	–
Simmons' citrate	–
Adonitol fermentation	–
Malonate utilization	–
Methyl red	+
Voges-Proskauer	–
Arginine dihydrolase	–
Lysine decarboxylase	–
Ornithine decarboxylase	*S. sonnei* -, other *Shigella* +
Urea hydrolysis	–
Phenylalanine deaminase	–
Salicin fermentation	–
Lactose fermentation	–
Sucrose fermentation	–
Xylose fermentation	–
Inositol fermentation	–
Catalase	+ (all species except *S. dysenteriae*)

Table 2.6 Differentiation of *Shigella* spp. and *E. coli* (V-variable)

Differentiation of *Shigella* spp. and *E. coli*		
Test	*Shigella* spp.	*E. coli*
Mucate fermentation	–	+
Acetate utilization	–	+
Christensen's citrate	–	V
Indole production	V	+
Lysine decarboxylase	–	V
Glucose (gas)	–	+
Salicin (fermentation)	–	V
Motility	–	+/–

viability. Therefore, transport to the laboratory should be as rapid as possible and preferably in suitable transport media, in order to protect the cells from drying out and from the toxic effect of oxygen. Culture analysis of *Campylobacter* spp. is usually performed by direct plating on selective media or by enrichment followed by cultivation on solid selective media. Enrichment may be required if the bacteria are present in low numbers or have been damaged by environmental stresses such as temperature changes, nutrient deprivation, dehydration, or exposure to atmospheric oxygen, whereas direct culture is recommended for samples in which a high level of campylobacter is suspected. A selective medium is necessary when a diverse flora is expected, such as in clinical samples from intestinal contents where large amounts of bacteria from the family Enterobacteriaceae are generally present. The selectivity of the medium is determined by the use of antibiotics. Cephalosporins are often used in combination with other antibiotics such as trimethoprim, vancomycin, amphotericin, and rifampicin, since most *Campylobacter* spp. are resistant to these antibiotics. Several media are described in the literature for the bacteriological culture of thermophilic *Campylobacter* spp. (Fernandes 2009). The main difference between these media is the amount and type of antibiotics used for inhibition of the contaminating flora. The media come from two main groups: blood-containing and charcoal-containing media. Blood components and charcoal remove toxic oxygen derivates such as peroxides, singlet oxygen, and superoxide ions, which can be formed when media are exposed to light. These products are toxic to campylobacter due to the fact that they lack the enzymes superoxide dismutase and peroxidase. Pre-enrichment in particular commonly uses Exeter Broth, Bolton Broth, Preston Broth, CEB (Campylobacter Enrichment Broth), and Park and Sanders Broth. Commonly used selective media with blood components for campylobacter include Skirrow agar and Preston agar, while selective media with charcoal include CAT agar (mCCDA) and Karmali agar. In most protocols, a loopful of the enrichment is streaked onto a campylobacter-selective agar plate to obtain single colonies. Some media are favorable for some *Campylobacter* spp., for example, some strains of *C. coli* are sensitive to polymyxin B and may therefore be inhibited in PEB and PA, and some strains of *C. coli* and a few strains of *C. jejuni* may be missed due to their sensitivity to cephalotin. As thermophilic *Campylobacter* spp. in some samples can be difficult to culture, there are other methods available for isolation and identification, for example, membrane filtration, enzyme immunoassay, and PCR-based methods.

Incubation of thermophilic *Campylobacter* spp. should preferably be performed at 42 °C to minimize the growth of contaminants. Most *Campylobacter* spp. are microaerophilic, but a few species show a range of oxygen tolerance, and some are almost anaerobic. Optimal growth is obtained in an atmosphere with 5–10% oxygen and 1–10% carbon dioxide, and the growth of some species is enhanced by the presence of hydrogen. Several methods are available to achieve the optimal gas mixture, by appropriate gas-generating envelopes such as CampyGen (Oxoid Basingstoke, UK) and Campypac (BBL Microbiology Systems, Cockeysville, MD) or airtight jars with valves for evacuation and filling of gas (Anoxomat system, Anoxomat, MART-Anoxomate system, Netherlands). *Campylobacter* spp. grow in a spreading manner; characteristic campylobacter colonies are grayish and slightly pink with a metallic sheen on blood-containing agars. On charcoal-based media, the campylobacter colonies are grayish to white with a metallic sheen.

Confirmation of *Campylobacter* spp. is mostly based on colony morphology, microscopic appearance, and the following phenotypic characteristics: motility, production of oxidase and catalase, and hippurate hydrolysis reaction. *Campylobacter* spp. are typically spiral or curved rods with slender corkscrew motility, but bacteria from older cultures are less motile and coccoid forms occur. Bacteria from *C. jejuni* can be distinguished from other *Campylobacter* spp. on the basis of the hydrolysis of hippurate, as this is the only hippurate-positive species. However, hippurate hydrolysis-negative *C. jejuni* does exist. Biochemical speciation may be supplemented or replaced by molecular methods. A variety of polymerase chain reaction (PCR)-based methods for identifying thermophilic *Campylobacter* spp. have been developed.

2.7.12 Determination of Total Fungi Count

Principle The method includes use of culture plates of dry medium supplemented with antibiotics, dye to enhance visualization of growth, and cold water-soluble gelling agent. Diluted or undiluted suspensions are added to the plate at a rate of 1 ml/plate and spread uniformly (30cm^2 growth area). Gelling agent is allowed to solidify, plates are incubated, and yeasts/mold are counted (AOAC 2000).

Materials
The yeast and mold plates contain nutrients supplemented with chlortetracycline, chloramphenicol, cold water-soluble gelling agent, and dye sensitive to presence of phosphatase (5-bromo-4-chloro-3-indolyl phosphate) that enhances visualization. Circular growth area of single plate contains 30 1 × 1cm^2 outlines on film base.

Procedure Aseptically prepare 1:10 or greater dilutions of food product with water. Pour 1 ml suspension to the plates and spread using sterile spreader. Incubate for 5 days at 20–25 °C. Count plates promptly after incubation. Yeast appears as blue-green or off-white small defined colonies, while mold colonies are blue, black, yellow, or green large diffused colonies.

Calculations

$$\text{Yeast / mold colonies} = \text{total no.of colonies} \times \text{dilution factor}$$

2.8 Trends in Quality Control and Assurance in Seafood Processing Industry

2.8.1 HACCP, ISO, and FAO (Boziaris 2013)

Hazard Analysis and Critical Control Point (HACCP) is an internationally recognized system for reducing the risk of safety hazards in food. An HACCP system requires identification of potential hazards and controlling these at specific points in the process. The hazards may include biological, chemical, or physical. Any company involved in the manufacturing, processing, or handling of food products needs to use HACCP to minimize or eliminate food safety hazards in their product. Presently, it has become an important criterion for all food processors (Fig. 2.11).

- *Develop a food safety team* – The management should appoint a multidisciplinary, knowledgeable, and experienced team from all areas (quality control manager, production manager, microbiologist, process engineer, people from different sections such as packaging, raw material purchase, distribution, etc.) to implement the food safety management system (FSMS). The members must have the knowledge and experience on the company's products, processes, equipment, and food safety hazards. The team leader convenes the group and directs the work of the team. Team leader should be an expert in product and process.
- *Raw materials, ingredients, and product contact surfaces description* – This should describe the biological, chemical, and physical characteristics of product, composition, formulated ingredients, additives, processing aids, origin, method of production, packaging and delivery methods, storage conditions and shelf life, preparation and handling before use or processing, and the acceptance criteria.
- *End product description* – Details such as product name, composition, biological, chemical, and physical characteristics related to food safety, intended shelf life and storage conditions, packaging, labeling and handling instructions, preparation and handling, modes of distribution, intended use, unintended use,

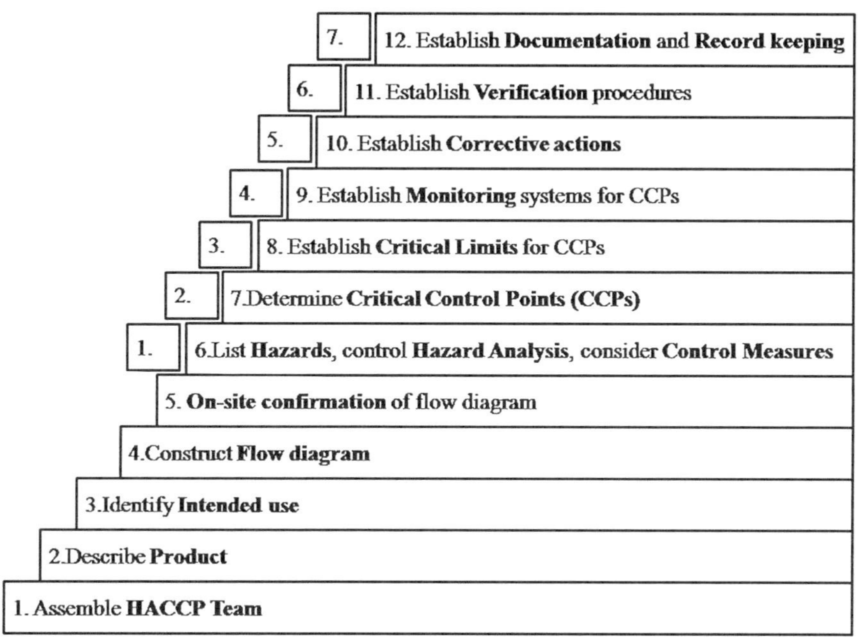

Fig. 2.11 12 Codex steps and seven principles of HACCP

expected mishandling and misuse, and vulnerable groups to specific food safety hazards need to be described.

- *The flow diagram and description* – This involves developing the flow chart depicting all the process steps involved to get the final product, from the raw material to dispatch of the final product.
- *Schematic and onsite verification* – A schematic diagram of the production unit showing all process routes, waste, and interactions between the personnel, wastes, and the product. Onsite verification is the confirmation, through walking along the process to confirm that all processes are represented in the scope.
- *Conduct a hazard analysis* – Effective hazard identification and hazard analysis are the keys to a successful HACCP plan. All real or potential hazards that may occur in each ingredient at each stage of the processing should be considered and preventive measures to apply to control these hazards. Hazards can either be as follows:

  Biological: foodborne bacterial pathogens (*Salmonella*, *Listeria*, and *E. coli*, viruses, algae, parasites, fungi, etc.)

  Chemical: commonly naturally occurring chemicals (cyanides in some root crops and allergenic compounds in peanuts), toxins produced by microorganisms (mycotoxins and algal toxins), and chemicals added to the commodity to get desired effects (fungicides or insecticides)

  Physical: contaminants such as broken glass, metal fragments, insects, or stones
- *Identify critical control point (CCP)* – A CCP is a point or a step in which more care is needed to prevent the hazard.
- *Establish critical limits for each CCP* – A critical limit is the minimum value of a hazard that makes product unacceptable.
- *Establish CCP monitoring requirements* – These are follow-up activities that help to maintain a hazard in acceptable levels.
- *Establish corrective actions* – These are actions taken when monitoring indicates a deviation from an established critical limit. These prevent the recurrence of the hazard.
- *Verification* – Once the HACCP plan has been drawn up and all the CCPs have been validated, then the complete plan must be verified at regular intervals by a trained personnel. Microbiological and/or alternative chemical tests are used to confirm that the plan is in control and the product is meeting customer specifications.
- *Documentation and record keeping* – The HACCP regulation requires that documents (hazard analysis, HACCP plan, monitoring of critical control points, critical limits, verification activities, the handling of processing deviations, etc.) are maintained. Implementation involves monitoring, verifying, and validating of the daily work that is compliant with regulatory requirements in all stages all the time.

The *International Standards Organization (ISO)* is a principal voluntary global leader in the establishment of guidelines for the establishment of private sector. ISO

22000 is a generic food safety management standard for the organizations involved directly or indirectly in food chain. It was published on 1 September 2005. ISO22000:2005 is first global food safety standard that harmonizes the voluntary international standards, employs proven management system principles (HACCP), enables a common understanding of what the system is, requires legal compliance checking, integrates Good Manufacturing Practices (GMP), and has internal and external monitoring. ISO 22000 Family Standards are as follows:

ISO 22000:2005	Food safety management system: Requirements for any organization in the food chain
ISO 22002-1:2009	Prerequisite programs on food safety – Part 1 – food manufacturing
ISO 22003:2007	Food safety management systems– Requirements for bodies providing audit and certification of food safety management systems
ISO TS 22004:2005	Food safety management systems – Guidance on the application of ISO 22000:2005
ISO 22005:2005	Traceability in the feed and food chain – General principles and basic requirements for system design and implementation
ISO 22006:2009	Quality management systems – Guidelines for the application of ISO 9001:2008 to crop production

The *Food and Agriculture Organization of the United Nations (FAO)* is the only international organization overseeing all aspects of the food chain, thus providing a unique 360° vision on food safety. Its long-standing partnership with the World Health Organization (WHO) enhances this perspective. Through complementary mandates, both the organizations cover a range of issues to support global food safety and protect consumer health. FAO addresses food safety issues along the food production chain, while WHO oversees and maintains strong relationships with the public health sector.

FAO assists member countries in food safety considerations by:

- Strengthening national food control regulatory systems and enhancing member countries participation in Codex
- Providing independent scientific advice through the Joint FAO/WHO Expert Committee on Food Additives (JECFA) and Joint FAO/WHO Expert Meetings on Microbiological Risk Assessment (JEMRA) expert bodies to support standard settings of Codex
- Enhancing food safety management along food chains to prevent diseases and trade disruptions
- Promoting food safety emergency preparedness to build resilient agri-food chains (Emergency Prevention System for Food Safety, EMPRES)
- Developing online platforms for global interactions through networking, databases developments, and tools to support food safety management (International Food Safety Authorities Network, INFOSAN)

2.8.2 Seafood Hazards and Prophylaxis

Biological, chemical, and physical safety hazards may be identified with seafood (Table 2.7).

Prophylactic Measures

- Avoid preparing food far in advance of service.
- Avoid holding cooked foods/thawed foods at room temperature.
- Employ quick chill methods to cool food below 7.2 °C (for *Bacillus*) within 4 h of preparation.
- Store in shallow pans/small quantities <4 inches deep.
- Store hot foods >60 °C until served.
- Reheat food rapidly to>74 °C.
- Proper thermal process and rapid uniform cooling to <10 °C within 2–3 h (*Clostridium* spp.) or heavy salting/drying/fermentation/acidification pH <4.6.
- Prevent cross-contamination and maintain food processing areas, cleaning and sanitizing, and good personal hygiene practices.

Table 2.7 Different types of hazards

Biological	Chemical	Physical
Bacterial:	Marine biotoxins:	Glass
Bacillus cereus	Amnesic shellfish poisoning (ASP)	Metal fragments
Campylobacter jejuni	Diarrhetic shellfish poisoning (DSP)	
Clostridium botulinum	Neurotoxic shellfish poisoning (NSP)	
C. perfringens	Paralytic shellfish poisoning (PSP)	
Pathogenic *E.coli*	Ciguatera fish poisoning (CFP)	
Listeria monocytogenes	Gempylotoxin	
Salmonella spp.	Scombroid toxin	
Shigella spp.	Tetrodotoxin	
Pathogenic *Staphylococcus aureus*	Other chemical hazards	
Vibrio cholerae	Aquaculture drugs	
V. parahaemolyticus	Allergens	
V. vulnificus	Chemical contaminants	
Yersinia enterocolitica	Food additives	
Viral:		
Hepatitis A virus		
Norwalk virus		
Parasites:		
Anisakis simplex		
Pseudoterranova decipiens		
Diphyllobothrium latum		

- The FDA has established three freezing processes (<6inches) to kill parasites: freezing and storing at −20 °C or below for 7 days or freezing at −35 °C or below for 15 h or freezing at −35 °C or below and storing at −20 °C for 24 h.
- Follow HACCP and ISO22000 strictly.
- Timely detection, determination, and documentation; ensure quality assurance by providing useful information.

References

Ahmed, J., Ramaswamy, H. S., Kasapis, S., & Boye, J. I. (2016). *Novel food processing: Effects on rheological and functional properties*. Boca Raton: CRC Press.

AOAC Official Method 997.02. (2000). *Yeast and mold counts in foods*.

Borda, D., Nicolau, A. I., & Rasper, P. (Eds.). (2018). *Trends in fish processing technologies*. CRC Press.

Bourne, M. C. (1990). Basic principles of food texture measurement. In *Dough rheology and baked product texture* (pp. 331–341). Boston: Springer.

Boziaris, I. S. (Ed.). (2013). *Seafood processing: technology, quality and safety*. Hoboken: Wiley.

Castellini, M. A., & Somero, G. N. (1981). Buffering capacity of vertebrate muscle: correlations with potentials for anaerobic function. *Journal of Comparative Physiology, 143*(2), 191–198.

Chandrasekaran, S., Ramanathan, S., & Basak, T. (2013). Microwave food processing—A review. *Food Research International, 52*(1), 243–261.

Cortez-Vega, W. R., Pizato, S., & Prentice, C. (2014). Nutritional quality evaluation of surimi and kamaboko obtained from mechanically separated chicken meat. *Nutrition & Food Science, 44*(6), 483–491.

Eklund, T. (1989). Organic acids and esters. In G. W. Gould (Ed.), *Mechanisms of action of food preservation procedures* (pp. 161–200). London: Elsevier Applied Science.

Fernandes, R. (Ed.). (2009). *Microbiology handbook: Fish and seafood* (p. 258). Royal Society of Chemistry.

Gómez, B., Munekata, P. E., Gavahian, M., Barba, F. J., Martí-Quijal, F. J., Bolumar, T., et al. (2019). Application of pulsed electric fields in meat and fish processing industries: An overview. *Food Research International, 123*, 95.

Hall, G. M. (2012). *Fish processing technology*. Springer.

Huss, H. H. (Ed.). (1995). *Quality and quality changes in fresh fish* (Vol. 348). Rome: FAO.

Jeong, H. S., Chung, H., Song, S. H., Kim, C. I., Lee, J. G., & Kim, Y. S. (2015). Validation and determination of the contents of acetaldehyde and formaldehyde in foods. *Toxicological Research, 31*(3), 273.

Jiao, Y., Tang, J., Wang, Y., & Koral, T. L. (2018). Radio-frequency applications for food processing and safety. *Annual Review of Food Science and Technology, 9*, 105–127.

Kleeberg, U., & Klinger, W. (1982). Sensitive formaldehyde determination with Nash's reagent and a 'tryptophan reaction'. *Journal of Pharmacological Methods, 8*(1), 19–31.

Koutchma, T. (2008). UV light for processing foods. *Ozone: Science and Engineering, 30*(1), 93–98.

Krishnamurthy, K., Khurana, H. K., Soojin, J., Irudayaraj, J., & Demirci, A. (2008). Infrared heating in food processing: an overview. *Comprehensive Reviews in Food Science and Food Safety, 7*(1), 2–13.

Lakshmanan, P. T., Mathew, S., Anandan, R., Asha, K. K., & Chatterjee, N. S. (2013). *Laboratory training manual-biochemical analysis of seafood*. CIFT, Government of India.

Mahapatra, B. K., Pailan, G. H., Subhendu, D., Sardar, P., & Munilkumar, S. (Eds.). (2013). *Fish processing and value added fish products*. CIFT, India.

Manual of methods of analysis of food. (2015). FSSAI, Government of India.

Mohan, C. O., Ravishankar, C. N., & Srinivasagopal, T. K. (2008). Effect of O2 scavenger on the shelf-life of catfish (*Pangasius sutchi*) steaks during chilled storage. *Journal of the Science of Food and Agriculture, 88*(3), 442–448.

Nielsen, S. S. (Ed.). (2010). *Food analysis* (pp. 139–141). New York: Springer.

Panasiuk, R., Amarowicz, R., Kostyra, H., & Sijtsma, L. (1998). Determination of α-amino nitrogen in pea protein hydrolysates: A comparison of three analytical methods. *Food Chemistry, 62*(3), 363–367.

Pranaitytė, B., Jermak, S., Naujalis, E., & Padarauskas, A. (2007). Capillary electrophoretic determination of ammonia using headspace single-drop microextraction. *Microchemical Journal, 86*(1), 48–52.

Pundhir, A., & Murtaza, N. (2015). Hurdle technology-An approach towards food preservation. *International Journal of Current Microbiology and Applied Sciences, 4*, 802–809.

Raman, M., & Mathew, S. (2015). Physiochemical and textural alterations in Indian squid (Loligo duvauceli) mantle during frozen storage and cooking. *Journal of Aquatic Food Product Technology, 24*(5), 454–467.

Rehbein, H., & Oehlenschlager, J. (2009). *Fishery products: Quality, safety and authenticity*. Wiley.

Sandhya. (2010). Modified atmosphere packaging of fresh produce: Current status and future needs. *LWT-Food Science and Technology, 43*(3), 381–392.

Shukla, S., Choi, T. B., Park, H. K., Kim, M., Lee, I. K., & Kim, J. K. (2010). Determination of non-volatile and volatile organic acids in Korean traditional fermented soybean paste (Doenjang). *Food and Chemical Toxicology, 48*(8–9), 2005–2010.

Sun, D. W. (2014). *Emerging technologies for food processing*. Elsevier.

Surendran, P. K., Thampuran, N., Nambiar, N. V., & Lalitha, K. V. (2006). *Laboratory manual on microbiological examination of seafood* (2nd ed.). CIFT, Government of India.

Venugopal, V., Shahidi, F., & Lee, T. C. (1995). Value-added products from underutilized fish species. *Critical Reviews in Food Science & Nutrition, 35*(5), 431–453.

Wang, H., Zhang, F., Cao, J., Zhang, Q., & Chen, Z. (2012). Comparison of Chromatographic and Titrimetric Methods for the Determination of the [alpha]-amino Nitrogen in Standard Solution and Fish Protein Hydrolysates. *Journal of Food Research, 1*(4), 174.

Zhao, Y. M., de Alba, M., Sun, D. W., & Tiwari, B. (2018). Principles and recent applications of novel non-thermal processing technologies for the fish industry—A review. *Critical Reviews in Food Science and Nutrition*, 1–15.

Water/Ice: Assessment of Quality

3

Water is the most important component in fish processing industry and in ice-making. The contamination of water used in seafood industry (sources) will increase the pathogen load in fish posing a serious health hazard to the consumers. Hence, it is urgent to meet the basic requirements of the water/ice quality and should meet the drinking water standards. The World Health Organization (WHO) has issued guidelines for drinking water quality and has identified that very stringent standards cannot be used universally; and hence, a range of guideline values for more than 60 parameters were elaborated. The water/ice quality parameters have to be qualitatively and quantitatively analyzed and monitored regularly for each source of water supply such as at the point of collection (harbor) to the final use (until packaging). The qualified technicians are to perform the analysis. The water sampling and analysis should be certified by ISO-certified laboratories. The different sources of water used in seafood industry could be bore wells, municipal mains, water tanks, reservoirs, and harbor basin water. The contamination of bore wells and municipal mains may occur from corroded casings, sewage, etc. The bore well water needs to be tested every month. Similarly, the tanks and reservoirs need to be checked regularly for the residual chlorine levels. Periodic scrubbing may assist in preventing deposition. The municipal mains, water tanks, and reservoirs have to be tested half yearly. Harbor basin water has to be tested yearly, more accurately at the peak of the dry season and the wet season, when the chances of contamination are high.

The testing procedures include:

(a) Physical tests
(b) Chemical tests
(c) Bacteriological tests

S. Mathew et al., *Fish and Fishery Products Analysis*,
https://doi.org/10.1007/978-981-32-9574-2_3

3.1 Physical and Chemical Characteristics of Water

3.1.1 Determination of pH

Principle The ionization of water results in the formation of hydrogen ion (H^+) and hydroxyl ion (OH^-). Change in the concentration of these would alter the condition of the water. Hence, pH is used as the measure of hydrogen ion concentration. It is an indicator of relative acidity or alkalinity of water. Values of 9.5 and above indicate high alkalinity, while values of 3 and below indicate acidity. Low pH values help in effective chlorination but cause problems with corrosion. Values below 4 generally do not support living organisms. Drinking water should have a pH between 6.5 and 8.5 (APHA 1998).

The pH value of a solution has been defined as the negative log of the hydrogen ion concentration:

$$pH = -\log\left(\left[H^+\right]\right)$$

Procedure The pH meter is set on a flat surface and to calibrate, the electrode of the pH meter is dipped in the buffer solution with known pH value. After calibration, the electrode is washed by dipping into distilled water to get rid of the adhering buffer. This is then gently wiped with a tissue paper and dipped in the sample solution. The reading is noted. The electrode is washed again by dipping in distilled water, and the pH of the sample is measured two more times.

The pH of the given water sample indicates that the sample is acidic/alkaline/neutral in nature.

3.1.2 Determination of Temperature

Principle Temperature of water is dependent on atmospheric and ground condition. It is measured using thermometers in degrees Fahrenheit or Celsius (Centigrade). Different types of thermometers are commercially available for measuring temperature (as listed below) including digital thermometers.

Low temp (°C)	High temp (°C)	Reading (°C)	Length (mm)	Material
−1	51	0.1	460	Mercury
−1	101	0.1	610	Mercury
−1	210	0.1	610	Mercury
−10	110	1	300	Mercury
100	650	2	405	Mercury

(continued)

Low temp (°C)	High temp (°C)	Reading (°C)	Length (mm)	Material
200	1200	5	405	Mercury
−10	500	2	405	Mercury
20	750	5	405	Mercury
20	930	5	405	Mercury
−35	50	1	305	Spirits
−10	260	1	305	Spirits
0	300	2	305	Spirits
20	500	2	405	Spirits
−1	101	0.2	450	Spirits
−1	201	0.5	430	Spirits
−50	50	1	305	Spirits

Water temperature is generally 2–3 °C higher or lower than the ambient temperature. Testing of temperature has, therefore, no practical significance for drinking water. However, it is to be known that for potable water, a temperature of 10 °C is highly desirable, whereas temperatures above 25 °C are considered to be objectionable (Adams 2017).

Procedure Immerse the thermometer in the sample to the proper depth for a correct reading. Record the temperature to the nearest fraction of a degree Celsius (°C) before removing. Record the temperature of water sample at different depth at different time intervals.

3.1.3 Determination of Color

Principle Chemically, pure water has a blue tint. Several surface waters are colored due to disintegration of organics (humic and peaty material), metallic salts, or colored clays. This color is considered as *apparent color* due to the presence of suspended matter, while the *true color* is derived from dissolved inorganic and organic matters. Samples can be centrifuged and/or filtered to remove turbidity in order to measure true color. Water which obtains its color from natural organic matter usually poses no health hazard. However, because of the yellowish brown appearance of such waters, the consumers may not find the water aesthetically acceptable. Many processing industries require low colored water. According to the World Health Organization (WHO), drinking water should meet the standards of <5 color units (1 color unit is defined as a 1 mg/l concentration of platinum in the form of potassium chloroplatinate, K_2PtCl_6) to be highly desirable. The maximum permissible color for domestic supplies is 20 ppm (APHA-AWWA-WPCF 1989).

Procedure There are two methods for the measurement of water color.

(a) Measurement of color is by comparison of the sample with standard color solutions using a spectrophotometer. A straight line calibration curve is initially developed by plotting absorbance versus platinum-cobalt color standard. The absorbance of the sample is measured and the corresponding concentration is read from the calibrated curve. For turbid waters, pretreatment has to be carried out to remove turbidity prior to measuring the true color. The color value depends on the pretreatment methods (centrifugation or filtration) and pH.
(b) The color of a sample is compared with the color of the glass tubes called Neosler tubes, containing standard solutions of different color intensities. Small color intensities are measured by tintometer. The instrument is lighted from inside and contains an eyepiece with two holes. A slide of standard colored water is seen through one hole, and the slide of water to be tested is seen from the other hole. The colored slide is replaced by another till both the slides match. This standard color intensity represents the intensity of color of the test water sample.

3.1.4 Determination of Turbidity

Principle Turbidity in water is caused by suspended materials which absorb and scatter light. These colloidal and finely dispersed turbidity-causing materials do not settle under quiescent conditions and are difficult to remove by sedimentation. Turbidity is also often used as indicative evidence of the possibility of bacteria being present.

Materials

Solution A: Dissolve 1 g hydrazine sulfate $(NH_2)_2.H_2SO_4$ in distilled water and dilute to 100 ml in a volumetric flask. Caution: hydrazine sulfate is a carcinogen; avoid inhalation, ingestion, and skin contact. Formazine (formazin) suspensions can contain residual hydrazine sulfate.

Solution B: Dissolve 10.0 g hexamethylenetetramine, $(CH_2)_6N_4$, in distilled water, and dilute to 100 ml in a volumetric flask.

In a flask, mix 5.0 ml solution A and 5.0 ml solution B. Let stand for 24 h at 25 ± 3 °C. This results in a 400-NTU suspension. Transfer stock suspension into an amber glass for storage. Make dilutions from this stock suspension. The stock suspension is stable for up to 1 year when properly stored.

Procedure Turbidity measurements are performed using nephelometric instruments/HACH turbidimeter and are expressed as nephelometric turbidity units (NTU). It can also be measured visually. The nephelometric apparatus is designed to measure forward scattering of light at 90° to the path of an incandescent light beam. Suspended particles present in a water sample reflect a portion of the incident light off the particle surface. The light reflected at 90° is measured by a photoelectric detector and is compared against light reflected by a reference standard. No interference exists for the turbidity test. A standard unit is that which is produced by 1 mg of finely divided silica in 1 L of distilled water.

Dilute 400 NTU primary standard suspensions with high-quality distilled water immediately before use. Working standards are prepared by dilution of the following quantities of the stock formazin turbidity suspension (nominal 400 NTU) to 200 ml. Secondary standards are standards provided by the manufacturer for calibration.

Dilute to 200 ml	Resultant NTU
10 ml	20
5 ml	10
2 ml	4
0.5 ml	1
0.2 ml	0.4
0.0 ml	0

The sample is filled in a clean sample cell and placed in the cell holder and held in place by the spring clip. Align the dot on the sample cell with the raised mark on the spill ring around the cell holder opening. The sample cell is covered with the light shield and turbidity is read. Degassing is necessary to avoid interference with air bubbles. It can also be measured in formazin turbidity units (FTU) when formazin polymer is used as a reference turbidity standard suspension in place of silica suspension. In some case Jackson turbidity units (JTU) are also used.

The turbidity of the given water sample is expressed as mg/l or ppm.

3.1.5 Determination of Odor and Taste

Principle The dissolved organic materials, inorganic salts, and dissolved gases impart tastes (T) and odor (O) to water. These are subjective tests which may be expressed upon description rather than quantitative results. Although tastes and odor occur together, there are certain compounds (sodium chloride) which impart only taste with no odors. For drinking purpose, water must not contain any undesirable or objectionable T/O. However, some tastes imparted by dissolved oxygen and dissolved carbon dioxide are generally desirable. In order to remove the taste and odor from the raw water, it is essential to ascertain chemical composition of water. Occasionally, the extent of taste and odor of a sample is measured quantitatively by threshold odor number or simple threshold odor. The threshold odor numbers represents the dilution ratio at which the odor is hardly detectable (APHA-AWWA-WPCF 1989).

Procedure Water sample to be tested is gradually diluted in the ratio of 1:4 with distilled water. The mixture at which the detection of odor by human observation is lost is determined. The number of times the sample is diluted represents the threshold odor number.

3.1.6 Determination of Alkalinity

Principle The pH and alkalinity are key water quality parameters in food/seafood processing. Alkalinity is a measure of the acid-neutralizing capacity of dissolved substances in water. Many compounds (carbonates, bicarbonates, and hydroxides) contribute to the alkalinity of water at different pH conditions. In pure water, water molecules dissociate into equal amounts of hydrogen and hydroxyl ions (10^{-7} moles/L).

The alkalinity of water is its quantitative capacity to neutralize acids. The three major forms of alkalinity are:

- Hydroxide alkalinity, [OH^-]
- Carbonate alkalinity, [CO_3^{2-}]
- Bicarbonate alkalinity [HCO_3^-]

The carbonates and bicarbonates react with H^+ to form carbonic acid (H_2CO_3) at pH 4.5. In nature, bicarbonates are the major form of alkalinity resulting from the reactions of CO_2 on calcium and magnesium rocks. Some $CaCO_3$ ($\leq$20 mg/l) go into solution as Ca^{2+} and CO_3^{2-} ions.

Procedure Calibrated pH meter is used for pH measurement. Samples are poured into a clean beaker, and rinsed probe is immersed in the sample (pH mode). Establish equilibrium between probe and sample by stirring to ensure homogeneity. Record the pH value. Rinse the electrode thoroughly with distilled water, and replace the bottle containing potassium chloride (4 M). The pH readings are taken thrice.

The sample is taken in 100 ml Erlenmeyer flask, and three drops of methyl orange indicator solution is added to it. Titrate this solution against 0.02 N H_2SO_4, constantly swirling the flask content until just after the color of the flask content changes from yellow to red. Record the titer volume, and calculate the total alkalinity. The experiments are repeated thrice.

Note: For samples whose initial pH is above 8.3, the titration is made in two steps. In the first step, the titration is conducted until the phenolphthalein indicator end point is reached (pH 8.3) with a color change from pink to colorless. During this first phase, the acid added to the sample reacts with $[OH^-]$ alkalinity and $[CO_3^{2-}]$ alkalinity, while in the second phase, the titration is continued until the methyl orange indicator end point is reached with a color change from yellow to red (pH 4.5). During this phase, the addition of acid changes the HCO^{3-} ions, initially present, as well as those produced by earlier reactions, into carbonic acid. The result of first step is known as *phenolphthalein alkalinity* and the overall titration is known as *total alkalinity* (the amount of acid required to react with all the hydroxide, carbonates, and bicarbonates in the sample), respectively. When the pH of a sample is less than 8.3, a single titration is made to the methyl orange end point. It is possible to determine the various components of alkalinity (hydroxide, carbonate, and bicarbonate fractions) from a combination of titration, pH measurements, and chemical equilibrium equations (APHA-AWWA-WPCF 1989).

Calculation

$$\text{Total alkalinity}\left(\text{mg}/1\,\text{CaCO}_3\right) = A \times N \times 1000 \times 50/\text{ml of sample}$$

where:

A = titer volume
N = normality of H_2SO_4

The total alkalinity of the given water sample is expressed as mg/l $CaCO_3$

3.1.7 Determination of Specific Conductance/Conductivity

Principle Conductivity is the ability of a material to conduct electrical current. The principle by which instruments measure conductivity is simple. The plates/wires are placed in the sample, a potential is applied across them (a sine wave voltage) and the current is measured. Conductivity, the inverse of resistance and is determined from the voltage and current (Ohm's law). The charged ion in the solution facilitates the conductance of electrical current; hence, the conductivity of a solution is proportional to its ion concentration. Ionic concentrations can alter the linear relationship between conductivity and concentration in some highly concentrated solutions. The basic unit of conductivity is the siemens (S) also referred to as *mho*. The cell geometry affects conductivity values; therefore, the standardized measurements are expressed in specific conductivity units (S/cm). The conductivity and salinity are interrelated; and the conductivity measurements together with temperature can be used to calculate salinity values using algorithms.

Specific conductance is one of the most frequently measured and useful water-quality parameters. It correlates the sum of dissolved major ion concentrations in water with a single dissolved-ion concentration (Hem 1970). Thus, it assists in analytical determination of major ions.

Specific conductivity (μS/cm)	Water quality
<250	Excellent
250–750	Good
750–2000	Permissible
2000–3000	Needs treatment
>3000	Unsuitable

Materials

Conductivity water: 0.01 M potassium chloride solution – Dissolved 745.6 mg anhydrous KCl in conductivity water and diluted to 1000 ml in a class A volumetric flask at 25 °C and store in a CO_2-free atmosphere. This is the standard reference solution, which at 25 °C has a conductivity of 1412 μS/cm. It is satisfactory for most samples when the cell has a constant value between 1 and 2 cm^{-1}. For other cell constants, use stronger or weaker KCl solutions as listed below. The KCl solutions <0.001 is unstable because of the influence of carbon dioxide on pure water. Store in glass-stopper borosilicate glass bottle until analysis.

KCl concentration (M) or equivalent/L	Equivalent conductivity($cm^2 ohm^{-1} equiv^{-1}$)	Conductivity, ksμ mho/cm
0	149.9	
0.0001	148.9	14.9
0.0005	147.7	73.9
0.001	146.9	146.9
0.005	143.6	717.5
0.01	141.2	1412
0.02	138.2	2765
0.05	133.3	6667
0.1	128.9	12,890
0.2	124.0	24,800
0.5	117.3	58,670
1	11.9	111,900

Procedure Rinse conductivity cell with at least three portions of 0.01 M KCl solution. Adjust temperature to 25.0° ± 0.1 °C. If a conductivity meter displays resistance, R ohms, measure resistance of this portion and note temperature. Compute cell constant, C.

$$C, cm^{-1} = (0.001412)(R_{KCl})[1 + 0.0191(t - 25)].$$

where:

R_{KCl} = measured resistance in ohms
t = observed temperature in °C

Conductivity meters indicate conductivity directly. Commercial probes commonly contain a temperature sensor. With such instruments, rinse probe three times with 0.01 M KCl. Adjusted temperature compensation dial. With probe in standard KCl solution, adjust reading to 1412 μ/cm. Thus, automatically internal cell constant is adjusted.

Thoroughly rinse cell with one or more portions of sample. Adjust temperature to about 25 °C. Measure sample resistance of conductivity and note temperature to ±0.1 °C.

Values of electrical conductivity:

Sample	μS/cm
Totally pure water	0.055
Typical DI water	0.1
Distilled water	0.5
Raw water	50–100
Domestic "tap" water	500–800
Potable water (max)	1055
Seawater	56,000
Brackish water	1,00,000

3.1.8 Determination of Salinity

Principle Salinity can be defined as the measure of the total amount of salts in grams dissolved per kg of the water. The levels of salt in various types of water from different sources vary extensively. Seawater has high degree of salinity, while freshwater has a very low amount of salinity. The salinity can be measured using refractometer (the degree of light refracted from the surface of the refractometer is an index and used to determine the salinity of a sample). Refractometer is based on the principle that the amount of refraction is caused by the density of the substance. The higher the density, the higher will be the refractive index. It consists of a prism and a beveled-clear plastic cover lens each with a known refractive index. Depending on the salinity of the water samples, it will refract the light through an angle when placed between two lenses of the refractometer that can be viewed against light source.

Procedure Salinity refractometry is a quick and precise method to measure the amount of salts in the water samples. Refractometer can be hand-held or Abbe's refractometer. In case of Abbe's refractometer, calibrate the instrument with water by turning on the light and the refractometer scale knob (to adjust the interface between illuminated and dark regions). Read the refraction index using the telescope scale. Temperature of the sample can be varied by using the slider. Experiment is repeated with different water samples (APHA-AWWA-WPCF 1989).

In case of hand refractometer, few drops of water is poured on the prism and closed with lid. The refractive index reading can be read.

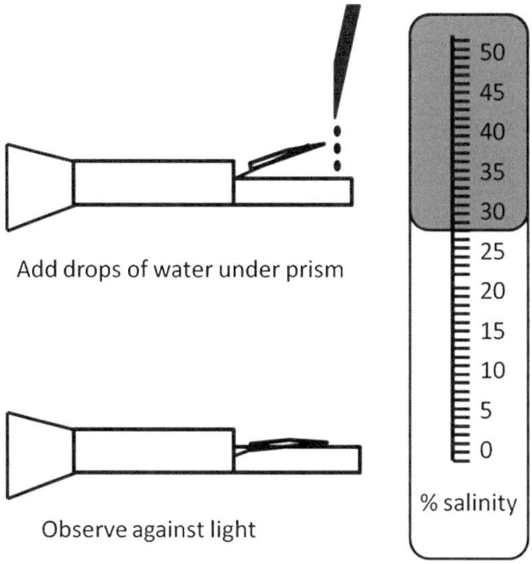

The salinity of the given water sample is expressed as g/kg.

3.1.9 Determination of Total Dissolved Solids

Principle Potable water should possess no taste and odor. Dissolved solids refer to minerals, salts, metals, cations, or anions dissolved in water. Total dissolved solids (TDS) comprise of inorganic salts (principally calcium, magnesium, potassium, sodium, bicarbonates, chlorides, and sulfates) and some small amounts of organic matter that are dissolved in water. The water used in processing industries should be free of total solids. The minerals dissolved in water tend to concentrate in the central part of ice blocks and give a cloudy appearance in the center which is undesirable. Hence, in general, total dissolved solids concentration is the sum of the cations (positively charged) and anions (negatively charged) in the water.

Cations with carbonates and chlorides	Effects on water
Cations combined with $CaCO_3$, $MgCO_3$	Associated with hardness, scale formation, bitter taste
Cations combined with NaCl, KCl	Salty or brackish taste, increase corrosiveness

Water quality	TDS/ppm
Fresh	<1000
Slightly saline	1000–3000
Moderately saline	3000–10,000
Very saline	10,000–35,000
Brine	>35,000

Procedure Stir the water sample thoroughly using a magnetic stirrer. Pipette a known volume and filter using Whatman filter paper (No. 1) in vacuum condition. Wash it thrice with 10 ml reagent-grade water. Transfer the total filtrate (with washings) to a pre-weighed evaporating dish and evaporate to dryness on a steam bath or in a drying oven (1 h at 180 ± 2 °C). Cool in a desiccator and weigh (APHA 1998; Sluiter et al. 2008).

Calculation

$$\text{Total dissolved solids}\left(\text{mg/L}\right)=\left(A-B\right)\times 1000/\text{sample volume}$$

A = weight of dried residue + dish in mg
B = weight of dish in mg

The total dissolved solids in the given water sample is expressed as mg/l (ppm).

3.1.10 Determination of Total Suspended Solids

Principle Total suspended solids are a measure of the undissolved solid matter in the water that remains on the surface of a filter after all the water has been evaporated. Suspended solids affect water quality.

Procedure A known volume of a well-mixed sample is filtered through a standard glass fiber filter having filter disk on the filter support screen of the filtration apparatus. Apply low to moderate vacuum and rinse the filter thrice with ≥30 ml distilled water. Collect the filter paper containing solid residue on the surface of the filter using forceps and place in an evaporating pan. The filter and residue is evaporated to a constant weight condition in an oven maintained at a temperature of 103–105 °C. The mass of the dried residue is determined and used to calculate the concentration of total suspended solids in the sample. This method is applicable for the measurement of total suspended solids in all natural waters and in raw, processed, and treated agricultural, municipal, and industrial wastewaters. The volatile suspended solids can be determined by igniting the filter paper in a pre-weighed crucible using muffle furnace (550 ± 50 °C) for 15 min (APHA 1998).

Calculation

$$\text{Total Suspended Solids}\left(\text{mg TSS/L}\right)=\left(A-B\right)/S\times 1000$$

where:

A = weight of filter + dried residue
B = filter weight
S = sample volume

The total suspended solids (TSS) in the given water sample is expressed as mg/l or mg TSS/l.

3.1.11 Determination of Hardness

Principle Hardness in water is defined as the characteristics that prevent formation of sufficient lather or foam when such water is mixed with soaps. Hardness due to presence of bicarbonates and carbonates of calcium and magnesium are known as *temporary hardness* as these could be removed by simple boiling or by adding lime to water. Hardness due to presence of sulfates, chlorides, and nitrates of calcium or magnesium that do not go by boiling and requires special treatment are termed as *permanent hardness*. Hardness is commonly defined as the calcium carbonate equivalent of calcium and magnesium ions present in water and is expressed in mg/l. It is also expressed as degree of hardness. Each British degree of hardness is equal to 14.25 mg/1 (https://www.canterbury.ac.nz/media/documents/science-outreach/magnesium_calcium.pdf).

$$\begin{aligned}&\text{Total hardness}\left(\text{mg}/1\right)\text{of CaCO}_3\\&\quad=\left\{\text{Ca}^{2+}\text{ in mg}/1\times\text{combining weight of CaCO}_3/\text{combining weight of Ca}^{2+}\right\}\\&\qquad+\left\{\text{Mg}^{2+}\text{ in mg}/1\times\text{combining weight of CaCO}_3/\text{combining weight of Mg}^{2+}\right\}\end{aligned}$$

The combining weight of Ca^{2+}, Mg^{2+}, and $CaCO_3$ are 20, 12, and 50, respectively.

$$\therefore\ \text{Total hardness}\left(\text{mg}/1\right)=\text{Ca}^{2+}\ \text{mg}/1\times 50/20+\text{Mg}^{2+}\ \text{mg}/1\times 50/12$$

Hardness of water

Range (mg/l)	Hardness level
0–50	Soft
50–100	Moderately soft
100–150	Slightly hard
150–200	Moderately hard
Over 200	Hard
Over 300	Very hard

Treating water with hardness below 150 mg/l is not economical.

When hardness is greater than the sum of carbonate and bicarbonate alkalinity, amount of hardness equivalent to the total alkalinity is called *carbonate hardness*.

Carbonate hardness (mg/l) = alkalinity

When, alkalinity > total hardness:

Carbonate hardness (mg/l) = total hardness

The amount of hardness in excess of this is called *noncarbonate hardness* (NCH). These are associated with sulfate chloride and nitrate ions.

Non-carbonate hardness, (mg/l) = total hardness-carbonate hardness

Procedure

The first method is calculation based method and the second method is titration method using EDTA.

(i) Calculation method

For this method, concentration of cations should be known and then all concentrations are expressed in terms of $CaCO_3$ using the following equation:

$$\text{Hardness}\left(\text{in mg/l as }CaCO_3\right) = \left[M^{2+}\left(\text{in mg/l}\right) \times 50\right] / \left(\text{E.Wt.of }M^{2+}\right)$$

where:

M^{2+} = mass of divalent ions (mg/l)
E. Wt. = equivalent weight of divalent ions (g/mole)

(ii) Titrimetric Method: This method uses ethylenediaminetetraacetic acid (EDTA), a chelating agent, which forms complex ions with $Ca^{2+}$and Mg^{2+} and other divalent ions causing hardness. The successful use of EDTA for determining hardness depends on presence of an indicator which can show presence of excess EDTA in solution or when all the ions present in solution has formed complex. Eriochrome Black T (EBT) (blue color solution) serves as an excellent indicator to show when all hardness ions have been consumed. When small amount of EBT is added to hard water with pH > 10, it combines with Ca^{2+} and Mg^{2+} ions to form weak complex ions (wine-red color solution). During the titration with EDTA, all free hardness ions are complexed, and, subsequently, EDTA disrupts the wine-red complex as it can form a stable complex with the hardness ions. At this stage, solution color changes from red wine color to blue color, indicating the end of the titration.

Ca-Hardness and Total Hardness can be determined by titration method. Separate samples are used for each measurement.

Total Hardness EDTA is standardized with magnesium chloride using EBT indicator in the presence of ammonia buffer. The end point is marked with permanent pink color change from blue. Calculate the number of moles used.

Taken 100 ml of the water sample and added 20 ml 0.005 mol/l EDTA solution, 10 ml of ammoniacal buffer, distilled water (50 ml) and 2–3 drops of EBT. Titrate it against 0.0025 mol/l magnesium chloride solution until a permanent pink color appears. Note down the titer volume used.

Ca-Hardness: Take 50 ml of the sample and add 1 ml sodium hydroxide solution (8%) and add pinch of Mercurex Powder. Titrate with standard EDTA solution until the light pink color of solution converts into light blue color.

Calculation

$$\text{Hardness}\left(\text{mg/l as CaCO}_3\right)=\left(V \times N \times 50 \times 1000\right)/\left(\text{SV}\right)$$

where:

V = volume of titrant (ml)
N = normality of EDTA
50 = equivalent weight of $CaCO_3$
SV = sample volume (ml)

The total hardness and Ca-hardness in the given water sample are expressed as mg/l as $CaCO_3$.

3.1.12 Determination of Chloride, Fluoride, Sulfate, Nitrite, and Phosphate

Chloride

*Principle:*Chlorides are mostly present as inorganic anion in natural waters (derived from natural mineral deposits, agricultural or irrigation discharge, sewage and industrial effluents) and widely range in the concentration. It is mainly due to sodium chloride, and to some extent, chlorides of calcium and magnesium as these are most stable in water. Most of the rivers and lakes have chloride concentration less than 50 mg/1. For potable water, the chloride content is restricted to 200 mg/l. High chloride content in water may harm metallic pipes and structures.

Procedure Take 25 ml sample in a conical flask and measure the pH. Add1.0 ml indicator solution and titrate against standard silver nitrate solution (N/71 or 0.0141 N) to pinkish yellow end point. The titer volume and pH of the sample are measured. [Note: The pH must be in the range of 7–8 as silver ions are precipitated as AgOH at high pH levels, and the chromate ions are converted to $Cr_2O_7^{2-}$ at low pH values].

Calculation

$$\text{Chloride Ion}\left(\text{mg/l}\right)=\left(A\times N\times 35.45\right)*1000/\text{volume of sample}$$

where:

A = titer volume
N = normality of silver nitrate

The chloride ion concentration in the given water sample is expressed as mg/l.

Fluoride
Principle: Fluoride occurs sometimes naturally in water, either through contamination or during treatment. Fluorination of water supplies to a level of 1 mg/l is safe (in public water, it ranges between 0.7 and 1.2 mg/l) and effective in reducing dental caries. Excess amounts of fluorine (>10 mg/l) consumption through water leads to fluorosis, which causes mottling of the teeth and sometimes bone damage both in children and adults. Groundwater passing through chalk or limestone under clay formations may have high concentration of fluoride. Since specialized treatment is to be applied to control higher concentration of fluoride, the treatment becomes quite costly, and it is better to go for some other source containing lesser fluoride content. Fluoride concentrations in water samples and appropriate aqueous standards are measured electrometrically using the ion-specific electrode. The limit of detection (LOD) of the electrode is 0.019 mg/l, a concentration far below the concentration typically found in water. The linear portion of the standard curve extends from 0.1 to 19,000 mg/l at the electrode (Danielle and Whitford 2014).

Materials
Total Ionic Strength Acetate Buffer I (TISABI): This buffer solution is added to all standards and water samples prior to analysis with the ion-specific electrode for the purpose of adjusting the pH and ionic strength of all solutions to uniform values.

To 700 ml of distilled water, add 57 ml of glacial acetic acid, 58 g of sodium chloride, and 0.30 g of sodium citrate. The solution will be warm and strongly acidic. Titrate this to pH 5.2 using 10 N sodium hydroxide and make up to a final volume of 1 l. Transfer the solution to a labeled plastic bottle for future use.

TISAB II: It differs from TISAB I only in that it contains CDTA (2-[[2-[bis(carboxymethyl)amino]cyclohexyl]-(carboxymethyl)amino]acetic acid), a chelator of aluminum and several other divalent cations that may form complexes with fluoride. It is recommended that TISAB II be used for samples suspected or known to contain significant amounts of aluminum which can complex fluoride ions and renders them undetectable by the fluoride electrode.

Stock Standards: Stock solution is prepared by dissolving 221.1 mg of sodium fluoride in 1 l water (fluoride concentration 100.0 mg/l). Serial dilute it to various concentrations 0.25–50 mg/l and store it at room temperature. Analytical standards: three sets of three analytical standards are prepared for each analytical run. These standards are used to generate the standard curve from which the concentrations of fluoride in the water samples are determined. The fluoride stock standards used are 0.50, 1.00 and 2.50 mg/l. The standards are prepared by adding 250 μl TISAB (I or II) to 500 μL of each stock standard solution. Similarly, the water samples are also prepared for analysis. One set of the three analytical standards is read at the beginning of the run, one set is read at or near the midpoint of the run, and one set is read at the end of the run.

Procedure All the standards and water samples are placed in beaker and mixed well. The fluoride and reference electrodes are placed in the standard solutions starting with the lowest and then progressively to higher standards. Rinse with distilled water and dry the electrodes between exposures to the solutions. This has to be repeated 3–5 times until stable readings are obtained. Then read the TISAB-buffered samples by placing the electrodes in 750 μL buffered solution. Record the pH (should be between 5 and 6). Lower pH values cause the formation of hydrofluoric acid which is not detected by the electrode. High pH values cause an overestimation of the fluoride concentration because the electrode responds equally well to hydroxyl ions.

Calculations

Enter the concentrations of the fluoride standards and corresponding millivolt readings in two separate columns. Then convert the standard values to natural logarithms (ln) and perform linear regression analysis with millivolts of the standards and samples on the X axis (independent variable) and the ln (natural logarithm) of the standards on the Y axis (dependent variable). Calculate the predicted values. The output will be the natural logarithms of the amounts of fluoride in the samples. Calculate the concentrations of fluoride in the samples by taking the antilogarithms.

The chloride ion concentration in the given water sample is expressed as mg/l.

Sulfates

Principle: The sulfates are found in appreciable quantity in all natural waters and may come from different sources. Dissolution of gypsum and other mineral deposits containing sulfates (due to the oxidation of sulfides, sulfites, and industrial effluents) are the main cause of its increase in water. Sulfates contribute permanent hardness of water imparting bitter taste and amalgamation with magnesium, and sodium has a laxative effect. The sulfates may undergo bacterial degradation under anaerobic conditions to produce hydrogen sulfide.

$$SO_4^{2-} + \text{organic matter} \rightarrow S^{2-} + H_2O + CO_2 \ (\text{anaerobic reaction})$$

$$S^{2-} + H^+ \leftrightarrow HS^-$$

$$HS^- + H^+ \leftrightarrow H_2S$$

The sulfates in water can be determined by:

- Gravimetric method with ignition of residue
- Gravimetric method with drying of residue
- Turbidimetric method

- *Gravimetric method with ignition of residue*

Principle Sulfate gets precipitated as barium sulfates in the presence of hydrochloric acid and barium chloride. This is carried out near the boiling temperature, and after digestion the precipitate is filtered, washed, ignited, and weighed as barium sulfates.

Procedure Take 250 ml water sample in a conical flask and adjust the acidity with HCl to 4.5–5 with methyl red as indicator. Add excess HCl, and heat the solution to boiling, and add little excess barium chloride under gentle stirring. Digest the precipitate at 80–90 °C (overnight), and filter using an ashless Whatman filter paper. Wash the precipitate with warm distilled water (free of chloride) and test for chloride using silver nitrate-nitric acid reagent. Place the precipitate and filter paper in a pre-weighed crucible dry, and ignite at 800 °C for 1 h. Cool in a desiccator, and the ash weight is represented as barium sulfate precipitate (https://biocyclopedia.com/index/enivronmental_science_engineering_laboratory_methodology/determination_of_sulphate_and_sulphide_in_water.php).

- *Gravimetric Method with Drying of Residue*

If organic matter is not present in the sample, first method can be done without igniting and instead drying the residue and weighing.

- *Turbidimetric Method*

Principle The turbidimetric method is based on the fact that barium sulfate tends to precipitate in a colloidal form, which is enhanced in presence of a sodium chloride-hydrochloric acid solution containing glycerol and other organic compounds. The absorbance of the barium sulfate solution is measured by a nephelometer or turbidimeter, and the sulfate ion concentration is determined by comparison with a standard curve.

Materials

Conditioning reagent: Mix 50 ml glycerol with a solution containing 30 ml concentrated hydrochloric acid, 300 ml distilled water, 100 ml 95% ethyl alcohol, and 75 g sodium chloride.

Barium chloride crystals.

Standard sulfate solution (100 μg SO_4/ml): Dissolve 147.9 mg anhydrous Na_2SO_4 in 500 ml distilled water and dilute to 1 l.

Procedure Measure 100 ml sample into a 250 ml Erlenmeyer flask, and add conditioning reagent (5 ml), and mix well using stirrer. Add excess barium chloride crystals, and stir at constant speed for 1 min. After stirring, pour an aliquot into the absorption cell of the photometer, and measure the turbidity at 30 s intervals for 4 min. Usually maximum turbidity occurs within 2 min, and the reading remains constant thereafter for 3–10 min. Prepare a calibration curve (absorbance versus concentration) using standards (0, 5, 10, 15, 20, 25, 30, 35, and 40 mg/l sulfate range), and determine the concentration of sulfates in the sample solution.

Calculation

$$\text{Sulphate}\left(\text{mg/l}\right)=\text{BaSO}_4\left(\text{mg}\right)/\text{volume of sample}\left(\text{ml}\right)\times 411.6$$

Sulfite

Principle: Sulfites occur in groundwater especially in hot springs and wastewater. Hydrogen sulfite escaping into the air from sulfite-containing wastewater causes odor. It is highly toxic and causes corrosion of sewers and pipes. Sulfites include H_2S, HS, and acid soluble metallic sulfites (present in the suspended matter). Iodine reacts with sulfite in acid solution, oxidizing it to sulfur; a titration based on this reaction is an accurate method for determining sulfites at concentration above 1 mg/l, if interferences are absent and if loss of H_2S is avoided.

Materials

Standard iodine solution (0.025 N): Dissolve 20–25g potassium iodide in a little water and add 3.2 g iodine. After dissolution, it is diluted to 1 l and standardized against 0.025 N sodium thiosulfate using starch as indicator.

Standard sodium thiosulfate (0.025 N): Dissolve 6.205 g of sodium thiosulfate in distilled water and make up to 1 l. Standardize this with 0.025 N potassium dichromate (1.226 g potassium dichromate in 1 l).

Starch indicator: Mix 5 g starch to 800 ml boiling water; stir thoroughly and dilute to 1 l.

Procedure Take 10 ml of standard iodine solution into a 500 ml flask, and add 10 ml distilled water. Following which, add HCl (6 N) and 200 ml sample. Titrate against standard sodium thiosulfate with starch as indicator, and record the end point (disappearance of blue color) (https://biocyclopedia.com/index/enivronmental_science_engineering_laboratory_methodology/determination_of_sulphate_and_sulphide_in_water.php).

Calculation

$$\text{Sulphite}\left(\text{mg/l}\right)=400\left(a-b\right)/\text{sample volume}\left(\text{ml}\right)$$

where:

a = iodine solution used
b = sodium thiosulfate solution used

The sulfite ion concentration in the given water sample is expressed as mg/l.

Nitrite

Nitrate and nitrite are naturally occurring ions and are part of the nitrogen cycle. The nitrate ions (NO^{3-}) are the stable form and chemically unreactive that can be reduced by microbial action. Nitrate is used mainly in inorganic fertilizers. It is also used as an oxidizing agent and in the production of explosives. The purified potassium nitrate is used for glassmaking. Sodium nitrite is used as food preservative, especially in cured meats. Nitrate is sometimes also added to food to serve as a reservoir for nitrite. Nitrate can occur naturally in surface and groundwater at levels that may not be harmful for human health. The nitrate deposits increase in water sources as a result from its location/improper construction, overuse of chemical fertilizers, or improper disposal of human and animal waste. Nitrite can also be formed chemically in distribution pipes by *Nitrosomonas* bacteria during stagnation of nitrate-containing and oxygen-deficient water or if the chlorination process is not well controlled.

Materials

Sulfanilamide solution: Dissolve 5 g of sulfanilamide in a mixture of 50 ml concentrated hydrochloric acid (HCl) and about 300 ml of deionized distilled water. Dilute to 500 ml with deionized distilled water. The solution is stable for several months.

N-(1-naphthyl)ethylenediaminedihydrochloride: Dissolve 0.5 g of the dihydrochloride in 500 ml of deionized distilled water. Store the solution in a dark bottle.

Primary standard: Dissolve in 0.345 g of sodium nitrite ($NaNO_2$) in 1 l of deionized distilled water.

Procedure

(a) Running standard: Dilute 5 ml of primary standard to 500 ml deionized distilled water. Use fresh.

(b) Make three nitrite standards: Pipette 2 ml of the running standard into three 50 ml plastic graduated cylinders with caps. Fill each to 50 ml with deionized distilled water.
(c) Make two reagent blanks: Fill two 50 ml plastic graduated cylinders with deionized distilled water.
(d) Nitrite standard and reagent blank analyses: Add 1 ml sulfanilamide to each Nitrite Standard and each Reagent Blank; and mix thoroughly (2–10 min). Add 1 ml of N-(1-naphthyl) ethylenediaminedihydrochloride solution to each Nitrite standard and each reagent blank and mix thoroughly. Allow to react for at least 10 min, but no more than 2 h. N-(1-naphthyl) ethylenediaminedihydrochloride reacts with the diazo compound to create a pink azo dye. This dye has an absorbance of 543 nm. Measure the absorbance at 543 nm using a spectrophotometer with distilled water as blank. Record the values.
(e) Water sample analyses: Fill plastic graduated cylinders with 50 ml water samples. Add 1 ml sulfanilamide and mix well. Add 1 ml of N-(1-naphthyl) ethylenediaminedihydrochloride solution and mix thoroughly. Measure the absorbance at 543 nm and record the values. Turbidity blank determination for each water sample is also recorded (APHA-AWWA-WPCF 1989).

Calculations

$$\text{Calibration}\left(\text{F}\right)\text{Factor,F}_{1\text{cm}}=2.00/\left[\left(\text{ABS}_{\text{stnd}}-\text{ABS}_{\text{rb}}\right)\times 10\right]$$

This value should be around 2, if not experiments with new standard has to be repeated.

$$\text{Corrected Sample Absorbance}=\text{ABS}_{\text{sample}}-\text{ABS}_{\text{rb}}-\text{ABS}_{\text{turbidity}}$$

$$\text{\\mgNO}_2\text{/l}=\left(\text{Corrected Sample ABS}\right)\times \text{F}_{1\text{cm}}$$

The nitrite ion concentration in the water sample is expressed as µg/l.

Phosphate

Principle: Phosphate is considered to be one the most important nutrients in natural water and is nontoxic to plants or animals. In fact, it is a plant nutrient which stimulates the growth of aquatic weeds and algae. The growth of algae in natural water will rarely occur at phosphate concentrations below 0.05 mg/dm^3. Drinking water may have a maximum allowable phosphate content of 0.3 mg/dm^3, while on average, raw sewage contains about 30 mg/dm^3.

The phosphate found in natural waters mainly exists as orthophosphate (PO_4^{3-}) and polyphosphates ($P_2O_7^{4-}$ and $P_3O_{10}^{5-}$). The polyphosphate are hydrolyzed to produce the orthophosphate. The principle of phosphate estimation involves the formation of molybdophosphoric acid, which is reduced to the intensely colored complex, molybdenum blue. The intensity is directly proportional to the concentration of phosphate in the solution. This analytical method is extremely sensitive and is reliable at a concentration of 0.1 mg of phosphorus per liter of water.

Materials

Combined reagent: Prepared by combining 500 ml of 2.5 M H_2SO_4, 50 ml potassium antimony tartrate solution (dissolve 1.371 g $K(SbO)C_4H_4O_6.0.5H_2O$ in about 400 ml distilled water and dilute to 500 ml), 150 ml ammonium molybdate solution (20 g $(NH_4)_6Mo_7O_6.4H_2O$ in 500 ml of water), and 300 ml ascorbic acid solution (made by dissolving 5.28 g of ascorbic acid in 300 ml of H_2O). The solution is prepared fresh every day.

Procedure Preparation of standard solutions: Prepare a 5-point calibration curve with concentrations ranging from 0 to 5 mg/l phosphate (including blank) from the stock standard solution. Take water samples in another volumetric flask (100 ml). Add 50 ml distilled water and mix thoroughly. Add 13 ml combined reagent, shake thoroughly, and allow for color development (30 min). Make up to 100 ml using distilled water, and absorbance is recorded at 880 nm (APHA-AWWA-WPCF 1989).

The phosphate ion concentration in the water sample is expressed as mg/l (ppm)

3.2 Toxic Components

3.2.1 Determination of Copper, Chromium, Cadmium, Zinc, Lead, Mercury, Iron, and Manganese

Principle Metals in water may be present in any forms such as dissolved metals (pass through 0.45 μm membrane filter), suspended metals (retained by 0.45 μm membrane filter), total metals (sum of dissolved and suspended), and acid extractable metals (retained in water after hot diluted mineral acid treatment). These metals can be determined by atomic absorption spectrophotometry (AAS) (Smith 1983). The principle, technical details of the instrument, and different types of the instrument are discussed in Chap. 5 (Table 3.1).

Materials

Cesium-lanthanum interference suppressant solution: Add 400 mi of conc. hydrochloric acid slowly and carefully to 47 g lanthanum oxide in a 2 l Pyrex beaker. Stir the mixture until the lanthanum oxide is dissolved, then add 12.7 g cesium chloride, and again stir until dissolved. Cool the solution and add 800 ml deionized distilled water. Dilute to 2 l with deionized distilled water. This solution contains 5 g/l Cs and 20 g/l La.

Cadmium, chromium, copper, iron, lead, manganese, and zinc standard solutions:

Secondary stock solution 1: Transfer 100 ml of the standard stock solutions (1000 mg/l) of chromium, copper, iron, lead, manganese, and zinc and 20 ml of

Table 3.1 Recommended wavelengths and flame gases

Element	Wavelength (nm)	Flame gases
Copper	324.7	Air-acetylene
Chromium	557.9	Air-acetylene or nitrogen oxide-acetylene
Cadmium	228.8	Air-acetylene
Zinc	213.9	Air-acetylene
Lead	217.0/283.3	Air-acetylene
Mercury	253.7	Cold vapor generation
Iron	248.3	Air-acetylene
Manganese	279.5	Air-acetylene

the standard stock solution (1000 mg/l) of cadmium to standard volumetric flask and diluted to 1 l with deionized distilled water. This solution contains 20 mg/l Cd and 100 mg/l, Cr, Cu, Fe, Pb, Mn, Zn.

Secondary stock solution 2: Transfer 100 ml of secondary stock solution 1 to 1 l standard volumetric flask and dilute with deionized distilled water. This solution contains 2 mg/l Cd and 10 mg/l Cr, Cu, Fe, Pb, Mn, and Zn.

Add the following quantities of secondary stock solutions 1 and 2 to 250 ml standard volumetric flasks, each containing 2.5 ml nitric acid and 25 ml cesium-lanthanum solution, and dilute to 250 ml with deionized distilled water.

Standard no.	Secondary stock solution no.	Volume of secondary stock solution added (m)	Concentration (μg/l)	
			Cadmium	Others
1		0	0	0
2	2	0.63	5	25
3	2	2.5	20	100
4	2	2.5	100	500
5	1	2.5	200	1000
6	1	5.0	400	2000
7	1	7.5	600	3000
8	1	10.0	800	4000
9	1	12.5	1000	5000

Procedure

Pretreatment of samples: Add 5 ml of cesium-lanthanum solution to a clean dry 50 ml standard volumetric flask and dilute.

Aspirate the standard and sample solutions and note the absorbance values obtained. Choice of either the air-acetylene or the nitrous oxide-acetylene flame for chromium determinations will be dependent on the sample constitution. Calculate the concentration of each metal ion with reference to the calibration curves obtained by plotting concentrations of the standard solutions versus the corresponding absorbance readings (https://www.epa.gov/sites/production/files/2016-06/documents/npwdr_complete_table.pdf).

Mercury **(Cold Vapor Generation)**

Materials

Sulfuric acid (500 ml/l)
Potassium permanganate solution (50 g/l)
Potassium persulfate solution (50 g/l)
Sodium chloride-hydroxylamine sulfate solution (120 g/l
Stannous chloride solution (100 g/l)
Secondary stock solution: Transfer1 ml of the standard stock solution (1000 mg/l) of mercury to 1I standard volumetric flask, and dilute with deionized distilled water. This solution contains 1 mg/lHg. Add the following quantities of secondary stock solution to 250 ml standard volumetric flasks, each containing 2.5 ml nitric acid, and dilute to 250 ml with deionized distilled water.

Standard no.	Volume of secondary stock solution added (ml)	Concentrations (μg/l)
1	0	0
2	0.5	2
3	1.5	6
4	2.5	10
5	5	20

Pretreatment of samples and standard solutions: Transfer 100 ml of each of the sample and standard solutions to a 250 ml Erlenmeyer flask. Add 10 ml sulfuric acid, 2.5 ml nitric acid, and 15 ml potassium permanganate solution to each flask and mix well. Allow to stand for 15 min (color of potassium permanganate is retained). Add 8 ml potassium persulfate solution to each flask, and heat for 2 on a water bath at 95 °C. Cool and add 6 ml sodium chloride-hydroxylamine sulfate solution or more to reduce the excess potassium permanganate until colorless.

Procedure Set the operating parameters of the instrument and connect the mercury vapor generation accessory to the instrument. Transfer 50 ml of the treated sample or standard solution to the reaction vessel of the accessory, and add 3 ml stannous chloride solution. Immediately insert the assembly in the reaction vessel, and stir the solution vigorously with the magnetic stirrer for 90 s. Record the absorbance obtained as peak height on the chart recorder. Calculate the concentration of mercury in each sample by reference to the calibration curve. Total (organic + inorganic) mercury is determined by this technique (https://www.epa.gov/sites/production/files/2016-06/documents/npwdr_complete_table.pdf).

3.2.2 Determination of Pesticides

Principle Pesticides are chemicals that kill or control pests such as weeds, rodents, fungi, insects, and nematodes. These are toxic. These are classified into four main categories, herbicides, fungicides, insecticides, and nematocides. These pose as a potential danger if they are consumed in large quantities. Pesticides reach water-bearing aquifers below ground from applications onto crop fields, seepage of contaminated surface water, accidental spills and leaks, improper disposal, and even through injection waste material into wells. Short-term high exposure can result in vomiting, stomach pain, eye irritation, acute gastrointestinal distress, headaches dizziness, and even seizures. Lower-level long-term exposure can cause liver and lung cancers, genetic mutations, and deformities in fetuses. Different screening methods can be used to analyze pesticides/residues in water samples. Recently several kits have been developed to make it convenient to analyze the pesticides.

Materials

Potassium hydroxide (KOH, 20%)
Hydrochloric acid (6 N)
Sodium chloride solution (5% (w/v)
Sodium sulfate solution (2% (w/v), pH 8.5–9.0)
GPC calibration solution (Prepare a solution containing 2.5 mg/ml corn oil, 0.05 mg/ml bis(2-ethylhexyl) phthalate (BEHP), 0.01 mg/ml methoxychlor, 0.002 mg/ml perylene, and 0.008 mg/ml sulfur, or at concentrations appropriate to the response of the detector).

Column descriptions

Column	Gel permeation (600–700 mm height × 25 mm Φ), packed with 70 g of 200–400 mesh SX-3 bio-beads
Syringe	10 ml with lucr fitting
Syringe filter holder	Stainless steel/ glass fiber/ fluoropolymer
Detectors	UV (254 nm), preparative or semi-preparative flow cell
Column	Reverse-phase high-performance liquid chromatograph, 100 × 4.6 mm, 5 Φ mm particle size
Injector, pump, autosampler, detector, fraction collector	
Column	GC
Injector	Split/splitless
Conditions	Helium gas as carrier, 2 min @ 200 kPa injector temperature 180–220 °C or temperature programmed maximum column temperature 300 °C GC temperature program initial temperature and hold 50 °C for 1 min initial ramp 50–180 °C @ 10 °C per minute second hold 180 °C for 0 min second ramp 180–200 °C @ 1.5 °C per minute third hold 200 °C for 2 min third ramp 200–295 °C @ 6 °C per minute final hold 295 °C for 1 min or until methoxychlor elutes Interface temperature 290 °C mass spectrometer conditions source temperature 250 °C Electron energy 35 eV trap current 500–900 ΦA mass resolution 8000 detector potential 340–400 V

Sample preparation: Following table suggests the sample preparation (stock and spikes), phases and suggested quantities for extraction of various sample matrices (Tables 3.2 and 3.3).

Continuous liquid/liquid extraction: Place 100–150 ml methylene chloride in each continuous extractor and 200–300 ml in each distilling flask. Pour the sample(s)/blank/aliquots into the extractors, rinse with 50–100 ml methylene chloride, and add to the respective extractors. Heat the flask until the methylene chloride is boiling, and adjust such that 1–2 drops of methylene chloride fall from the condenser tip into the water in 1 s. Extract for 16–24 h. Remove the distilling flask, estimate and record the volume of extract (100 ml), and pour the contents through a drying column containing 7–10 cm of granular anhydrous sodium sulfate into the concentration flask. Rinse the distilling flask with 30–50 ml of methylene chloride, and pour through the drying column. Add 1 ml of toluene "keeper" to the extract and concentrate.

Procedure

The organochlorine, organophosphorus, triazine, and pyrethroid pesticides in water samples can be detected by high-resolution gas chromatography/high-resolution mass spectrometry (HRGC/HRMS).

Extraction of aqueous samples (samples containing <1% solids): Stable isotopically labeled analogs of the pesticides are spiked into a 1 l sample. The sample is extracted at neutral pH with methylene chloride using separating funnel extraction (SFE) or continuous liquid/liquid extraction (CLLE). Extracts are macro-concentrated using rotary evaporation, a heating mantle, or an evaporator and injected into the HRGC/HRMS (20 μl). Prior to injection, the samples are cleaned up using an aminopropyl SPE column followed by a microsilica column and gel permeation chromatography (GPC) or solid-phase cartridge techniques. Extracts in which the organochlorine pesticides only are to be determined are cleaned up using silica gel, Florisil, or alumina chromatography. A labeled injection internal standard is added to each extract. The analytes are separated by the GC and detected by a high-resolution (≥8000) mass spectrometer. Two exact m/z's (mass/charge) for each pesticide are monitored throughout a predetermined retention time window. The individual pesticides are identified by comparing the GC retention time and ion abundance ratio of two exact m/z's with the corresponding retention time of an authentic standard and the theoretical or acquired ion-abundance ratio of the two exact m/z's. Quantitative analysis is performed using selected ion current profile (SICP) areas. For pesticides, for which a labeled analog is available, the GC/HRMS is multipoint calibrated, and the concentration is determined using the isotope dilution technique. Pesticides, for which a labeled analog is not available, are determined using the internal standard technique. The labeled compounds are used as internal standards, affording recovery correction for all pesticides. The quality of the analysis is assured through reproducible calibration (https://www.epa.gov/sites/production/files/2016-06/documents/npwdr_complete_table.pdf).

Table 3.2 Preparation of stock solutions and spikes

Pesticide	Stock (ng/ml)	Spiking solution (pg/ml)	In 20 μL extract (ng/ml; pg/μl)
Tecnazene	800	800	40
Hexachlorobenzene	800	800	40
Quintozene	1600	1600	80
Heptachlor	600	600	30
Alpha-BHC	1200	1200	60
Gamma-BHC (Lindane)	1200	1200	60
Beta-BHC	1200	1200	60
Delta-BHC	1200	1200	60
Aldrin	1200	1200	60
Dacthal	400	400	20
Octachlorostyrene	600	600	30
Oxychlordane	1200	1200	60
Heptachlor epoxide B	600	600	30
Trans-chlordane	600	600	30
Cis-chlordane	600	600	30
Trans-Nonachlor	800	800	40
Cis-Nonachlor	600	600	30
Endosulfan I (alpha)	600	600	30
Endosulfan II (beta)	600	600	30
Dieldrin	600	600	30
2,4′-DDD	600	600	30
4,4′-DDD	600	600	30
2,4′-DDE	600	600	30
4,4′-DDE	600	600	30
2,4′-DDT	600	600	30
4,4′-DDT	600	600	30
Perthane	600	600	30
Endrin	600	600	30
Endosulfan sulfate	600	600	30
Mirex	600	600	30
Methoxychlor	600	600	30
Endrin ketone	600	600	30
Desethylatrazine	800	800	40
Simazine	1600	1600	80
Atrazine	1600	1600	80
Ametryn	1600	1600	80
Metribuzin	400	400	20
Cyanazine	1600	1600	80
Hexazinone	2000	2000	100
Permethrin	800	800	40
Cypermethrin	4000	4000	200
Chlorothalonil	800	800	40
Diazinon	1600	1600	80

(continued)

Table 3.2 (continued)

Pesticide	Stock (ng/ml)	Spiking solution (pg/ml)	In 20 μL extract (ng/ml; pg/μl)
Disulfoton	8000	8000	400
Phorate	1600	1600	80
Methamidophos	1600	1600	80
Diazinon-oxon	1600	1600	80
Fonofos	1600	1600	80
Chlorpyrifos-methyl	2000	2000	100
Parathion-methyl	4000	4000	200
Pirimphos-methyl	1600	1600	80
Chlorpyrifos	1600	1600	80
Fenitrothion	1600	1600	80
Malathion	20,000	20,000	1000
Parathion-ethyl	1600	1600	80
Chlorpyrifos-oxon	1600	1600	80
Disulfotonsulfone	400	400	20
Azinphos-methyl	2000	2000	100
Captan	4000	4000	200
Phosmet (Imidan)	4000	4000	200
13C6-HCB	1800	1800	90
13C6-gamma-BHC	2600	2600	130
13C4-Heptachlor	1400	1400	70
13C6-beta-BHC	1600	1600	80
13C6-delta-BHC	1600	1600	80
13C12-Aldrin	1600	1600	80
13C10-Oxychlordane	1600	1600	80
13C10-Heptachlor-epoxide	1600	1600	80
13C9-alpha-Endosulfan	1600	1600	80
13C12-Dieldrin	1600	1600	80
13C10-t-Chlordane	1600	1600	80
13C10-t-Nonachlor	1600	1600	80
13C12-p,p-DDE	1600	1600	80
13C12-Endrin	1600	1600	80
13C9-beta-Endosulfan	1600	1600	80
13C10-c-Nonachlor	1600	1600	80
13C12-o,p-DDT	1600	1600	80
13C12-p,p-DDT	1600	1600	80
13C8-Mirex	1600	1600	80
13C12-Methoxychlor	1600	1600	80
Azinphos-methyl-d6	1600	1600	80
Diazinon-d10	1600	1600	80
13C6-Fonofos	1600	1600	80
13C3-Atrazine	1600	1600	80
13C6-Permethrins	1600	1600	80
13C12 PCB 52	1600	1600	80

Table 3.3 Pesticides and their mass and charge ratios and retention time

Function	Pesticide	m1 (1)	m2 (1)	m1/m2 Ratio	Tolerance (+/−)	RT (min)
1	Methamidophos	93.9642	94.9721		0.35	09:01
2	HCB	283.8102	285.8072	1.25	0.25	15:55
2	Tecnazene	258.8761	260.8732	0.78	0.35	14:44
2	13C6-HCB	289.8303	291.8273	1.25	0.25	15:54
2	Phorate	260.0128	262.0086	6.92	0.35	16:11
3	Desethylatrazine	172.0390	174.0360	3.11	0.35	16:50
3	Alpha-HCH	218.9116	220.9086	2.08	0.25	16:35
4	Atrazine	215.0938	217.0908	3.08	0.35	18:01
4	Simazine	201.0781	203.0752	3.1	0.35	18:21
4	Fonofos	246.0302	247.0336		0.35	18:25
4	Gamma-HCH	218.9116	220.9086	2.08	0.25	18:16
4	Quintozene	236.8413	238.8384	1.56	0.35	17:39
4	13C3-Atrazine	218.1038	220.1009	3.08	0.35	18:00
4	13C6-Fonofos	252.0503	253.0537	1000	0.35	18:25
4	13C6-gamma-BHC	222.9346	224.9317	0.77	0.25	18:15
4	Diazinon-d10	282.1074	314.1638	1000	0.35	17:32
4	Disulfoton	274.0285	275.0318		0.35	18:34
4	Diazinon	276.0698	304.1011		0.35	17:44
4	Diazinon-oxon	273.1004	288.1239		0.35	17:55
5	Aldrin	262.8569	264.854	1.56	0.25	21:17
5	Beta-BHC	218.9116	220.9086	2.08	0.25	19:27
5	Delta-BHC	218.9116	220.9086	2.08	0.25	21:01
5	Heptachlor	271.8102	273.8072	1.25	0.25	19:37
5	13C12-Aldrin	269.8804	271.8775	1.56	0.25	21:15
5	13C6-beta-BHC	222.9346	224.9317	0.77	0.25	19:26
5	13C6-delta-BHC	222.9346	224.9317	0.77	0.25	21:00
5	13C4-Heptachlor	276.8269	278.824	1.24	0.25	19:36
5	Chlorothalonil	263.8816	265.8786	0.78	0.35	21:08
5	Chlorpyrifos-methyl	285.9261	287.9232	1.44	0.35	21:26
5	13C12-PCB-52	301.9626	303.9597	0.78	0.25	21:51
6	Octachlorostyrene	270.8443	272.8413	0.63	0.25	23:18
6	Ametryn	227.1205	228.1238		0.35	22:41
6	Dacthal	298.8836	300.8807	0.78	0.35	23:18
6	Metribuzin	198.0701	199.0735		0.35	23:04
6	Parathion-methyl	263.0017	264.0051		0.35	22:28
6	Pirimiphos-methyl	276.0572	290.0728		0.35	22:42
7	Oxychlordane	262.8569	264.8540	1.56	0.25	24:11
7	13C10-Oxychlordane	269.8804	271.8775	1.56	0.25	24:09
7	Chlorpyrifos	313.9574	315.9545	1.44	0.35	23:33
7	Chlorpyrifos-oxon	269.9490	271.9462	1.54	0.35	24:30
7	Fenitrothion	260.0146	277.0174		0.35	24:07
7	Malathion	283.9942	285.0020		0.35	24:12

(continued)

Table 3.3 (continued)

Function	Pesticide	m1 (1)	m2 (1)	m1/m2 Ratio	Tolerance (+/−)	RT (min)
7	Parathion-ethyl	291.0330	292.0364		0.35	24:26
8	Heptachlor-epoxide	262.8569	264.8540	1.56	0.25	24:14
8	Alpha-Endosulfan	262.8569	264.8540	1.56	0.25	27:53
8	Dieldrin	262.8569	264.8540	1.56	0.25	30:34
8	o,p-DDE	246.0003	247.9974	1.56	0.25	28:07
8	p,p-DDE	246.0003	247.9974	1.56	0.25	30:38
8	13C12-Heptachlor-epoxide	269.8804	271.8775	1.56	0.25	25:11
8	13C9-alpha-Endosulfan	269.8804	271.8775	1.56	0.25	27:51
8	13C12-Dieldrin	269.8804	271.8775	1.56	0.25	30:31
8	13C12-p,p-DDE	258.0406	260.0376	1.56	0.25	32:36
8	13C10-t-Chlordane	269.8804	271.8775	1.56	0.25	26:39
8	13C10-t-Nonachlor	269.8804	271.8775	1.56	0.25	26:48
8	Cyanazine	240.0890	242.0861	3.06	0.35	28:13
8	c-Chlordane	262.8569	264.854	1.56	0.25	27:44
8	t-Chlordane	262.8569	264.854	1.56	0.25	26:41
8	t-Nonachlor	262.8569	264.854	1.56	0.25	26:50
9	Endrin	262.8569	264.854	1.56	0.25	32:56
9	c-Nonachlor	262.8569	264.854	1.56	0.25	33:19
9	o,p-DDD	235.0081	237.0052	1.56	0.25	32:21
9	13C12-Endrin	269.8804	271.8775	1.56	0.25	32:53
9	Captan	263.9653	265.9623	1.44	0.35	31:26
9	Disulfoton-sulfone.	213.0173	214.0251		0.35	32:49
9	Perthane	224.1520	223.1487		0.35	32:58
10	Beta-Endosulfan	264.8540	262.8569	0.64	0.25	34:32
10	Endosulfan-sulfate	264.8540	262.8569	0.64	0.25	36:54
10	o,p-DDT	235.0081	237.0052	1.56	0.25	33:59
10	p,p-DDD	235.0081	237.0052	1.56	0.25	34:31
10	p,p-DDT	235.0081	237.0052	1.56	0.25	35:54
10	13C9-beta-Endosulfan	269.8804	271.8775	1.56	0.25	34:30
10	13C10-c-Nonachlor	269.8804	271.8775	1.56	0.25	33:17
10	13C12-o,p-DDT	247.0484	249.0454	1.56	0.25	33:58
10	13C12-p,p-DDT	247.0484	249.0454	1.56	0.25	39:53
11	Endrin-ketone	247.8521	249.8491	0.63	0.25	39:47
11	Methoxychlor	227.1072	228.1106		0.35	39:44
11	Mirex	236.8413	238.8384	1.56	0.25	39:30
11	13C12-Methoxychlor	239.1475	240.1508		0.35	39:43
11	13C8-Mirex	241.8581	243.8551	1.56	0.25	39:29
11	13C6-Permethrins-Peak_1	189.1011	190.1045		0.35	24:33
11	13C6-Permethrins-Peak_2	189.1011	190.1045		0.35	24:21

(continued)

Table 3.3 (continued)

Function	Pesticide	m1 (1)	m2 (1)	m1/m2 Ratio	m1/m2 Tolerance (+/−)	RT (min)
11	Azinphos-methyl-d6	160.0511	161.0544		0.35	24:33
11	Hexazinone	171.0882	172.0916		0.35	39:38
11	Phosmet	160.0399	161.0432		0.35	40:55
11	Permethrins-Peak_1	183.0081	184.0843		0.35	42:04
11	Cypermethrins-Peak_1	163.0081	165.0052	1.56	0.35	43:52
11	Cypermethrins-Peak_2	163.0081	165.0052	1.56	0.35	44:03
11	Cypermethrins-Peak_3	163.0081	165.0052	1.56	0.35	44:11
11	Permethrins-Peak_2	183.0081	184.0843		0.35	42:21
11	Azinphos-methyl	160.0511	161.0544		0.35	42:39
11	Total cypermethrins	163.0081	165.0052	1.56	0.35	
11	Total permethrins	183.0081	184.0843		0.35	

Enzyme-Linked Immunosorbent Assays (ABRAXIS kit) is a convenient screening method for measuring single or closely related pesticide residues in water. The analysis is focused on only one or one family of pesticide. These are sensitive to levels down to ~0.05 μg/l (0.05 ppb). Test includes: Fill the test tube array with blanks, standards (known concentrations), and water samples. Add antibody solution specific to the pesticide and (in some kits) magnetic particles to each tube. Mix well; allow the antibodies to bind with pesticide molecules in the standards and samples. Rinse the tubes. The bound antibodies remain inside the tubes. (Kits using magnetic particles employ a powerful magnet tray to retain the bound antibodies in the tubes when washing). Add a color-developing solution that binds with the bound antibodies, read using a spectrophotometer, and compare with the standards.

PRO Lab has also developed a kit for testing pesticide residues in water.

3.2.3 Determination of Phenols

Principle The presence and levels of phenols in aquatic environments is of major concern as these are released as by-products during the production of plastics and dyes and in pulp and paper industries. These are also used in the production of a large variety of aromatic compounds, explosives, fertilizers, paint, paint removers, textiles, and drugs and as disinfectants and insecticides. Chlorinated phenols are formed on reaction with chlorine. These compounds are toxic and persistent. Prolonged oral and subcutaneous exposure causes damage to the lungs, liver, kidneys, and the genitourinary tract.

Materials

Gas chromatograph fitted with a flame ionization detector (FID), a glass sleeve injection port, a temperature programmer, and a compatible integrator or data station.

Operating parameters

Primary analytical column	2 m × 2 mm ID glass column packed with Carbopack F-TA, 60/80 mesh
Initial oven temperature	160 °C
Program rate	10 °C/min
Final oven temperature	220 °C
Final delay	5 min
Injector temperature	250 °C
Detector temperature	250 °C
Carrier gas	He or N_2
Carrier gas flow	20 ml/min
Injection sample size	2 μl
Analytical column	2 m × 2 mm ID glass column packed with 10% GP SP 2100, 100/120 in Supelcoport
Initial oven temperature	110 °C
Program rate	30 °C/min
Final oven temperature	200 °C
Final delay	10 min
Injector temperature	250 °C
Detector temperature	250 °C
Carrier gas	He or N_2
Carrier gas flow	20 ml/min
Injection sample size	2 μl

Standard phenol stock solution (100 μg/ml of each phenol): Weigh 0.1–0.11 g (± 0.0001 g) of each reagent grade phenol, o-, p-, and m-cresol and 3, 4 dimethylphenol using a plastic weighing tray, and quantitatively transfer to a 1 l volumetric flask containing 500–600 ml of NaOH solution. Shake well and make up to mark using NaOH.

Standard Phenol Working Solution (10 μg/ml): Pipet 10 ml of the phenol stock solution to a 100 ml volumetric flask and add NaOH to the mark and mix well. Prepare fresh prior to use.

Procedure To 9.0 ml of the phenol working standard, add I.0 ml of H_2SO_4. Check the pH (pH~1), and inject 2 ml into the gas chromatograph using a 10 ml syringe. Record the retention time and the peak area of each phenolic compound. Retain the chromatogram, and rinse the syringe with acidified NaOH solution between injections. All samples are refrigerated. If analyzed within 3 days, add 1 ml of 10% copper sulfate solution to each. If the sample is contaminated with heavy oil, extract the phenolic compounds using carbon tetrachloride as the extracting solvent. This procedure is not necessary if the sample is "clean." Measure and record the total liquid volume in each impinger. Transfer 9.0 ml aliquot to a 25 ml test tube, and add 1 ml of sulfuric acid. Cover the test tube with Parafilm, invert, and mix well. Check the pH, and inject into the gas chromatograph. The blank contains 1 ml H_2SO_4 and 9 ml NaOH solution in a 25 ml test tube. Record the retention time and peak area of the phenolic compounds, and compare with sample (Levaggi and Feldstein 1971).

Calculation

$$\text{Total } \mu\text{g of Phenol per impinger} = \text{Total } V_{(S)} \times PA_{(S)} \times \text{Concn}_{(STD)} / PA_{(STD)}$$

where:

Total $V_{(s)}$ = total volume of sample in the Impinger
$PA_{(S)}$ = peak area of the phenolic compound in the sample
$\text{Concn}_{(STD)}$ = mg/ml of the individual phenolic compound in the standard
$PA_{(STD)}$ = peak area of the phenolic compound in the standard

Total μg of phenol in the sample = μg in the 1st Impinger + μg in the 2nd Impinger
Total μg of o-, p-, and m-cresol in the sample = μg in the 1st Impinger + μg in the 2nd Impinger
Total μg of DMP in the sample = μg in the 1st Impinger + μg in the 2nd Impinger
ppm Phenol = Total μg Phenol in the Sample/(f∗) × Sample Volume (liters)
where:

(f∗) = 3.84 for phenol; 4.42 for o-, p-, and m-cresol; and 4.99 for DMP. The (f∗) values are μg/μl of the individual phenol at 25 °C and 760 mm Hg.

$$\text{Total ppm Phenols} = \text{ppm Phenol} + \Sigma(\text{ppm Cresols}) + \text{ppm DMP}$$

3.2.4 Determination of Oil/Grease

The concentration of oil and grease (OG) is an important parameter for water quality and safety. OG in water can cause surface films and shoreline deposits leading to environmental degradation and pose as threat to human health when discharged to surface or groundwaters. OG may interfere with aerobic and anaerobic biological processes leading to decreased wastewater treatment efficiency. OG in water is commonly determined by extraction into a nonpolar, hydrocarbon-free solvent followed by measurement of the infrared absorption spectrum of the extract. The absorption between 3000 and 2900 cm^{-1} by C-H groups in the OG is correlated to the concentration of OG, using tetrachloroethylene as standard. Tetrachloromethane and 1,1,2-trichloro-1,2,2-trifluoroethane (CFC-113) have serious environmental effects (ozone depletion).

Procedure A calibration stock solution of 100 mg/ml total OG is prepared by placing 0.55 ml of octanoic acid and 0.72 ml of isooctane in a 10 ml volumetric flask and filling to the mark with tetrachloroethene. This stock solution is diluted (2.5–50 ml with tetrachloroethylene). Calibration standards are prepared as in Table 3.4. A ref-

Table 3.4 Calibration standards

Standard	Volume of 5 mg/ml stock diluted to 10 ml	Concentration (mg/ml)	Equivalent concentration of OG (mg/ml) in 250 ml water sample extracted into a 50 ml volume of solvent
1	0.05	0.025	5
2	0.10	0.050	10
3	0.20	0.100	20
4	0.30	0.150	30
5	0.50	0.250	50
6	0.60	0.300	60
7	0.70	0.350	70
8	0.80	0.400	80

erence OG mixture and spiked samples are prepared. Volumetric concentrations of OG are converted to mass/volume concentrations by using values of 773 and 911 g/l (densities of isooctane and octanoic acid, respectively).

For the extraction of OG from water samples, 250 ml of water sample is acidified to pH 2 using 50% sulfuric acid and extracted serially thrice with ~15 ml tetrachloroethylene. These are dried by filtering through 1 g sodium sulfate and make up to 50 ml. The absorbance is measured using FTIR/NIR spectrophotometer. Spectra obtained for water sample over a range of 3200–2700 cm^{-1} were compared with the spectra for standards. The details of FTIR/NIR spectrophotometer are discussed in Chap. 5 (United States Environmental Protection Agency, Office of Water, Method 1664) (Table 3.4).

3.3 Organic Nutrient and Demand

3.3.1 Determination of Biological Oxygen Demand

Biochemical oxygen demand or BOD is a chemical procedure for determining the amount of dissolved oxygen needed by aerobic microorganisms to break down organic material present in the given water sample at certain temperature over a specific time period. It is expressed as mg O_2 consumed/l of sample during 5 days (BOD_5) of incubation at 20 °C. The rate of oxygen consumption is affected by temperature, pH, the presence of certain kinds of microorganisms, and the type of organic and inorganic material in the water. High BOD means that there is less of oxygen to support life and indicates organic pollution. There are two stages of decomposition in the BOD test, a carbonaceous stage and a nitrogenous stage:

- The carbonaceous stage, or first stage, represents that portion of oxygen demand involved in the conversion of organic carbon to carbon dioxide.

- The nitrogenous stage, or second stage, represents a combined carbonaceous plus nitrogenous demand, when organic nitrogen, ammonia, and nitrite are converted to nitrate. Nitrogenous oxygen demand generally starts after about 6 days.

Materials

Calcium chloride solution (27.5 g/l)
Ferric chloride solution (0.25 g/l)
Magnesium sulfate (22.5 g/l)
Phosphate buffer: Dissolve 8.5 g KH_2P0_4, 21.75 g K_2HP0_4, 33.4 g $Na_2HP0_4.7H_20$, and 1.7 g NH_4Cl in approximately 500 ml distilled water and dilute to 1 l. The pH should be 7.2. Store at 4 °C in refrigerator.

Procedure Check the pH of the water sample (pH 6.5–7.5), as needed neutralize samples with 1 N sulfuric acid or 1 N sodium hydroxide. Do not dilute the sample with the acid or base by more than 0.5% (1.5 ml in a 300 ml BOD bottle). Prepare the suitable dilutions for water samples as shown.

Estimated BOD_5 (mg/l)	Suggested sample volumes (ml)
< 5	200, 250, 300
< 10	100, 150, 200
10–30	25, 50, 100
30–60	15, 25, 50
60–90	10, 15, 25
90–150	5, 10, 15
150–300	3, 5, 10
300–700	1, 3, 5
700–1500	0.5, 1, 3
1500–2500	0.25, 0.5, 1

Use less than 3 ml for strong industrial wastes, 3–15 ml for raw and settled wastewater, and 15–75 ml for biologically treated effluent. Transfer the required volume to 300 ml BOD bottle and dilute. Stopper the bottle without leaving air bubbles, also ensure overflow. Record the initial dissolved oxygen using a calibrated DO meter (initial DO). Record the pressure. Incubate the BOD bottle with the sample for 5 days at 20 °C in a BOD incubator. Repeat the experiments similarly for blanks and duplicates. Fill seals with dilution water to reduce evaporation. Pour off the excess water, and determine the DO (final DO) (Standard Methods for the Examination of Water and Waste Water 1976).

Calculations

$$\text{BOD mg/l} = (\text{Initial DO} - DO_5) \times \text{Dilution Factor}$$

$$\text{Dilution Factor} = \text{Bottle Volume}(300\,\text{ml})/\text{Sample Volume}$$

Winkler Titration Technique

Materials

Manganous sulfate solution (480 g/l).
Alkali-iodide-azide reagent: Dissolve 500 g NaOH and 135 g NaI in distilled water, and add 10g NaN_3 dissolved in 40 ml, and make up to 1 l.
Standard sodium thiosulfate titrant (0.025 N).
Starch solution (0.5%).

Procedure Siphon three portions of aerated dilution water to three BOD bottles, avoiding atmospheric O_2. To two of the three BOD bottles, add 1 ml $MnS0_4$ solution, followed by 1 ml alkali-iodide-azide reagent. Stopper carefully, to exclude air bubbles. When precipitate has settled to about half the bottle volume, carefully remove the stopper, and add 1 ml sulfuric acid. Re-stopper and mix by gentle inversion until the iodine is uniformly distributed throughout the bottle. Transfer 203 ml of sample into a 500 ml conical flask, and titrate with 0.025 N sodium thiosulfate to a pale straw color. Add 1–2 ml of starch solution, and continue to titrate to first disappearance of the blue color. (200 ml of original dilution water is equal to 203 ml of dilution water plus reagents.) Titrate two of the three samples.

The BOD of the given water samples is expressed as mg/l (ppm).

3.3.2 Determination of Chemical Oxygen Demand

Principle The Chemical Oxygen Demand (COD) determines the quantity of oxygen required to oxidize the organic matter in a waste sample, under specific conditions of oxidizing agent, temperature, and time. The test results are independent measurement of organic matter in the sample. The method could be applied to domestic and industrial waste samples having an organic carbon concentration > 50 mg/l. The presence of sufficient concentration of dissolved oxygen is critical in maintaining water quality.

Materials

Standard potassium dichromate solution (0.250 N)

Sulfuric acid reagent: Conc. H_2SO_4 containing 23.5 g silver sulfate per 4.09 kg bottle

Standard ferrous ammonium sulfate (0.25 N)

Standardization: To approximately 200 ml of distilled water added 25 ml of 0.25 N potassium dichromate and 20 ml sulfuric acid. Cooled and titrated against ferrous ammonium sulfate using 3 drops of ferroin indicator. The end point is sharp color change, going from blue-green to reddish-brown.

Phenanthroline ferrous sulfate (ferroin) indicator solution: Dissolve 1.48 g of 1, 10 (ortho) phenanthroline monohydrate, together with 0.70 g of $FeSO_4.7H_2O$ in 100 ml water.

Procedure Collect samples in glass bottles, and maintain at pH 2 by adding sulfuric acid. Reflux 50 ml sample or an aliquot diluted to 50 ml with 1 g of $HgSO_4$ and 5 ml conc. sulfuric acid; mix well until the mercuric sulfate has dissolved. Place in an ice bath and slowly added 25 ml of potassium dichromate with swirling. Then add 70 ml sulfuric acid-silver sulfate solution to the cooled reflux flask with swirling motion. Reflux for 2 h, and allow it to cool. Wash down the condenser with 25 ml of distilled water. If required transfer the mixture to an Erlenmeyer flask. The washing is repeated three or four times. Dilute the acid solution to about 300 ml with distilled water, and allow the solution to cool to about room temperature. Add 8–10 drops of ferroin indicator, and titrate the excess dichromate with 0.25 N ferrous ammonium sulfate solution to the end point. Simultaneous blank is also run (Standard Methods for the Examination of Water and Waste Water 1976).

Calculation

Calculate the COD in the sample in mg/L as follows:

$$\text{COD} = 8 \times 1000 \times \text{DF} \times M \times \left(V_B - V_S\right) / \text{volume of the sample}\left(\text{ml}\right)$$

where:

DF = dilution factor (if applicable)
M = molarity of the standardized ferrous ammonium sulfate solution
V_B = titer value for blank
V_S = titer value for sample

The COD of the given water sample is expressed as mg/l or ppm.

3.3.3 Determination of Nitrates

Principle Nitrate is the most common form of nitrogen found in water. Other forms of nitrogen include nitrite and ammonia. In water, nitrate has no taste or scent. The

Maximum Acceptable Concentration (MAC) for nitrate in drinking water is 10 mg/l (laboratory). Generally the ground water has high nitrate concentration because of the percolating sewage, industrial waste, chemical fertilizers, leaches from solid waste landfills, septic tank effluents, etc. The low acidity in the infants' intestine permits the growth of nitrate-reducing bacteria that converts the nitrate to nitrite, which is absorbed in the bloodstream. The nitrite has a great affinity for hemoglobin than the oxygen, and it replaces oxygen in the blood. The deficiency of oxygen causes suffocation, and the skin color becomes blue, so it is termed as blue baby disease. The medical name is *methemoglobinemia*. This disease is a fatal and should have nitrate concentration above 50 ppm. Nitrate can be removed from drinking water by ion exchange, reverse osmosis, and electrodialysis.

Procedure Prepare standard nitrate solution (0, 5, 10, 15, 20 … 35 ppm), and read spectrophotometrically at a wavelength of 220 nm to obtain absorbance/transmittance. Distilled water with no nitrate is taken as blank. The concentration of nitrate is calculated from the standard curve (concentration versus absorbance) and expressed as ppm.

3.4 Microbiological Parameters

3.4.1 Determination of Most Probable Number (MPN)

The plating methods have limitations to detect low numbers of bacteria in water or food, because the inoculum size is only 1 ml per petri dish. Pathogenic or indicator bacteria may not be present in sufficiently large numbers in water or food to be detected by plating methods. For example, in the case of potable water, permitted levels of coliforms are less than 1 per 100 ml. In such cases, MPN methods are used, where larger volumes of sample can be used for inoculation. MPN is only a statistical approximation of the test bacteria in the given sample and not the actual number. MPN is defined as "that bacterial density which, if it had been actually present in the sample would, more frequently than any other, have given the observed analytical results." Usually MPN method is used to detect coliform bacteria in water or food. MPN method for coliforms is a three-step process, and all the media used are liquid media (Surendran et al. 2006).

3.4.2 Determination of Total Coliforms in Water (Five-Tube MPN Method)

Procedure

Step I (for presumptive total coliforms) – Arrange the DS and SS MacConkey broth tubes in a test tube rack. 50 ml of the given sample is inoculated into 50 ml DS MC broth; 10 ml each of the water sample is inoculated into five tubes DS MC broth,

1 ml each into five tubes of SS MC broth. Make a decimal dilution of the water sample by adding 1 ml of the water to 9 ml of normal saline, and add 1 ml each of the first dilution (equivalent to 0.1 ml of the original sample) to the next five tubes of SS MC broth. Label appropriately. Incubate in the incubator at 37 °C for 24 h. After 24 h, observe the tubes for growth and gas production. If the color of the MC broth has turned toward yellow with gas production in Durhams' tubes, the reaction is noted as positive. Note the results as number of positives in each set of 50 ml, 10 ml, 1 ml, and 0.1 ml tubes. Compare the results with standard MPN table (for 5 tube MPN), and note the MPN values for presumptive total coliforms.

Step II (for confirmed total coliforms) – Inoculate one loopful of culture from the positive tubes of step I to BGLB 2% broth, and mark with the corresponding label. Incubate at 37 °C in a serological water bath for 24 h. Note growth and gas production. Compare with five-tube MPN table, and obtain the results as MPN confirmed total coliforms.

Step III (for fecal coliforms and E.coli) – From the positive tubes of step II, inoculate one loopful each to EC broth and tryptone broth (indole medium). Label appropriately. Incubate at 44.5 °C ± 0.5 °C for 24 h.

EC broth: Note growth and gas pr0oduction. Note the number of positives in each set. Compare with five-tube MPN tables. The values obtained gives MPN fecal coliforms.

Tryptone broth: Test for indole production by adding four drops of Kovac's indole reagent. A pink or red color at the top layer indicates a positive test for indole. Note the number of positives in each set. Compare with five-tube MPN tube table. Coliform bacteria which produce gas in EC broth and indole in tryptone broth at 44.5° ± 0.5 °C are considered as *E. coli*. Note: MPN method can be used to detect coliforms and *E. coli* in food samples also, provided the expected number of these bacteria in the food products are less than 10^3per gram. Otherwise the results may not be meaningful.

3.4.3 Determination of Standard Plate Count

The sampling, spread plating, isolation, purification, and identification details are given in Chap. 2.

3.4.4 Determination of Total Coliforms/Fecal Coliforms

This is discussed in Chap. 2.

3.4.5 Determination of Biofouling and Biofilm Formation

Biofouling is an unwanted deposition and growth of living organisms on surfaces. It is an assemblage of microbial cells attached to the surface and encapsulated in a film or slime layer of extracellular polymers. An indirect procedure is to count the number of viable microorganisms by classical microbiological methods in the samples. These methods are time-consuming and retrospective, as it can take days before the results are available. There are several advanced methods available for detecting and studying biofilms in research laboratories such as microscopic, spectroscopic, and immunological techniques. However, these methods are complicated and require sophisticated instrumentation and highly specialized personnel (Agle 2007).

Visual observation: Organic sediments appear as whitish or yellow coatings under UV light because the molecular configuration of organic material will cause sediments to fluoresce when illuminated by UV light. Microbial contamination on accessible surfaces can be evaluated by swab methods such as the ATP swab test. ATP (adenosine triphosphate) is the principal energy carrier of all living organisms, and measurement of ATP bioluminescence is a widely used rapid method for monitoring contamination. After swabbing of a surface, the probe is measured in a portable instrument. The method is easy to use, but the cost per analysis is relatively high, and swab tests are not particularly useful when large areas need inspection. In addition, the detection limit of the ATP test is relatively high. It should also be realized that a biofilm may stick so tight to a surface that swabbing only removes a small part of the film.

The rinse water can be tested by ATP bioluminescence or by measuring total organic carbon (TOC) using TOC analyzer. TOC comprises both microbial organisms and organic sediment, and the method is widely used. Thermal combustion ion chromatography could be used to detect TOC in water (Fung et al. 1996). For complete oxidation, potassium persulfate (0.25%) is added to oxidize nonvolatile organic compounds in solution and cupric oxide heated at 900 °C to convert volatile organic compounds to carbon dioxide. These are scrubbed in a 20 ml solution of 50 mM KOH with 10 drops of butanol. The sample is filtered (0.45 μm); and the carbonate anion obtained is determined by non-suppressed ion chromatography using 0.6 mM potassium hydrogen phthalate as eluent. Separations are carried out on a 150 mm (l) × 4.1 mm (ID) Hamilton PRPX100 column at room temperature using the ion chromatograph at an eluent flow rate of 1 ml/min with indirect UV detection at 272 nm (Tarver 2009; Mortensen 2014).

3.4.6 Determination of Pathogens

Determination and biochemical tests for the identification of *E. coli*, *Staphylococcus aureus*, *Salmonella typhi*, and *Vibrio cholera* are given in Chap. 2.

3.5 Biological Parameters

3.5.1 Determination of Phytoplankton and Zooplankton

Planktons are drifters in the ocean. They are unable to swim against a current and include both animals and plants. Planktons are extremely important in the marine

Table 3.5 Plankton size

Category	Size
*Ultra*plankton	< 0.005 mm
*Nano*plankton	0.005–0.06 mm
*Micro*plankton	0.06–0.2 mm
*Meso*plankton	0.2–0.5 mm
*Macro*plankton	0.5–2 mm
*Mega*plankton	> 2 mm
Phytoplankton: free-floating autotrophs and holoplankton – organisms that spend their entire life spans as plankton	Size
Diatoms	0.005–1 mm
Dinoflagellates and *Zooxanthellae*	0.05–0.2 mm
Zooplankton: free-floating heterotrophs and holoplankton – organisms that spend their entire life spans as plankton	Size
Copepod	1–10 mm
Foraminiferans	0.1–1 mm
Radiolarians	0.1–1 mm
Dinoflagellates	0.05–0.2 mm

ecosystem. The planktons are either holoplanktonic (floating organisms for entire life) or meroplanktonic (floating organisms for only part of life cycle, benthic or nektonic for the rest of life cycle) (Appendix 3.6). These are collected using a fine mesh plankton net towed through the water. Based on their size these are classified into following (Table 3.5):

Methods of Collection Phytoplanktons can be sampled with conical tow nets, pumps, Clarke-Bumpus plankton sampler, depth-integrating samplers, and water-sampling bottles (Von Dom bottles or water samplers with closing mechanism). First three are also used for zooplankton sampling. The sampling success depends on the selection of suitable gear, mesh size of netting material, time of collection, water depth of the study area, and sampling strategy (Goswami 2004) (Figs. 3.1, 3.2 and 3.3).

For quantitative plankton sampling and to know the actual amount of water passed through the net, flow meter is used, which helps in calculating the volume of water filtered. Based on the net shape and the study area, the volume differs.

Fixation and Preservation The fixation of samples should be carried out immediately in 4–5% buffered formalin (borax buffer) and fixatives (2–5%). These are then preserved in 70% ethanol or 40% isopropanol. Glycerin prevents shrinkage of the specimens.

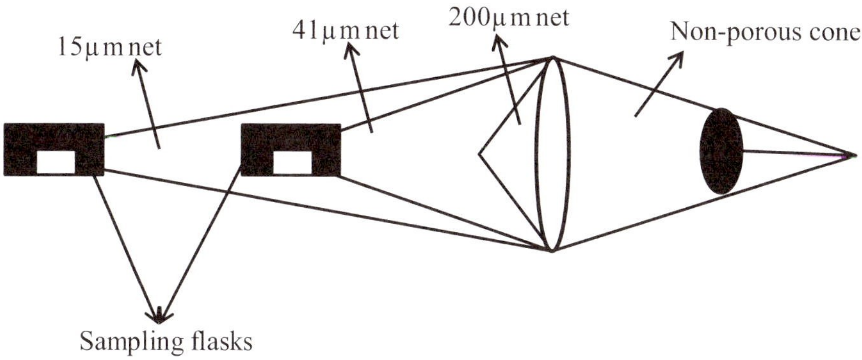

Fig. 3.1 Conical Tow Nets

Fig. 3.2 Clarke-Bumpus plankton sampler (FAO Aquaculture photo library)

Fig. 3.3 Pumps

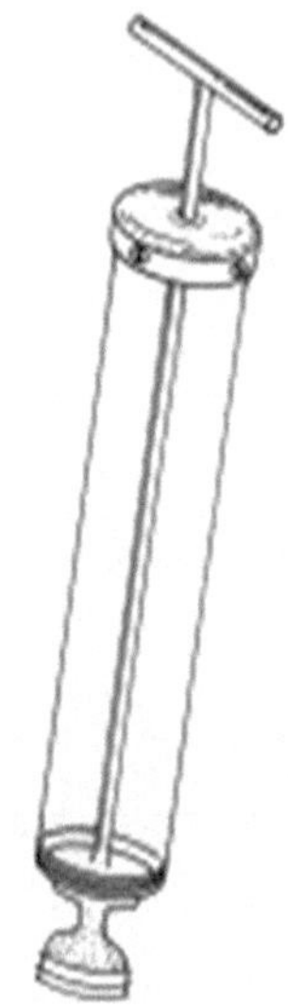

Analysis The basic analysis consists of measurements of biomass (standing stock), enumeration of common taxa, and species.

Biomass: The term denotes the live weight or the amount of living matter present in the specimen. It is estimated by the following methods:

1. Volumetric (displacement volume and settling volume) method
2. Gravimetric (wet weight, dry weight and ash free dry weight)method
3. Chemical method (measurement of constituents such as carbon, nitrogen, phosphorus and biochemical elements such as protein, lipid, and carbohydrate)

Faunal enumeration: Aliquots or subsampling using Folsom plankton splitter and used for taxa identification, dry weight estimation, abundance of species, and as reference collection. This is followed by counting and species identification.

Phytoplanktons: Soft Algal Sample Analysis

Procedure: The plankton samples are mixed gently for 1 min. Samples are added to the chamber with a syringe (<10 ml) or macropipettor (10 ml or more). The sample bottle should be inverted at least once between each addition as larger organisms settle quickly and remain at the bottom. The chamber is then topped with a round glass top plate. Algae samples are allowed to settle onto the base of the settling chamber, and oil immersion is used for identification. The time recommended for complete sedimentation varies with the height of the chamber, i.e., 8 cm/day to 4 cm/day depending on accuracy required in enumeration. Approximate settling times necessary are as follows:

100 ml	100 h
50 ml	50 h
25 ml	25 h
10 ml	10 h
5 ml	5 h
2 ml	2 h

The chamber of settled material is scanned, and the dominant (four or five most common organisms) as well as subdominant taxa are determined. This is to give the analyst an idea of the sample composition as well as to insure that the sample is evenly settled. Enumeration and identification are done by scanning parallel strips of 10 mm per strip (area 2 mm^2). A minimum of three strips is required (no less than 250 "live" organisms). The area counted is recorded as it is needed for cells per ml calculation. The number of "live" cells are identified and enumerated to the lowest taxonomic rank possible. At least 10 specimens of each taxa are measured for cell volume calculations.

Diatom Sample Analysis

Procedure: Initial steps are carried out under the hood. A specified volume (500 ml) of uniformly mixed sample is poured into a beaker. Mix well gently. Add 20 ml of concentrated HNO_3 to digest organic matter in the sample. Place beaker on a hot plate, and concentrate sample to approximately 20 ml by heat evaporation. Allow sample to cool, and transfer to a 300 ml beaker. Rinse with distilled water, and transfer to the digested sample. Adjust the volume to 150 ml with distilled water oxidized with 25 ml of 30% H_2O_2. Accelerate the process by adding a few grains of $K_2Cr_2O_7$. Concentrate to 10 ml by heat evaporation. Cool and centrifuge (1500 rpm for 30 min). Remove the supernatant, and add 10 ml distilled water to the tube, gently vortex and centrifuge again. Repeat ten times. The cleaned sample is spread on warm cover slip and fixed for analysis. These are then identified and enumerated (2 frustules = 1 diatom cell). A minimum of 500 frustules are counted.

Calculations

$$\text{Cells per ml} = C \times \text{TA}/L \times W \times V \times S$$

where:

C = cell count
L = length of strip (mm)
W = width of strip (mm)
V = volume of chamber (ml)
S = number of strips counted
TA = total area of chamber bottom (mm^2)

3.6 Radioactive Elements

3.6.1 Alpha Emitter and Beta/Photon Emitter

Principle Radioactivity in water could be natural (weathering of rocks containing radioactive minerals and fallout of cosmic-ray produced nuclides) or artificial (nuclear power industry, nuclear weapons testing, and applications of nuclear materials and devices). The principal radionuclides introduced naturally into surface and groundwaters are uranium, radium-226, radium-228, radon, potassium-40, tritium, and carbon-14. All except last two are derived from radioactive minerals. Radioactive elements including uranium, thorium, and actinium and radioactive daughters resulting from these decay series are important primarily for reasons of health and as potential energy sources. The alpha particles are toxic when inhaled or ingested, while beta particles pass through body and may cause harm. The effectiveness of the contaminants can be reduced using reverse osmosis, aeration, ion exchange, granular activated carbon, distillation, etc etc (Fons-Castells et al. 2017).

Procedure For absolute gross alpha and gross beta measurement, the detectors must be calibrated to obtain the ratio of count rate to disintegration rate. Americium-241 (used for alpha activity in the collaborative test of this method) has higher alpha particle energy (5.49 MeV) than those emitted by the naturally occurring uranium and radium-226 radionuclides. However, it has similar energy levels of the alpha particles emitted by naturally occurring thorium-228 and radium-224. Strontium-90 and cesium-137 are used as standards for gross beta activity. Standard solutions of each of these radionuclides are readily available. Cesium is volatile at elevated temperatures (>450 °C). Sample weight stability is essential to measure gross alpha and gross beta radionuclides to ensure the accuracy of the self-absorption counting efficiency factor to be used for the samples. Strontium-90 in equilibrium with its daughter yttrium-90 is the prescribed radionuclide for gross beta calibrations. Separate alpha and beta particle self-absorption graphs showing water sample residue weight (mg) vs the efficiency factor (dpm/cpm) are prepared using standard alpha and beta emitter solutions and tap water. For the alpha graph standard, alpha activity is added to varying size aliquots of tap water, such that the aliquot residue weight is varied between 0 and 100 mg (for a 2-inch counting planchet). A similar graph is prepared with standard beta activity and tap water aliquots, varying the residue weight between 0 and 300 mg (for a 2-inch planchet). If water sample aliquot volumes are used, then it must contain 100 mg of dried water solids. Tap water aliquots with added americium-241 or strontium-90 standard should be acidified with 16 N nitric acid, evaporated to a small volume in a beaker on a hot plate, transferred quantitatively in 5 ml portions or less to a tarred counting planchet, evaporated to dryness, and finally dried at 105 °C for 2 h. Weight-stable aliquot residues should then be alpha and/or beta counted until at least 10,000 total counts have been accumulated. A single set of reference standards prepared in this way can be used for each counting instrument for separate graph preparations and can be stored for verification whenever needed.

Large amounts of water samples are collected in plastic bottles and adjusted to pH 2 using 1 N nitric acid. Transfer to an aliquot of a water sample of a volume size that contains no more than 100 mg (for alpha only or alpha and beta determination) or 200 mg (for beta only determination) of total water solids. Evaporate the aliquot to near dryness on a hot plate. The water samples containing chlorides are converted to nitrates (using 16 N nitric acid) before it is transferred to a stainless steel planchet. Dry the sample residue in a drying oven at 105 °C for at least 2 h; cool in a desiccator; weigh; and count. Store the sample residue until ready for counting. The hygroscopic residues may be flamed to a dull red heat with a Meeker burner for a few minutes to convert the nitrate salts to oxides before weighing and counting, for accurate reading. Count for alpha and beta activity at their respective voltage plateaus (Friedlander et al. 1965).

Liquid scintillation counting (LSC) is a commonly used technique for the determination of alpha and beta emitters. However, LSC has poor resolution, and the continuous spectra for beta emitters hinder the simultaneous determination of several alpha and beta emitters from the same spectrum. The alpha (^{nat}U, ^{241}Am and ^{226}Ra) and beta emitters (^{40}K, ^{60}Co, $^{90}Sr/^{90}Y$, ^{134}Cs, and ^{137}Cs) are used in LSC for counting the radionuclides in water samples. Samples are dissolved or suspended in a "cocktail" containing aromatic solvent (toluene or benzene and small amounts of other additives known as fluors). The beta particles transfer energy to the solvent molecules which in turn transfer energy to fluors. Excited fluor molecules dissipate energy by emitting light of different wavelengths. Samples are placed in small transparent or translucent (glass) vials and loaded to the instrument. The counter has two photomultiplier tubes (PMT) connected to a coincidence circuit, which assures that genuine light reaches the PMT, while other noises are ignored (Fig. 3.4).

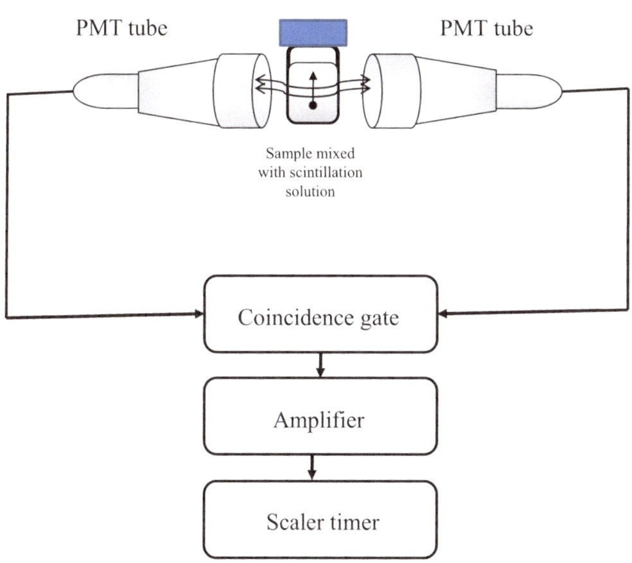

Fig. 3.4 Liquid scintillation counter

Calculation

$$\text{Alpha radioactivity}\left(\text{picocuries per liter,pCi/l}\right)=A\times 1000/2.22\times C\times V$$

where:

A = net alpha count rate (gross alpha count rate minus the background count rate) at the alpha voltage plateau
C = alpha efficiency factor, read from graph of efficiency versus mg of water solids per cm^2 of planchet area, (cpm/dpm)
V = volume of sample aliquot, (ml)
2.22 = conversion factor from dpm/pCi

$$\text{Beta radioactivity}\left(\text{pCi / l}\right)=B\times 1000\,/\,2.22\times D\times V$$

Beta particles are also measured in mrem (milli roentgen equivalent man)where:

B = net beta count rate (gross count rate minus the background count rate at the beta voltage plateau)
D = beta efficiency factor, read from the graph of efficiency versus mg of water solids per cm^2 of planchet area, (cpm/dpm)
V = volume of sample aliquot, (ml)
2.22 = conversion factor from dpm/pCi

When counting beta radioactivity in the presence of alpha radioactivity by gas-flow proportional counting systems (at the beta plateau), alpha particles are also counted. Since alpha particles are more readily absorbed by increasing sample thickness than beta particles, the alpha/beta count ratios vary with increasing sample thickness. Therefore, it is necessary to prepare a calibration curve by counting standards containing americium-241 with increasing thickness of solids on the alpha plateau and then on the beta plateau, plotting the ratios of the two counts vs density thickness. The alpha amplification factor (E) from that curve is used to correct the amplified alpha count on the beta plateau. When significant alpha activity is indicated by the sample count at the alpha voltage plateau, the beta activity of the sample can be determined by counting the sample at the beta voltage plateau and calculating the activity from the following equation:

$$\text{Beta}\left(\text{pCi/l}\right)=\left(B-\text{AE}\right)\times 1000/2.22\times D\times V$$

where:

E = alpha amplification factor, read from the graph of the ratio of alpha counted at the beta voltage/alpha counted at the alpha voltage vs sample density thickness.

Appendices

Appendix 3.1 Drinking Water Quality Standard

Chemicals of Health Significance as Described by World Health Organization Guidelines (WHO) for Drinking Water Quality in Fourth Edition (2011)

Parameter	Unit	Latest WHO 4th edition (2011) guideline value
Acrylamide	μg/l	0.5
Alachlor	μg/l	20
Aldicarb	μg/l	10
Aldrin and dieldrin	μg/l	0.03
Antimony	mg/l	0.02
Arsenic	mg/l	0.01 (A,T)
Atrazine and its chloro-s-triazine metabolites	μg/l	100
Barium	mg/l	0.7
Benzene	μg/l	10
Benzo[a]pyrene	μg/l	0.7
Boron	mg/l	2.4
Bromate	μg/l	10 (A,T)
Bromodichloromethane	μg/l	60
Bromoform	μg/l	100
Cadmium	mg/l	0.003
Carbofuran	μg/l	7
Carbon tetrachloride	μg/l	4
Chlorate	μg/l	700 (D)
Chlordane	μg/l	0.2
Chlorine	mg/l	5 (C)
Chlorite	μg/l	700 (D)
Chloroform	μg/l	300
Chlorotoluron	μg/l	30
Chlorpyrifos	μg/l	30
Chromium	mg/l	0.05 (P)
Copper	mg/l	2
Cyanazine	μg/l	0.6
2,4-D (2,4-dichlorophenoxyacetic acid)	μg/l	30
2,4-DB (2,4-dichlorophenoxybutyric acid)	μg/l	90
DDT (Dichlorodiphenyltrichlorethane) and metabolites	μg/l	1
Di(2- ethylhexyl)phthalate	μg/l	8
Dibromoacetonitrile	μg/l	70
Dibromochloromethane	μg/l	100

(continued)

Parameter	Unit	Latest WHO 4th edition (2011) guideline value
1,2-Dibromo-3-chloropropane	μg/l	1
1,2-Dibromoethane	μg/l	0.4 (P)
Dichloroacetate	μg/l	50 (D)
Dichloroacetonitrile	μg/l	20 (P)
1,2-Dichlorobenzene	μg/l	1000 (C)
1,4-Dichlorobenzene	μg/l	300 (C)
1,2-Dichloroethane	μg/l	30
1,2-Dichloroethene	μg/l	50
Dichloromethane	μg/l	20
1,2-Dichloropropane	μg/l	40 (P)
1,3-Dichloropropene	μg/l	20
Dichlorprop	μg/l	100
Dimethoate	μg/l	6
1,4-Dioxane	μg/l	50
Edetic acid	μg/l	600
Endrin	μg/l	0.6
Epichlorohydrin	μg/l	0.4 (P)
Ethylbenzene	μg/l	300 (C)
Fenoprop	μg/l	9
Fluoride	mg/l	1.5
Hexachlorobutadiene	μg/l	0.6
Hydroxyatrazine	μg/l	200
Isoproturon	μg/l	9
Lead	mg/l	0.01 (A,T)
Lindane	μg/l	2
MCPA (4-(2-Methyl-4-chlorophenoxy) acetic acid)	μg/l	2
Mecoprop	μg/l	10
Mercury	mg/l	0.006
Methoxychlor	μg/l	20
Metolachlor	μg/l	10
Microcystin-LR	μg/l	1 (P)
Molinate	μg/l	6
Monochloramine	mg/l	3
Monochloroacetate	μg/l	20
Nickel	mg/l	0.07
Nitrate (as NO^{3-})	mg/l	50
Nitrilotriacetic acid	μg/l	200
Nitrite (as NO_2^-)	mg/l	3
N-Nitrosodimethylamine	μg/l	0.1
Pendimethalin	μg/l	20
Pentachlorophenol	μg/l	9 (P)
Selenium	mg/l	0.04 (P)

(continued)

Parameter	Unit	Latest WHO 4th edition (2011) guideline value
Simazine	μg/l	2
Sodium dichloroisocyanurate (as cyanuric acid)	mg/l	40
Styrene	μg/l	20 (C)
2,4,5-T (2,4,5-trichlorophenoxy acetic acid)	μg/l	9
Terbuthylazine	μg/l	7
Tetrachloroethene	μg/l	40
Toluene	μg/l	700 (C)
Trichloroacetate	μg/l	200
Trichloroethene	μg/l	20 (P)
2,4,6-Trichlorophenol	μg/l	200 (C)
Trifluralin	μg/l	20
Trihalomethanes		The sum of the ratio of the concentration of each to its respective guideline value should not exceed 1
Uranium	mg/l	0.03 (P)
Vinyl chloride	μg/l	0.3
Xylenes	μg/l	500 (C)

According to WHO Drinking-water Quality 4th edition (2011)
A = provisional guideline value because calculated guideline value is below the achievable quantification level
C = concentrations of the substance at or below the health-based guideline value may affect the appearance, taste, or odor of the water, leading to consumer complaints
D = provisional guideline value because disinfection is likely to result in the guideline value being exceeded
P = provisional guideline value because of uncertainties in the health database
T = provisional guideline value because calculated guideline value is below the level that can be achieved through practical treatment methods, source protection, etc.

Appendix 3.2 – Drinking Water Quality Standard (Rodger et al. 2017)

Parameter	Limits
pH at 25 °C	8.2–8.8
Color	Not exceeding 5 Hazen units
Turbidity	Not exceeding 1.5 NTU
Iron as Fe	Not exceeding 0.1 mg/l
Manganese as Mn	Not exceeding 0.05 mg/l
Aluminum as Al	Not exceeding 0.10 mg/l
Free residual chlorine	0.5–1.5 mg/l
Fluoride as F	± 10% of nominal level (current 0.5 mg/l)
Taste and odor	Unobjectionable
Total coliforms and *E.coli* (no./100 mL)	Absent

Appendix 3.3 Drinking Water and Human Health Quality Criteria (IRMA 2018)

Metals/Metalloids	Units	Criteria	Source
Aluminum	μg/l	100	CAN, WHO
Antimony	μg/l	6	USA, CAN
Arsenic	μg/l	10	USA, CAN, AUS, EU, SAF, WHO
Barium	μg/l	1000	CAN, PER
Beryllium	μg/l	60	AUS
Cadmium	μg/l	5	USA, CAN, EU, SAF, CHI, PER
Chromium (Total)	μg/l	50	CAN, AUS, EU, WHO, SAF, CHI, PER
Copper	μg/l	1000	USA, CAN, AUS
Iron	μg/l	300	USA, CAN, AUS, SAF, CHI
Lead	μg/l	10	CAN, AUS, EU, SA, WHO, CHI, PER
Manganese	μg/l	50	USA, CAN, EU, SAF
Mercury	μg/l	1	CAN, AUS, EU, SAF, PER, PHI
Molybdenum	μg/l	50	AUS
Nickel	μg/l	20	AUS, EU, CHI, PHI
Radium 226/228	Bq/l	13.5	CAN, AUS
Selenium	μg/l	40	WHO, PER
Silver	μg/l	100	USA, AUS
Thallium	μg/l	2	USA
Uranium	μg/l	30	USA, WHO
Zinc	μg/l	3000	AUS, SAF, PER
Non-Metals/Ions	Units	Criteria	Source
Alkalinity (as CaCO3)	mg/l	–	
Ammonia	mg/l	0.5	AUS, EU, PER
Chlorine	mg/l	5	AUS, WHO
Chloride	mg/l	250	AUS, USA, CAN
Cyanide (Free or WAD)	μg/l	80	AUS
Fluoride	mg/l	1.5	CAN, AUS, EU, WHO, PER
Hydrogen Sulfide (as S^{2-})	mg/l	*	
Nitrate (asNO^{3-})	mg/l	45	CAN, USA, CHI
Nitrite (asNO^{2-})	mg/l	3.3	CAN, USA, CHI
pH (standard units)	s.u.	6.5–8.5	USA, CAN, AUS, CHI, PHI
Sulfate	mg/l	400	Value between CAN, PER and USA, WHO, CHI
Total Dissolved Solids	mg/l	500	USA, CAN

Abbreviations for Sources/Standards: *AUS* Australia, *CAN* Canada, *CHI* China, *EU* European Union, *IFC* International Finance Corporation, *PER* Peru, *PHI* Philippines, *SAF* South Africa, *USA* United States, *WHO* World Health Organization of the United Nations

*A limit for hydrogen sulfide is not included because the methods available for analyses are presently well below the method reporting limit (the lowest amount of an analyte in a sample that can be quantitatively determined with stated, acceptable precision and accuracy under stated analytical conditions, i.e., the lower limit of quantitation). However, if there is some reason to believe that sulfide is present, then it should be measured

Appendix 3.4 EPA Standards for Toxin Metals, Pesticides, and Radioactive Components in Drinking Water

Contaminant	Maximum contaminant level (mg/L)
Acrylamide	–
Alachlor	0.002
Alpha/photon emitters	15 picocuries per liter (pCi/L)
Antimony	0.006
Arsenic	0.010
Asbestos (fibers >10 micrometers)	7 million fibers per liter (MFL)
Atrazine	0.003
Barium	2
Benzene	0.005
Benzo(a)pyrene (PAHs)	0.0002
Beryllium	0.004
Beta photon emitters	4 millirems per year
Bromate	0.010
Cadmium	0.005
Carbofuran	0.04
Carbon tetrachloride	0.005
Chloramines (as Cl_2)	MRDL = 4.0[1]
Chlordane	0.002
Chlorine (as Cl_2)	Maximum residual disinfectant level = 4.0
Chlorine dioxide (as ClO_2)	Maximum residual disinfectant level = 0.8
Chlorite	1.0
Chlorobenzene	0.1
Chromium (total)	0.1
Copper	Action level = 1.3
Cyanide (as free cyanide)	0.2
2,4-D	0.07
Dalapon	0.2
1,2-Dibromo-3chloropropane (DBCP)	0.0002
o-Dichlorobenzene	0.6
p-Dichlorobenzene	0.075
1,2-Dichloroethane	0.005
1,1-Dichloroethylene	0.007
cis-1,2-Dichloroethylene	0.07
trans-1,2,Dichloroethylene	0.1
Dichloromethane	0.005
1,2-Dichloropropane	0.005
Di(2-ethylhexyl) adipate	0.4
Di(2-ethylhexyl) phthalate	0.006
Dinoseb	0.007

(continued)

Contaminant	Maximum contaminant level (mg/L)
Dioxin (2,3,7,8-TCDD)	0.00000003
Diquat	0.02
Endothall	0.1
Endrin	0.002
Epichlorohydrin	–
Ethylbenzene	0.7
Ethylene dibromide	0.00005
Fluoride	4.0
Glyphosate	0.7
Haloacetic acids (HAA5)	0.060
Heptachlor	0.0004
Heptachlor epoxide	0.0002
Hexachlorobenzene	0.001
Hexachlorocyclopentadiene	0.05
Lead	Action level = 0.015
Lindane	0.0002
Mercury (inorganic)	0.002
Methoxychlor	0.04
Oxamyl (Vydate)	0.2
Pentachlorophenol	0.001
Picloram	0.5
Polychlorinated biphenyls (PCBs)	0.0005
Radium 226 and radium 228 (combined)	5 pCi/L
Selenium	0.05
Simazine	0.004
Styrene	0.1
Tetrachloroethylene	0.005
Thallium	0.002
Toluene	1
Total Trihalomethanes (TTHMs)	0.080
Toxaphene	0.003
2,4,5-TP (Silvex)	0.05
1,2,4-Trichlorobenzene	0.07
1,1,2-Trichloroethane	0.005
Trichloroethylene	0.005
Uranium	30 μg/l
Vinyl chloride	0.002

Appendix 3.5 Water Quality (Indian Standard Drinking Water Specification (BIS-10500: 1991))

Parameters	Desirable-tolerable	If no alternative source available, limit extended
Essential characteristics		
Turbidity (NTU unit)	Less than 5	10
Color (Hazen scale)	Less than 5	25
Odor	Unobjectionable	–
Taste	Agreeable	–
pH	6.5–8.5	No relaxation
Total hardness (as $CaCO_3$) mg/l, *Max*	300	600
Iron (as Fe) mg/l, *Max*	0.3	1.0
Chlorides (as Cl) mg/l, *Max*	250	1000
Residual, free chlorine, mg/l, *Min*	0.2	–
Fluoride (as F) mg/l, *Max*	1.0	1.5
Desirable characteristics		
Dissolved solids mg/l, *Max*	500	2000
Calcium (as Ca) mg/l, *Max*	75	200
Magnesium (as Mg), mg/l, *Max*	30	100
Copper (as Cu) mg/l, *Max*	0.05	1.5
Manganese (as Mn) mg/l, *Max*	0.1	0.3
Sulfate (as SO_4) mg/l, *Max*	200	400
Nitrate (as NO_3) mg/l, *Max*	45	100
Phenolic compounds (as C_6H_5OH) mg/l, *Max*	0.001	0.002
Mercury (as Hg) mg/l, *Max*	0.001	No relaxation
Cadmium (as Cd), mg/l, *Max*	0.01	No relaxation
Selenium (as Se), Mg/l, *Max*	0.01	No relaxation
Arsenic (as As), mg/l, *Max*	0.05	No relaxation
Lead (as Pb), mg/l, *Max*	0.05	No relaxation
Zinc (as Zn), mg/l, *Max*	5	15
Anionic detergent (as MBAS) mg/l, *Max*	0.2	1.0
Chromium (as Cr^{6+}) mg/l, *Max*	0.05	No relaxation
Polynuclear aromatic hydrocarbons (as PAH) g/l, *Max*	–	–
Mineral oil mg/l, *Max*	0.01	0.03
Pesticides mg/l, *Max*	Absent	0.001
Radioactive materials:		
(a)Alpha emitters Bq/l, *Max*	–	–

(continued)

Parameters	Desirable-tolerable	If no alternative source available, limit extended
(b)Beta emitters pci/l, *Max*	–	–
Alkalinity mg/l, *Max*	200	600
Aluminum (as Al), mg/l, *Max*	0.03	0.2
Boron, mg/l, *Max*	1	5
Bacteriological Standards		
Water entering the distribution system: The coliform count in any sample of 100 ml should be zero. Any sample of the water entering the distribution system that does not conform to this standard calls for an immediate investigation into both the purification process and the method of sampling		
Water in the distribution system should satisfy all the three criteria indicated below:		
The *E. coli* count in 100 ml of any sample should be zero		
Coliform organisms present in any sample should be less than 100 per ml		
Coliform organisms should not be detectable in 100 ml of any two consecutive sample or 95% of the samples collected throughout the year. If coliform organisms are found, re-sampling should be done. The repeated presence of 1–10 coliform organisms in 100 ml or the appearance of higher numbers in any sample necessitates the investigation and removal of the source of pollution		
Unpiped water supplied: If water is supplied by wells, bore holes, and springs, it should satisfy the following criteria:		
Fecal coliforms should be zero per 100 ml of water		
If fecal coliform occurs frequently and sanitary protection cannot be improved, an alternate source must be found, if possible		
Emergency water supplies: Fecal and other coliform number per 100 ml of water should be zero. In case of failure to meet above guideline, boiled water should be used		
Virological quality of drinking water		
The drinking water should be free from virus, which causes diseases to the consumers. It is desirable to examine the raw water sources and treated water for the presence of virus. But it is costly and requires elaborate facilities. In general, virological examination is done for plague forming unit (PFU) of virus. If PFU is zero, water can be considered as safe to drink. However, large volumes of drinking water are required to detect virus (e.g., 10 liters per PFU level). In practice, 0.05 mg/l of free residual chlorine for 1 h is sufficient to inactivate virus		

Appendix 3.6 Different Types of Phytoplankton and Zooplanktons)

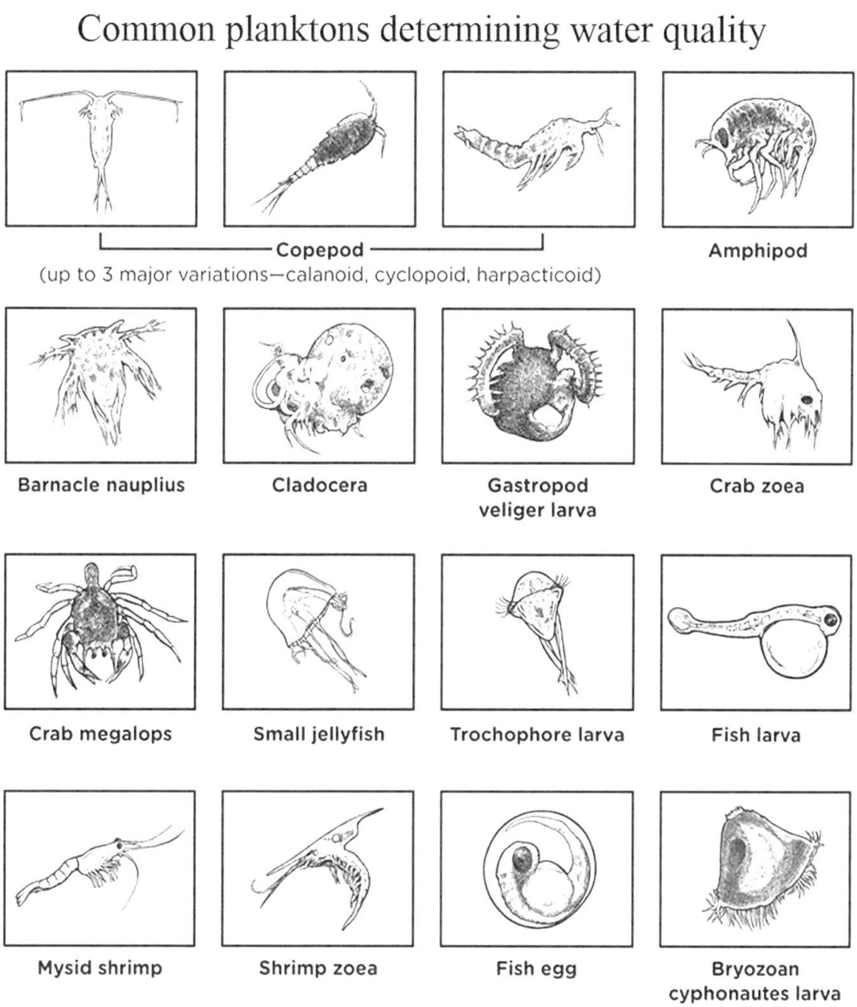

References

Adams, V. D. (2017). *Water and wastewater examination manual*. New York: Routledge.

Agle, M. E. (2007). *Biofilms in the food industry. Biofilms in the food environment* (H. P. Blaschek, H. H. Wang, & M. E. Agle, Eds., Vol. 3). Oxford: Wiley Blackwell.

APHA. (1998). *Standard methods for the examination of water and wastewater* (20th ed.). Washington, DC: American Public Health Association, American Water Works Association and Water Environmental Federation.

APHA-AWWA-WPCF. (1989). *Standard methods for the examination of water and wastewater* (17th ed.). Washington, DC: American Public Health Association.

Danielle, R., & Whitford, G. M. (2014). *Laboratory Procedure Manual. Fluoride in water*. NHANES 2013–2014.

Fons-Castells, J., Tent-Petrus, J., & Llauradó, M. (2017). Simultaneous determination of specific alpha and beta emitters by LSC-PLS in water samples. *Journal of Environmental Radioactivity, 166*, 195–201.

Friedlander, G., Kennedy, J. W., Miller, J. M., & Shore, B. W. (1965). Nuclear and radiochemistry. *Physics Today, 18*, 82.

Fung, Y. S., Wu, Z., & Dao, K. L. (1996). Determination of total organic carbon in water by thermal combustion-ion chromatography. *Analytical Chemistry, 68*(13), 2186–2190.

Goswami, S. C. (2004). *Zooplankton methodology, collection & identification-a field manual*. Dona Paula: National Institute of Oceanography.

Hem, J. D. (1970). *Study and interpretation of the chemical characteristics of natural waters* (2nd ed.). U.S. Geological Survey Water-Supply Paper 1473, p. 361).

http://shodhganga.inflibnet.ac.in/bitstream/10603/27485/2/appendix.pdf

http://site.iugaza.edu.ps/t77328/files/Experiment-6.pdf

http://www.cee.ntu.edu.sg/aboutus/CEELabs/Env/Documents/LabManu/CV2701%202A-5.pdf

https://biocyclopedia.com/index/enivronmental_science_engineering_laboratory_methodology/determination_of_sulphate_and_sulphide_in_water.php

https://clu-in.org/download/ert/1808-r20.pdf

https://responsiblemining.net/wp-content/uploads/2018/08/IRMA_WATER-QUALITY-TABLES_2018.pdf

https://www.canterbury.ac.nz/media/documents/science-outreach/magnesium_calcium.pdf

https://www.epa.gov/sites/production/files/2016-06/documents/npwdr_complete_table.pdf

https://www.perkinelmer.com/lab-solutions/resources/docs/APP_OilandGreaseinWaterbyMid-Infrared.pdf

IRMA STANDARD v.1.0. (2018 June). www.responsiblemining.net

Levaggi, D. A., & Feldstein, M. (1971). *Determination of phenols and trimethylamine in industrial effluents*. NASA Technical Reports.

Mortensen, B. K. (2014). *Detection of biofilms, cracks, and pores in stainless steel tanks in the food industry*. http://www.bactoforce.com/downloadsmedia/

Rodger, B. B., Andrew, D. E., & Eugene, W. R. (2017). *Standard Methods for the Examination of Water and Wastewater* (23rd ed.). Washington, DC: APHA.

Sluiter, A., Hames, B., Hyman, D., Payne, C., Ruiz, R., Scarlata, C., et al. (2008). Determination of total solids in biomass and total dissolved solids in liquid process samples. *National Renewable Energy Laboratory, Golden, CO, NREL Technical Report No. NREL/TP-510-42621*, pp. 1–6.

Smith, A. (1983). *A laboratory manual for the determination of metals in water and wastewater by atomic absorption spectrophotometry*. CSIR Technical Guide K63, pp 1–29, ISBN.0 7988 2482 4.

Surendran, P. K., Thampuran, N., Nambiar, N. V., & Lalitha, K. V. (2006). *Laboratory manual on microbiological examination of seafood* (2nd ed.). Cochin: Central Institute of Fisheries Technology.

Tarver, T. (2009). Biofilms – A threat to food safety. *Journal of Food Technology, 63*(2), 47.

United States Office of Water Environmental Protection Agency (4303). (2010 February). Method 1664, Revision B: n-Hexane Extractable Material (HEM; Oil and Grease) and Silica Gel Treated n-Hexane Extractable Material (SGT-HEM; Non-polar Material) by Extraction and Gravimetry.

US Environmental Protection Agency. (2003). *Standard operating procedure for phytoplankton analysis* (p. 41). Chicago: US Environmental Protection Agency.

World Health Organization. (2004). *Guidelines for Drinking Water Quality* (3rd ed., Vol. 1). Geneva: Recommendations, World Health Organisation.

Toxicants: Assessment of Quality

4

4.1 Analysis of Food Additives

Food is inevitable to sustain life, and as a result, contaminated food becomes a very common source of toxicant exposure to humans. Several naturally occurring and synthetic contaminants find their way into food. More than 2500 chemical substances are added to foods to modify it. They are grouped under the regulatory term "food additives" and are defined as chemical substances deliberately added to foods in known quantities for purposes of assisting in the processing of foods and preservation of foods or in improving the flavor, texture, or appearance of foods. Food additives can be used directly or indirectly. Direct additives are those that are intentionally added to foods for a specific purpose, while indirect additives are those to which the food is exposed during processing, packaging, or storing. If a substance is added to a food for a specific purpose, it is referred to as a direct additive. Food additives and preservatives are used in today's food supply to prevent foodborne illness, enable the transportation of food to areas that otherwise wouldn't be possible, and for the efficient manufacture of products to consistently meet the established quality standards from batch to batch. The US Food and Drug Administration evaluates the safety of food additives and determines how they may be used in the food supply. If an additive is approved, the FDA issues regulations that may include the types of foods in which it can be used, the maximum amounts to be used, and how it should be identified on food labels. This section looks into the analysis of some of the commonly used intentional and unintentional additives.

4.1.1 Synthetic Food Additives and Adulterants

Synthetic food additives can be added intentionally to food to perform certain technological purposes which consumers often take for granted. They are needed to ensure that processed food remains safe and in good condition throughout its journey from factories or industrial kitchens, during transportation to warehouses and

S. Mathew et al., *Fish and Fishery Products Analysis*,
https://doi.org/10.1007/978-981-32-9574-2_4

shops, and ultimately to consumers. WHO encourages national authorities to monitor and ensure that food additives in food and drinks produced in their countries comply with permitted uses, conditions, and legislation. National authorities should oversee the food business, which carries the primary responsibility for ensuring that the use of a food additive is safe and complies with legislation.

Boric Acid and Borates

Boric acid (H_3BO_3) and borax ($Na_2\ B_4\ O_7 \cdot 10H_2O$) are common boron-containing compounds. Boric acid and borax occur naturally in water and soil and are used in a wide range of consumer products, including preservatives and pesticides. Boric acid and borax have long been used as additive in various foods. Since boric acid and borax are effective against yeasts, and to a much lesser extent, against molds and bacteria, they can be used to preserve food products. In addition, both additives can be used to increase the elasticity and crispiness of foods as well as prevent shrimps from darkening. In the early 1900s, some scientists started questioning the safety of the use of large amounts of boric acid and borax in foods. By the mid-1920s, many countries began legislating against the use of these additives in foods due to their toxicity. Nowadays, their uses in food are not permitted in many countries such as Mainland China, Australia, New Zealand, and Canada. The presence of boric acid can be determined by following the procedure given below (Jacobs 1999a).

Principle Boric acid and turmeric react to yield a characteristic red color. The production of this color forms the base of the qualitative test for borates. The volumetric method (quantitative method) for the determination of boric acid is done after ashing. This method might be erroneous when applied to samples of food containing large proportions of fat, or phosphates, or small proportions of boric acid such as those found naturally in certain fruits.

Procedure Turmeric test – Prepare strips of turmeric paper by soaking filter paper in an alcoholic extract of good turmeric. Allow the paper to dry spontaneously and then cut into strips. If the sample is a solid, prepare a paste of the material with water. Otherwise use the sample as it is and add 7 ml of hydrochloric acid for 100 g of sample used. Stir vigorously and dip into the mixture a piece of the prepared turmeric paper. Allow the paper to dry spontaneously. If a red color develops which is changed by ammonium hydroxide solution to a dark blue green and black and again to red by acid, boric acid is present.

Quantitative determination – Boil the acidified ash with methyl alcohol in a 300 ml Kjeldahl flask and carry the vapors over into a vertical, double-surface condenser to which the Kjeldahl is inclined at a 30° angle. The exit tube of the condenser is elongated. It dips below the surface of alkaline methyl alcohol into a conical flask. The flask stands on a tripod and gauze and is heated with a burner. It is closed by a rubber stopper, the condenser exit, and another wide glass tube passing through a second hole in the stopper of the Kjeldahl flask down into the methyl alcohol. Immediately above the stopper of the conical flask is a side tube sealed onto the condenser outlet which runs close to the condenser. It carries a trap bulb and a thistle funnel.

Moisten with 10 ml of 2 N sodium hydroxide solution. If much fat is present, remove it by washing with petroleum ether, decanting the petroleum ether layer, and repeating the process until nearly fat-free. Evaporate the water and, if petroleum ether was used, the petroleum ether on a steam bath, and ash the sample. Transfer the ash to the Kjeldahl flask with minimum water. Wash the dish with dilute sulfuric acid and dissolve the ash in the flask in a further quantity of acid by warming. This dissolves the lumps of ash and removes most of the carbon dioxide. Add methyl red indicator solution and run in 30% sodium hydroxide solution until the color changes to yellow. Concentrate the liquid to 1–2 ml with continuous stirring. After cooling, add 1 ml of methyl red solution and 60 ml of methyl alcohol followed by addition of sulfuric acid drop by drop till the solution is strongly acid to the indicator. Attach the flask to the apparatus and the receiving flask containing 0.5 ml sodium hydroxide solution. Add few drops of phenolphthalein solution. Heat the Kjeldahl flask, and when enough alcohol has been collected in the receiving flask, heat strongly, so that the vapor bubbles vigorously through the acid liquid in the Kjeldahl flask. Adjust the flame so that 15–20 ml of methyl alcohol remains in the flask. Add a further 0.5 ml of the sodium hydroxide solution through the thistle funnel, if during the distillation the color of the phenolphthalein in the conical flask is discharged.

Replace the flame beneath the Kjeldahl flask by a beaker of cold water after 30 min and distill as much of the methyl alcohol as possible up into the Kjeldahl flask for subsequent recovery. Remove the receiving flask followed by washing the condenser tube, both inside and out, with water. Boil the remaining methyl alcohol off, making the solution just acid to methyl red with 0.1 N sulfuric acid. Boil for an additional 5 min to remove carbon dioxide. Adjust the acidity with 0.05 N sodium hydroxide solution after cooling, until it is just acidic to methyl red. Add more phenolphthalein followed by addition of 1 g of mannitol. Carry the titration to the phenolphthalein color change. The result is usually of the order of 0.1 ml of 0.05 N equivalent to 0.0031 g boric acid.

Sulfur Dioxide and Sulfites

Sulfur dioxide and sulfites (E220-E228) are generally used for their preservative effects. The maximum permitted amount of sulfur dioxide or sulfites in breakfast sausages and burger meat (with a minimum vegetable and/or cereal content of 4%) is 450 mg/kg (total of all sulfites). Manufacturers of these products should be aware that seasoning mixes and other ingredients used in the manufacture of these products may already contain these additives.

Principle The presence of sulfites as such in a material is not easy to detect qualitatively within the foodstuff because of the possible presence of other reducing agents. This method is based on the bleaching action of sulfur dioxide on iodine. The sulfur dioxide is carried over the iodine in a foam and current of gas caused by carbon dioxide.

Procedure Place 20–50 g of the food material rubbed up to a paste with 20 ml of water, if solid, in a wide-mouth, 8 ounce bottle. Add 20 ml of a solution containing 2% sodium hydroxide solution and 3% sodium carbonate solution. Then acidify with 30 ml of hydrochloric acid (1:1). When the foam has decreased, suspend in the bottle a piece of starch-potassium iodide paper slightly blued with iodine vapor or by dipping into a dilute solution of chloramine-T. The starch-potassium iodide paper is suspended by inserting it into a slit made in a cork, provided with a vent, that fits the wide-mouth bottle. If sulfur dioxide is present, the paper is decolorized. Measure of the amount of sulfur dioxide present is obtained by the rapidity with which the starch-iodide paper is decolorized.

Rapid quantitative method – The determination of sulfur dioxide or sulfites is generally made on a quantitative basis. The following method is comparatively short, practical, and does away with the sometimes-unnecessary refinements of distillation in an atmosphere of carbon dioxide or some other inert gas. Exactly the same setup issued as that needed for a Kjeldahl ammonia distillation. For this, transfer 50–100 g of the material to an 800 ml Kjeldahl flask. Add 250 ml of water and 15 ml of 20% phosphoric acid to make certain that the mixture is distinctly acid. Connect the flask by means of a trap to a condenser equipped with an adapter dipping into and under the level of a dilute solution of iodine in a beaker. Distill over about 150 ml. If the iodine is decolorized by the liberated and distilled sulfurous acid, sulfites are present. It is possible that other materials may cause the decolorization. Add more iodine solution immediately and continue the distillation until no further decolorization takes place. Care must be taken to see that the distilled solution does not foam over or that the iodine solution is sucked back. Remove the beaker, wash the adapter, catch the washings in the beaker, and boil off the excess iodine. Make acid with hydrochloric acid and add 10% barium chloride solution until no further precipitation takes place. If no precipitate is formed, sulfites are absent. If a precipitate is formed, digest on a hot plate for an hour or leave in a warm place overnight and then filter through ashless filter paper. Wash thoroughly with water. Transfer the filter paper and precipitate to a weighed quartz or platinum crucible previously heated to the same temperature as that to which the ash will be heated and burn to a white ash in a muffle furnace. Remove from the muffle, place in a desiccator to cool, and weigh when cool. The additional weight is due to barium sulfate. Calculate the amount of sulfur dioxide present, and report in parts per million of the original sample.

Care must be taken in the interpretation of results, especially for low amounts of sulfur dioxide. If any doubt arises, a more precise determination in an oxygen-free atmosphere should be made. Such a method is that of Monier-Williams 11 which is applicable in the presence of other volatile sulfur compounds but is not in the presence of nitrates and nitrites.

Monier-Williams method – Connect an 800 ml Kjeldahl flask to a sloping reflux condenser, the lower end of which is cut off at an angle. Pass carbon dioxide from a generator or tank through a sodium carbonate solution. Connect the upper end of the reflux condenser to the bottom of a small flask which is connected in turn with a

Peligot tube. One Peligot tube has been found to be sufficient to catch traces of sulfurous acid swept through the small receiving flask. The receiving flask contains 15 ml of pure neutral 3% hydrogen peroxide solution, and the Peligot tube contains 5 ml of the same reagent. Hydrogen peroxide generally contains free sulfuric acid. In order to free the peroxide from sulfate, start with 30% hydrogen peroxide solution, dilute somewhat, and neutralize with barium hydroxide solution, using bromophenol blue solution as indicator. After the reagent has settled in the cold, filter from the barium sulfate, and determine its strength by a permanganate titration, and finally adjust to a 3% strength. The bromophenol blue indicator in the hydrogen peroxide remains unaffected for some time.

Connect the apparatus, introduce into the flask 300 ml of water and 20 ml of hydrochloric acid, and boil for a short time in a current of carbon dioxide. Then add the food to be tested, adapting the procedure to the sort of food. Add liquids directly by means of a dropping funnel and solids by rapid transfer directly to the Kjeldahl flask. After introducing the food, boil the mixture for 1 h, or 1.5 h in the case of dried foods, in a slow current of carbon dioxide, stopping the flow of water in the condenser just before the end of the distillation. This will cause the condenser to become hot and drive over the residual traces of sulfur dioxide retained in the condenser. When the delivery tube just above the receiving flask becomes hot to the touch, remove the stopper connecting the delivery tube to the upper end of the condenser. Wash the contents of the Peligot tube into the receiving flask, and titrate the liquid at room temperature with 0.1 N sodium hydroxide solution, using bromophenol blue as indicator. The sodium hydroxide must be standardized with this indicator. Bromophenol blue is unaffected by carbon dioxide and also gives a distinct color change in cold hydrogen peroxide solution. One ml of 0.1 N sodium hydroxide solution is equivalent to 3.2 mg of sulfur dioxide, so that titration of small quantities of sulfur dioxide requiring less than 0.5 ml of sodium hydroxide is not accurate. A gravimetric determination may be made after titration, the precipitation of barium sulfate being carried out at room temperature. After allowing the supernatant liquid to settle, filter, and wash the residual barium sulfate three times by decanting with boiling water. Determine a blank on the reagents, both by titration and gravimetrically, and correct the results accordingly.

Formaldehyde

Formaldehyde is a common adulterant in fish. Traders and suppliers use it to extend the storage life of fresh or chilled fish and artificially improve the sensory attributes. Formaldehyde diluted in water is known as formalin. Consumption of fish adulterated with formalin can cause health conditions such as abdominal discomfort, vomiting, renal injury, etc. There are many well-known tests for formaldehyde. Some of these may be performed directly on the food sample, whereas others can only be made on an alcoholic extract or on the distillate from phosphoric acid solution of the food material. Among these tests are the sulfuric acid-ferric chloride reaction of Hehner, the hydrochloric acid reaction of leach, the phenylhydrazine hydrochloride reaction, the phenylhydrazine hydrochloride plus sodium nitroprusside reaction, etc. The following tests are adequate.

Materials

Fuchsin-sulfite reagent –Dissolve 0.5 g pure fuchsin in 400 ml of warm water. Cool, add 2 g of anhydrous sodium bisulfite, and stir until dissolved. Add 4 ml of concentrated sulfuric acid. Transfer to a brown bottle and allow to stand at least overnight before use. This reagent will not deteriorate over long periods of time if kept in a refrigerator.

Rosaniline-sulfite reagent – Dissolve 0.2 g of rosaniline, or an equivalent weight of its salt, in 120 ml of hot water, cool, and add this to a solution of 2 g of sodium bisulfite in 20 ml of water. Finally, add 2 ml of concentrated hydrochloric acid and dilute the entire mixture to 200 ml. This solution should become colorless or nearly so after standing. If it is protected from air, no deterioration occurs.

Procedure Schiff's test – To about 10 ml of the sample, made acid with at least 7 ml of hydrochloric acid to 100 ml of sample or prepared solution or mixture, add 1 ml of fuchsin-sulfite reagent, and allow to stand. If a pink to deep violet color develops on standing, formaldehyde is present. This test can be performed directly on most food products; however, if the color may be masked, or it is desired to make the test more sensitive, it should be performed with 10 ml of a distillate.

Chromotropic acid reaction – When a solution containing formaldehyde is heated with an acid solution of chromotropic acid, 1,8-dihydroxynaphthalene-3,6-disulfonic acid, or its sodium salt, a magenta to violet colored solution is formed. To a few drops of a distillate, or a colorless solution, if that is available, add 5 ml of freshly prepared chromotropic acid solution, prepared by dissolving 50 mg of chromotropic acid orits sodium salt in 100 ml of 75% sulfuric acid, and heat in a water bath for 10 min at 60 °C.

Weinberger dimetol method – Prepare a 5–10% solution of dimethylcyclohexanedione in alcohol. This reagent is also known as methone, dimetol, and dimethylhydroresorcin. To an aqueous solution of the aldehyde, add sodium chloride and neutralize or make faintly acid with acetic acid. Add a few drops of the reagent to the cold solution and stir vigorously. If formaldehyde or any other aldehyde is present, a precipitate will form. Filter off the precipitate, and recrystallize from hot water or alcohol and determine the melting point. Dimethylformaldehyde melts at 187 °C.

Formic Acid

Formic acid is the simplest carboxylic acid. A major use of formic acid is as a preservative and antibacterial agent in livestock feed. It is applied on silage to promote the fermentation of lactic acid and to suppress the formation of butyric acid. Formic acid arrests certain decay processes and causes the feed to retain its nutritive value longer, and so it is widely used to preserve winter feed for cattle. In the poultry industry, it is sometimes added to feed to kill *E. coli* bacteria. Formic acid analysis can be done as described below.

Procedure Steam distill about 50 g of the sample made acidic with 15 ml of 20% phosphoric acid until about 200 ml of the distillate has been collected. Test the distillate for formaldehyde by one of the above tests. If formaldehyde is absent, add to another portion of the distillate sulfuric acid (1:4) and some magnesium filings. The hydrogen produced will reduce the formic acid, if present, to formaldehyde which may then be detected by one of the above methods.

Quaternary Ammonium Compounds

The quaternary ammonium compounds are widely used as surface disinfectants. Within the past decades, there has been interest displayed in the use of quaternary ammonium compounds as food preservatives. Many of these chemicals are surface-active agents. A large number of these surface-active agents have been developed for use as detergents and are generally classified into three groups: (1) the cationic compounds, such as the alkylbenzyldimethylammonium chlorides in which the hydrophobic group is in the cation; (2) the anionic compounds, such as sodium lauryl sulfate in which the hydrophobic group is in the anion; and (3) the unionized compounds, such as the polyethers and polyglycerol esters. The detergents of the cationic group are more effective germicidal agents than the anionic group, and for this reason, this is the group in which the food analyst is most interested.

Principle The alkylbenzyldimethylammonium chlorides form colored salts with bromophenol blue (tetrabromophenolsulfonphthalein) and with bromothymol blue (dibromothymolsulfonphthalein) which are readily extracted from alkaline aqueous solutions by a number of organic solvents, especially the chlorinated solvents. Thus, bromophenol blue, either in its acid form or as the sodium salt, is insoluble in ethylene dichloride; nor does it form salts extractable from alkaline solution with the common primary, secondary, and tertiary amines and alkaloids. In the Auerbach method, the quaternary ammonium-dye salt in carbonate solution is extracted with ethylene dichloride, and the intensity of the color of the extract is measured with the aid of a photoelectric colorimeter.

Procedure Auerbach method preparation of extract – Place 10 g of the sample in a 250 ml beaker, add 100 ml of acetone, and mix with a rod. If a layer of undissolved oil separates on the bottom of the beaker, add sufficient acetone to dissolve it. If gummy material separates, knead it with the rod to press out the oil. Filter with the aid of a Buchner funnel using a 589 SS white ribbon filter or equal. Wash two or three times with 15–20 ml of acetone. Transfer the acetone to a separatory funnel and wash out the suction flask with an equal volume of water, adding the washings to the separatory funnel. Wash the suction flask with 75 ml of petroleum ether and transfer the petroleum ether to the separatory funnel. Shake for 1 min, allowing to separate. Draw off the aqueous acetone layer and discard the petroleum ether phase. Extraction of the undesired fat can be done with two additional 50 ml portions of petroleum ether. Transfer the washed aqueous acetone layer to a beaker. Evaporate on a steam bath till the odor of acetone is gone and the volume has been reduced to less than 75 ml. Cool the solution, transfer to a 100 ml volumetric flask, complete to volume with water, and mix. Transfer 25 ml to a separatory funnel, add 3 ml of bromophenol blue solution, add 1 ml of hydrochloric acid (1:1), and proceed as directed below.

Add 50 ml of ethylene chloride, with the aid of a pipette equipped with a safety pipetter, to the separatory funnel containing the samples prepared as detailed above and shake for 3–4 min. Allow to stand until the layers separate and are clear and draw off the lower layer into a second separatory funnel containing 10 ml of 1% sodium carbonate solution, 10 g Na_2CO_3per liter, and shake for 3–4 min. When the two layers separate, observe the lower layer. If it is blue, the quaternary ammonium compound is present. The depth of color indicates whether it can be estimated photometrically. Draw off the lower layer into a glass-stoppered flask containing 1–2 g of anhydrous granular sodium sulfate, allow to stand for 30 min, and read in an appropriate cell using all adequate light filter. If the color is too deep, acidify the contents of the second separatory funnel with 1–2 ml of hydrochloric acid (1:1), shake until the contents become yellow, and transfer back to the first separatory funnel. Add a second 50 ml portion of ethylene dichloride to the first separatory funnel, shake 3–4 min, and allow to stand until the lower layer is clear.

Estimate a suitable volume of the extract to be used as an aliquot. Transfer this aliquot with a pipette to a 50 ml volumetric flask and fill to the mark with ethylene dichloride. Then continue with the method as detailed above. The clarified colored solution can be decanted into a Klett-Summerson photocolorimeter tube and read in the instrument, using a No. 54 filter, or an equivalent device and filter may be used.

Color Additives

Coloring is added to food to replace colors lost during preparation or to make food look more attractive. It includes both synthetic substances and substances derived from natural sources. The FDA continually monitors reports of problems that may be related to color additives and takes action when necessary. The FDA can issue a warning letter to the manufacturer, detain products before they are shipped to stores, issue import alerts, or even seize products that are found to be unsafe or to contain color additives that are prohibited, misused, or not properly identified as ingredients. The color additives include D&C Green no. 6, D&C Yellow no. 11, Orange SS, etc. They can be broadly classified into oil-soluble and water-soluble dyes based on their solubility and are determined separately.

Oil-Soluble Dyes

Materials

1. Florisil column – Fill tube to height of 10 cm, tapping to pack and remove air. Wash with petroleum ether and drain to top level of column.
2. Alumina column – Add 50 ml petroleum ether to a closed tube, add 18 g alumina, and work plunger to break lumps and remove air. Drain to top level.

3. Magnesia column – Mix equal weights of MgO and Celite 545. Fill tube to 10 cm height and wash with petroleum ether and drain. Compress column with slight pressure.
4. Silica acid column – Add 10 cm of mixture of equal weights silicic acid column and Celite 545 to column, using suction. Tamp and smooth upper surface, and wash with n-hexane using pressure.

Procedure Isolated dye solution is prepared in suitable concentration of $CHCl_3$. Determine spectrophotometric curve of this solution. Compare this curve with that of known dyes in $CHCl_3$ solutions determined on the same instrument under similar conditions. If spectrophotometric data cannot be correlated with that of known color, unknown color may be a mixture. In such cases proceed as below (AOAC 1995).

Dilute 10 ml oil-based sample with 10 ml petroleum ether and place on Florisil column. Elute with petroleum ether. Discard colorless portion and begin collection when color appears. Continue elution until eluate (no. 1) is colorless. Set eluate (no. 2) aside, start elution with alcohol-ether (1 + 3), and watch eluting colors. Change receivers when eluate color changes. Identify receivers by position in scheme. Usually first eluate is yellow from natural color of base oil and has no distinctive spectrophotometric curve. Discard this eluate. Next eluate is D&C Yellow No.11. When alcohol-ether eluate is colorless, begin elution with CH_2CN. This will elute last trace of D&C Yellow No.11, also, Toluidine Red.

Evaporate individual alcohol-ether and CH_3CN eluates to dryness. Dissolve the residues in $CHCl_3$ and dilute to volumes appropriate for spectrophotometer. Scan between 350 and 700 nm and relate with curves of known colors. Separation of D&C Violet No.2 and D&C Yellow No.11 may not be complete, but colors can be identified because adsorption peaks are widely separated. These colors may be separated by extracting D&C Yellow No.11 with 70% alcohol from solution of their mixture in petroleum ether. Evaporate original ether eluate (No. 2) to remove all ether, add petroleum ether eluate (No. 1), and evaporate to 15 ml. Transfer carefully to alumina column. When all solution enters column, wash with 50 ml petroleum ether and discard eluate. Add two 10 ml portions $CHCl_3$. If $CHCl_3$eluate is colorless, discard it. Continue elution with alcohol-$CHCl_3$(1:3) till eluate is colorless. Completely evaporate the solvent and dissolve residue in petroleum ether.

Carefully add solution to MgO column, dropwise at side of tube, with pipet. Apply slight pressure till the solution passes into adsorbent. Then wash column with 25 ml petroleum ether and discard petroleum ether wash. Elute with $CHCl_3$ and watch for colors, collecting separate fractions. First fraction may contain D&C Green No. 6 and Hexyl Blue. Second may contain Orange SS and Oil Red XO. Continue to colorless eluate and alter receivers; then elute with alcohol-$CHCl_3$(1 + 3), changing receivers as different colors appear (Oil Red OS, Yellow AB and OB, and D&C Red No. 17). Evaporate individual solutions, dissolve each in $CHCl_3$, and scan from 350 to 700 nm. Compare curves those from known colors. If curve for blue-green portion does not conform to known color, use the following separation.

Evaporate $CHCl_3$ and dissolve residue in n-hexane. Put on silicic acid column and elute with n-hexane benzene (1 + 1). Collect eluate until colorless and continue elution with benzene until eluate is colorless. If curve for D&C Red No.17 has minimum at 385 nm, Yellows AB and OB may be present. Separate yellows from Red No.17 as follows: evaporate $CHCl_3$. Dissolve residue in minimum volume petroleum ether. Put on MgO column and elute with alcohol-$CHCl_3$(1:3). Collect separate fractions as color changes. Determine color present by evaporating solvent in fraction, diluting to volume with $CHCl_3$, and scanning from 350 to 700 nm. Compare curves with known color on same chart.

Water-Soluble Dyes

Materials

Preparation of column – Lightly tamp glass wool plug into constricted end of chromatographic tube of 100 cm long × 2.54 cm diameter. Prepare thin aqueous slurry of 40 g powdered cellulose and pour into column. Let liquid drain as cellulose settles and add more slurry as needed until all is added. When liquid level drops almost to top of absorbent bed, add 20% NaCl solution. Just before last of this solution enters adsorbent, close constricted end of column. Column may be used immediately or may be stored for several weeks before use.

Procedure Prepare a neutral solution of the dye in concentration suitable for the spectrophotometric analysis. Divide solution into three portions, and to one portion, add few crystals of NH_4CH_3COO. To second portion add HCl to make it 0.1 N. To third portion add NaOH solution to make it 0.1 N. Determine spectrophotometric curves of the three solutions and compare with corresponding curves of known dyes determined under same conditions on same instrument. If spectrophotometric data of unknown color cannot be correlated with that of a known color, unknown color may be a mixture. In such cases subject unknown color solution to chromatography.

Chromatographic separation – To neutral aqueous solution of the color, add enough NaCl to make 20% solution. Pour solution into column so that adsorbent bed is not disturbed; then open constricted end of tube. When last of solution is ready to pass into adsorbent bed, add few ml 20% NaCl solution. If any color moves down column at moderate rate, continue washing with 20% NaCl solution. If all color remains at or near the top of column, change to 10% NaCl solution. If this solution fails to move any color down column, change to 5% NaCl solution. Continue lowering NaCl concentration by half until concentration is found which moves color down the column at moderate rate.

Continue adding appropriate concentration of NaCl solution until color is eluted and collected. If color separates into two or more bands as it progresses down the column, collect each band separately. In some cases, it may be necessary to change to still more dilute NaCl solution to elute upper bands of color. If ≥ 2 bands of color are found, examine each spectrophotometically. If this procedure gives no indication that > 1 color is present, it may be assumed that the color is not a mixture.

Antioxidants

Antioxidants are additives added in foods to prevent or slow down damage caused by free radicals. The analytical method for determining the presence of antioxidants is given below (AOAC 1995).

Materials

Florisil column for cleanup of BHT extract – Insert glass wool plug into chromatographic tube 20 (od) × 250 mm with Teflon stopcock, and add approximately 12 g Florisil. Wash twice with 15 ml portions petroleum ether. Second washing is to be done when liquid level drops just above top of Florisil and then close stopcock. The column should not become dry.

Procedure Propyl gallate (PG) – Approximately 30 g fat (melted) or oil is weighed. Dissolve in 60 ml petroleum ether followed by transfer to 250 ml separator. Add 15 ml H_2O and shake gently for 1 min. Drain aqueous phase into 125 ml separator. Leave any emulsion in organic phase. Extraction of petroleum ether may be repeated with two additional 15 ml portions of H_2O. Add 15 ml ether to combined aqueous extracts and agitate for 1 min. After discarding aqueous phase, evaporate ether to dryness in small beaker. To the residue add 4 ml 50% alcohol, swirl, and add 1 ml NH_4OH. If solution chances to rose, PG is present. Color is unstable and disappears after few minutes.

Nordihydroguaiaretic acid (NDGA) – Petroleum ether solution is extracted (a) by shaking 2 min with 20 ml CH_3CN. Let the layers separate out and later drain CH_3CN into 1 L separator. Extraction is repeated with two additional 30 ml portions of CH_3CN and discards petroleum ether. Dilute combined CH_3CN extracts with 400 ml H_2O, add 2–3 g NaCl, and shake 2 min with 20 ml petroleum ether. Let the layers separate. Drain diluted CH_3CN solution for further extraction. Combine petroleum ether extracts in 100 ml beaker. This is used for BHA and BHT tests.

Ether – Petroleum ether in the ratio 1:1(50 ml) is added to diluted CH_3CN and shake 2 min. Allow the layers to separate. Discard CH_3CN followed by evaporating ether just to dryness in a small beaker. Add 4 ml 50% alcohol, whirl, and then add 1 ml 1% $Ba(OH)_2$solution and mix. Solution turns blue and fades rapidly, if NDGA is present.

Butylated hydroxyanisole (BHA) – Take 1/3 of combined petroleum ether solution kept aside for BHA-BHT tests. Evaporate this solution just to dryness in beaker, using gentle heat, under air. Add 2.5 ml alcohol to dissolve residue. Dilute with 2.5 ml H_2O. Swirl, add 1 ml Ehrlich reagent, immediately add 1 ml 1 N NaOH, and swirl again. BHA is present, if solution turns red purple.

Butylated hydroxytoluene (BHT) – Remaining 2/3 combined petroleum ether solution is passed through Florisil column and elute with 150 ml petroleum ether.

Collect eluate in 200 ml beaker followed by evaporation till it is dry. Add 2 ml dianisidine solution and blend. Add 0.8 ml of 0.3% $NaNO_2$ solution. Mix, and let stand 5 min; then pour into a small separator. Add 0.5 ml $CHCl_3$, swirl vigorously for 30 s, and let it separate. BHT is present, if $CHCl_3$ turns pink to red. Confirm BHT by comparing spectrophotometric curve of colored $CHCl_3$ extract with control prepared from reference standard BHT.

4.1.2 Analysis of Antibiotics

Aquaculture is marked by unregulated use of antibiotics and limited attention to waste management in policy and practice. The analytical method for detecting antibiotic sulfonamide is given below:

Thin Layer Chromatographic Screening Method (AOAC 1995)
Principle: Sulfonamide is extracted with ethyl acetate after addition of sulfapyridine as internal standard. Extracts are cleaned up by partitioning between organic and aqueous solvents and chromatographed on silica gel TLC plates. Developed plates are treated with fluorescamine and scanned by fluorescence densitometer.

Procedure Sample extraction – Accurately weigh 2.5 g homogenized liver or muscle into 50 ml centrifuge tube. Add 100 μ 1 internal standard solution. Prepare three control samples (using tissue known to be free of sulfonamides) and fortify with internal standard (0.1 ppm) and each sulfonamide of interest, one control each at 0.05, 0.10, and 0.20 ppm. Wait 15 min and then add 25 ml ethyl acetate. Blend muscle sample 1 min and centrifuge. For liver samples, cap tube tightly, shake tube 20 min on horizontal shaker, and centrifuge. Transfer ethyl acetate to clean tube and discard tissue. Add 10 ml glycine buffer to extract, mechanically shake 5 min, and centrifuge. Vacuum-aspirate and discard organic phase. Adjust pH of aqueous phase to 5.2–5.3 by adding 2 ml (1 + 1) mixture of 2 M pH 5.25 phosphate buffer and 1.75 M HCL. Check pH and make final adjustment with either additional buffer or 0.1 N NaOH. Add 5 ml hexane, mechanically shake 5 min, and centrifuge. Aspirate and discard hexane phase. Remove any solid or emulsified material remaining at interface. Add 10 ml CH_2Cl_2, shake 5 min, centrifuge, aspirate, and discard aqueous phase. Add 10 μl diethylamine to CH_2Cl_2 extract and concentrate just to dryness under stream of N_2 at 40 °C. During evaporation, occasionally rinse tube walls with CH_2Cl_2. Redissolve residue in 100 μl methanol and mix 30s on vortex mixer. Let it stand 5 min before chromatography to let insoluble oils settle.

Chromatography – Apply 20 μl portion of sample to pre-absorbent spotting zone of TLC plate. Avoid using lane at each side of plate and spot the three fortified control samples at intervals across plate to minimize effects of across plate variations. Develop plate 1 cm in methanol followed by two developments, 6 cm and 12 cm, in CH_2Cl_3-tert-butanol (80 + 20). Between each development, dry plate 1 min at 110°. Maintain development tank temperature at 30–33° for optimum resolution of multiple sulfonamides. Derivatize compounds by quickly dipping plate in fluorescamine solution. Bands are visible after 15–30 mins. Scan each lane and obtain its response scanning curve. For each sample and standard (each lane), determine ratio of response for sulfonamide of interest to response for internal standard (SPY). For each sulfonamide, calculate slope and intercept of standard curve; concentration v/s response ratio, using linear regression.

4.1.3 Analysis of Pesticide

Principle Chlorinated pesticides and polychlorinated biphenyls (PCBs) are extracted from prepared fish sample with petroleum ether, cleaned up on Florisil column, and determined by GC against reference standards (AOAC 1995).

Procedure Extraction – Weigh 20 g thoroughly ground and mixed sample into metal blender cup. Moisten 40 g granular Na_2SO_4 with petroleum ether and add to sample. Mix sample, using stirring rod, let stand 20 min, and mix again. Add 10 ml petroleum ether to sample and blend 1–2 min (Lourdes blender in series with rheostat set at 40–60%, or equivalent, may be used). Centrifuge balanced sample cup 1–2 min at 2000 rpm to obtain clear petroleum ether extract. Place glass wool plug in funnel, overlay with granular Na_2SO_4, and place funnel in 250 ml volumetric flask. Decant petroleum ether extract through Na_2SO_4into volumetric flask. Mix sample again with stirring rod, add 100 ml petroleum ether, and extract as before. Repeat using 70 ml petroleum ether. Dilute to volume with petroleum ether. Transfer 25 ml aliquot to tard 1 ml flat bottom extraction flask. Place flask on steam bath to evaporate solvent, leave additional 30 min on steam bath, remove, and cool. Weigh flask and determine % fat in fish.

For fish containing < 10% fat, transfer 25 ml, and aliquot to 125 ml Kuderna-Danish concentrator. For fish containing >10% fat, take aliquot containing not > 2 mg fat. Add several granules of 20–30 mesh carborundum and concentrate to ca 3 ml on steam bath. Let cool and remove Snyder column. Rinse concentrator with two 1 ml portions of petroleum ether, and using only current of air, concentrate sample to 3 ml for transfer to Florisil column.

Florisil cleanup – Use 4 g Florisil adjusted for lauric acid value. Add Florisil to 300 × 10 mm ID chromatographic tube, and add $Na_2SO_4$to height 2 cm above Florisil. Completely open stopcock tap tube to settle adsorbent and mark tube 1 cm above Na_2SO_4 layer. Add 20–25 ml petroleum ether wash to Florisil column; as solvent level reaches mark, place 125 ml K-D flask under column. Using disposable Pasteur pipet, transfer 3 ml sample to column and wash tube with 1 ml petroleum ether and add wash to column. Solvent level must not go below mark. Temporarily close stopcock if necessary. Add 35 ml petroleum ether-ether mixture (94: 6) and elute PCBs and DDT and its analogs. When solvent reaches mark, change K-D flask and add 35 ml petroleum ether-ether (85:15) to elute compounds such as dieldrin and endrin. Add few granules of carborundum to first concentrator, attach Snyder column, and carefully concentrate on steam bath. Remove Snyder column and evaporate solvent under air to appropriate volume for GC determination.

Gas chromatography – Instrument consisting of on-column injection system, all-glass column in oven controlled to ±0.1°, electron capture, and thermionic detectors. Column is made of glass, 1.85 m × 4 mm id packed with 10% DC-200 (weight/weight) on solid support.

Weigh 2 g Dow Corning 200 silicone fluid into beaker. Dissolve in $CHCl_3$ and transfer to 300 ml Morton-type flask, using total of 100 ml $CHCl_3$. Add 18 g solid support to flask. Swirl and let stand for 10 min. Place flask on rotary evaporator and remove solvent slowly with intermittent rotation, using 50° water bath and slight vacuum. Foaming may occur initially. When solids appear damp, increase vacuum. Remove last traces of $CHCl_3$ without rotation or by air-drying. Use only free-flowing material to fill column. Use care at all stages of column preparation to prevent fracturing solid support. Condition column at 250–260° with N_2 flow of ca 100 ml/min ≥ 48 h or until endrin exhibits single peak.

Electron capture detector (ECD) – Concentric design. For use with dc voltage supply and ^3H source. Determine detector operating characteristics as follows: Apply dc voltage to detector. After system becomes stable (overnight), determine current-voltage relationship at various voltages between 200 and 0 V. Slightly lower, stable standing curves may be obtained after detector has been at operating temperature several hours. This is probably due to loss of some easily removed radioactive material. Determine and plot response-voltage relationship at 1×10^{-9}amp full-scale sensitivity for 1 ng injections of heptachlor epoxide at same voltages used in obtaining current-voltage curve. Select as operating voltage that voltage at which heptachlor epoxide causes an 40–50% full-scale recorder deflection. Check linearity of system from 0.2 to 2.0 ng heptachlor epoxide.

Other electron capture detector may be substituted for the voltage concentric design ^3H detector, which is no longer marked. Constant current, variable frequency ^{63}Ni electron capture detectors are acceptable substitutes when operated at conditions to produce stable reproducible linear responses. Optimum conditions may produce more sensitive response than from ^3H detector. Use Ar-CH_4 gas, as recommended for most ^{63}Ni detectors.

Concentration techniques – Weigh 25–50 g fish sample. Thoroughly ground and mix sample into high-speed blender. (If fat content is known or can be estimated, adjust sample size so that maximum of ca 3 g fat will be extracted). Add 100 g anhydrous Na_2S0_4 to combine with water present and disintegrate sample. Alternately blend and mix with spatula until sample and Na_2SO_4 are well mixed. Scrape down sides of blender jar and break up caked material with spatula. Add 150 ml petroleum ether and blend at high speed 2 min. Decant supernate petroleum ether through 12 cm Buchner, fitted with two sharkskin papers, into 500 ml suction flask. Scrape down sides of blender jar and break up caked material with spatula. Re-extract residue in blender jar with two 100 ml portions petroleum ether and blend 2 min each time. (After blending 1 min, stop blender, scrape down sides of blender jar, and break up caked material with spatula; continue blending 1 min).

Scrape down sides of blender jar and break up caked material between extractions. Decant supernate petroleum ether from repeat blending through Buchner and combine with first extract. After last blending, transfer residue from blender jar to Buchner and rinse blender jar and material in Buchner with three 25–50 ml portions of petroleum ether. Immediately after last rinse, press residue in Buchner with bottom of beaker to force out remaining petroleum ether. Pour combined extracts through column of anhydrous Na_2SO_4 and collect eluate in 500 or 1000 ml Kuderna-Danish concentrator with plain tube. Wash flask and column with small portions of petroleum ether and evaporate most of petroleum ether from combined extracts and rinses in Kuderna-Danish concentrate. Transfer fat solution to tared beaker, using small amounts of petroleum ether. Evaporate petroleum ether at steam bath temperature under current of dry air to obtain fat. When petroleum ether is completely removed, weigh and record weight of fat extracted.

Calculation Record weight of fat taken for cleanup. (Weight of fat from cleanup/weight of fat extracted) × (weight of original sample = weight of sample analyzed). If it is known that ≤3 g fat will be extracted from particular sample, do not isolate and weigh fat before CH_3Cl partitioning. Determination is then on basis of weight of original sample.

4.1.4 Analysis of Heavy Metals

Principle The determination of metals in foods resolves itself into the problem of determining those metals in the presence of organic material. The organic material must to be destroyed or removed before the estimation of the metal is made and then the usual qualitative or quantitative method applied (Jacobs 1999a).

Procedure Preparation of ash – there are two general methods for the destruction of interfering organic matter. The first is called the "wet ash" or acid-digestion method, and the second is ordinary ashing by means of heat with, or without, the aid of an "ash aid."

Wet ash or acid digestion – depending on the type of food stuff and whether the metallic contamination is throughout the product (might be possible in the case of fish) or whether it is exclusively on the outside, weigh 100–200 gms of the product. Place the weighed portion in one or more 800 ml Pyrex Kjeldahl flasks. Add 50 ml of nitric acid and then carefully add 20 ml of sulfuric acid. Heat cautiously so that no excessive foaming takes place. Add nitric acid in small portions until all the organic matter is destroyed. This point is reached when no further darkening of the solution occurs on continued heating after the production of a clear solution and copious fumes of sulfur trioxide. Cool and add 75 ml of water and 25 ml of a saturated solution of ammonium oxalate to aid in the expulsion of nitrogen fumes. Evaporate again to the appearance of sulfur trioxide fumes. Cool, dilute with water, transfer to a 500 ml or liter volumetric flask, and make to volume. Use aliquot portions for the analysis detailed below.

Ash by ignition – weigh a representative portion of the material to be analyzed and transfer to a porcelain dish or casserole of convenient size. This will be from 5 to 200 gms depending on the amount of metal in the food. Add 2 to 5 ml of a solution of aluminum nitrate and calcium nitrate, 40 g of $Al(NO_3)_3$ $9H_2O$ + 20 g of Ca $(NO_3)_2$ $.4H_2O$ into 100 ml of water. An excess of lime or magnesia can be used as a substitute. Dry in a thermostatically controlled oven at 1000 °C. Char the material, controlling swelling, if any, by playing the flame from a glass jet over the sample. Ash in a muffle overnight, if possible, at not over 4500 °C. If a clean ash cannot be obtained, cool the dish and add more ash-aid solution, that is, aluminum nitrate-calcium nitrate solution, or 2–3 ml nitric acid, if permissible, dry, and re-ash. Do not add nitric acid if an alkaline ash is necessary, as, for example, if arsenic is to be estimated. Dissolve in an appropriate solvent when the ashing process is complete and proceed with one of the methods detailed below.

In general, the qualitative demonstration of the presence of heavy metals in foodstuffs can be made by using the group sulfide precipitations. For this, transfer the wet ash to a beaker, dilute, and neutralize. Add 1 ml of hydrochloric acid (1:3) to each 10 ml of solution and warm to 500 °C. Add an equal volume of saturated hydrogen sulfide solution and allow to stand for 10 min at 350 °C. If no color or precipitate is produced, the following metals may be considered absent in more than traces: silver, arsenic, antimony, tin, copper, mercury, bismuth, thallium, and cadmium. Lead may remain behind as $PbSO_4$, lead sulfate, in the wet-ash process; hence, if any residue remains, it should be investigated separately with ammonium acetate solution.

If the mixture of sample solution and hydrogen sulfide is now filtered and made alkaline with ammonium hydroxide, any precipitate which forms may be due to one of the following: aluminum, chromium, zinc, manganese, iron, cobalt, and nickel. This portion of the test is not of much value because many food materials contain iron, which would give the test.

Arsenic

Principle: The Reinsch test is a simple, although not very sensitive, one. It is based on the deposition of arsenic from solution as a copper arsenide. This test may vary and often be applied directly without previous destruction of organic material.

Procedure Place 200 ml of the liquid food or beverage, or of a mixture of water and the solid food, in a casserole or similar container and acidify with 1 ml of arsenic-free hydrochloric acid. Then evaporate to one-half its volume. Add 15 ml more of hydrochloric acid and also a piece of pure burnished copper foil. Keep the liquid simmering for an hour and replace the water lost by evaporation from time to time. If at the end of this time the copper foil remains bright, arsenic is absent. If the copper has a black or brown deposit, remove it and wash well with water, alcohol, and ether, and dry. Place the foil in a subliming tube and heat over a low flame. If a sublimate is present, examine under a microscope. Arsenic forms tetrahedral crystals in contradistinction to mercury. Antimony, silver, and bismuth will also give a deposit but will not sublime.

Antimony

Antimony may get into foods that are cooked or processed in enamelware, the enamel of which was made with antimony compounds, or it may contaminate foods that are covered with antimony-bearing tinfoil. Acids, such as citric, may at times extract sufficient antimony to have an emetic or even more harmful effect. It may be detected quantitatively by the Reinsch and other tests. The Reinsch test is detailed in the section, "Arsenic."

Lead

Lead is considered far more toxic than formerly, and the determination of small amounts is important in biochemical estimations as an indication of lead poisoning. Lead sometimes gets into foods which are processed in lead lined tanks or pipes or that are stored in leaded containers or in containers that are soldered.

Procedure Sulfide method – The following method based on the precipitation of lead as the sulfide is a comparatively simple one and is adaptable to all types of foodstuffs. For determining lead, ash an appropriate amount of the material in a porcelain or silica dish, and after obtaining a clean ash as directed in a preceding section, dissolve the ash in 10 ml of water and 1 ml of nitric acid. Filter and make the colorless filtrate alkaline with ammonia. The phosphate precipitate will contain practically all of the lead, and the copper will be in the filtrate. Filter, wash well, and dissolve the precipitate in 5 ml of dilute acetic acid. Make up to 50 ml in a Nessler tube. Add 5 ml of hydrogen sulfide solution and match the color with standard lead solutions treated the same way. The lead sulfide precipitate may be stabilized using gums, Arabic or ghatti, and may then be estimated colorimetrically against a standard as detailed in the method for antimony.

Dithizone method –These methods have the ability of detecting very small quantities of lead and are based on the formation of a red precipitate of a lead-dithizone complex, which is soluble in chloroform or carbon tetrachloride when an ammoniacal cyanide solution of dithizone is added to a solution containing lead.

Dithizone is the short name for diphenyl thiocarbazone (Fig. 4.1). It is the type of reagent which is best used for estimation of low concentrations. It forms green solutions in chloroform. The lead complex has a red color and is soluble in chloroform but is practically insoluble in dilute ammonia, whereas dithizone itself is

Fig. 4.1 Diphenyl thiocarbazone

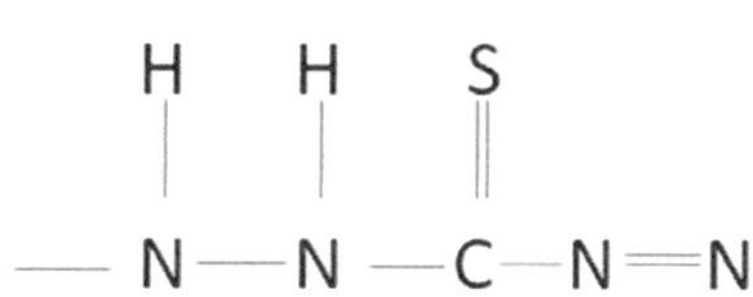

soluble in this solvent. Upon these factors, the various methods for the isolation and subsequent determination of lead depend.

The nature of the reaction that takes place between dithizone and a metallic salt and the structure of the resulting compound are not definitely known, but it is probable that the hydrogen of the NH group is adjacent to the phenyl group. An enol complex may also be formed in which an additional metal ion is attached to the sulfur atom with a double bond between the central carbon atom and a nitrogen.

Mercury

This metal generally comes in contact with food products because it is a component of some insecticides and fungicides. It is also the metallic constituent of many organic antiseptics and coloring matters, as, for instance, Merthiolate and Mercurochrome.

Procedure Mercury may be detected directly in foodstuffs by applying the Reinsch test. This is carried out in a manner similar to the test for arsenic. A piece of pure burnished copper is immersed in a mixture of the material to be analyzed, with 1/5 its volume of hydrochloric acid, and the mixture is allowed to simmer for a number of hours. A bright lustrous mirror is formed on the copper in the presence of mercury. The piece of foil is washed with water, alcohol, and ether, dried, placed in a subliming tube, and heated. The mercury will deposit in the cool part of the tube and may be identified under the microscope.

Electrolytic method – If small quantities are present, the mercury may be electrolytically determined, either in a nitric acid electrolyte or using an alkaline sulfide electrolyte. The mercury may be deposited on a gauze cathode while employing rotating anodes with 2.5 amperes of current. The cathode is washed with water and dried in a desiccator. Do not dry mercury deposits in an oven. The cathode may be placed in a subliming tube and then heated. The mercury will sublime and deposit in the cool part of the tube. A more detailed method is the following.

The sample should be transferred to a 500 ml Kjeldahl flask and boiled off the alcohol very rapidly for 5–10 min. Cool and add 10 ml of concentrated sulfuric acid and then 4 g of potassium permanganate. Wash down the neck of the flask with water. Allow the sample to digest at a temperature just below boiling for 2 h. Decolorize with oxalic acid, for which 3 g is generally required, warming if necessary. If the reaction mixture is not cold, add the oxalic acid in small amounts during this step. The solution is transferred to a 250 ml wide-mouth glass-stoppered centrifuge bottle, add 1 ml of 0.5% copper sulfate solution, and then pass in hydrogen sulfide for about 0.5 h. Stopper the flask and permit the precipitate to settle overnight. Wash by centrifuging. Pass chlorine gas into the centrifuge bottle containing the washed mercury and copper sulfides after the addition of 5 ml of water. Solution is generally affected in 15 min. Aspirate air through the flask in order to remove the excess chlorine.

Transfer this solution to a 50 ml beaker. Saturated oxalic acid solution (2 ml) and saturated ammonium oxalate solution (5 ml) are added to this. Plate out the mercury, using a pure gold cathode, 1 by 3 cm. and ¼ mm thick. Keep the voltage at about 1.3–1.5 volts. At this voltage, 18–24 h is required for complete deposition. The gold electrode can be easily made in the laboratory. Cut a piece of gold foil to the specified dimensions and weld it to a platinum wire by heating the wire and foil in position on an anvil with a Bunsen flame, finally tapping gently with a small hammer. Care must be taken not to melt the gold by excessive heating. Wash the electrode upon which both copper and mercury have deposited with water, alcohol, and ether, successively, dry in a desiccator, and weigh on a microbalance.

After weighing, place the electrode in a Pyrex combustion tube, pass through a stream of hydrogen, heat the tube carefully, and drive off the mercury. After cooling in a desiccator, weigh again. The difference in weight represents mercury. Never dry the original mercury deposit in an oven.

Cadmium

It is considered poisonous in small amounts. Cadmium may be detected in the wet-ash solution after the total expulsion of nitric acid by precipitation as the sulfide, by separation from copper, if necessary, and reprecipitation as the sulfide.

Principle A method for the detection and estimation of cadmium is based on the separation of cadmium from other metals as the sulfide, with its subsequent estimation by the amount of yellow color produced in a solution containing cadmium when viewed under a mercury-arc lamp. In concentrations of less than 0.1 mg of cadmium in 50 ml of solution, differences in the yellow color of cadmium sulfide are indistinguishable in ordinary light, while under the quartz mercury-vapor lamp, the yellow color is perceptible in concentrations as low as 0.01 mg per 50 ml of solution.

Procedure Sulfide method – Add sufficient nitric acid to the sample to cover it and heat gently. After the solid material has dissolved, add 10 ml of concentrated sulfuric acid and add, when necessary, small amounts of nitric acid until oxidation is complete. The method described for arsenic may also be used. Dilute to 75 ml and add the equivalent of 0.5 mg of copper and 2 g of sodium citrate. The copper is added to act as an entrainer or collector. Neutralize the acid solution for the first precipitation with ammonium hydroxide and adjust the concentration of hydrogen ion to pH 3 by means of the indicators thymol blue (thymolsulfonphthalein) and bromophenol blue (tetrabromophenolsulfonphthalein). Saturate with hydrogen sulfide solution for 5–10 min, add 1 drop of 5% aluminum chloride solution, and allow the solution to stand for 6–12 h. Filter, dissolve the precipitate in nitric acid and hydrochloric acid, and carefully evaporate to dryness. Repeat the precipitation as sulfide twice more, omitting the addition of sodium citrate the last time and adjusting the pH to two by means of dilute potassium hydroxide. Carefully evaporate the final solution of chloride to dryness, dissolve it in water, and make up to a convenient exact volume in a volumetric flask. Transfer an aliquot portion of this prepared solution to a Nessler tube for final reading. To each tube add five drops of 10%

potassium cyanide, water, and 5 ml of hydrogen sulfide water, and make to volume. Mix thoroughly and compare under a flood of ultraviolet light with standards similarly prepared. The solution should exhibit a bright clear yellow color under the mercury arc. Dark or turbid solutions may indicate incomplete removal of iron.

For larger quantities of cadmium, the following method may be used. Neutralize the wet-ash solution, make slightly acid but sufficiently so as to hold all the zinc in solution, and pass in hydrogen sulfide. Yellow cadmium sulfide is precipitated. If copper is present, it may be separated as follows: redissolve the sulfides in either sulfuric or hydrochloric acid. Add an excess of sulfurous acid to make certain that no oxidizing medium exists and then add N ammonium thiocyanate, $Cu_2(SCN)_2$. Filter and wash with cold water. Collect the filtrate and the washings. Precipitate the cadmium in the filtrate with hydrogen sulfide and estimate in a manner similar to lead by comparing with standard cadmium sulfide precipitates in Nessler tubes. This method has a large error, at times, since cadmium sulfide is often contaminated with a basic salt in the hydrogen sulfide precipitation.

Copper

Copper often occurs in foods that have been processed in copper kettles. The copper is oxidized to copper carbonate which is soluble in organic acids. Thus, tomato products often are contaminated with copper. Copper salts are sometimes added to intensify the green color of chlorophyll. The views on copper contamination have undergone changes, and it is now accepted that minute amounts of this metal are needed for proper human metabolism.

Principle In iodide thiosulfate method, the copper is separated from the other materials by means of a sulfide precipitation, and it is subsequently estimated by iodine liberated in the cupric-cuprous iodide reaction.

Procedure Weigh 100 g of the sample into a porcelain dish, about 9–10 cm, and add 5 ml of a mixture of 5 ml of sulfuric acid and 95 ml of 95% alcohol. Burn to a white ash. Add 100 ml of water after cooling and hydrochloric acid until slightly acid. Try to dissolve all of the ash. Filter and, if necessary, burn again. Pass in hydrogen sulfide for 15–20 min. Filter through quantitative paper. Wash. Place the filter paper plus the precipitate directly into a 100 ml squat flask with a wide-mouth, and add 3–4 ml of sulfuric acid and 6 ml of nitric acid. Place glass hooks on the flask and cover with a watch glass. If the resultant mixture is dark, add nitric acid until it is clear. Evaporate to 1 or 2 ml, cool, and add 30 ml of water and an excess of bromine water. Place the flask on the steam bath until the solution is colorless, after which cool and add ammonium hydroxide solution. In case iron is present, filter and then evaporate off the ammonium hydroxide solution. Make acid with acetic acid and titrate with 0.01 N sodium thiosulfate solution in the presence of about 5 g potassium iodide and starch solution. Factor on 0.01 N thiosulfate solution divided by 2 multiplied by the titration equals the mg of copper per 100 g.

Standard curve: a standard curve can be prepared. Standardize a 1% solution of the quaternary ammonium preparation by the ferricyanide method. Then ascertain the maximum and minimum quantities of the material that will produce in 50 ml of ethylene dichloride using the method detailed, colors having densities within the range of the instrument used. Prepare a set of three or more standards containing, in 50 ml, appropriate quantities of the quaternary covering the range with a wedge photometer. Standards containing 0.0, 0.1, 0.2, and 0.25 mg per 50 ml are adequate.

Detection of Metals by Inductively Coupled Plasma Spectrometer

ICP, abbreviation for inductively coupled plasma, is one method of optical emission spectrometry. The ICP is for elemental analysis in water and solution systems. It has a solid-state CID detector, which can simultaneously analyze sample for multiple elements. The design has been optimized to offer excellent resolution over the entire spectrum from 166 to 847 nm enabling access to all wavelengths and minimizing spectral interference.

Principle Plasma is a stream of highly ionized gas containing an equal number of electronics and positive ions. Plasma is electricity conductive. It is affected by a magnetic field. When plasma energy is given to an analysis sample from outside, the component elements (atoms) are excited. When the excited atoms return to low-energy position, emission rays (spectrum rays) are released, and the emission rays that correspond to the photon wavelength are measured. The element type is determined based on the position of the photon rays, and the content of each element is determined based on the rays' intensity.

To generate plasma, first, argon gas is supplied to torch coil, and high-frequency electric current is applied to the work coil at the tip of the torch tube. Using the electromagnetic field created in the torch tube by the high-frequency current, argon gas is ionized, and plasma is generated. This plasma has high electron density and temperature (10,000 K), and this energy is used in the excitation-emission of the sample. Solution samples are introduced into the plasma in an atomized state through the narrow tube in the center of the torch tube. The steps leading to the emission are desolvation, vaporization, atomization, and ionization.

Procedure: Sample should be homogeneous, representative of bulk, free of suspended particles, and free flowing. It was introduced in the form of droplets/fine spray. Sample should be digested in microwave digestion. Take 0.25–0.50 g of sample to vessel. Add 8 ml nitric acid followed by slow addition of 2 ml H_2O_2. Keep for 10 min. Close the vessel and keep in microwave digestion chamber for digestion. After digestion the sample should be made up to 100 ml. After every digestion the vessels should undergo a blank run in microwave digestion by adding acid only. Digested sample can be introduced to ICP (inductively coupled plasma) spectrometer for analysis.

4.1.5 Analysis of Metallothioneins

Metallothioneins (MTs) are a class of ubiquitously occurring low-molecular-weight cysteine- and metal-rich proteins containing sulfur-based metal clusters. High-performance liquid chromatography (HPLC) and capillary electrophoresis (CE) are able to separate different MT isoforms; however, UV detection has relatively poor sensitivity and no metal-specific capabilities. HPLC followed by metal-specific detection using atomic absorption spectrometry (AAS) or inductively coupled plasma atomic emission spectrometry (ICP-AES) was employed for the detection and determination of metal-binding proteins as an early hyphenated technique. HPLC systems with the combination of two different types of chromatography, such as size-exclusion column chromatography (SEC) and ion-exchange chromatography, or tandem connection with SEC are used.

From the physiological point of view, it is very difficult to classify the MTs. This has led to the development of numerous methods for MT quantification. Since MT has no known catalytic function, measurement of their concentration is purely based upon quantitative assay of the protein itself. A number of methods have been employed such as electroanalytical technique, UV-VIS spectrophotometry, metal saturation assay, and immunological assay such as enzyme-linked immunosorbent assay (ELISA) and radioimmunoassay (RIA). The basis for spectrophotometric assay is the ability of absorption of radiation in combination with mercaptons. Some of the major methods for analyzing the MTs are discussed below.

Procedure Chromatographic method (Satomi et al. 2014) – MT has got small molecular mass which makes them suitable candidate for chromatographic quantification. If metal content is low in marine environment as detected by atomic absorption spectroscopy, then ion-exchange or gel chromatography is mainly employed. MT is size-specific; therefore, gel chromatography is the best technique for estimation. The column packing agent or gel for separation of MT should possess the pore size of 10–100 nm. Silicates or organic polymers have the pore size between this range, which is why they are proven to be the best packing agents for this process. The mobile phase is always either water or buffer because they are neutral with respect to interaction with MT and prevent metal dissociation from MT. Furthermore, ion-exchange chromatography has also been proven to be a good detection technique. In ion-exchange chromatography, as MT has got high affinity for stationary phase, a high ionic strength buffer is utilized to elute out the MT from stationary phase. High ionic strength buffer alters the three-dimensional structure of MT. To overcome this copolymeric styrene, divinylbenzene is used as column packing material. If metal content is very high in the marine environment, then MT quantification is done by fluorescence detection using high-performance liquid chromatography (HPLC). Supelcosil LC-18 (0.46 cm × 25 cm, 5 μm particle) is used as a column, fluorescence light detector is used as a detector with excitation wavelength 382 nm, emission wavelength is 470 nm, and sample injection volume is 20 μL. Mass spectrometry is most often used to identify MTs from fractions obtained from chromatography techniques.

Electrophoresis – Electrophoresis has become best alternative to chromatographic techniques as it can provide high resolution, high efficiency, and rapid analysis from small sample size of MT. MT can be separated and detected by both native and denatured sodium dodecyl sulfate polyacrylamide gel electrophoresis(SDS-PAGE). For proper separation of MT, the gel should have 15–17% acrylamide in it. Furthermore, gradient gel is also considered as proper gel for MTs separation. Native electrophoresis uses tris-glycine as buffer. Coomassie blue and silver staining are used for quantification of MT-1 and MT-2 isoforms of MT.

4.2 Analysis of Biotoxins

Toxins are specific chemical products of microbes, plants, and some animals that are poisonous to other organisms. It is a metabolic product of an organism which alters the normal metabolism of host cell producing deleterious effects. The power of an organism to produce toxins is called toxigenicity which is genetically controlled by many species. If the toxin producers are present in the fishes/shellfishes kept for consumption, it is most likely that the consumer will be a victim of intoxication. This section discusses the important biotoxins present in fish feeds, finfishes, and shellfishes that pose a threat to humanity.

4.2.1 Analysis of Aflatoxin

Aflatoxins are fungal metabolites that contaminate foods and feeds, causing adverse health effects in humans and animals. AFB1 is one such fungal metabolite found in agricultural products, including rice, peanuts, and cereals. All these ingredients are used for producing fish feeds and due to that reason is a threat in aquaculture. *Aspergillus flavus* and *Aspergillus parasiticus* are the main producers of AFB1. Food grains may be contaminated by AFB1 at a number of stages but is most common when crops are exposed during harvesting and storage, provided that hot and humid weather conditions and improper and unsanitary storage exist for a prolonged period.

Aflatoxin is categorized under mycotoxins. The technical term for poisoning by aflatoxin mycotoxins is aflatoxicosis. This usually occurs from eating food contaminated with aflatoxin mycotoxins. Aflatoxin mycotoxins are toxic to humans and even more toxic to animals. They also cause cancer in humans and animals.

Symptoms

Symptoms of aflatoxicosis are liver damage, disruption of food digestion, absorption or metabolism, liver cancer, mental impairment, abdominal pain, vomiting, convulsions, edema, pulmonary edema, hemorrhaging, coma, and death. The aflatoxin LD50 rate (the dosage level that causes 50% of a group to die) for animals is between 0.5 and 10 mg/kg of the animal's weight.

Mode of Action

AFB1 is a genotoxic hepatocarcinogen that presumptively causes cancer by inducing DNA adducts, leading to genetic changes in the target cells, which then cause DNA strand breakage, DNA base damage, and oxidative damage that may ultimately lead to cancer. DNA adducts arise from chemical modification of the bases in DNA or amino acids in proteins by toxic carcinogenic chemicals. Approximately half of human cancers are due to a mutated *Tp53* gene. The mutations affecting *p53* are diverse by their nature and position. For example, mutations such as the transversion in codon 249 [guanine (G) to thymine (T)], which causes an arginine (R) to serine (S) substitution, are present in 50% of HCCs. This may be due to consumption of AFB1-contaminated food.

Detection

The occurrence of aflatoxins in fish feeds can be determined by the following method.

Aflatoxin B1 standard solution – A stock standard solution of aflatoxin B1 (AFB1) should be prepared by dissolving 1 mg of pure crystalline AFB1 in toluene: acetonitrile (9:1, v/v) at concentration of 80 μg mL-1. The concentration of the AFB1 stock solution should be determined by measuring the UV absorbance at 348 nm and calculated by using the molar extinction coefficient ε of 19,300. The concentration of working standard solution in toluene: acetonitrile (9:1, v/v) was 0.8 μg ml^{-1}. The standard and working solutions should be kept frozen (−18 °C).

Aflatoxin B1 extraction – The samples weighing approximately 1 kg should be ground and homogenized, and a subsample of 50 g should be removed for analysis in duplicate. 50 g of each sample should be extracted with 270 ml methanol and 30 ml 4% (w/v) potassium chloride. Samples should be blended at high speed for 5 min and filtered; 150 ml of the filtrate should be collected. After that, 150 ml 10% (w/v) copper sulfate and 5 g diatomaceous earth were added, followed by moderate stirring and filtration. 150 ml of the filtrate should be transferred to a separation funnel with 150 mL of distilled water, and AFB1 should be extracted twice with 10 ml chloroform. The chloroform extracts were collected and submitted to solvent evaporation in water bath at 60 °C (Hoeltz et al. 2010).

Chromatography – The extracts should be resuspended in 100 μl chloroform and submitted to thin-layer chromatography (TLC). Three, 6, and 9 microliter aliquots of sample extract and 2, 4, and 6 microliter aliquots of AFB1 standard solution (0.8 μg ml^{-1}) should be spotted 1 cm apart on precoated silica gel plates ($10 \times 10\ cm^2$). The spots should be dried, and the plates developed in a saturated chamber with chloroform: acetone (99:1, v/v). Aflatoxins spots should be observed under long-wave ultraviolet light (λ = 366 nm) and identified by comparison with AFB1 standard. Confirmatory tests for AFB1 should be carried out using trifluoroacetic acid.

4.2.2 Finfish Toxins

Marine organisms harbor hundreds of toxic secondary metabolites. Many of these compounds demonstrate extremely high toxic activities. Toxins produced by marine microorganisms may be transferred through food chains into various shellfish, fish, and other macroorganisms. Accumulation of toxins in these marine animals makes them dangerous for consumption.

Ciguatera Toxicity
Edible fishes sporadically turn toxic in tropical waters when they become contaminated with potent toxins of microalgal origin. Toxicity results in fatal human poisoning. Pacific Islanders suffered from poisoning since the early 1999. Hence, in tropical areas it is high risk for the people ingesting saltwater fish.

Two types of toxins have been distinguished based on water solubility, those insoluble in water (ciguatoxins) and those soluble in water (maitotoxin). Maitotoxin is one of the most toxic and complex unprecedented marine toxin. Maitotoxin toxicity exceeds that of palytoxin by several times and that of any other nonprotein toxins. Its LD_{50} is approximately 50 μg/kg^{-1} following intraperitoneal administration to mice.

Symptoms
Symptoms of ciguatera toxicity in humans include gastrointestinal, nervous, and cardiovascular disorders, nausea, headache, and dizziness. In severe cases, respiratory paralysis and death have been observed. Activation of intestine stabbing and burning pain in response to mild cooling following ciguatoxin activation is also observed.

Mode of Action
The molecular mechanism of action of ciguatera poisoning includes its interaction with calcium-permeable, nonselective ion channels (caNSC) and its capability to increase calcium influx and increase sodium ion outflow from excitable and non-excitable cells. Moreover, activation of calpain-1 and calpain-2 proteases along with membrane permeability disturbance induces oncotic cell death. The toxins are produced by dinoflagellates in the *Gambierdiscus* genus. Ciguatera-causing microalgae of this genus are benthic and are distributed in tropical and temperate waters. Microalgae containing the toxins contaminate fish via the food chain. Ciguatoxin was isolated from contaminated fish (eel – *Gymnothoraxjavanicus)* and cultivated dinoflagellate *G. Toxicus*. It was also isolated from a species of the genera *Ostreopsis* and *Thecadinium.* This toxin is highly toxic with LD_{50} values in the range of 0.25–2.3 μg/kg^{-1}when intraperitoneally administered to mice. It can activate voltage-gated sodium channels and block potassium channels when administrated in large doses. These actions damage nerve conduction.

Detection
Extraction of *Gambierdiscus* cells (Lewis et al. 2016) – Cell pellets should be sonicated twice for 1 min in methanol/water/hexane (2:1:1) 10 ml per 1 million cells. After the disruption of cells, the sample should be centrifuged at 3200 × g for 10 min. The supernatant is later transferred to a clean separatory funnel. The remaining pellet should be resuspended in 10 ml of methanol/water/hexane (2:1:1),

sonicated 2 × 1 min, centrifuged at 3200 × g for 10 min, and the supernatant added to the same separatory funnel. The hexane layer should be discarded followed by drying of the remaining methanol/water solution under N_2 gas. This residue should be then dissolved in 10 ml of dichloromethane (DCM). The redissolved residue is placed into a clean separatory funnel and extracted twice with 5 ml of 60% aqueous methanol. The methanol:water fraction containing maitotoxin (MTX) and the DCM fraction containing ciguatoxin (CTX) should be collected in separate 20 ml scintillation vials and dried under ultra-high purity nitrogen.

Fractionation of *Gambierdiscus* extracts – The MTX and CTX extracts should be resuspended in 0.5 ml of 30% acetonitrile (ACN). It is fractionated and eluted at 0.7 ml min⁻initially for 5 min at 5% acetonitrile (ACN)/0.1% formic acid followed by a linear gradient from 5 to 90% ACN/0.1% formic acid over 60 min. 70 × 1 min fractions should be collected. 25% (v/v) of these fractions should be freeze-dried and resuspended in 15 μl physiological salt solution containing 0.1% bovine serum albumin (BSA) just prior to FLIPR assay analysis.

Measurement of CTX- and MTX-like activity – Human-derived neuroblastoma cells, SH-SY5Y, express a range of voltage-gated sodium channel subtypes. Sensitive direct detection of MTX-induced calcium influx and indirect detection of CTX activity is possible by the use of these cells. It is done by measuring enhancement of veratridine-induced calcium influx. SH-SY5Y cells should be maintained at 37 °C/5% CO_2 in RPMI media comprising 15% fetal bovine serum (FBS) and 2 mM L-glutamine. Cells should be routinely passaged at a 1:5 dilution every 3–5 days using 0.25% trypsin/EDTA. SH-SY5Y cells should be seeded at 120,000 cells/well in 40 μl of culture medium on black-walled 384-well imaging plates and cultured until 90–95% confluent monolayers are obtained. Cells are then loaded for 30 min at 37 °C with 20 μl of calcium-4 no-wash dye in PSS containing 0.1% BSA. The dye is absorbed into the cells' cytoplasm during this 3-0min incubation period. Plates holding the loaded cells should be then transferred to the fluorescent plate reader with a 470–495 nM excitation filter and 515–575 nM emission filter. Camera gain and intensity should be adjusted to yield a minimum baseline fluorescence of 1000 arbitrary fluorescence units (AFU). After 10 s baseline measurement should be recorded. 10 μl of sample should be added to each well containing the loaded cells using injection manifold. Fluorescence should be recorded every second for 300 s thereafter.

Responses are normalized to baseline, to compute responses induced by MTX-like activity. The maximum increase in fluorescence for reads 10–300 should be determined. For P-CTX and CTX-like activity, the maximum increase in AFU over baseline (reads 300–310) should be determined for reads 310–600. By determining the peak AFU for each CTX fraction, results should be normalized. In the absence of a suitable standard, to establish the relative levels of MTX, results should be normalized to the *Gambierdiscus/Fukuyoa* extract that produced the largest increase in AFU (100% activity) of the samples tested. Each fraction from both MTX and CTX should be measured in triplicate.

Scombroid Poisoning/Histamine Food Poisoning
Histamine has been implicated in the toxicity of scombroid and even non-scombroid fishes. Formation of histamine in tuna, sardine, horse mackerel, and anchovy is reported by many workers. When decomposition progresses in a fish, the amount of biogenic amines also increases. As with other biological indicators such as TMA, DMA, TVB-N, and K value, amines are also used as a potential freshness index in fish and shellfish. The formation of putrescine, cadaverine, and histamine and loss of spermidine and spermine was observed as decomposition of tuna progressed. These amines might serve as quality indicators of tuna.

The Food and Drug Administration established a guidance level for histamine of 5 mg/100 g (50 ppm) for assuring the safe consumption of scombroid or scombroid-like fish and recommended the use of other data to judge fish freshness, such as the presence of other biogenic amines associated with fish decomposition. A maximum average histamine content of 10 mg/100 g (100 ppm) has been established in the European Community (EC) for acceptance of tuna and other fish belonging to the *Scombridae* and *Scomberesocidae* families. The EC has suggested that in the future, a maximum of 300 ppm for total biogenic amines in fish and fish products may be an appropriate legal limit. In addition to this, cadaverine and putrescine could be used as freshness indices for fish and shellfish, respectively. 0.5 mg/kg has been suggested as threshold limit of cadaverine in fish.

Symptoms
Symptoms of histamine toxicity typically begin within 5–30 min after eating spoiled fish, although there are cases when symptoms are delayed for as long as 2 h. These symptoms may include flushing of the face and body, nausea, burning in the mouth, headache, faintness, sometimes with blurring vision, abdominal cramps, diarrhea, wheezing or other breathing problems, and swelling of the face and tongue. Symptoms typically last a few hours or a day. In rare cases, symptoms can persist for a few days.

Mode of Action
Histamine exerts its effects by binding to histamine receptors on cell's surfaces. There are four types of histamine receptor: H1, H2, H3, and H4. The binding of histamine to these receptors stimulates them to produce functional responses. The H1 histamine receptor plays an important role in allergic response and is widely distributed throughout the peripheral nervous system, particularly the smooth muscle, where its activation causes vasoconstriction. Activation of the H1 receptor also causes blood vessel dilation, increased vessel permeability, stimulation of sensory nerves in the airways, and bronchoconstriction. In addition, activation of this receptor promotes the chemotaxis of eosinophils, which can lead to nasal congestion, sneezing, and rhinorrhea. Once activated in the cerebral cortex, the H1 receptor can inhibit potassium channels in neuronal cell membranes, depolarizing the neurons and increasing neuronal excitation. The H2 receptor is found on the parietal cells within the stomach, heart, and, to a limited extent, in immune cells and vascular smooth muscle. Activation of the H2 receptor stimulates vasodilation and release of

the gastric acids required for digestion. Physical responses to the H2 receptor include a decrease in the chemotaxis and activation of neutrophils and basophils, stimulation of suppressor T cells, lymphocyte proliferation, and natural killer cell activity. Combined activation of H1 and H2 receptors contributes to rhinorrhea and swelling of the nasal airways. The H3 histamine receptor is a presynaptic autoreceptor found on nerve cells that contain histamine. It is widely distributed throughout the central nervous system, with the greatest expression found in the cortex, caudate nucleus, thalamus, hypothalamus, olfactory tubercle, and hippocampus. The diverse distribution of the H3 receptor throughout the cortex suggests this receptor is able to modulate many neurotransmitters such as dopamine, GABA, acetylcholine, and norepinephrine in the central and peripheral nervous systems. The H4 histamine receptor is mainly found on immune cells and tissues including peripheral blood leukocytes, the spleen, bone marrow, and thymus. It is also found in the colon, lung, liver, and epicanthus. Stimulation of this receptor also mediates the chemotaxis of eosinophils and upregulates adhesion of molecules.

Detection

Preparation of the fish samples – According to Frank et al. (1981), the concentration of histamine can vary considerably even between different portions of a single fish (Fig. 4.2). Therefore, in order ascertain uniform histamine level, samples from the same region should be taken in all cases.

Biogenic amine analysis reagents – Biogenic amines standards (putrescine dihydrochloride, cadaverine dihydrochloride, spermidine trihydrochloride, spermine tetrahydrochloride, histamine dihydrochloride, and agmatine sulfate) can be purchased from Sigma-Aldrich, UK. HPLC grade acetonitrile is used as solvent A, and deionized ultrapure millipore water is used as solvent B (RiOs and Elix millipore water purification system).

Preparation of standard amine solution – Putrescine dihydrochloride (182.9 mg), cadaverine dihydrochloride (171.4 mg), spermidine trihydrochloride (175.3 mg),

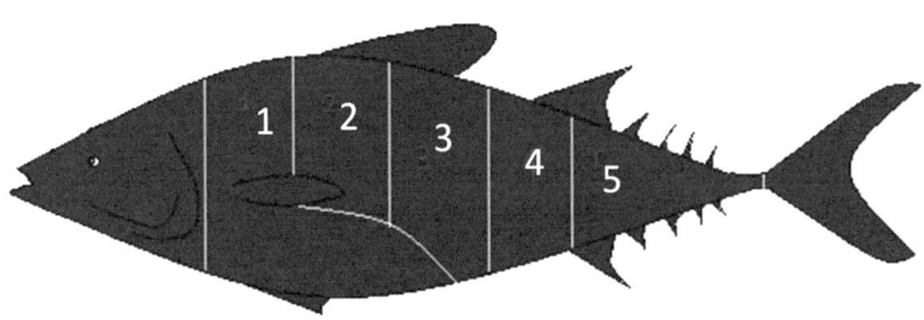

Histamine mg/kg	1020	70	16	45	29

Fig. 4.2 Within – fish variation in histamine concentrations within a fish. (Frank et al. 1981)

spermine tetrahydrochloride (172.0 mg), histamine dihydrochloride (165.7 mg), and agmatine sulfate (175.4 mg) are dissolved separately in 10 ml HPLC grade water. A composite standard comprising all the above biogenic amines can also be used. The final concentration of free base for each amine should be 10 mg/ml solution.

Sample preparation – Fish muscle (5 g) from samples should be taken and transferred to a centrifuge tube. The sample should be homogenized with 20 ml 6% TCA (trichloroacetic acid) for 3 min, centrifuged at 12000 g for 10 min at 4 °C, and filtered through Whatman No.1 filter paper. This should be made up to 50 ml with distilled water and was stored at −30 °C until further analysis.

Derivatization procedure: (Ozogul et al. 2002) – A stock solution should be prepared by dissolving 2% benzoyl chloride in acetonitrile to enhance the reaction with amines. For derivatization of standard amine solutions, 50 μl should be taken from each free base standard solution (10 mg/ml) and 2 ml of TCA extract for fish samples.

One ml of 2 M sodium hydroxide should be added, followed by 1 ml benzoyl chloride (2%), and vortex mixed for 1 min. The reaction mixture was left at room temperature (24 °C) for 30 min. The benzoylation should be stopped by adding 2 ml of saturated sodium chloride solution, and the solution should be extracted two times with 2 ml of diethyl ether. The upper organic layer should be transferred into a clean tube and evaporated to dryness in a stream of nitrogen. The residue should be dissolved in 500 μl of acetonitrile, and 20 μl aliquot should be injected for HPLC analysis.

Apparatus – Quantitative determination of the biogenic amines can be conducted using a Waters HPLC system with a binary pump model M 515, a 600 gradient mixer solvent delivery system, a dual λ absorbance UV/VIS detector model 2487, and a C 18 symmetry column (5 μM particle size, 4.6 mm id × 250 mm length column) with a flow rate of 1.5 ml/min. Data analysis can be performed using EMPOWER 2 chromatography software.

Chromatographic conditions – Chromatographic separation makes use of continuous gradient elution with acetonitrile (eluent A) and HPLC grade water (eluent B). The gradient will start at 80% acetonitrile and will decreased to 20% and finally increase to 80% in 16 min (Table 4.1.). The total separation time will take less than 7 min, and the gradient can be run for 20 min to ensure full separation. Detection should be monitored at 254 nm.

Table 4.1 HPLC gradient profile for separation of biogenic amines

	Time (min)	Flow rate ml/min	Acetonitrile (%A)	Deionized millipore water (% B)	Curve
1	–	1.5	80	20	
2	10	1.5	80	20	6
3	15	1.5	20	80	6
4	16	1.5	20	80	6
5	20	1.5	80	20	6

A calibration curve for each of the amines in the range of 0–100 μg/ml should be prepared. Correlation of peak areas of individual amines standards and composite standards with known concentration can be calculated after injecting each of the standard amine solutions.

Pufferfish Poisoning/Tetrodotoxin

Tetrodotoxin (TTX) is of microbial origin and represents a polyhydroxylated polycyclic bipolar ion with orthoester and guanidine fragments. The toxin accumulates in the liver, gonads, and sometimes the skin of the puffer fish. Puffer fishes belong to the family Tetraodontidae and hence the name tetrodotoxin. Some toxic fish like *Takifugu (Speroides rubripes)* are used to prepare delicious dishes in Japan. Mortal poisoning occur almost annually following human consumption of these delicacies. The toxin is considered to be produced by certain bacteria (*Vibrio alginolyticus* and *Pseudoalteromonas tetraodonis*). This explains the isolation of TTX from different biological sources.

Symptoms

Symptoms occur generally 4–5 h after consuming the pufferfish. It begins with numbness and tingling around the mouth, salivation, nausea, and vomiting. Symptoms may progress to paralysis, loss of consciousness, and respiratory failure and can lead to death.

Mode of Action

TTX is an extremely potent neurotoxin that enters the body by injection of TTX-containing marine animals. Tetrodotoxin selectively inhibits the TTX-sensitive neuronal sodium channels and blocks action potentials in nerves and muscles. TTX enters the outer vestibule of channel pores and attaches to amino acids residues that control permeation. Toxicity is especially high following intravenous application of the toxin (LD_{50} ~330 μ/kg^{-1}). Human death is mainly caused by paralysis of the diaphragm and thus cessation of breathing. TTX in sub-toxic doses have been used to ease pain in clinical trials. Its analgesic effect was sustained in 50% of patients with severe cancer-related pain.

Detection

TTXs can be analyzed by the methods described by McNabb et al. (2014). All biological material should be stored frozen at −18 °C until required for subsampling and testing.

Apparatus – (a) Ultra-performance LC (UPLC) system consisting of an acquity ultra-performance liquid chromatography coupled to a Premier triple quadrupole mass spectrometer, (b) sample homogenization unit, (c) handheld blender, (d) Strata-X RP SPE cartridges (200 mg, 8B-S029-FBJ; Phenomonex, Auckland, NZ), and (e) Strata-X-C strong cation-exchange SPE cartridges (200 mg, 8B-S029-FBJ-S, Phenomonex).

Reagents – (a) Acetone, (b) acetonitrile, (c) ammonia solution (25%), (d) ammonium acetate, (e) formic acid, (f) TTX, (g) glacial acetic acid, (h) hydrochloric acid (0.18 and 5 M HCl solutions prepared from concentrated acid), (i) methanol, (j) Milli-Q-filtered water system, and (k) sodium hydroxide (10 and 2.5 M NaOH solutions prepared from solid pellets). All liquids were mixed on a volumetric basis and usually expressed as a percentage by volume.

Calibration solutions – 200 μg/mL TTX.-1 mg solid TTX dissolved in 5 mL Milli-Q water containing 0.1% acetic acid. (b) 1 μg/mL TTX.-5 μL of 200 μg/mL TTX aqueous solution (prepared as described above) diluted with 995 μL methanol containing 0.1% acetic acid. These stock solutions were stored frozen for up to 6 months. Mobile phases – Filtered to 0.2 μm.

Sample extraction – Homogenize sample using a blender. Weigh 1 g subsample into 50 mL polypropylene tube. Add 9 mL 0.1% acetic acid. Homogenize using an ULTRA-TURRAX at 17000 rpm for 1 min. Centrifuge at 3000 × g for 5 min. Transfer 1–15 mL polypropylene and add 9 mL of 0.1% acetic acid in methanol. Place in freezer for at least 1 h and then recentrifuge (3000 × g for 5 min) to precipitate solids. Dilute samples further with 0.1% acetic acid in methanol 1:4 for environmental (low level) samples and 1:100 for highly positive samples. The overall dilutions range from 1:400 to 1:10000 from the 1 g starting sample.

Calibration and QC – TTX standards were prepared from a 200 μg/mL stock solution diluted with methanol containing 0.1% acetic acid. Calibration standards were prepared at 100, 50, 10, 5, 1, and 0.5 ng/mL to test for linearity of the LC/MS/MS response curve. Samples were run at suitable dilutions to fall within this linear concentration range, and one standard was run after every sixth sample to ensure a consistent LC/MS response.

LC/MS/MS apparatus – (1) Waters Premier MS coupled to a Waters Acquity UPLC system. (2) Thermo GOLD aQ 50 × 2 mm column (Thermo Fisher Scientific Australia Pty Ltd., Scoresby, Victoria, Australia). (b) Mobile phases, (1) stock buffer solution, 500 mM formic acid, and 40 mM ammonia; (2) TTX A, stock buffer diluted to 10% with Milli-Q water containing 5% acetonitrile; and (3) TTX B, stock buffer diluted to 10% using acetonitrile. (c) LC mobile phase parameters (Table 4.1). LC injector parameters: mode, partial loop with overfill, weak wash 600 μL, strong wash 400 μL, column temperature 40 °C, overfill factor 3, injection volume 2.5 μL. (d) MS parameters. Capillary 0.75 kV, cone 40 V, extractor 5 V, RF lens 0.0 V, source 100 °C, desolvation 1100 L/h and 450 °C, cone gas flow 100 L/h, and collision gas flow 0.3 mL/min. (e) Ions monitored, m/z (collision energy, eV). 320.20 > 60.20 (32), 320.20 > 162.10 (37), 320.20 > 302.10 (24), dwell time 50 ms.

Calculations of results – MassLynx software (Version 4.1, Waters MicroMass) can be used to integrate the MRM peaks. The concentrations of TTX in sample extracts (ng/mL) were calculated from the TTX area responses using the linear six-point calibration produced from the TTX standard dilutions.

$$\frac{\text{mgTTX}/\text{kg} = \text{ngTTX}/\text{mL} \times 10(\text{volume}) \times 40(\text{dilution factor})/1(\text{weight of sample})}{1000}$$

Sardine Poisoning/Clupeotoxicism

Herrings, anchovies, and related species apparently become poisonous after consuming certain planktonic organisms, such as toxic dinoflagellates. These fishes belong to Clupeidae family and hence the name clupeotoxicism. The intoxication is rare and is similar to ciguatera or paralytic shellfish poisoning. The toxin acts very rapidly and produces a very high mortality (Deshpande 2002).

There are a few scattered reports of clupeotoxicism. In the Philippines, two outbreaks were reported in 1953 and 1955, with 14–17 deaths. The common Philippine sardine *tamban (Sardinella longiceps)* was the culprit. Three cases of sardine (*harengula* spp.) poisoning were also reported in Tarawa in the Gilbert Islands in the South Pacific with the death of all three victims. Poisoning resulting in five deaths caused by another sardine (*Clupea venenosa*) were also reported in the Fiji Islands in the South Pacific.

Clupeotoxic fish are listed in Table 4.2. Most thrive in tropical waters. Clupeotoxicism is unpredictable and can be a serious public health problem in certain parts of the world. The violent and often lethal nature of the poisoning makes it dangerous. It has been associated with toxic dinoflagellate blooms. The viscera are observed as the most toxic part of the fish. It is most likely to be toxic during the warmer months. The actual source and nature of the poison have never been identified. The toxin may be obtained from the food chain, particularly from certain types of dinoflagellates. *Pyrodinium bahamense*, the red tide dinoflagellate, is suspected as a possible source of the toxin.

Table 4.2 Fishes exhibiting clupeotoxicism

Species	Common name
Family Clupeidae	
Anondostomachacunda	Shirt-finned gizzard shad
Clupanodonthrissa	Sprat or thread herring
Clupea sprattus	Sprat
Dussmieriaacuta	Round herring
Harengulahumeralis	Red-ear sardine
H.ovalis, H. Zunasi, Sardinella fimbriata	Sardine
S.longiceps, S.performata, S.sindensis	
Ilisha africana	Herring
Macrura ilisha	Sable fish
Nematalosa nasus	Gizzard shad or long-finned gizzard
Opisthonema oglinum	Atlantic thread herring
Family Elopidae	
Megalops cyprinoids	Tarpon
Family Engrulidae	
Engraulis enchrasiculus	Anchovy
E.japonicus, Thrissinabaelama	

Symptoms

The symptoms and signs of clupeotoxicism are discrete and usually violent. A sharp metallic taste is the first indication of intoxication and may be present immediately after ingestion of the fish. This is flowed by dryness of the mouth, nausea, vomiting, malaise, abdominal pain, and diarrhea. A feeble pulse, tachycardia, chills, and cold may convoy the gastrointestinal upset, vertigo, a drop in blood pressure, cyanosis, and other evidences of vascular collapse.

Within a very short period, a variety of neurological disturbance such as nervousness, dilated pupils, violent headaches, numbness, tingling hypersalivation, muscular cramps, respiratory distress, progressive muscular paralysis, and convulsions rapidly ensure coma and death. Death may occur in less than 15 min.

There are no precise statics available regarding the mortality rate of clupeotoxicism. But, arbitrating from the documented case reports, the fatality rate appears to be very high. The victims generally die within minutes to hours. It is believed that clupeotoxicity in some instance may be associated with ciguatera poisoning. This has not been documented, however.

Hallucinogenic Fish Poisoning

Several species of fish are claimed to produce hallucinogenic effects when consumed. *Sarpa salpa*, a species of seabream, is commonly claimed to be hallucinogenic. These widely distributed coastal fish are normally found in the Mediterranean and around Spain and along the west and south coasts of Africa. Occasionally they are found in British waters. They may induce hallucinogenic effects that are purportedly LSD-like if eaten. In 2006, two men who apparently ate the fish experienced hallucinations lasting for several days. The likelihood of hallucinations depends on the season. *Sarpa salpa* is known as "the fish that makes dreams" in Arabic (Kim 2014).

Other species claimed to be capable of producing hallucinations include several species of sea chub from the genus Kyphosus. It is unclear whether the toxins are produced by the fish themselves or by marine algae in their diet. Other hallucinogenic fish are *Siganus spinus*, called "the fish that inebriates" in Reunion Island, and *Mulloidichthys flavolineatus*, called "the chief of ghosts" in Hawaii.

Symptoms

The first symptoms usually seen are loss of balance and coordination, and generalized malaise primarily effects the central nervous system. GI symptoms are generally mild and include nausea, abdominal pain, and diarrhea. Within a few hours, specific signs of poisoning occur including delirium, visual and auditory hallucinations, depression, and feeling of impending death with reactive tachycardia and hyperventilation. Patients report terrifying nightmares.

Mode of Action

The active agent(s) that cause hallucinations in humans, and the origin of these agents, are not clear. Some authors think they could come from toxins associated with macroalgae that accumulate in the flesh of the fish. Toxins from the green algae

Caulerpa prolifera in the Mediterranean Sea appear to be implicated, as is the seagrass *Posidonia oceanica*. When herbivores eat seagrass leaves, they ingest algal epiphytes and toxic dinoflagellates that live on the seagrass leaves. Dreamfish might contain the hallucinogen DMT.

4.2.3 Shellfish Toxins

Shellfish poisoning can occur after eating clams, mussels, oysters, scallops, and crustaceans contaminated by toxins. Herbivorous dinoflagellates are the primary transvectors that accumulate the toxins in their digestive organs and soft tissues. Shellfish poisoning is more common during red tides, when sea waters turn a reddish color because of the presence of large numbers of dinoflagellates. These dinoflagellates produce at least 12 toxins, which are heat and acid stable. Some of the toxins, the symptoms caused by their poisoning, and their detection methods are discussed in this section.

Red Tides

Filter-feeding mollusks such as oysters, clams, and mussels can become toxic to humans during the red tides. The phenomenon of red tide is caused by the fast growth of microscopic and single-celled algae, which are usually not harmful. Unfortunately, a small number of species produces potent toxins that can be transmitted throughout the food chain. This affects and even kills zooplankton, shellfish, and eventually humans that feed on them either directly or indirectly.

On-site refrigeration and transportation have taken the problem far from the incident point. The human poisoning associated with shellfish and toxic blooms are paralytic (PSP), amnesic (ASP), neurotoxin (NSP), and diarrhetic (DSP) shellfish poisonings. Their occurrence is extremely rare. Still, regulations are in place effectively to protect the consumers from shellfish toxins.

The methods most widely used for the assay of naturally occurring toxins are mouse time to death bioassay (MBA) and high-performance liquid chromatography (HPLC). These methods are slow, expensive, and not sufficiently rugged for routine use in the field.

Paralytic Shellfish Poisoning (PSP) Toxins

Ingestion of mollusks contaminated with saxitoxin cause paralytic shellfish poisoning (PSP). Saxitoxin was first isolated from the butter clam, *Saxidomus giganteus*. Later, the toxin was found in many edible mollusk species and pufferfish. PSP toxin infiltrates into edible mollusk from dinoflagellates *Alexandrium*, *Gymnodium*, and *Pyrodinium* spp. through the food chain. Toxins can accumulate in certain body compounds of mollusks, where they are well kept for several weeks.

Symptoms

Ingestion of mollusks contaminated with PSP results in the following symptoms. Five to 30 min from consumption, there is slight perioral tingling progressing to numbness which spreads to face and neck to moderate cases. In severe cases, these symptoms spread to the extremities with incoordination and respiratory difficulty. There are medullary disturbances in severe cases evidenced by difficulty swallowing, sense of throat constriction, speech incoherence or complete loss of speech, as well as brain stem dysfunction. Within 2–12 h, in very severe cases, there is complete paralysis and death from respiratory failure in absence of ventilatory support. After 12 h, regardless of severity, victims start to recover gradually and are without any residual symptoms within a few days.

Other symptoms include headache, dizziness, nausea, vomiting, rapid pain, and anuria. There is no loss of consciousness, and the reflexes are unaltered except may be pupillary size, and sight may be temporarily lost. As opposed to tetrodotoxin poisoning, there is rarely significant hypotension. Symptomatology is essentially identical for Pacific and Atlantic cases, although gastrointestinal symptoms may be more prominent in the Atlantic.

The overall mortality was about 8.5–9.5% in two large series. However, the Guatemalan 1987 outbreak on Pacific Coast had a case fatality rate of 14%, which was even higher in young children (50%). It is possible that children may be more sensitive to PSP toxins than adults. In addition, the access to emergency medical services in acute cases is crucial to the prognosis.

The differential diagnosis of this clinical scenario of an acute gastrointestinal illness with recent shellfish ingestion would be bacterial or viral gastroenteritis. The neurologic manifestations are more consistent with poisoning by other marine toxins such as NSP and pufferfish poisoning or even recent organophosphate pesticide poisoning.

Detection (AOAC 1995)

Sample preparation – Independent of the type of determination method, PSP toxins have to be quantitatively extracted from the sample materials. The standard method using 0.N hydrochloric acid as extraction solvent was suggested by the Association of Official Analytical Chemist (AOAC). The sample and the extraction solvent are heated for 5 min at 100° C. The hereby N-sulfocarbomoyl toxins are converted into the respective carbamate toxins, and B1-, B2-, and C1-C4 toxins are no longer present in the extracts.

However, the natural toxin pattern could be observed by extracting samples with acetic acid. Mix fresh sample (1 g) with acetic acid (0.03 N, 3 ml) using an ULTRA-TURRAX. Centrifuge the mixture for 10 min (2980 g) and pass through a 0.45 μm nylon filter. A second sample (1 g) should be extracted with hydrochloric acid (0.2 N, 3 mL); mixing, centrifugation, and filtration should be executed as described above. The extracts obtained from procedures were injected separately into the high-performance liquid chromatography (HPLC) system. After separate injection of the different extracts into the HPLC system, determination of N-sulfocarbamoyl toxins possible by calculating the peak height increases for carbamoyl toxins formed.

Spectrophotometric assay – PSP toxins which exhibit neither ultraviolet (UV) absorption nor fluorescence were oxidized in alkaline solution to fluorescent pyrimidinopurins. After acidification, the intensity of the fluorescence of the oxidation products should be measured in the solution. Individual PSP toxins differ both in toxicity and fluorescence intensity after oxidation. As a group, the carbamate toxins (STX [saxitoxin]), NEO, and GTX1–4 [gonyautoxins]) are the most toxic, the N-sulfocarbamoyl toxin (B1, B2, and C1–4) are the least toxic, and the decarbamoyl toxins have intermediate toxicity. However, no correlation exists between toxicity and fluorescence intensity. Therefore, a chromatographic separation of the PSP toxins was suggested prior to the determination of the fluorescence. Total toxicity is calculated from the individual toxin concentrations and the absolute toxicities for each of the PSP toxins using equations:

$$T = \Sigma\left[C(i)T(i)D(\mathrm{CF}):10\right]$$

where T = total toxicity (in micrograms of STX/100 g), $C(i)$ = concentrations of individual toxins (in micromoles), T (i) = toxicity of individual toxins (in mouse units per micromoles), D = dilution of extract (in milliliters per gram of meat), and CF = coercion factor (in micrograms of STX per mouse unit).

Biological detection methods – Biological methods included here are those assays where the end point is related to the toxin-receptor interaction, that is, binding of saxitoxin molecules (s) to site 1 in the α-VDSC. The mouse bioassay (MBA) is official method for the determination of the PSP in seafood, but this is neither specific nor sensitive; it requires a continuous supply of mice, and results are affected by test conditions such as animal strain and sample extract preparation. Other methods include fluorimetric assay and liquid chromatography. The latter requires expensive equipment for pre- or post-column analyte oxidation. Additionally, samples must be analyzed one at a time, and so the method is unsuitable for routine on-site testing.

Immunochemical methods have advantages in terms of both sensitivity and speed and are therefore of increasing importance in food control as rapid screening test. Because of the highly specific antigen-antibody interaction, several laboratories have attempted to develop an immunoassay for PSP.

Neurotoxic Shellfish Poisoning (NSP) Toxins

Neurotoxic shellfish poisoning is due to the accumulation of microalgae toxins known as brevetoxins and analogous natural products in shellfish and other marine organisms. Brevetoxin-1 is a natural polyether product which usually bears a system of fused oxygen- containing rings.

Symptoms

People ingesting toxins do not die. However, they exhibit gastrointestinal and neurological symptoms such as nausea, vomiting, slurred speech, loss of coordination, and paresis of facial muscles. Neurological symptoms can lead to partial paralysis. Immune systems of humans and other animals are weakened by brevetoxins. It also increases the danger of inflections and other secondary diseases.

Mode of Action

Brevetoxins are fixed to voltage-gated sodium channels in nerve cells, and this activates them. This gives rise to sodium influx and membrane depolarization which induces poisoning. This always often leads to hospitalization.

Detection

There are three distinct methodologies for the assessment of brevetoxins in environmental and biological samples (Hua 1995). These include the mouse bioassay (MBA) test which is still considered the "gold standard" for shellfish toxicity assessment. For the test, intraperitoneal injections of an extracted sample of the putatively contaminated shellfish are done to elicit a toxic response. Results are expressed as mouse units (MUs) per 100 g of shellfish. US guidelines state that shellfish demonstrating toxicity values equal to or greater than 20 MUs per 100 g of shellfish are considered toxic.

Recent methods have been developed using in vitro methods for the detection of brevetoxins. The enzyme-linked immunosorbent assay (ELISA) for brevetoxin is an antibody-based bioassay. It is based on the activity of goat anti-brevetoxins polyclonal antibody which has high affinity for brevetoxins. It allows for the detection and quantification of brevetoxin seven at low levels.

High-performance liquid chromatography (HPLC) can also be used for analyzing brevetoxin. It utilizes standards for the comparison of elution times. There are limitations due to co-eluding complexes when it is used alone. But when paired with mass spectrophotometry (MS) detectors, a conclusive identification can be made. This method is used most frequently on samples from seawater to quantify the presence of red tide toxins and to estimate the concentrations of the various brevetoxins.

Receptor-binding assay (RBA) is another method currently used for identifying brevetoxins. An unknown quantity of non-radiolabeled brevetoxin functionally competes with radiolabeled brevetoxin for the site 5 receptor of the VGSC in this method. Preparation of the biological receptor is from rat brain crude membranes. Using a range of PbTx-3 concentrations, competition curves for assay calibration are generated. Results obtained are expressed as equivalent of PbTx-3 concentrations.

Diarrhetic Shellfish Poisoning (DSP) Toxins

Diarrhetic shellfish poisoning (DSP toxicity) is a result of several microalgae toxins, including okadaic acid, dinophysistoxins, pectenotoxins, and yessotoxins. All these contain polyether structures and can accumulate in shellfish.

Okadaic acid was first isolated from the sponges (*Halichondria okadai*). It enters sponges from dinoflagellates (*Prorocentrum* and *Dinophys* sp.). Okadaic acid easily penetrates into planktivorous mollusks and other marine animals because of the symbiotic relationship of *Prorocentrum* sps. with different marine invertebrates and macrophytic algae.

Mode of Action

Okadaic acid inhibits protein serine/threonine phosphates by interacting with the PP1 and PP2A catalytic domains of these enzymes. Both human and animal enzymes are inhibited by okadaic acid. It is also known as powerful tumor promoter. European mollusks contain okadaic acid, whereas mollusks from Canada and Japan primarily accumulate dinophysis toxins. Dinoflagellate species pose a major threat to shellfish aquaculture due to their diarrhetic toxins (Reguera et al. 2012). Pectenotoxins-1 and pectenotoxins-2 are usually found with okadaic acid. Biological activity of pectentoxins relates to their capability to depolymerize actin filaments. Pectentoxin-2 exhibits a significant cytotoxic effect against tumor cells.

Detection

The common DSP assays are biological-based techniques. Immunoassays were also established using the RIA and ELISA methods. ELISA has a short assay time with good quantification and little cross-reactivity to other DSP toxins. HPLC techniques will detect OA and DTXs by the spectrophotometric detection of 9-antharyldiazomethane derivates with levels of sensitivity at approximately 15 μg OA/100 g shellfish tissue. It has been shown that up to 50% of the toxin can be lost in the derivatization and extraction procedure. Relatively rapid radioactive protein phosphatase (PP) assay was developed by Honkanen. It can be used to detect OA in oyster extracts.

Amnesic Shellfish Poisoning (ASP)/Domoic Acid Poisoning

Domoic acid was obtained from the red algae *Chondria armata*, and it was identified as a neurotoxic agent causing amnesic shellfish poisoning (ASP toxicity). Later it was found that domoic acid is produced by some strains of diatoms, such as *Pseudo-nitzschia* spp., *Nitzschia navis-varingica*, and *Amphora coffeaeformis*. *Pseudo-nitzschia multiseries* loses its ability to produce domoic acid in axenic culture. When bacteria from original toxic cultures are added to the culture, the alga recovers production. This experiment suggests that association between bacteria and microalgae might be necessary for synthesis of domoic acid.

Domoic acid from algal blooms may accumulate from algal blooms in shellfish and finfishes such as sardines and anchovies. Through the food chain, it can be transferred into seabirds, sea lions, and other animals, causing illness and sometimes death depending on the amount of toxic fish they eat. Toxicosis from this substance is also called DAP toxicity or domoic acid poisoning.

Symptoms

The first incident of the ASP poisoning documented in 1987 in Canada, after consumption of contaminated blue mussels, *Mytilus edulis*, resulted in the death of four people and more than 100 exhibiting symptoms of acute toxicosis. Short-term memory loss is the most prominent clinical symptoms described and is the reason for the name amnesic shellfish poisoning (Pulido 2008). Elderly men are more susceptible to both the severity of illness and to memory loss when compared with women and younger men.

Mode of Action

Both kainic and domoic acids are excitatory amino acids. They are analogs of glutamate, a neurotransmitter in the brain, which activates glutamate receptors (GluRs). Domoic acid induces activation of ionotropic GluRs on both sides of the synapse. It results in permanent stimulation of hypothalamic neurons and leads to their destruction.

Detection

Mouse bioassay is not sufficiently sensitive to detect the regulatory level of 20 μg DA/g of edible tissue. Methods most commonly employed for determining the (HPLC) is accompanied with a variety of sample extraction techniques. Liquid chromatography with ultraviolet absorbance detection (LC-UVD) is currently the reliable technique for the determination of DA in shellfish. Domoic acid can be extracted from shellfish tissues by the AOAC hot acid method or by blending with aqueous methanol. The detection limit of the toxin is 10–80 ng/mL in an extract solution. A rapid, sensitive, and disposable amperometric immunoassay (ELISA) method has been developed for determination of DA in various marine samples.

Spectrophotometric study – The spectrophotometric ELISA should be performed in a 96-well microplate. This immunoassay should be done in an indirect competitive assay involving antibodies against DA conjugate and DA to bovine albumin serum (BSA-DA) for coating and horse anti-goat IgG alkaline phosphatase conjugate (IgG-AP) for the detection. 4-nitrophenyl phosphate (4-NPP) is the enzyme substrate. The operative range can be calculated by the equation:

$$Y = d + \left[(a - d) / (1 + X / c)^{b} \right]$$

where "*d*" and "*a*" are the asymptotic values from maximum and minimum (higher concentrations and zero concentrations), *c* corresponds to the analyte concentration (X), which give $Y = d + (a - d)/2$, *b* gives the slope of the curve. The linear range for such an assay is between 1 and 185 ng/mL of DA.

Azaspiracid Toxins

Azaspiracid were isolated from contaminated mollusks. These compounds are produced by a microalgal species *Azadinium spinosum*. It causes severe and protracted diarrhea in humans, although they have never been included in the DSP group. They are established to be open-state blockers of human ether-a-go-go-related gene (*h*ERG) potassium channels. Azaspiracid have high cytotoxicity, teratogenic, and probably carcinogenic activities. The toxicity experienced following ingestion of these compounds is due to these physiological effects, including injuries to gastrointestinal and lymphoid systems. Azaspiracid are highly toxic to humans, and its LD_{50} is ~ 110–200 μ/kg (Ito et al. 2002).

Detection

Refer to lipid extract preparation (Mudadu et al. 2017) in the coming section "Yessotoxin."

4.2.4 Other Toxins

Apart from the commonly reported food poisoning, there are other types of seafood poisoning like escolar poisoning, whelk poisoning, clupeotoxin, etc. which have become a threat in the recent years. This section discusses about such toxins. For many of these poisoning, detection method is yet to be standardized.

Escolar Poisoning

Escolar is a large deepwater fish often caught as a bycatch of tuna fishing. It contains approximately 20% indigestible waxy esters (by body weight) which can have a laxative effect, called keriorrhea, in those who eat them. Associated symptoms can be varied and include mild and rapid passage of oily yellow or orange droplets, severe diarrhea with nausea and vomiting, abdominal cramps, and headache. A recent report describes a cluster of illness associated with escolar that was attributed to both its laxative effects and scombrotoxin ingestion (Leask et al. 2004). To reduce the chances of scombroid fish poisoning, fish, especially of the *Scombroidei* suborder and escolar, should be appropriately refrigerated.

Although escolar is not a well-recognized vehicle for scombroid fish poisoning in the medical literature, it was reported in 15 out of 218 scombroid outbreaks in CDC's database for 1990–2001. Of these 15, 12 (80%) were reported in 2001. The recent increase in the number of escolar-associated outbreaks probably reflects the increasing popularity of this fish: annual landings of escolar increased from 11.1 metric tons in 1991 to 65.6 metric tons in 2002. Fish of the *Scombroidei* suborder (including tuna and mackerel) are much more common vehicles for scombroid fish poisoning than escolar. Other fish that have been implicated in scombroid fish poisoning outbreaks include mahi mahi, bluefish, and marlin. When susceptible fish are not promptly and continuously refrigerated, bacteria can metabolize histidine and degrade the protein of fish flesh to produce scombrotoxin, which consists of histamine and other amines. The presence of 50 ppm or more of histamine is indicative of decomposition whether or not there is gross evidence of this, such as bad odor. When illness results, histamine levels in implicated fish have commonly been at least 200 ppm and often greater than 500 ppm. Histamine is heat resistant and can, therefore, cause illness even when fish is properly canned or cooked thoroughly.

Symptoms

Patients experienced from 1 to 18 symptoms that started immediately after eating the fish to as long as 2 h afterward. Acute symptoms lasted from 15 min to 24 h after onset. The most common symptoms reported were headache (67%), facial flushing (62%), palpitations or a rapid pulse rate (57%), nausea (48%), dizziness (48%), and diarrhea (41%).

Detection

Escolar poisoning is due to the presence of histamine. It can be detected as described in Sect. 4.2.2 (Finfish toxins: Scombroid poisoning/histamine food poisoning).

Whelk Poisoning

Numerous cases of food poisoning associated with consumption of whelks (Neptunea spp.) have been reported. The poisoning is caused by ingestion of meat containing tetramine, $(CH_3)_4N^+$, which occurs naturally. High levels of tetramine occur in the salivary glands of many species of whelks, including Neptunea species. In Korea, hospitalized patients who were presumed to have tetramine poisoning due to ingestion of whelks have been reported sporadically, but no causative agent or origin have been identified in any of these cases. In March 2005, a toxin-type food poisoning incident occurred in Korea. Seventeen persons experienced gastrointestinal or neurological illness or both after eating whelks identified as *Neptunea intersculpta*, based on features of the empty shells collected after the incident. A meat extract obtained using 0.1 N HCl and administered to mice to look for signs of paralytic shellfish poisoning caused intoxication symptoms consistent with tetramine poisoning. It was hypothesized that tetramine might have contributed to the outbreak. Tetramine can be analyzed by using the following method.

Symptoms

Symptoms of tetramine poisoning include neurological and gastrointestinal effects, but no fatal cases have been recorded. The lag time between ingestion and symptoms is short.

Detection

Chemicals – Tetramethylammonium chloride, n-hexane, and methanol (both analytical grade and HPLC grade), pyridine, and formic acid. Deionized water is required for preparing the required chemicals.

Samples should be stored at 220uC until analysis. Fresh whelk specimens should be transported to the laboratory in crushed ice. Fresh specimens can be dissected into meat, salivary gland, and midgut gland portions for tetramine analysis.

Extraction of tetramine – For LC-MS-MS analysis, tetramine should be extracted from all tissues with methanol. Tetramine should be extracted from 10 g of meat and midgut gland homogenates using five times the volume of methanol and from the salivary gland sample using ten times the volume of methanol (relative to sample weight). The extracts should be centrifuged at 3800 g for 10 min, and the precipitates should be repeatedly extracted in same manner as used for the homogenates. The supernatants should be evaporated under reduced pressure to dryness, and the dried materials should be suspended in 20 ml of deionized water and defatted three times with an equal volume of n-hexane. The aqueous layer should be further evaporated under reduced pressure to dryness, and dried material should be dissolved in deionized water: 10 ml for the boiled meat implicated in the incident, the meat from purchased specimens, and the midgut gland tissue samples, and 50 ml for the salivary glands. Each sample solution (diluted with deionized water if necessary) should be filtered through a 0.2 mm pore-size membrane. These filtered solutions should be used for analysis of tetramine concentration by LC–MS-MS.

LC-MS-MS analysis – LC-MS-MS analysis should be performed using the method of Kawashima et al. (2004) with slight modifications. Separation should be performed on a cation-exchange Nucleosil 100-10SA column (0.46 by 25 cm) maintained at 35uC. The mobile phase for LC separation consists of 0.03 M pyridine-formic acid buffer (pH 3.1) containing 20% methanol. The total run time should be 30 min at a flow rate of 1 ml/min, and the injection volume should be 10 ml. The sample tray temperature in the Surveyor AS Plus should be set at 7uC. A triple-quadrupole mass spectrometer (Thermo Electron) should be used for mass detection. The eluate from the column should be introduced into the MS-MS system at a split ratio of 7:3 (waste volume:introduced volume) using an electrospray ionization source. The mass spectrometer should be operated in the positive ion detection mode with the spray voltage set at 4.0 kV. The nitrogen sheath gas pressure should be set at 60 arbitrary units, and the auxiliary gas pressure should be set at 12 arbitrary units. The heated capillary temperature was 340uC. A collision energy of 35 V should be used with argon at a pressure of 1.5 mTorr for collision-induced dissociation. Tetramine should be confirmed by Q1-Q3 (MS-MS) fragment ion scan mode to locate m/z 30–150. Quantification of tetramine should be performed by selected reaction monitoring of the molecular ion of m/z 74 and the fragment ion of m/z 58.

Yessotoxin

Yessotoxins (YTXs) are sulfated polyethers belonging to lipophilic phycotoxins family. They were first isolated in Japan in 1986, and afterward, production of YTXs has been demonstrated worldwide. In Italy the presence of YTXs in mussel was reported for the first time in hepatopancreas of mussels bred in Northern Adriatic Sea. Potentially YTX producers are the dinoflagellates *Protoceratium reticulatum*, *Lingulodinium polyedra*, and *Gonyaulax spinifera*. The toxins can be accumulated by filtering within the edible bivalve mollusks, like mussels, scallops, and clams; this permits their entry into the food chain. YTXs were initially included in the Diarrhetic Shellfish Poisoning (DSP) toxin group due to their occurrence with okadaic acid (OA) and other lipophilic toxins, but it was demonstrated that they have a different mechanism of action and not produce diarrheal neither in humans nor in mice. For these reasons, nowadays they are considered in different groups and are independently regulated.

Symptoms

The toxicological information for YTXs is limited and includes mainly studies on their acute toxicity in mice, even because no human cases of intoxication have been reported to date. Related YTX symptoms and toxicity effects after oral assumption are rather unknown in humans, so consumption of YTX-contaminated shellfish still remains an unidentified health risk. In contrast, mice showed discoordination described by jumping and restlessness, due to cerebellar cortical alterations as a result of damaged motor function.

Detection

The Regulation EC No 15/2011 established that liquid chromatography-mass spectrometry (LCMS/MS) method will be used as the reference method for marine biotoxins. Commission Regulation No 786/201 increased the permitted limits of YTXs in live bivalve mollusks from 1 to 3.75 mg eq YTX/Kg e.p.

Lipid extract preparation (Mudadu et al. 2017) – 100 g of mussels should be mixed with 300 ml acetone, filtered, and re-extracted with 200 ml of acetone. The acetone supernatant should be mixed and evaporated with a rotavapor, yielding a residue which should be recovered with 30 ml of dichloromethane and 60 ml of methanol (60% v/v) and subjected to partition by the separating funnel. The methanol phase should be stored, and the dichloromethane phase should be re-extracted twice with 60 ml of methanol. For both extracts, solvent should be evaporated and the resultant residue suspended in aqueous solution of 1% Tween 60.

Mouse bioassay – For the investigation of OA-toxin group, pectenotoxin group (PTXs), and Azaspiracid group (AZAs), 1 ml of dichloromethane extract should be injected intraperitoneally into albino mice Swiss strain weighing between 19 and 21 g. The death of two out of three injected mice, within a 24 h observation period, constituted a positive result. For the specific determination of YTXs, the methanolic extract should be injected in mice. The death of two out of three mice, within a 5 h (with a 6 h observation period), constituted a positive result.

In the use of Harmonised Standard Operating Procedure by MBA, the death of two out of three injected mice within 6 h observation, with a YTX-specific symptomatology, constituted a positive result. The positive samples should be subjected to LC-MS/MS method. It is based on the extraction of OA-toxin group, PTXs, AZAs, and YTXs by homogenized mussel tissues with methanol 90%. Extract should be filtered and directly analyzed by LCMS/ MS. To determine the total content of lipophilic toxins, an alkaline hydrolysis is necessary. After hydrolysis, extract should be filtered and analyzed by LCMS/MS. Chromatographic separation should be performed by gradient elution.

Botulism

Botulism is a serious and rare intoxication caused by neurotoxins produced by *Clostridium botulinum*. There are three forms of botulism – foodborne botulism, infant/intestinal botulism, and wound botulism. Foodborne botulism results from ingestion of preformed neurotoxins, commonly from food placed in anaerobic storage. Commonly intoxicated products are inadequately processed.

Causal organism – *Clostridium botulinum* is an anaerobic bacteria, commonly found in soils and aquatic habitats throughout the world. It is straight to slightly curved, gram positive, motile, anaerobic rods, 0.5–2.0 μm in width and 1.6–22.0 μm in length, with oval and subterminal spores. It produces a family of serologically related seven neurotoxins (BoNT, botulinum toxin): A, B, C1, D, E, F, and G which are able to cause food poisoning. All the seven BoNt are the enterotoxins because their main route of entry is the intestinal lining. Another clostridial toxin, C2 cytotoxin, is also an enterotoxin but not a neurotoxin. Among all the BoNTs, only A, B, E, and F usually cause human botulism, and BoNT/D affect mainly birds and mammals.

Lethal dose – The lethal dose of botulinum toxin for humans is not known but can be estimated from primate studies. By extrapolation, the lethal amounts of crystalline type A toxin for a 70 Kg human would be approximately 0.09–0.15 μg intravenously or intramuscularly, 0.70–0.90 μg inhalationally, and 70 μg orally.

Mechanism of Action

At neuromuscular (N-M) junction release of acetylcholine (Ach) is mediated by assembly of synaptic fusion complex that allows membranes of synaptic vesicles to fuse with neuronal cell membrane. Synaptic fusion complex is a set of SNARE (soluble N-ethylmaleimide-sensitive factor attachment protein receptor) proteins, which include (i) synaptobrevin, (ii) SNAP-25 (synaptosome-associated protein of 25 kDa), and (iii) syntaxin.

After membrane fusion, acetylcholine is released into synaptic cleft and then bonded with the receptors present on the surface of muscle cell. This binding stimulates the muscle cell to initiate cascade of reaction that in turn results in muscle contraction. But whenever there is exposure to the botulism toxin, it does the action in the following steps: (i) systemic absorption, (ii) binding to the nerve terminal and internalization, (iii) penetration of the endosome and release of toxin inside the nerve cell at presynaptic terminal, and (iv) synaptic poisoning.

The toxin reaches the lymphatic channels and then the blood stream either by absorption through the upper gastrointestinal tract (food borne and infantile botulism) or through tissue absorption (as in the case of wound botulism). Then the toxin circulates in the blood until it reaches cholinergic synapses of the peripheral nervous system. But the cholinergic synapses of the central nervous system are not involved because the toxin is unable to cross the blood-brain barrier. Now the toxin binds to neuronal cell membrane at nerve terminal and enters neuron by receptor-mediated endocytosis (internalization). After internalization, an acid-induced conformational change in the neurotoxin's translocation domain is believed to allow it to penetrate the endosome by forming a pore. The membrane interaction and pore formation are believed to facilitate the passage of the catalytic domain across the membrane into cytosol.

Pectenotoxin

Okadaic acid usually accompanied by pectenotoxins-1 and pectenotoxins-2. These toxins were first identified in sponges and later obtained from the microalga *D.acuminta*. According to Leira et al., biological activity of pectentoxins is related with their ability to depolymerize actin filaments. Pectentoxin-2 exhibited a significant cytotoxic effect against tumor cells.

Detection

Refer to lipid extract preparation (Mudadu et al. 2017) in section "Yessotoxin."

Maitotoxin

Maitotoxin is one among the two types of toxins identified in ciguatera poisoning. For details refer to section "ciguatera toxicity."

Pinnatoxins

Pinnatoxins were initially isolated and characterized from Japanese *Pinna muricata*. They belong to the spiroimine group of toxins and are dangerous. These are produced in nature by dinoflagellates (*Vulcanodinium rugosum*) and then accumulate in shellfish. The LD_{50} of pinnatoxin F is 13 μg/mg^{-1}. Pinnatoxins are found to activate calcium channels. After their total synthesis, it was established that pinnatoxins inhibit nicotinic acetylcholine receptors selectively for the humans neuronal alpha 7 subunit.

Detection

Solvents and reagents – Acetonitrile (LC-MS grade) and methanol (HPLC grade) should be used. Ammonium formate (LC-MS quality) and formic acid (analyses quality) should be obtained. PnTX-G reference material (CRM) should be obtained and diluted in methanol to calibration levels from 0.5 to 100 ng ml^{-1}.

Extraction – At least 100 g of whole mussel or clam tissue should be homogenized with a blender. Methanol (9 ml) should be added to aliquots of shellfish tissue (2 g), and toxins should be extracted using a high-speed homogenizer (Polytron) at 15000 rpm for 2 min. Samples should be then centrifuged at 3700 g for 8 min at 4 °C, and supernatants transferred to 20 mL volumetric flasks. This extraction should be repeated twice, and volumetric flasks should be made up to 20 ml with MeOH. Aliquots of extracts were filtered (0.2 μm membrane filters) and analyzed by LC-MS/MS.

LC-MS/MS – Analysis of pinnatoxins can be performed on a UFLC coupled to a triple-quadrupole mass spectrometer. Chromatography should be performed with a HyperClone MOS C8 column (50 × 2.0 mm, 3 μm) with a C8 guard column (4 × 2.0 mm, 3 μm, Phenomenex). A binary mobile phase should be used, phase A (100% aqueous) and phase B (95% aqueous acetonitrile), both containing 2 mM ammonium formate and 50 mM formic acid. The flow rate should be 0.2 ml min^{-1} and injection volume should be 5 μL. The column and sample temperatures were 25 °C and 4 °C, respectively. A gradient elution was employed, starting with 30% B, rising to 95% B over 2.5 min, held for 5 min, then decreased to 30% B in 0.1 min, and held for 2.5 min to equilibrate the system. The LC-MS/MS system should be used in positive ionization and multiple reaction monitoring (MRM) mode. All toxins can be quantified against the PnTX-G standard from NRC, assuming that all analogs (pinnatoxins and pteriatoxins) have the same response factor as PnTX-G.

Confirmatory high-resolution mass spectrometry – Accurate mass data can be acquired on a QTOF 6540 Agilent instrument equipped with a dual ESI source coupled to an Agilent 1200 HPLC system. Separation can be carried out on an Agilent ZORBAX RRHD which extend C18 column (50 × 2.1 mm, 1.8 μm) at 40°°C with a flow rate at 0.5 mL min^{-1} using a linear gradient elution from 10% B to 100% B over 5.5 min. 100% B should be held for 2 min before returning to 10% B at 7.5 min. The column should be then equilibrated for 2.5 min with 10% B prior to the next run. Mobile phase A should be 100% water containing 50 mM formic acid and 2 mM ammonium formate, and B should be 95% acetonitrile with 5% water with 50 mM formic acid and 2 mM ammonium formate.

The instrument should be operated in positive mode, performing full-scan analysis over m/z 100–1700 at one spectrum s^{-1} and targeted MS/MS analysis at five spectra s^{-1}. Capillary voltage and fragmentor voltage should be 3500 V and 200 V, respectively. Resolution achieved is typically 25,000 to 40,000, depending on molecular weight.

Microplate receptor-binding assay for pinnatoxin G-comparison of natural PnTX-G with that obtained by chemical synthesis should be performed by using a microplate-receptor binding assay (Aráoz et al. 2012).

Palytoxins
Palytoxin, one of the most potent marine toxins to cause poisoning in humans and is 10,000 times more toxic than potassium cyanide. It was first isolated from the Hawaiian zoanthid *Palythoa toxica*. The LD_{50} is approximately 0.30 μg/kg^{-1}. Palytoxin and closely related ostreotoxins (osterocines) are probably biosynthesized by the benthic dinoflagellates of the *Ostreopsis* genus, which live on sandy and rocky substrates or macrophytes in tropical, subtropical, and, as recently established, temperate coastal zones. In fact, osterotoxins were isolated from the culture of the microalga *Ostreopsis siamensis*, and their structural similarity to palytoxins was established. Toxins in this structural group are known to be the cause of food poisoning due to their accumulation in crabs sea urchins and saltwater fish. Occurrence of *Ostreopsis* spp. is presently increasing in the Mediterranean Sea, probably in response to climate change. As a result, there is a real danger of numerous human poisonings, because of bioaccumulation of palytoxins and related toxins in both filter-feeding bivalve mollusks and sea urchins. The increase in *Ostreopsis* spp. and their bioaccumulation of palytoxin is a threat to human population.

Symptoms
The most common complication of severe palytoxin poisoning is rhabdomyolysis. This involves skeletal muscle breakdown and the leakage of intracellular contents into the blood. Other symptoms in humans are bitter/metallic taste, abdominal cramps, nausea, vomiting, diarrhea, mild to acute lethargy, tingling, slow heart rate, kidney failure, impairment of sensation, muscle spasms, tremor myalgia, cyanosis, and respiratory distress. In lethal cases palytoxin usually causes death by cardiac arrest via myocardial injury.

Mode of Action
The sodium/potassium/adenosine triphosphatase (Na^+/K^+-ATPase pump), a transmembrane enzyme, maintains the resting potential, and the gradient ion concentrations on both sides of biological membranes of nerve and muscle cells are the main molecular targets palytoxins. The toxin turns the pump in cardiocytes into an ion channel. It is followed by an increase in sodium conduction and influx of calcium ions. It was also established that palytoxin forms a dimeric supramolecule in aqueous solutions, and this molecule interacts with Na^+/K^+-ATPase pump.

4.3 Analysis of Filth (Jacobs 1999b)

Presence of filth or decomposition is the most common cause of detention of fish products imported into the United States (FDA 2002). Out of 4527 detentions (April–June 2002), 443 of the detained products were fish or seafood products. Of the 443 detentions, half (213) were detained because of filth and/or spoilage (FDA 2002).

4.3.1 Examination for Insects and Rodent Contamination

Insect, rodent, or mold contamination can be judged adequately by a visual examination of the untreated product in some cases. If so, the following method of examination should be employed. *Rodent excreta pellets* may readily be distinguished from the nondescript filth that may be found either around a food packing establishment or in the food itself. However, care should be taken that *cockroach excreta*, which is somewhat same size as small mouse pellets, is not mistaken with that of mice and rats. Rodent excreta may be identified by its size and shape by the fact that it has a mucous-like coating and by embedded rodent hair. In infested plant, rodent excreta are usually found along the walls, behind sacks, or in any storage space that is not regularly cleaned. Since rodents need both food and water, their pellets are usually found adjacent to such sources. Suspected rodent excreta pellets always should be checked for the presence of rodent hairs.

In addition to feces, urine may be voided by rats, mice, and cats in food processing establishments. The urine stains appear as somewhat darkened soiled areas on or about the food, or on the snacks, boxes, or other containers. Urine fluoresces under ultraviolet light. When there is any doubt, the suspected boxes are to be removed to the laboratory for a confirmatory examination for the presence of urea.

Animal hairs may be divided into four general classes: (1) guard hairs, or over hairs, which are coarse, heavy hairs that in the living animal act as protection for the softer hairs; (2) soft, under, or fur, hairs which are soft fine hairs used to keep the animal warm; (3) intermediate or curly over hairs which have the characteristics of both guard and fur hair alternating along their length; and (4) tactile hairs such as the vibrissae of animals. Tactile hairs are rarely found in food products.

All animal hairs, whether they are fur or guard hairs, exhibit the following structures. They are covered with a cuticle composed of overlapping keratinous scales and sometimes are of value in identifying some types of hairs, but in cat and rodent hairs, this feature is of limited importance; inside the cuticle the hair contains a cortex. This area may or may not contain some pigment granules, and in some hairs, it may appear to be evenly stained with color because of the presence of very fine pigmentation. The size and density of the pigment granules, as well as the size of the cortex and relative sizes of the cortex to the underlying medulla, sometimes are of diagnostic value; inside the cortex, the hair contains a medullary shaft. It is composed of various loosely arranged cells, interspersed with air spaces in some hairs, and in other hairs, the loosely arranged cells are not noticeable, and the cortex appears to be a tube with a hollow medulla.

The color effect of hair is based on two different factors, air spaces, and pigment granules. The air spaces when examined microscopically with transmitted light appear to be black because the light is reflected with hem and does not pass on through. If the air spaces are filled with balsam or some other substance, the hair appears almost completely clear; if some true pigment is present, the pattern of distribution of the pigment may be distinctly seen. When pigment is present in the cortex, it is present as true pigment granules, but in the medulla, it may be present as true pigment granules or confused with air spaces. In the examination of a hair or hair fragment to determine whether it is a cat or rodent, the first character to note is medulla. If the medulla is multiple-rowed and discontinuous, it is the guard hair of a rodent; if it is smooth or erose (continuous), it is the guard hair of a cat. It is only when the hair has a single rowed discontinuous medulla that difficulty in differentiation is encountered. This type of medulla occurs in the fur hairs of both rodents and cats. Fortunately along the length of the hair of these animals, there are a number of constrictions or internodes.

The internodes of common house mouse and brown rat are very prominent, while those of cat are much less prominent. Unfortunately, the internodes of some rats are not overly pronounced. For example, approximately 10% of the last internodes toward the proximal end of hairs of brown rats are as indistinct as the more distinct one of the cat. The internodes of the black rat similarly are about are indistinct as those of cat and, therefore, a frequently difficult to differentiate. However, the black rat is common only in a few coastal localities in this country, and this factor is of a rodent, but if the hair fragment is short and the internode is either indistinct or lacking, further study may be necessary.

In general cat hairs are larger in diameter than either rat or mouse hairs, but when individual measurements are made, it is found that considerable overlapping occurs. The cortex of cat hair is thicker generally than that of rodent hairs, and the relative size of the cortex to the medulla often shows the cortex of cat hairs to be relatively thick, whereas that of rodents often is relatively thin. Again, actual measurements have shown that some overlapping of the character occurs, and it should be used only in conjunction with other features.

Cortex pigment is of more common occurrence in the hairs of cats than it is in those of rodents. Usually, when large numbers of cortex pigment granules are present, it is the hair of a cat rather than of a rodent. When pigment is present in the cortex of a rodent hair, there will invariably be some in the medulla, and when pigment is present in the cortex only, the hair is definitely of a cat. Most of the cat furs have more densely concentrates pigment granules than most rat or mouse hairs. At 200–400 X this pigment is visible as a blotched darkening of the cortex. At 800–1000 X (oil immersion objective) the pigmentation resolves itself into distinct and separate granules. Hair swollen in sodium hydroxide solution or chlorine (hypochlorite) solution and then examined using the oil immersion objective shows the granules even more plainly. In cat hairs they are arranged in short broken parallel rows, whereas in rat and mouse hairs, the rows are usually widely separate and less obviously parallel. Granules are ovate-elongate to round in both types of hair, although there may be more elongate granules in cat hairs.

The flattened cuticular scales of cut hairs are much smaller and finer than those of mice. In mice the cuticular scales of guard hairs are ovate to crenate, and typically small flattened scales are not found. In rats they are ovate elongated to flattened, the ovate elongate scales being the most pointed of any of these three.

When hairs are treated with 10% sodium hydroxide solution on a microscope slide, the hair swell, and within about half an hour, additional diagnostic characters are visible at 400–500X. After several hours the hairs more or less disintegrate and are no longer usable for study. When the hairs are swollen by sodium hydroxide, the cuticular scales are enlarged and stand out against the swollen cortex. Within the cortex the medullary shaft appears as individual segments variously shaped and arranged.

The following characters apply to fur hairs treated with sodium hydroxide solution:

Rodent fur hairs – the medulla segments of rodent fur hairs, if pigmented, are compactly pigmented; usually one end of the segment is clear with the compacted pigment at the other although not in frequently both ends of the segment are clear with the compacted pigment in the middle (Table 4.3.).

Individually, the pigment granules are black or extremely dark brown, although occasionally light brown granules occur. Segments often are longer than are wide and have an hourglass-like constriction in the middle. The segments are relatively far apart when compared to the segment in cat hairs. The cortex as a rule is not heavily pigmented, but the pigment granules are rather large and appear compact.

When examined under the microscope, they appear as though, if pushed together, adjacent segments would fit similarly to a ball-and-socket joint. They do not touch one another, although they are relatively close together and a straight line drawn across the tips of the concave end of one of the segments will in many cases touch the convex end of the adjacent one. Pigment granules often are rather abundant in the cortex of cat hairs. Sometimes the cortex appears stained with colors because of dispersed very fine pigment granules. In any case, it seldom or never has the clearance of the rodent cortex, and this is due in part to faint parallel longitudinal lines that occur in the cortex of the cat hairs. Rodent hairs seldom if ever contain these longitudinal lines.

Guard hairs (cat or rodent) – in addition to the compound medulla of the untreated rodent guard hair and the continuous medulla of the cat guard hair, some additional observations on the swollen hairs after treatment with sodium hydroxide solution also appear diagnostic. The medulla of the rodent guard hair breaks up into two or more rows of irregularly shaped blocks that appear to have been fitted closely together until forced apart by the chemical action. These blocks may or may not contain pigment. The medulla of the cat guard hair, on the other hand, breaks up into thin plates which appear to have been pressed one upon another along the length of the medulla. The individual plates often do not cover the full circumference of the medulla, but, when they do, they appear as a great number of wafer-like discs not unlike the chamber pith of certain plants.

Table 4.3 Pest hair identification

Character	Rodent	Cat
Medulla fur hair	Single rowed, discontinuous	Single rowed, discontinuous
Medulla guard hair	Multiple rowed, discontinuous	Smooth or erose, continuous
Internodes	Prominent or few indistinct	Indistinct
Diameter	Usually smaller	Usually larger
Cortex size in relation to medulla	Usually narrow	Usually wide
Cortex pigment	Usually sparsely scattered or absent	Usually abundant
	None when none in medulla	May be present when medulla is unpigmented
	Pigment granules in scattered rows	Granules are often in dense parallel rows
Cuticular scales of guard hairs	Ovate, elongate to flat. So very ovate elongate present	Crenate to flat. Some very flat present
	If present, usually compact, black, or dark brown; often localized at one end of the segment with other end clear or at both ends with center clear	If present, composed of loosely scattered granules. Granules brown or reddish; seldom black or dark brown
In NaOH		
Medulla pigment	Usually long in direction of hair elongation	Usually wider than long
Shape of medulla segment	Usually have hour glass constriction.	Sometimes have hourglass-like constructions
Distance apart of adjacent medulla segments	Usually relatively far	Usually relatively close
Contact of adjacent segments	Often squared across	Usually ball-and-socket effect
Cortex appearance	Usually has a clear gelatin like appearance	Usually not clear and with faint parallel, longitudinal lines
Cortex-in hydrochloride, NaOH, or untreated	Usually not heavily pigmented	Usually abundantly pigmented with dense pattern of pigment granules arranged in broken parallel rows

4.3.2 Filth Recovery Methods

The types of filth most commonly met in food products are usually insoluble pieces of insects, rodent filth, or molds. Certain soluble components of urine may be detected by microchemical methods, and relatively insoluble substance such as uric acid from bird or insect excreta may be precipitated and its identity confirmed, but the majority of substances that are encountered are pieces of plant or animal material. Occasionally glass and sand or quartz are found, and these usually require special examination.

Insects fragments, rodent hairs, and molds can stand rather vigorous treatment and still retain their microscopic characteristics although hairs will dissolve in strong alkali and be lost. The most satisfactory way to recover filth from food is to retain it on a filter paper after dissolving the food and filtering it. Such a process is possible in only a limited number of food products. Sugars, some heart candies, and a few other materials such as dextrose-molt preparations fall into this group. By far the great majority of separations are made by some procedures involving a differential of wetting, specific gravity, size, solubility, and/or appearance of the filth and the food involved. Often a combination of steps is necessary before a satisfactory separation is obtained. As is the case with chemical analysis, each precipitation or extraction is incomplete to some small extend, and the fewer the manipulations involved, other factors being equal, the better the recovery.

Heavy materials such a sand can be retained as a sediment by using heavier-than-water organic solvents with such materials as peanut butter and ground spices. Canned leafy vegetables can be floated in strong salt solutions while adhering soil settles out. In either case it is necessary to work the soil free from the plant material, for when the bulk of the separations rises to the top, there is a tendency for small particles to become entrapped and consequently not settle out.

Sieving – in certain instances, a size separation offers a rapid means of separating filth from food. Adult insects, larvae, insect eggs, and excreta, as well as rodent excreta, bits of metal etc may be shifted from foods, or food filth separations may be accomplished by washing the food (or filth as the case may be) through a screen. A screening operation can be used for a preliminary separation before other operations are carried out.

Wildman trap flask methods – since for most foods neither a plain water solution no sieving will give adequate separation, other procedures must be used. When gasoline is mixed with an aqueous mixture containing insects or insect's fragments, the insects float up with the gasoline layer. This principle has been utilized repeatedly for filth extractions.

Specific gravity and oil wetting – part of this effect is brought about when the hydrophobic and oliophilic insect cuticle is wet by oil, which is lighter than the liquid in which the food is soaking, and as the oil droplets rise, the insects are carried up. Some of the oil wets the smooth insect cuticle; some clings to external processes, and water is repelled. Most plant tissue will settle out. This explanation is only part of the picture, for in some instances, the separation appears to be due mainly to the low-specific gravity of the insects. This is particularly noticeable when the insects or fragments are dried out and contains air. Both the relative densities of the liquid, plant material and insects, and the increased differential caused by the oil increment are utilized in the flour method where the insects are floated from the white wheat flour in a saturated salt medium aided by gasoline.

Maggot Recovery

Fly eggs and maggots (and at least some nematodes) respond differently to oil-water notation than do other insects. They settle out, while other insects float. Two theories have been advanced to explain this; both may be play a part. While it is known that certain maggot parts, such as spiracles, are hydrophobic and oliophilic, it may be that the reminder of maggot body surfaces does not attract oil. The maggots encountered in foods generally live in a watery medium, and perhaps they are too full of water and thus are too heavy to be floated. Because of this peculiarity, they are separated by a process opposite to that used for other insects and insect fragments. They are permitted to settle out while the sample is floated. Consequently, instead of gently stirring gasoline into mixture, as is carried out in a Wildman trap, the mixing is carried out vigorously in a separatory funnel so that the sample is caught in the rising gasoline and air and carried up and away from the maggot and eggs which settle. Since even small fly eggs and maggots can be retained on a 10XX blotting cloth, a mixture containing such filth can be filtered on the cloth rather than on a filter paper. By its use much of the extraneous matter is lost through the cloth, and the microscopic examination is simplified.

Removal of Interfering Components Before Extraction of Filth

The above-discussed sedimentation, sieving, and floatation of filth constitute virtually ideal conditions from the analyst's point of view. In practice many foods present difficulties which make it necessary to examine each by individual procedure. At times the troublesome elements become so troublesome that they must be removed before a filth extraction can be attempted.

Starch and protein digestions – Starch can be hydrolyzed by different methods. Boiling, boiling in dilute acids, or enzymatic digestions all have been used. Proteins can be broken down by comparable methods, and they along with starch can be hydrolyzed with pancreatin. Enzymatic digestions have been most helpful with bakery products where the enzymes are used to release filth from the food and to produce a liquid mixture which can be extracted or filtered. The digestions are accomplished readily if the right conditions are maintained. Pancreatin is the most useful enzyme. Its pH and temperature requirements are rather easily fulfilled, while it acts on carbohydrates, fats, and proteins. It is sometimes necessary to boil or otherwise soften the food to prepare it for digestion, but at other times, a dilute pancreatin solution can simply be soaked into the food, and digestion will start immediately.

Fat and oil removal – Fats and oils are troublesome under some condition and of minor importance in others. Chloroform or carbon tetrachloride used for the sedimentation of heavy filth will remove enough of the oil from light foods, so that any subsequent floatation for light filth can be carried out without further attention to the oils. In other instances petroleum ether should be used to remove fats and oils. Bakery products with large amounts of shortening can be digested with pancreatin more readily if most of the shortening is first removed, by soaking in an organic solvent and decanting off the oil bearing solvent, to permit more complete penetration by water and the enzymes. This principle of oil removal is one of the simplest in the filth method. At times, sodium hydroxide solution can be added to emulsify the oil, and while they are not removed, the soapy emulsion formed can be handled more readily than could the original droplet of fat.

Treatments to avoid with hair and insect fragments – Fortunately insect fragments, hairs, and mold will withstand rather vigorous treatment and yet retain their identity. However, hairs are susceptible to the action of alkaline solution; and strong alkalies, such as sodium hydroxide or potassium hydroxide when used hot even in

1% solution, will dissolve rodent hair. Hence when rodent hairs are to be recovered, the use of these and other relatively strong alkaline reacting substances, such as trisodium phosphate, sodium carbonate, and ammonium hydroxide, should be avoided.

Hairs are much more resistant to the action of acids, although the use of hot sulfuric acid and nitric acid at even 5 % should be avoided. Rodent hairs will hold up for 15–40 min in boiling 5% hydrochloric acid, the degree of attack on the hairs depending upon the protection given them by a particular food. Phosphoric acid and other relatively weak acid are much safer. After treatment with acid, the hairs may be soft or brittle and should not be subjected to vigorous mixing action. On a filter paper, they cannot be teased or turned readily without breaking. A further difficulty introduced by severe acid treatment is that the characteristics of morphological pattern may be altered somewhat; and, in identifying hairs by the sodium hydroxide swelling technique, the treatment they have been subjected to must be taken into account since a characteristics pattern may not be obtained. It is believed that rodent hairs processed in acid foods, such as tomatoes, apples, and grapefruit, give us slightly modified pattern when swollen in 1% sodium hydroxide solution as compared with the unprocessed hair.

Insect fragments will withstand any chemical action to which it has been found necessary to subject the food. Alkali will clear insect fragment and remove much of the pigments, especially when used in strong concentrations. Even fragments boiled in saturated aqueous potassium hydroxide solution can be identified as insect part almost are readily as the original untreated pieces. However, it must be remembered that strong sodium hydroxide or potassium hydroxide solutions soften insects, and after boiling in aqueous 10% sodium hydroxide solution for 5–10 mints, they are more subject to mechanical breakage. For all practical purpose in filth work, insect fragments and insects may be considered to be unaffected by acids, even though they can be carbonized by extended boiling in strong sulfuric acid.

Solution and filtration – Sometimes it is possible to complete a separation by simply dissolving the food and leaving the filth in a beaker or on a filter paper without further treatment. Boiling in dilute acid solution will dissolve the water-soluble portions and hydrolyzes a starchy coating or dissolve any carbonites that may be present in the coating. The acid treatment also serves to soften and disperse the food so that it is accessible to other reagents. However, the boiling must be stopped before caramelization of the carbohydrates has gone too far. Acetone can be added to the water to facilitate the solution of the chicle component by an organic solvent such as turpentine or chloroform. By attacking each component separately, it is possible to filter the food through a No. 100 or 150 sieve or bolting cloth.

Washing the filth from the food – Washing is advantageous with such products as nut meals and dried fruits where either plain water or water and detergents can be used to remove and concentrate a surface contamination for microscopic observation.

Centrifuging and other treatment – On occasion centrifuging may solve an otherwise difficult separation problem. In general procedures involving few transfer of material and few manipulations will give better recoveries than more complicated methods. Each manipulation introduces a point at which materials may be lost. This does not mean that a separation in a 100 ml beaker can be more efficiently carried out than in as 400 ml beaker, for at times it is necessary to dilute menstruum in order to loosen adhering filth. Whenever a method is under question, it should be checked by repeated runs with known amounts of material added.

4.4 Emerging Pathogens (Duffy 2009)

The foodborne agents that have emerged in the past three decades include bacteria, viruses, parasites, biotoxins, and a prion. They often emerge from animal reservoirs; 70% are sustained in animal populations and affect humans only incidentally. Many were identified in the course of outbreak investigations, when both their pathogenicity and association with food could be determined. Some common bacterial foodborne pathogens are adapted to particular reservoirs, making targeted control strategies feasible.

4.4.1 Campylobacter

They are adapted to birds and particularly to poultry, among which they are commensal intestinal flora, and contaminate poultry meat by cross-contamination at slaughter. Some strains can also colonize cattle and are transmitted via raw cows' milk.

Campylobacter jejuni/coli, a common cause of illness in the Unites States, contaminates at least four of chicken breasts at retail. As a result, poultry is a major source of *Campylobacter* infections in humans. People become infected by consuming inadequately cooked poultry or other foods that become cross-contaminated via contact with poultry. Even infants riding in shopping carts containing raw poultry are at increased risk.

4.4.2 *E. coli* O157:H7

Shiga toxin – producing *E. coli* O157:H7 can colonize the perirectal glands of ruminants and transfer from hides and feces to meat during the slaughter process. *E. coli* O157:H7 infection has emerged as an important cause of human illness ranging from simple diarrhea, to hemorrhagic colitis, to hemolytic-uremic syndrome, characterized by hemolytic anemia, thrombocytopenia, and renal injury. It was first recognized as a human pathogen in foodborne outbreaks associated with ground beef in 1982. Still beef remains the most frequently identified vehicle for *E. coli* O157:H7 infections, followed by produce-associated outbreaks. Reducing these infections further will depend on preharvest interventions to decrease the shedding of *E. coli* O157 by cattle before they come to slaughter. *E. coli* O157 is common among cattle, particularly in the summertime, and reducing carriage may be achieved using a suite of interventions, including vaccines, probiotics, and bacteriophage treatments, and microbicidal agents such as sodium chlorate.

4.4.3 Salmonella

Strains of *Salmonella* serotype Enteritidis that spread around the world in the 1980s colonize the periovarian tissues of the hen's reproductive tract, where they come in contact with the egg yolk as it forms and contaminates the internal contents of normal-appearing eggs. If the egg is fertilized, these *Salmonella* then colonize the reproductive tissues of the chick embryo and reach the next generation, while in the unfertilized table egg, *Salmonella* can multiply in the yolk and infect the eater of a less than fully cooked egg.

Shell eggs and poultry are common sources of human *Salmonella* serotype Enteritidis (SE) infections, which cause 6000–7000 laboratory-confirmed illnesses annually in the United States. In 2010, approximately 1900 more laboratory-confirmed illnesses with this outbreak strain were reported than were expected.

Another emerging challenge for food safety officials is multidrug-resistant *Salmonella* infections. In the United States, many outbreaks of multidrug-resistant *Salmonella* infections have been investigated in recent years. Because infections with multidrug-resistant *Salmonella* may be associated with an increased risk of hospitalization and antibiotic treatment failure, these illnesses are especially concerning. These outbreaks of multidrug-resistant salmonellosis highlight the importance of preharvest food safety programs to reduce the need for antibiotic usage in animals and for considering further measures to reduce contamination with multidrug-resistant *Salmonella*.

4.4.4 Identifying Reservoirs

The reservoirs where these pathogens persist, and the pathways by which they reach humans, are revealed in outbreak investigations. By epidemiological methods, the illnesses in an outbreak can often be associated with consuming a particular food, the food vehicle of infection. Between 2003 and 2008, the food vehicles identified in 1565 outbreaks reported to the Centers for Disease Control and Prevention (CDC) with specific food vehicles are a broad spectrum of animal- and plant-derived foods.

FoodNet, the network for active surveillance of infections often transmitted through foods in 10 sentinel sites around the country, led by the CDC and supported by the Food Safety and Inspection Service of the US Department of Agriculture (USDA) and the Food and Drug Administration (FDA), provides accurate tracking of what is diagnosed in human clinical laboratories, overcoming local variation in reporting requirements. By 2010, FoodNet data showed that the incidence of *E. coli* O157 infections had declined by 44% since the baseline period of 1996–1998, that of *Campylobacter* by 27%, and that of *Listeria* infections by 38%, while those caused by *Salmonella* had not decreased at all.

4.4.5 Detection of Emerging Pathogens

There are many methods to characterize and compare microorganism at a strain level. They are referred to as typing and subtyping methods. Typing method is classified into phenotypic or genotypic. Phenotypic methods include phage typing, stereotyping and biotyping, and antibiograms. However, genotypic techniques are increasingly used as they are more discriminatory. It also yields substantially more information about the isolate.

Obtaining a genetic profile/fingerprint of the isolate and to then compare the genetic fingerprints for each isolated strain is the most discriminatory approach to categorically relate bacteria. Some of the genetic fingerprinting approaches are described below.

Pulsed-Field Gel Electrophoresis

Pulsed-field gel electrophoresis (PFGE) is one of the most commonly applied genetic fingerprinting approaches. It is a method for separating large DNA molecules in a unique manner. Restriction enzyme is employed to make a limited number of cuts in the bacterial chromosome. This provides a unique chromosomal restriction pattern or "fingerprint" for each individual bacterial isolate. When a fingerprint is generated, computer software is used to compare the profiles with that of other bacterial isolates and established how closely they are related. As PFGE methodology, software, etc. are highly standardized internationally, it has facilitated the development of national and international database of generated PFGE profiles. The largest of these is called "PlusNet" (www.doc.gov/pulsenet/) which is a North American database coordinated by the Centers for Disease Control and Prevention (CDC) which contains thousands of PFGE profiles for pathogens that can be downloaded from the PulseNet website. There are a number of smaller databases in Europe which are generally administrated by national pathogen reference laboratories. The European Center for Disease Prevention and Control (ECDC) is planning to establish an EU-wide "PulseNet Europe" database.

Restriction Fragment Length Polymorphism (RFLP)

RFLP analysis looks into certain types of sequence polymorphism. They are mainly point mutations which can be base exchanges, base deletions, or insertions. The ability of restrictions enzymes, the "endonucleases," to cut double-stranded DNA according to a certain succession of bases, in a process called digestion, is the basis for RFLP technique. It can be done by digestion of DNA samples followed by analysis using standard gel transfer hybridization procedures. The restriction digestion of a PCR-amplified DNA segment that contains a variably present restriction site is another method followed. The technique requires prior knowledge of the DNA sequence flanking that restriction site (Duffy 2009). RFLP technique has been applied to differentiate psychrophilic and psychrotrophic *Clostridia* and lactic acid bacteria associated with vacuum-packed cooked meat product spoilage.

Multi-locus Sequence Typing (MLST)

This technique exploits the unique nucleotide sequence in microorganism by amplifying and sequencing segments of several housekeeping genes. There will be so-called mutations which can be base (A, T, G, and C) exchanges, base deletions, or insertions, and the sequences from different isolates can be compared to establish relatedness between isolates. *Multi-locus sequence typing* has been applied to a wide range of foodborne pathogenic bacteria, and there is a database (http://www.mlst.net) of profiles which also contains specific protocols. It is particularly useful for subtyping *Campylobacter* spp.

Multi-locus Variance Analysis (MLVA)

Majority of the bacterial genomes contain tandem duplications of short DNA sequence. It is termed as "variable-number tandem repeats" (VNTR). Repeat unit size and repeat sequence can vary when multiple loci are examined in a number of different isolates of an individual microbial species. The number of repeat units of locus is a strain-defining parameter. Therefore, there is an isolate specificity in the number of repeats per locus when different strains of a given bacterial species are compared. Multiple-locus VNTR analysis (MLVA) is a subtyping method targeting these repeats. It has emerged as a powerful tool for characterization of bacterial pathogens. It is used effectively for tracing *Salmonella* and VTEC and is a timely, modern, and versatile bacterial typing methodology.

Microarrays

The next generation of typing and tracking technology is likely to be microarrays. These can consist of a range of specific gene sequence (nucleotides) immobilized in surface such as microtiter plates, microbeads, micro-wells, or Eppendorf's tubes (Duffy 2009). The underlining principle microarray analysis is the hybridization of nucleic acid strands on the microarray with the corresponding complimentary sequence in the target microorganism. There is the capacity to put a wide range of specific genes, such as virulence genes, into this format or whole genome information. Microarrays can be used to detect and characterize bacteria but also to investigate gene expression. The analysis of the information generated where tens of thousands of nucleotides are involved is tremendous and requires highly specialist bioinformatics. The cost of microarrays currently prohibits their widespread use for routine purpose, but as they are most widely used, cost will continue to drop, and their use will correspondingly increase.

References

AOAC. (1995). *Official methods of analysis* (16th ed.). Washington DC: Association of official analytical chemists.

Aráoz, R., Ramos, S., Pelissier, F., Guérineau, V., Benoit, E., Vilariño, N., Botana, L. M., Zakarian, A., & Molgó, J. (2012). Coupling the *Torpedo* microplate-receptor binding assay with mass spectrometry to detect cyclic imine neurotoxins. *Analytical Chemistry, 84*, 10445–10453.

Deshpande, S. S. (2002). Seafood toxins and poisoning. In *Handbook on food toxicology* (pp. 695–696). New York, ISBN 0-8247-0760-5: Marcel Dekker.
Duffy, G. (2009). Detecting and tracking emerging pathogenic and spoilage Bacteria from farm to fork. In *Food microbiology and food safety series, safety of meat and processed meat* (p. 447). Fidel Toldra: Springer, Chapter 17.
FDA (Food and Drug Administration). (2002). *Import refusal reports for OASIS*. http://www.fda.gov/ora/oasis/ora_oasis_ref.html
Frank, H. A., Yoshinaga, D. H., & Nip, W. K. (1981). Histamine formation and honeycombing during decomposition of skipjack tuna (Katsuwonuspelamis), at elevated temperatures. *Marine Fisheries Review, 43*, 9–14.
Hoeltz, M., Welke, J. E., Noll, I. B., & Dottori, H. A. (2010). Photometric procedure for quantitative analysis of Aflatoxin B1 in peanuts by thin-layer chromatography using charge coupled device detector. *Quím. Nova, 33*(1), 43–47.
Hua, Y. (1995). High performance liquid chromatography-electrospray ionization mass spectrometry for the determination of Brevetoxins in 'Red tide' algae. *Analytical Chemistry, 67*, 1815–1823.
Ito, E., Satake, M., Ofuji, K., Higashi, M., Harigaya, K., McMahon, T., & Yasumoto, T. (2002). Chronic effects in mice caused by oral administration of sublethal doses of azaspiracid, a new marine toxin isolated from mussels. *Toxicon, 40*, 193–203.
Jacobs, M. B. (1999a). Filth and decomposition in foods, chapter XX. In M. B. Jacobs (Ed.), *The chemical analysis of food and food products* (3rd ed., p. 845). CBS publishers and distributors: New Delhi.
Jacobs, M. B. (1999b). Chemicals food poisoning. In M. B. Jacobs (Ed.), *The chemical analysis of food and food products* (3rd ed., p. 260). New Delhi: CBS publishers and distributors.
Kawashima, Y., Nagashima, Y., & Shiomi, K. (2004). Determination of tetramine in marine gastropods by liquid chromatography/electrospray ionization–mass spectrometry. *Toxicon, 44*, 185–191.
Kim, S. (2014). *Seafood science: Advances in chemistry technology and applications* (p. 504). Boca Raton: CRC press.
Leask, A., Yankos, P., & Ferson, M. J. (2004). Fish, so foul! Foodborne illness caused by combined fish histamine and wax ester poisoning. *Communicable Diseases Intelligence, 28*, 83–85.
Lewis, R. J., Inserra, M., Vetter, I., Holland, W. C., Hardison, D. R., Tester, P. A., & Wayne, L. R. (2016). Rapid extraction and identification of Maitotoxin and Ciguatoxin-like toxins from Caribbean and Pacific Gambierdiscus using a new functional bioassay. *PLoS One, 11*(7), e0160006. Published online 2016 Jul 28. https://doi.org/10.1371/journal.pone.0160006.
McNabb, P. S., Taylor, D. I., Ogilvie, S. C., Wilkinson, L., Anderson, A., & Hamon, D. (2014). First detection of tetrodotoxin in the bivalve Paphies australis by liquid chromatography coupled to triple quadrupole mass spectrometry with and without precolumn reaction. *Journal of AOAC International, 97*(2), 325–333.
Mudadu, A. G., Lorenzoni, G., Bazzoni, A. M., Bazzardi, R., Tedde, G., Arras, I., Sanna, G., Santucciu, C., Marongiu, E., & Virgilio, S. (2017). Yessotoxin detection in bivalve molluscs: A case study from coastal mussel farms (Sardinia, Italy). *Italian Journal of Food Safety, 6*(4), 7015. https://doi.org/10.4081/ijfs.2017.7015.
Ozogul, F., Taylor, K. D. A., Quantick, P., & Ozogul, Y. (2002). Biogenic amines formation in Atlantic herring (Clupea harengus) stored under modified atmospheric packaging using a rapid HPLC method. *International Journal of Food Science & Technology, 37*, 515–522.
Pulido, O. M. (2008). Domoic Acid Toxicologic Pathology: A Review. *Marine Drugs, 6*(2), 180–219. https://doi.org/10.3390/md20080010.

Reguera, B., Velo-Suarez, L., Raine, R., & Park, M. G. (2012). Harmful Dinophysis species: A review. *Harmful Algae, 14*, 87–106.

Satomi, K., Kunihiko, N., Akira, N., Hiroshi, K., & Hiroshi, S. (2014). Simple analysis method for Metallothionein-1, -2 and -3 in the brain by one-step size-exclusion column HPLC on-line coupling with inductively coupled plasma mass spectrometry. *Journal of Analytical & Bioanalytical Techniques*. https://www.omicsonline.org/open-access/simple-analysis-method-for-metallothionein-1-2-and-3-in-the-brain-2155-9872.1000224.php?aid=33335

Techniques Used in Fish and Fishery Products Analysis

5

5.1 Instruments Used for Physiochemical Analysis

5.1.1 pH Meter

The pH meter is an electronic device for measuring the pH (acidity or alkalinity) of a sample. It was only in 1909 that the concept of pH (hydrogen ion concentration) was introduced by Soren Peter Lauritz Sørensen and an electrode for pH measurements was realized in the year 1922 by W.S. Hughes (Nikhil et al. 2016).

5.1.1.1 pH Meter Probe

The probe is a key part of the pH meter. It is a rodlike structure usually made up of glass. At the bottom of the probe, there is a bulb, which is sensitive part of a probe that contains the sensor.

The pH electrode is typically an ion-selective electrode (ISE) that responds to one specific ion in solution, the hydrogen ion. The most common forms of pH electrode are the glass electrode (Fig. 5.1a) and the combination electrode (Fig. 5.1b) which contains an inbuilt calomel reference electrode. Typical glass pH electrodes are made of thin glass porous membrane sealed at the end (the glass bubble) of a hard glass tube containing 0.1 M hydrochloric acid, into which is immersed a silver chloride-coated silver wire. The pH of the internal solution varies from 1.0 (by using 0.1 M HCl) to 7.0 (different buffers are used by different manufacturers). This silver/silver chloride electrode acts as an internal reference that generates a constant potential. The porous membrane is typically 0.1 mm thick, the outer and inner 10 nm consist of a hydrated gel layer that contains exchange-binding sites for hydrogen or sodium ions. On the inside of the membrane, the ion exchange sites are predominantly occupied by hydrogen ions from the hydrochloric acid, while on the outside, the ion exchange sites are occupied by sodium and hydrogen ions. The bulk of the membrane is a dry silicate layer in which all ion exchange sites are occupied by sodium ions. Most of the coordinated ions in both hydrated layers are free to diffuse into the surrounding solution, while hydrogen ions in the test solution can

S. Mathew et al., *Fish and Fishery Products Analysis*,
https://doi.org/10.1007/978-981-32-9574-2_5

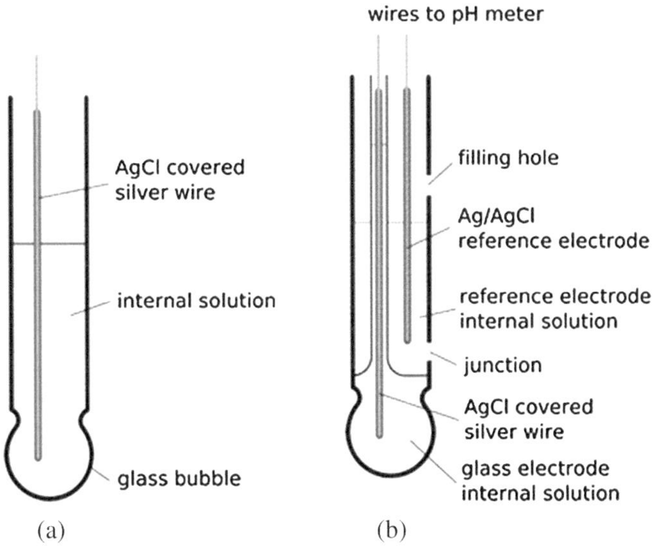

Fig. 5.1 Common pH electrodes (**a**) glass pH electrode (**b**) pH combination electrodes. (Source: http://www.ph-meter.info/pH-electrode-construction)

diffuse in the opposite direction replacing bound sodium ions in a process called ion-exchange equilibrium. Any other types of cations present in the test solution are unable to bind to the exchange sites, thus ensuring the high specificity of the electrode. Note that hydrogen ions do not diffuse across the dry glass layer but sodium ions can. Thus effectively the membrane consists of two hydrated layers containing different hydrogen ion activities separated by a sodium ion transport system. The glass surface is protonated by both internal and external solution until equilibrium is achieved. The adsorbed protons on either sides of the glass are responsible for the potential difference.

Most of the commercially available pH electrodes are combination electrodes, which have both a glass H^+ ion-sensitive electrode and an additional reference electrode that are conveniently placed in a single casing. The pH meters having separate pH electrodes and reference electrodes are still used in some specific applications as they allow higher precision. Nevertheless, most combination electrodes are adequately precise and much more convenient to use.

The construction of a combination electrode is demarcated by the necessary processes while measuring the pH (i.e., difference of potentials between sides of glass in the glass electrode), such that there is a closed circuit through the internal and external solutions. The reference electrode must be kept isolated from the solution such that they do not cross-contaminate, so as to ensue precise and reliable results (Fig. 5.1b). This is achieved by making a small hole in the electrode body, which is blocked by a porous membrane, or ceramic (asbestos in older models) wick. Internal solution flows very slowly through the junction, and such electrodes are referred to as flowing electrodes. The internal solution is gelled, so as to minimize the leaking in a gel electrode.

5.1.1.2 Principle of pH Meter or pH Sensor

The pH meter basically works on the fact that interface of two liquids produces an electric potential which can be measured. In other words, if there is a gradient of hydrogen ion activity across the membrane, this will generate a potential size of which is determined by the hydrogen ion gradient across the membrane. Moreover, since the hydrogen ion concentration on the inside is constant (due to the use of 0.1 M hydrochloric acid), the observed potential is directly dependent upon the hydrogen ion concentration of the test solution. This potential is explained by the Nernst equation and is directly proportional to the pH difference between solutions on either sides of the glass. The observed potential across the membrane is therefore given by the equation:

$$E = E^0 + \frac{0.059\ \text{V}}{Z} \log_{10} \frac{a_{\text{Ox}}}{a_{\text{Red}}}$$

where E^0 = standard cell potential, V = volt, Z = No. of electrons transferred in a half reaction, "a" is the chemical activity of the relevant species, a_{Ox} is the chemical activity of the oxidized form, and a_{Red} is the chemical activity of the reduced form.

In practice, a small junction or asymmetry potential ($E*$) is also created in part as a result of linking the glass electrode to a reference electrode. Since the precise composition of the porous membrane varies with time so also does the asymmetry potential. This contributes to the need for the frequent recalibration of the electrode commonly using two standard buffers of known pH. For each tenfold change in the hydrogen ion concentration across the membrane (equivalent to a pH change of 1 in the test solution), there will be a potential difference change of 59.2 m V across the membrane. The sensitivity of pH measurements is influenced by the prevailing absolute temperature.

5.1.1.3 Other Ion-Selective Electrodes (ISEs)

The principle of operation of these ion-selective electrodes (ISEs) is very similar to that of the pH electrode in that permeable membranes specific for the ion to be measured are used. The present-day bioanalytical ion-sensitive electrodes are discussed in Sect. 5.17 (Biosensors). Some of the commonly used sensors include:

(i) Glucose electrode utilizes glucose oxidase to oxidize glucose in the test sample to generate hydrogen peroxide that is reduced at the anode causing a current to flow that is then measured amperometrically.
(ii) Oxygen electrode measures molecular oxygen in solution rather than an ion.
(iii) Ammonium ion-selective electrodes (NH_4^+ -ISEs) monitor changes in the concentration of ammonia produced.
(iv) Optical sensors use the enzyme luciferase to measure ATP by generating light and detecting it with a photomultiplier.

Electrodes exist for the measurement of many other ions such as Li^+, K^+, Na^+, Ca^+, Cl^-, and NO_3^- in addition to H^+. These electrodes lack absolute specificity, and their selectivity factor is conveyed as selectivity coefficient that expresses the ratio of the response to the competing ions relative to that of the desired ion. Most ISEs have a good linear response to the desired ion and a fast response time.

5.1.2 Moisture Meter

5.1.2.1 Measurement of Moisture Content

A moisture meter is used to quantify the water content/percentage of water in a given substance. Moisture meters are the class of instruments that are able to measure the trace amounts of moisture in solids, gases, and hydrocarbon liquids. Controlling moisture is vital as water vapor above certain levels can lead to condensation, corrosion, contamination, or other issues. Each material and application is affected differently by moisture, and each has a range of moisture most suitable for it. Only a moisture meter can let you know when moisture levels are ideal. Moisture meter accomplishes moisture measurement by employing the relationship which exists between the moisture conditions and depends on the conductivity measurements, capacitance measurement, and electrical resistivity. These meters are equipped with sensors known as transducers which help in converting one kind of energy into another type of energy. A novel moisture meter sensor depends on the principles of frequency measurement (the meter operates on the principle of nondestructive measurement) and time-domain measurement (the meter works on the principle of change in the time duration of passage of the pulse through the material).

5.1.2.2 Online Moisture Measurement

All continuous moisture analyzers utilize secondary measurement principles and must be calibrated against a primary reference technique. They have the advantage of continuous or rapid sampling measurement and may be used for real-time process monitoring and control. A continuous moisture analyzer will provide information between laboratory sampling, even if left uncalibrated. In this form, the instrument is a useful set point controller, where the process set point may be adjusted after each laboratory sample. There are numerous online moisture measurement techniques. Dielectric measurement and near-infrared reflectance are the two that have proven to be accurate and reliable in many industries.

Dielectric Technique

This method relies on the relatively high dielectric of water relative to most solids. Numerous techniques have been developed to determine dielectric, including radio frequency, microwave, and time-domain reflectometry. The various mechanical sensor designs suit a wide range of process conditions and can be used in high-temperature environments.

The dielectric (capacitance) technology is used in many moisture meters. It is based on the principle that a functional relationship exists between the moisture content of the sample and its dielectric constant. Moisture meters based on the dielectric principle typically incorporate a test cell in the form of an electrical capacitor, that is, two conductors separated by an insulator. When the cell is empty, only air separates the two conductors, and the insulator is air. In a dielectric (capacitance) technology, the sample displaces most of the air, when it is placed between the conducting surfaces of the test cell. This creates a change in the electrical characteristics of the capacitor due to the dielectric properties of the sample. A moisture meter senses the change in the electrical characteristics and predicts the moisture content of the sample. Because the bulk density and the temperature of the sample also affect the electrical characteristics of the sample-filled test cell, the meter must measure these parameters and apply the necessary corrections.

A "dielectric material" (or dielectric) is an electrically insulating material that will be polarized under an electric field, and the phenomenon is called dielectric polarization. The dielectric properties of the material provide valuable information about the storage and dissipation of electric and magnetic fields in materials and also provide insight into the feasibility of using the material in potential applications.

True dielectric moisture instruments are rare, and even the lowest-cost moisture instruments have several advantages over conventional methods in that it can be used to measure nonhomogeneous products since it is a penetrating measurement. It is very reliable, robust, and comparatively inexpensive compared to other online techniques. Dielectric property is a characteristic of plant materials, vegetables, fruits, seafood, and other food materials mainly due to the structure of the biomaterials and the large amount of water content.

Infrared Technique

The near-infrared reflectance, NIR or IR, technique is a widely used technology for online moisture testing. Its popularity is due in large part to the ease with which it can be applied. A light source (typically quartz halogen bulb) is collimated and filtered into specific wavelengths. The filters, mounted in a rotating wheel, chop the light into a series of pulses of specific wavelength. The filtered beam is directed onto the surface of the product to be measured. A portion of light is reflected back to a detector (usually lead sulfide) such that a specific wavelength of light is absorbed by water. If filters are chosen such that one wavelength will be absorbed by water (sample beam) and one wavelength will be unaffected by water (reference beam), then the amplitude ratio of the two reflected wavelengths will be proportionate to the amount of water in the product.

5.1.3 Hygrometer

A hygrometer is an instrument used to measure water activity.

5.1.3.1 Measurement of Water Activity (a_w)

Measurement of water activity (a_w) is performed by determining the relative humidity of the air surrounding the sample when the air and the sample are at equilibrium. The sample must be in an enclosed space so as to achieve this equilibrium. Once this occurs, the water activity of the sample and the relative humidity of the air are equal. The measurement taken at equilibrium is called an equilibrium relative humidity, or "ERH."

5.1.3.2 Types of Hygrometers

During the past, the measurement of water activity of foodstuffs was a cumbersome experimentation. New instrument technologies have vastly improved speed, accuracy, and reliability of measurements. Commercially available water activity instruments are of two types: the chilled mirror dew point technology and the electrical resistance or capacitance meter. Each has its own advantages and disadvantages and varies in its precision, speed and repeatability, stability in calibration, linearity, and ease of use.

Chilled Mirror Dew Point Hygrometers

The temperature at which dew forms on a clean surface is directly related to the vapor pressure of the air. The mirror is cooled until the dew point temperature is measured by means of an optical sensor. This temperature is then used to find the relative humidity of the chamber using psychrometric charts. This method is theoretically the most accurate (+/− 0.003 a_w) and often the fastest.

Dew point hygrometers work by placing a mirror over a closed sample chamber that contains a mirror, an optical sensor, an internal fan, and an infrared temperature sensor. The relative humidity of the air in the chamber is the same as the water activity of the sample, at equilibrium. A thermoelectric (Peltier) cooler specifically controls the mirror temperature. An optical reflectance sensor detects the precise point at which condensation appears first. Then, a beam of infrared light is directed onto the mirror and reflected back to a photo detector that detects the change in reflectance when condensation occurs on the mirror. A thermocouple attached to the mirror accurately measures the dew-point temperature. The internal fan allows air circulation, which reduces vapor equilibrium time and controls the boundary layer conductance of the mirror surface. An additional thermopile sensor (infrared thermometer) measures the sample surface temperature. Both the dew point and sample temperatures are then utilized to determine the water activity. The instrument repeatedly determines the dew-point temperature until vapor equilibrium is reached. Generally, calibration is not necessary, since the measurement is based on temperature determination. However, monitoring using a standard salt solution ensures proper functioning of the instrument. The mirror is easily manageable and can be cleaned in a few minutes during fine-tuning.

Capacitance Sensor Hygrometers

Capacitance sensor instruments use a sensor made from a hygroscopic polymer membrane and associated circuits that give a signal relative to the ERH (equilibrium relative humidity). As the membrane adsorbs water, its ability to hold charge increases, and the capacitance is measured. This value is roughly proportional to the water activity and is determined by a sensor-specific calibration. The ERH is equal to the water activity of the sample, only as long as the temperatures of the sample and the sensor are the same. Since these instruments relate an electrical signal to relative humidity, the sensor must be calibrated with known salt standards. The capacitance hygrometers are not affected by most volatile chemicals, but they can be affected by residual water in the polymer membrane (hysteresis). They do not require cleaning, but are less accurate than dew point hygrometers. Accurate measurements with this type of system require good temperature control.

5.1.3.3 Evaluating the Instrumentation

When evaluating water activity measurements, precision and accuracy are, of course, important considerations. But equally important to consider is how susceptible the sensor is to contamination and how frequently calibration is required. Also, when comparing water activity instruments, it should be ensured to evaluate precision and accuracy over the entire range of water activities most commonly found in specific products.

The chilled mirror dew point instruments are advantageous in terms of accuracy, speed, and ease of use. Measurement is achieved in less than 5 min, in the range of 0.030–1.000a_w, with a resolution of ±0.001a_w and accuracy of ±0.003a_w. However, the capacitance sensors have the advantage of being inexpensive, but are not typically as accurate or as fast as the chilled mirror dew point instruments. Capacitive instruments measure over the entire water activity range 0–1.00 aw, with a resolution of ±0.005a_w and accuracy of ±0.015a_w. Some commercial instruments can measure in 5 min, while other electronic capacitive sensors may require 30–90 min to attain equilibrium relative humidity conditions.

5.1.3.4 Water Activity: Accepted and Approved

Water activity is an important aspect for many food products, as it predicts stability with respect to physical properties, degradative reactions, microbial growth, and perishability. The US Food and Drug Administration ensues the incorporation of water activity principle in the definition of non-potentially hazardous foods (potentially hazardous foods are foods with a finished equilibrium pH greater than 4.6 and a water activity greater than 0.85). The FDA uses this and other criteria for quality control (e.g., the thermal destruction of *Clostridium botulinum* to prevent botulism).

The nonlinear relationship of water activity (a_w) and moisture content in a sample is known as moisture sorption isotherm curve. These isotherms are substance and temperature specific (Refer Sect. 5.2) and can be used to forecast product stability at different storage conditions.

5.1.4 Torry Meter

Torry meter is typically a fish freshness meter that measures changes in the electrical properties of a sample. It can be used as an indicative sign of quality deterioration in fish and therefore provides a means of measuring postmortem changes or the level of spoilage. The development of the Torry meter was carried out at the Torry Research Station, Scotland.

5.1.4.1 Instrumentation

The Torry meter is a compact device, equipped with a sensing unit that can be applied directly on to the skin of the fish. The base of the Torry meter instrument is a sensing head, consisting of a pair of electrodes which passes an alternating current through the fish, and as a result, a voltage is sensed in the inner pair. The phase angle between the current and voltage is measured and converted electronically to allow digital measurement on a convenient scale in the range 0–18. The phase angle, and hence the meter reading, decreases with spoilage. Since the phase angle depends on temperature, as well as freshness, the reading displayed by the instrument is corrected to the value it would show at a reference temperature of 0 °C.

Generally, the measuring electrodes are supplemented by two auxiliary electrodes that are in conjunction with one of the carbon electrodes and help in ensuring whether there is proper contact with the experimental fish and not with that of air or ice. The current passing through the fish is approximately 1 milliamp, and hence do not harm the operator or affect the fish. The meter may also be powered by rechargeable batteries within the instrument.

5.1.5 Microkjeldhal Apparatus

The Kjeldahl method has been almost universally applied technique to determine nitrogen content (AOAC 2000) (Fig. 5.2).

5.1.5.1 Protein Analysis Using Microkjeldhal Method

The Kjeldahl method or Kjeldahl digestion is used in analytical chemistry as a method for the quantitative determination of nitrogen contained in organic substances and the nitrogen contained in the inorganic compounds ammonia (NH_3) and ammonium (NH_4^+), excluding detection of nitrate (NO_3^-). The present-day Microkjeldahl method to analyze organic nitrogen was originally developed by Johann Kjeldahl in 1883.

Modern Kjeldahl apparatus combine digestion and distillation/titration in an automated method. The early Kjeldahl digestions were performed using stone fume hoods and gas mantles as a heat source. Then, around 1930s, they were replaced by classical Kjeldahl digestion and distillation apparatus. The Kjeldahl digestion apparatus is available freestanding, appropriately combined with fume removal system, or can be positioned under a fume hood.

Kjeldahl apparatus for distillation are available with different levels of automation, ranging from relatively manual models to highly automatic Kjeldahl distillation units. Modern Kjeldahl apparatus for distillation is designed to accept straight digestion tubes, directly from block digesters; hence, the digested sample does not need to be transferred.

Steam distillation is much more rapid than classical Kjeldahl distillation, typically taking 3–5 min. Entry-level models dispense sodium hydroxide under control of a push button. Many models include a timer to control the length of distillation. Some units automate the entire Kjeldahl nitrogen process once the digestion flask is

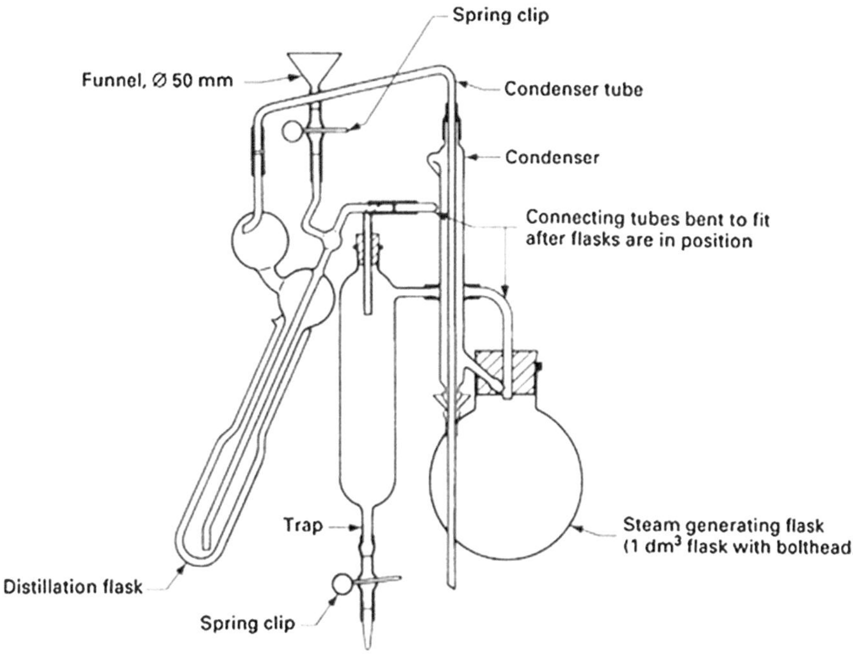

Fig. 5.2 The Kjeldahl distillation apparatus is used to determine organic nitrogen (N_2) and protein content in foods (https://www.iso.org/obp/ui/#iso:std:iso:1656:ed-3:v1:en:fn:3)

in place, and others automatically titrate to end point after Kjeldahl distillation and calculate and display the results in a printed report.

The nitrogen content in foods includes the protein as well as the nonprotein nitrogen (NPN) constituents. The nitrogen content of proteins actually varies from about 13 to 19%, upon equating to nitrogen conversion factors ranging from 5.26 (1/0.19) to 7.69 (1/0.13), based on Jones (1941).

5.1.6 Soxhlet Apparatus

Franz Ritter von Soxhlet (1848–1926) was a German agricultural chemist who invented the Soxhlet extractor in 1879. Soxhlet is also known as the first scientist who fractionated the milk proteins albumin, globulin, casein, and lactoprotein and also the first scientist to describe lactose, the disaccharide sugar present in milk (Fig. 5.3).

The Soxhlet apparatus is simple to set up and has a reservoir flask, an extraction chamber, and a condenser (Fig. 5.3) that features ground glass joints for easy dismantling and cleaning. This inert all-glass system can be used for extracting semivolatile compounds (e.g., fat/oil) from a solid or semisolid sample – seafood/meat, for example – into an organic extraction solvent, such as petroleum ether, diethyl ether, hexane, acetone, benzene, or alcohol.

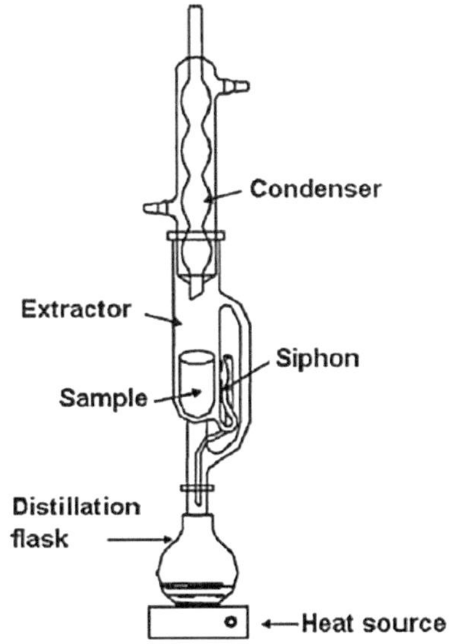

Fig. 5.3 Soxhlet extraction apparatus (Wang and Weller 2006)

5.1.6.1 Semicontinuous Extraction of Lipids

The total lipid content of a food is commonly determined by organic solvent extraction methods, preferably using a Soxhlet apparatus and is a type of semicontinuous extraction process (AOAC 2007). It involves transferring the partially soluble components of a sample to the liquid phase using a Soxhlet extractor. The sample taken in a filter paper/cellulose thimble is placed into the main chamber of the Soxhlet extractor. The extraction solvent inside the boiling flask is first evaporated and recondensed in the distillation column. The solvent then condenses down onto the solid sample material requiring extraction. The chamber containing the sample material (inside the thimble) is connected to the boiling flask below that allows the chamber to fill to a point in the siphon arm, at which it will empty its contents onto the boiling reservoir flask by a syphoning mechanism. The extracted compounds will accumulate in the boiling reservoir flask below, and the process is allowed to continue until all extraction is done. The lipid sample is recovered from the solvent by repeated distillation. Fat content is then measured gravimetrically.

Ideal solvents for fat extraction should have a high solvent power for lipids and low or no solvent power for proteins, amino acids, and carbohydrates. They should evaporate readily and leave no residue, have a relatively low boiling point, and be nonflammable and nontoxic in both liquid and vapor states (David and Wayne 2014).

5.1.7 Muffle Furnace

Muffle furnace is oven-type heat treatment equipment used to change physical properties of samples at very high temperature ranging from 500 °C to nearly

1700 °C. These laboratory furnaces are widely used in scientific experiments mainly for ash determination (dry ashing method) that can be accomplished by incineration at high temperature (525 °C or higher). Ash content denotes the total mineral content in a sample. Generally, it is a part of proximate analysis for nutritional evaluation of foods (Sect. 5.1.2). Determining the ash content may be important for several reasons. Several models of muffle furnaces ranging from small bench top to large-capacity units are available for various applications including coal analysis, leaves carbonization, lime calcination, etc.

5.1.7.1 Instrumentation

Typically, muffle furnace is oven-type equipment that can reach very high temperatures (up to 1700 °C). The instrument is equipped with a high-temperature heating coil in an insulated chamber. They can be metallic or ceramic, and main ones used are NiCr, FeCrAl alloys, silicon Carbide/SiC, or $MoSi_2$. Advances in materials for heating elements, such as molybdenum disilicide ($MoSi_2$), can now produce working temperatures up to 1800 °C. An insulating material effectively acts as a muffle, preventing heat from escaping. The furnace chamber is heated by electric resistance elements and is insulated with ceramic fiber insulation. Generally, good thermal insulation is achieved by refractory bricks (ceramic) and refractory fiber/wool, usually made of SiO_2 or AlO_3. The controller is located under the furnace chamber and is well insulated from the heat generated in the furnace chamber. A door safety switch removes power to the heating elements whenever the furnace door is opened. The temperature is controlled by one of three types of controllers. Heating to the desired temperature is done by conduction, convection, or blackbody radiation from electrical resistance heating elements. A metal thermocouple senses the temperature in the chamber and transmits this information to the temperature control in millivolts. The control section consists of a temperature controller, a current controller, a transformer, a contactor (relay), and a circuit breaker. The temperature controller senses the furnace temperature (by means of a current controller), which controls the current to the heating elements by adjusting the magnitude of electrical current (Fig. 5.4).

5.1.7.2 Crucible Types

Crucible is a container in which food samples, metals, or other substances may be melted or subjected to very high temperatures for ash content/mineral analysis (Table 5.1).

The conventional dry ashing in a muffle furnace often takes several hours. Microwave muffle furnaces are now available for ashing the samples in a few minutes, typically reducing the analysis time by 97%. Microwave muffle furnaces that can reach temperatures of up to 1200 °C are now available for both dry and wet ashing methods. Microwave systems are equipped with automatic warm up and cool down programs, exhaust systems enabling air circulation, and a scrubber for neutralization of the fumes. Conventional crucibles made of porcelain, platinum, quartz, and quartz fiber can also be used in a microwave muffle furnace.

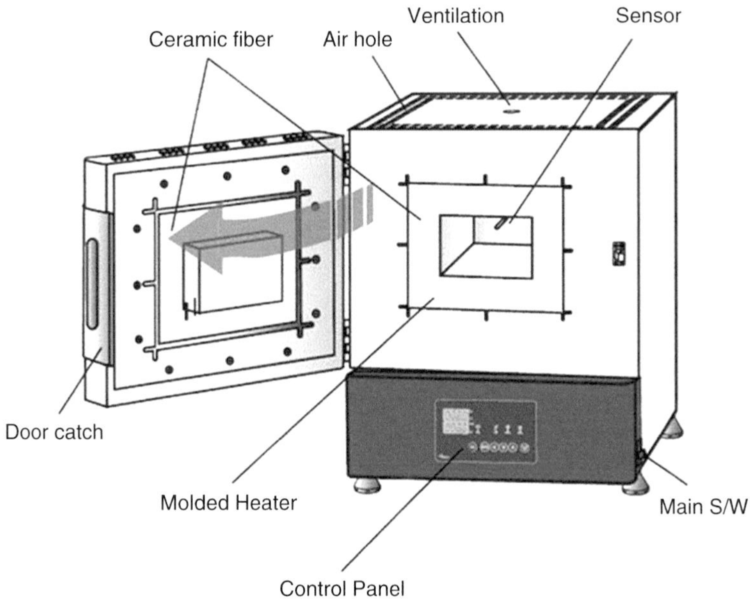

Fig. 5.4 Schematic diagram of a muffle furnace (https://79.twizer.co/diagram-of-a-muffle-furnace.html)

Table 5.1 Different types of crucibles

Crucible type	Properties
Porcelain crucibles	Porcelain crucibles are the most commonly used crucible type and are comparatively inexpensive. They resemble quartz crucibles in their properties, but will crack with rapid temperature changes
Vycor R brand crucibles	These are stable to up to 900 °C
Pyrex R Gooch crucibles	These crucibles are stable only up to 500 °C. Ashing at a lower temperature of 500–525 °C may result in less decomposition of carbonates giving slightly higher ash values. There is also a possibility of the loss of volatile salts
Steel crucibles	Even though steel crucibles are inexpensive and resistant to acids and alkalis, they are composed of chromium and nickel that are possible sources of contamination giving error values
Quartz crucibles	At high temperatures, quartz crucibles are resistant to acids and halogens, but not alkali
Quartz fiber crucibles	Quartz fiber crucibles reduce ashing times significantly since they are porous, allowing air to circulate around the sample. They are unbreakable and disposable, making them ideal for solids and viscous liquids. Quartz fiber crucibles can withstand temperatures up to 1000 °C and cools rapidly, virtually eliminating the risk of burn injury
Platinum crucibles	Platinum crucibles are the most expensive for routine use, but they are probably inert making them the best crucibles

5.2 Chromatography

Mikhail Semyonovich Tsvet (1872–1919) was a Russian botanist who developed the method of chromatography early in 1903, for extracting plant pigments using a mobile phase solvent, over a stationary phase glass column packed with chalk ($CaCO_3$) and alumina particles. The term "chromatography" in Greek, literally indicates "chroma" meaning "color" and "graphy" meaning "to write or record." Mikhail S. Tsvet is now popularly renowned as the "father of chromatography." Chromatography involves the separation of a sample (or sample extract) dissolved in a mobile phase (that is a gas, a liquid, or a supercritical fluid), when forced through an immobile, immiscible stationary phase. The phases are chosen such that components of the sample have different solubilities in each of the phase. A component with affinity/solubility in the mobile phase travels faster along the mobile phase, while those compounds with lesser affinity get retarded. As a result of these differences in mobilities, sample components get separated from each other as they travel through the stationary phase.

Techniques such as HPLC (high-performance liquid chromatography) and GC (gas chromatography) use columns – narrow tubes packed with stationary phase, through which the mobile phase is passed. The sample gets transported through the column along with the continuous supply of mobile phase in a process termed, elution. The average rate at which an analyte moves through the column is determined by the time it spends in the mobile phase.

5.2.1 HPLC (High-Performance Liquid Chromatography)

High-performance liquid chromatography is typically a highly automated form of the conventional column chromatography methods. HPLC is a powerful tool in analytical chemistry for the separation, identification, and quantification of components in a mixture. The acronym HPLC was coined by Professor Csaba Horváth in 1970, indicating the high-pressure flow generated for liquid chromatography (LC) in packed columns. Instead of a solvent being allowed to drip through the column under the influence of gravity, the mobile phase is forced through the column at high pressures of up to 400 atmospheres in an HPLC, making it faster.

Based on the modes of separation, HPLC is of two types: the normal phase and the reversed phase (RP). In reverse-phase HPLC, the stationary phase is nonpolar (e.g., C_{18}-bonded silica), while the mobile phase is polar, usually being water and polar organic solvents. RP-HPLC can be used to separate, identify, and or quantitate components in mixtures of soluble organic components based on their hydrophobicity. Compounds with the most hydrophobicity elute later in the chromatogram, and those with the least hydrophobicity elute earlier. Elution can be done in isocratic and gradient modes to optimize the separation of analytes. In a normal phase chromatography method, the stationary phase is polar, while the mobile phase is a nonpolar component (e.g., hexane, chloroform, etc.). Compounds with greater polarity elute later in the chromatogram, and those with the least polarity elute earlier. A number of HPLC applications have been employed in various fields including clinical analysis, pharmaceutical analysis, and polymer and phenolic compound analysis in addition to food analysis (Corradini 2011).

5.2.1.1 Instrumentation

HPLC system includes five main parts: (i) pump and solvent reservoir, (ii) injection port, (iii) column, (iv) detector, and (v) data acquisition software (Fig. 5.5).

(i) *Solvent Reservoir and Pump*

Mobile phase reservoir carries the mobile phase solvents, for isocratic (a single solvent container) or a gradient elution (two or more solvent containers), based on the chromatic conditions. The mobile phase solvents should be ideally pure and particularly filtered (using 0.45 μm filters) and degassed before use. The solvent reservoirs should be inert to mobile phase solvents, buffer salts, and solutes.

The role of the pump is to provide the pressure that propels the mobile phase solvents through the stationary phase (column) at a specific flow rate (normally 1–2 ml/min). They are made of stainless steel, titanium, and resistant minerals. Typically, pumps can achieve pressure ranges from 2000 to 6000 psi. An ideal pump should be solvent compatible and resistant to corrosion, capable of constant flow delivery independent of back pressure, and should also possess low dead volume for minimum problems on solvent change over.

The main types of pumps used in HPLC (or in LC) are constant pressure pumps and constant flow pumps. Constant pressure pumps provide consistent continuous flow rate through the column. Constant flow pumps include two types: "the reciprocating piston pumps" and "the syringe-type pumps." Reciprocating piston pumps deliver solvent through reciprocating motion of a piston in a hydraulic chamber. The main drawback of a reciprocating pump is that it produces a pulsing flow. With a flow-sensitive detector, such as micro-adsorption detector, a pulse damping system must be used, and detector sensitivity is reduced. While in syringe-type pumps, a

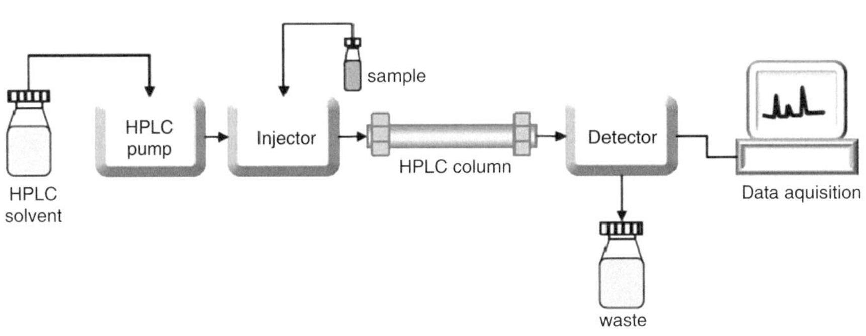

Fig. 5.5 Schematic representation of an HPLC apparatus. (Source: https://laboratoryinfo.com/hplc)

constant flow rate is delivered to a column by a motorized screw arrangement and is suitable for small bore columns. Pump operates in two different modes: isocratic and gradient elution. The isocratic pump delivers a constant single mobile phase composition, whereas gradient pump delivers variable mobile phase comprising two or more mobile phase solvents, enabling better resolution of the analytes.

A degasser assembly between the sample reservoir and pump enables the removal gas bubbles in mobile phase solvents that may block column performance.

(ii) *Injection Port*

An injection port functions for sample introduction into the stream of mobile phase. Manual and automated injector systems are available, which can withstand the high pressure of the mobile phase. An injection port (Rheodyne) valve serves to fill samples in an external loop, on application of samples using the syringe. Clockwise rotation of the valve rotor spaces the sample-filled loop (typical sample volumes 5–200 μL) into the mobile-phase stream, with subsequent injection of the sample into the column through a low-volume, cleanly swept channel. Valve injection allows the rapid, reproducible, and essentially operator-independent delivery of a wide range of sample volumes (μL to ml), at varying pressures up to 6000 psi. Manually operated valves are less expensive compared to automated versions. A minor disadvantage of most sample valves is that the sample loop must be changed accordingly, to attain various sample volumes, which can easily be achieved by an expert in a few minutes.

(iii) *Column*

Column is considered to be the most important part of the chromatograph, and the stationary phase packing material enables the separation of analytes, based on various chromatographic principles. It comprises of a metal housing (usually made of stainless steel to withstand high pressure) in a tube shape which is packed with tiny beads of stationary phase packing material that have an affinity for the analytes that are being analyzed (Refer Table 5.2). The column is also provided with a filter which lets the solvent to flow through but keeps the beads from leaving the column. HPLC columns are available in various modes including analytical (having an internal diameter (i.d.) of 1.0–4.6 mm and length ranging from 15 to 250 mm) and preparative columns (i.d. > 4.6 mm; lengths 50–250 mm). Capillary (. 0.1–1.0 mm i.d; varying lengths) and nanoanalytical (i.d. < 0.1 mm) columns are also available commercially.

(iv) *Detectors*

At typical HPLC analytical concentrations, the solutions of many compounds are colorless. An appropriate detector has the ability to detect the presence of a compound and send its corresponding electrical signal to a computer data station (Corradini 2011). The detector contains a flow cell that detects each separated

Table 5.2 Commonly used stationary phase column packing materials

Compound	Abbreviation	Chemical structure	Application
Aminopropyl	NH_2	Si-$(CH_2)_3$-NH2	Weak anion exchanger
Cyanopropyl	CN	Si-$(CH_2)_3$-CN	Compounds with double bonds and aromatic groups
Dihydroxypropyl	Diol	Si-$(CH_2)_3$-OCH(OH)-CH_2-OH	Polar biomolecules, pesticides, pharmaceutical metabolites
Octadecyl	C_{18}	Si-$(CH_2)_{17}$-CH_3	C_{18} alkyl: steroids, fatty acids, pharmaceuticals, phthalate, environmental pollutants
			C_{18} polar: water-soluble vitamins, organic acids, proteins and peptides
Octyl	C_8	Si-$(CH_2)_7$-CH_3	Water-soluble molecules, steroids, peptides, catecholamines, sedatives
Pentafluorophenyl	PFP	Si-$(CH_2)_3$-C_6F_5	Halogenated, polar, and isomeric analytes
Phenyl	C_6H_5	Si-$(CH_2)_3$-C_6H_5	Aromatic and moderately polar compounds

compound band against a background of mobile phase. A choice is made among different types of detectors, depending on the characteristics and concentrations of the compounds that are separated and analyzed.

UV-VIS Detector UV-VIS detector is the most commonly used detector, with specificity to light-absorbing functional groups present on a particular compound or class of compounds on the analyte. UV-VIS absorbance detectors are near universal as many solutes absorb UV light. They measure the amount of light transmitted through a solution according to the Beer's law: $A = \epsilon cl$. Some compounds which do not have such light-absorbing groups can give suitable response after post-column derivatization, thereby introducing light-absorbing entities. Approximate detection limit is 0.1–1.0 ng.

Photo Diode Array Detector Incorporation of large number of diodes which serve as detector elements makes possible the simultaneous monitoring of component mixtures at different wavelengths. This provides benefit of time saving and cost reduction on expensive solvents.

Fluorescence Detector Fluorescence detection offers greater sensitivity than a UV-VIS detector. However, the occurrence of naturally fluorescent compounds is meager compared to light-absorbing compounds. The limitation can be overcome by post-column derivatization using fluorescent dyes, even though they are very expensive. Fluorescence detectors are very sensitive with an approximate detection limit of 0.001–0.01 ng. They excite the eluate with a laser and measure fluorescence.

Chemiluminescence Detector Chemiluminescence detector offers a simple, sensitive, and comparatively low-cost method to quantify a wide variety of compounds. Chemiluminescence is simply the emission of light in an electronically excited state as a result of a chemical reaction. There are two types of chemiluminescence reactions: direct and indirect. In direct chemiluminescence, the reaction between compounds X and Y forms a product or intermediate (Z∗) that returns to its ground state (Z) by the ejection of a photon. In the case of indirect chemiluminescence, instead of Z∗ returning to the ground state by photon ejection, it undergoes an energy transfer reaction with a suitable fluorophore, which then exhibits a fluorescence emission.

Refractive Index Detector (RID) The RI detectors are sensitive to all analytes that have a different refractive index than the mobile phase. Light passes through a cell with pure solvent and is directed to a photocell by a deflection plate. When solute with a different RI enters the cell, the beam is deflected, and the photocell output changes. The response is dependent on changes in refractive index of eluting compounds in the mobile phase. The deflection of light is measured by Snell's law: $n_1 \sin\theta_1 = n_2 \sin\theta_2$. Gradient programming is not possible due to resultant alterations in the refractive index of mobile phase. This detector is less sensitive than UV-VIS detector with an average detection limit of 100–1000 ng. Temperature control is also necessary as it has high-temperature sensitivity. Typical applications are in size exclusion chromatography.

Electrochemical Detector Electrochemical detectors are very selective and sensitive. It measures the electron flow generated at the surface of electrodes during an oxidation or reduction reaction. It is, however, sensitive to changes in composition or flow rate of mobile phase.

Evaporative Light-Scattering Detectors These detectors are considered to be universal as they respond to any analyte that is significantly less volatile than the mobile phase constituents. Eluate is formed into a uniform dispersion of droplets, imparting a fine mist of solid particles upon solvent evaporation. After removal of mobile phase by passing through a heated zone, the solute molecules are detected by light scattering depending on molecular sizes. Light-scattering detectors are useful for detection of high-molecular-weight molecules.

Mass Spectroscopic Detector Mass spectroscopy offers very high sensitivity and selectivity. Detection is based on fragmentation of molecules by electric fields and separation on basis of mass-to-charge ratios of fragmented molecules. LC-MS technique has opened up new application areas due to advantages of resolution and sensitivity.

Data Acquisition Software

It includes a computer work station that controls all the modules of the HPLC instrument. Additionally, it collects the detector signals and performs to determine various quantitative and qualitative operations on the chromatogram thereby assisting in sample identification and quantitation. A chromatogram is a representation of the separation that has chemically/chromatographically occurred in the HPLC system. A series of peaks rising from a baseline is drawn on a time axis. Each peak represents the detector response for a different compound, and it is plotted by the computer data station.

5.2.1.2 The Retention Factor, Peak Width, and Resolution

The time space between sample injection and the analyte reaching the detector at the end of a chromatographic elution is termed the *retention time* (t_R). Each analyte in a sample will have a different retention time. The time taken by the mobile phase to pass through the column is termed t_M (Fig. 5.6).

Since the separation is based on adsorption, the quantitative measure of the tendency of a particular type of molecule to adsorb to the stationary phase is important and measured in terms of the retention factor (k'), also known as the capacity factor. The retention factor is often used to describe the migration rate of an analyte inside the column. The retention factor for an analyte, X, is defined as

$$k_X' = t_R - t_M / t_M$$

t_R and t_M can be easily obtained from a chromatogram. When the retention factor is less than one, elution of the analyte is so fast that accurate determination of the retention time becomes difficult. Retention factors greater than 20 indicate that the elution takes a very long time. The ideal retention factor is between one and five.

The separation factor, α for two different analytes A_1 and A_2, is the ratio of the two components,

$$\text{i.e.,}\ \alpha = A_2 / A_1$$

Due to the nature of the chromatographic equipment, a dilution of the injected sample occurs when the sample molecules migrate through the column. As a consequence, the zone containing the sample molecules broadens continually during its

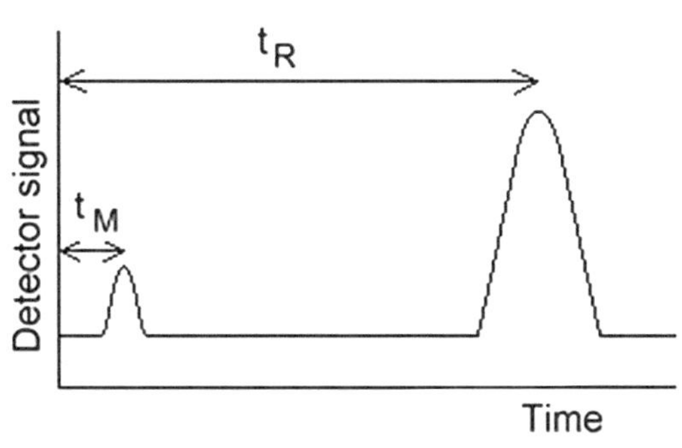

Fig. 5.6 Schematic representation of retention time and peak width

passage through the column. The detector will register a peak with a certain width. The height of a theoretical plate is indicated by the measure of the peak width. Resolution is merely the measure of the separation quality between two different kinds of molecules. It is composed of the two previous concepts: peak width and retention factor of the two components. Retention in a chromatographic column is achieved because the retained molecule adsorbs reversibly to the stationary phase. On a molecular level, this means the time an analyte spends in the stationary phase. The elution of the molecules is the result of random molecular movement and that caused by the flow of mobile phase. The sample components get separated based on the distribution coefficient on the stationary and mobile phase.

5.2.2 Gas Chromatography (GC)

Gas chromatography (GC) or the gas liquid chromatography (GLC) is a commonly used chromatographic technique for the identification, quantification, and quality control of samples. The technique separates highly complex mixtures primarily based upon the difference in polarity and boiling point/vapor pressure. The technique was invented by Archer, J. P. Martin and Richard, L. M. Synge in 1941. Contrasting other chromatographic techniques, a mobile phase is an inert carrier gas, such as helium, argon, or nitrogen. The stationary phase consists of a solid support (silica, alumina, or carbon) packed column (gas-solid chromatography) or utilizes a high- boiling polymer as the liquid stationary phase (gas-liquid chromatography), coated directly onto the walls of a small-diameter column tube. Gas-solid chromatography is commonly used for the separation of permanent gases and low-boiling hydrocarbons. The samples to be analyzed should be thermally stable and reasonably volatile (Lehotay and Hajslova 2002). GC applications are frequently employed in the analysis of essential oils, pesticides, petroleum, oil, fatty acids, steroids, and biological fluids.

Typically, a gas chromatograph consists mainly six parts: (i) carrier gas (mobile phase), which is in a high-pressure cylinder attached with pressure regulators and flow meters, (ii) a sample injection port, (iii) the stationary phase – column, (iv) separate thermostated compartments/column oven, (v) detectors, and (vi) data analysis software and recorder (Fig. 5.7).

(i) Carrier gas supply: The mobile phase carrier gas sweeps the sample onto the column, where the stationary phase is present. Inert gases, viz., helium, argon, or nitrogen, are commonly used as the carrier gas. The highly reactive and explosive hydrogen gas is also used as carrier gas in some specific analyses, in addition to its employability in a flame ionization detector. The gas supply is regulated to the correct pressure (or flow) and then fed to the column. Modern GC instruments have Electronic Pneumatic pressure controllers, whereas older instruments may have manual pressure control via regulators. The mobile phase utilizes an inert gas (e.g., nitrogen) for packed columns and helium or argon for capillary columns.

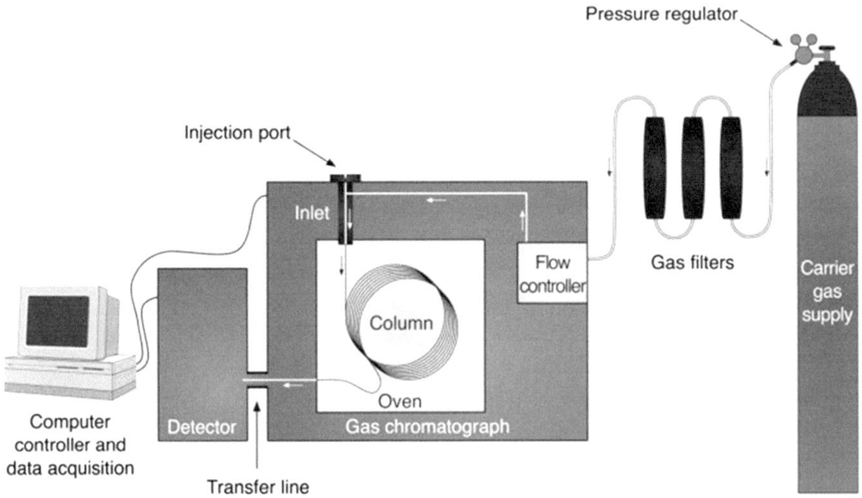

Fig. 5.7 Schematic representation of a GC instrument (Eric et al. 2008)

(ii) Injector: The sample is introduced via an injection port into the column inlet. Different types of injectors include the split/splitless, Programmed Thermal Vaporizing (PTV), cool on-column (COC), etc. The COC injector introduces the sample into the column as a liquid to avoid thermal decomposition or improve quantitative accuracy.

(iii) Column: The column is typically made of stainless steel and is between 1 and 4 meters long with an internal diameter of up to 4 mm, and it is coiled up so that it will fit into a thermostatically controlled oven. The separation of samples into its constituent components takes place inside the GC column. Columns vary in length and internal diameter depending on the application type and can be either packed or capillary. Packed conventional columns are made of a gently coiled glass or stainless steel, having 1–3 m column length and 2–4 mm internal diameter. They are packed with stationary phase coated onto an inert silica support. Commonly used stationary phases include the polyethylene glycols (Carbowax 20M, very polar), methylphenyl- and methyl vinyl silicone gums (OV17 and OV101, medium and nonpolar, respectively), Apiezon L (nonpolar), and esters of adipic, succinic, and phthalic acids. Beta-Cyclodextrin-based phases are available for chiral separations. The most commonly used support is Celite (diatomaceous silica), modified by silanization with hexamethyldisilazane. The support particles have a large surface area and an even size applicable for majority of applications. The capillary (open tubular) columns are made of high-quality fused quartz and are 10–100 m long and 0.1–1.0 mm internal diameter. They are of two types known as WCOT (wall-coated open tubular) and SCOT (support-coated open tubular), also known as PLOT (porous layer open tubular) columns, for adsorption. In WCOT columns, the stationary phase is directly coated (0.1–5 mm) as a thin layer onto the walls of the capillary, while in SCOT columns, the support matrix is bonded to the inner walls of the capillary column and the stationary phase coated onto the support. Commonly used stationary phases include polyethylene glycol (CP wax and DB wax, very polar) and methyl and phenyl-polysiloxanes (BP1, non-polar; BP10, medium polar). The capacity of SCOT columns is considerably higher than that of WCOT columns. The operating temperature for all types of column needs to be compatible with the stationary phase. Too high a temperature results in excessive column bleed owing to the phase being volatilized off, contaminating the detector and giving an unstable recorder baseline. The column temperature range is chosen to give a balance between peak retention time and resolution (Wilson and Walker 2010).

(iv) Column oven: The column, as well as injector and detector connections, harbors in a thermostatically controlled oven. The oven heats rapidly to give excellent thermal control. A hanger or cage is usually included to support the GC column and also to prevent it from touching the oven walls, as it may damage the column. For isothermal operation, the GC is held at a steady temperature during the analysis. In temperature-programmed gas chromatography, the oven temperature is increased according to the temperature program during the analysis. The oven is cooled using a fan and vent arrangement usually at the rear of the oven. The separation takes place as the mixture travels along the column, and upon exit from the column, they enter a detector, which provides an electronic signal proportional to the amount of the eluting analytes.

(v) Detectors: The GC detector responds to physicochemical property of the analytes, amplifies the response, and generates an electronic signal for the data system to produce a chromatogram. Based on the fate of the analytes, GC detectors can be classified into two main types: nondestructive and destructive. Photo ionization detector (PID), thermal conductivity detector (TCD), and electron capture detector (ECD) are examples of nondestructive detectors, whereas flame ionization (FID), nitrogen-phosphorus (NPD), flame photometric (FPD), and electrolytic conductivity (Hall/ELCD) detectors are destructive type.

(vi) Data analysis software and recorder: The data system is used to perform various quantitative and qualitative operations on the chromatogram thereby assisting in sample identification and quantization.

5.2.2.1 Nondestructive Type: GC Detectors

Electron Capture Detector (ECD)

Electron capture detectors are particularly sensitive to halogenated compounds. Gas entering the detector is ionized by high-energy electrons emitted from a radioactive source (i.e., ^{63}Ni, ^{3}H). The electrons formed are attracted to an anode, producing a baseline current. When an analyte molecule with a high electron affinity enters the detector, it captures some of these electrons. The detector responds by varying the frequency of voltage pulses between the anode and the cathode to maintain a constant current. This frequency is converted to a voltage which is directly proportional to the concentration of the analyte (< 5 fg/s).

Materials in the gas stream containing atoms of certain types capture electrons from the stream and measurably reduce the current. The most important of the capturing atoms are the halogens – fluorine, chromine, bromine, and iodine. Therefore, this type of detectors is particularly useful with chlorinated pesticides. The detector contains a cavity that comprises two electrodes and a radiation source that emits radiation (i.e., ^{63}Ni, ^{3}H). The electron and the carrier gas (methane plus an inert gas) collisions produce a plasma-containing electrons and positive ions. If a compound is present that contains electronegative atoms, those electrons will be "captured" to form negative ions, and the rate of electron collection will decrease. The detector is extremely sensitive for compounds with atoms of high electron affinity (10^{-14} g/s), but has a relatively small linear range (~10^2–10^3).

Thermal Conductivity Detector (TCD) Thermal conductivity detectors compare the heat-conducting ability of the exit gas stream to that of a reference stream of pure carrier gas, accomplished by passing gas streams over heated filaments (thin tungsten wires) in thermal conductivity cells. Measured changes in filament resistance of the cells are reflected as temperature changes caused by increments in thermal conductivity. The temperature difference between the sample cell filaments and the reference is monitored using a Wheatstone bridge circuit. This resistance change is monitored and registered continuously on a recorder. This detector is less sensitive than the FID (10^{-5} to 10^{-6} g/s, linear range, 10^3–10^4) but appropriate for preparative applications, since the samples are not destroyed. The detection is based on the comparison of the two gas streams, containing only the carrier gas, and the carrier gas plus analyte. A carrier gas with high thermal conductivity (e.g., helium or hydrogen) is used to maximize the temperature difference (and therefore the difference in resistance) between two tungsten filaments. The large surface-to-mass ratio permits a fast equilibration to a steady state.

Photo Ionization Detector (PID) PIDs rely upon ionization as the basis of detection. The PID is comprised of an ultraviolet lamp that emits photons that gets absorbed by the analyte in an ionization chamber. When sampled gas absorbs energy from a PID lamp, the gas becomes excited, and its molecular content is altered. The compound loses an electron (e^-) and becomes a positively charged ion. Once this process occurs, the substance is considered to be ionized, and the ions produced during this process are collected by electrodes. The generated current provides a measure of the analyte concentration. Because only a small fraction of the analyte molecules are actually ionized, this method is considered as a nondestructive method, allowing it to be used in conjunction with other detectors to confirm analytical results.

The PID is used mostly to detect volatile organic compounds (VOCs) in soil, sediment, air, and water. The PID can be used to detect volatile organic compounds such as formaldehyde, methane, benzene, toluene, xylene, hydrocarbons, and also certain inorganics such as ammonia and hydrogen sulfide. The PID is not suitable for the detection of semi-volatile compounds and may give false-positive readings for water vapor. High concentrations of methane can also hinder performance. Rapid variations in temperature at the detector, strong electrical fields, and naturally occurring compounds, such as terpenes in wooded areas, may affect instrument response. The PID must be recalibrated frequently. Detection limits for most PIDs are in the parts per million range. Thus, they are unsuitable for most vapor intrusion indoor air investigations, where screening or action levels are normally in the parts per billion range.

5.2.2.2 Destructive Type: GC Detectors

Flame Ionization Detector (FID) In a flame ionization detector, the eluate is ignited in a mixture of hydrogen and nitrogen (in air). Positive ions and electrons are produced in the flame, due to the presence of organic compounds as analytes. Elemental carbon atoms (except for carbonyl and carboxyl carbons) produce CH radicals that are thought to produce CHO^+ ions in the flame: $CH + O \rightarrow CHO^+ + e$. Electrons flow from the anode to the cathode, where they neutralize CHO^+ in the flame and produce a small, measurable current. This current is the detector signal. FID is very sensitive toward organic molecules, viz., hydrocarbons (1–2 pg/s, linear range, 10^6–10^7), but not sensitive to detect carrier gases, such as nitrogen, or highly oxidized materials, such as CO_2, carbon monoxide, as well as NO_x, H_2S, sulfur dioxide, and water. If proper amounts of hydrogen/air are mixed, the combustion does not afford any or very few ions resulting in a low background signal. If other carbon-containing components are introduced to this stream, cations will be produced in the effluent stream. The more carbon atoms in the molecule, the more fragments are formed and the more sensitive the detector is for this type of compounds. There is no exact relationship between the number of carbon atoms and the size of the signal. Hence, the individual response factors for each compound have to be experimentally determined for each instrument. This type of destructive detector technique is not suitable for preparative GC. In addition, several gases are usually required to operate a FID, i.e., hydrogen (as a fuel for ignition), oxygen or compressed air (to assist the flame ignition), and a suitable carrier gas.

Nitrogen-Phosphorus Detectors (NPDs) The NPD is a type of flame thermionic detector (FTD) and is also known as thermionic-specific detector (TSD). They are specific to nitrogen or phosphorus-containing compounds and are commonly used in pharmaceutical, food, environmental, and forensic applications. Alike the flame ionization detector, a NPD uses a hydrogen/air flame through which the sample is passed. However, a NPD uses a rubidium or cesium chloride alkali bead which is heated by a coil, over which the carrier gas mixed with hydrogen flow. The partially combusted nitrogen and phosphorus in a sample material are adsorbed onto the surface of the beads. The hot bead emits electrons by thermionic emissions which are collected at the anode and provides the background current which is then measured. A make-up gas can be used in GC-NPD at the column exit, to increase the flow rate into the detector.

The major limitation of GC-NPD is that the carrier gas with low water and oxygen impurities is required, since water and oxygen can interact with the stationary phase and cause significant problems such as high baseline noise and column bleed in the output gas chromatogram which both reduces analyzer sensitivity and decreases column life. The NPD is also extremely sensitive to hydrocarbon impurities in the hydrogen and air supply for the flame. Hydrocarbon impurities can cause increased baseline noise and reduce the detector sensitivity. But, the NPD response to normal hydrocarbons is approximately 100,000 times less than nitrogen or phosphorus-containing compounds.

Flame Photometric Detectors (FPD) The flame photometric detector (FPD) allows sensitive and selective measurements of volatile halogen, sulfur, phosphorus, and nitrogen-containing compounds. The flame photometric detector measures the intensity of light with a photometric circuit. In principle, the flame photometric detector (FPD) is identical to the FID with the addition of an S or P filter and a photomultiplier tube. The detection principle in a flame photometric detector is the formation of excited sulfur (S_2*) and excited hydrogen phosphorus oxide species (HPO*), in a reducing flame. The decomposed sample then enters into a second flame that supplies the energy for exciting these fragments, and further upon relaxation light is emitted. The specific emission that results after application of an S or P filter is converted into a current and amplified by the photomultiplier tube. The photomultiplier tube measures the characteristic chemiluminescent emission from these species. The optical filter can be changed to allow the photomultiplier to view light of 394 nm for sulfur measurement or 526 nm for phosphorus. The detector response to phosphor is linear, whereas the response to sulfur depends on the square of the concentration. Sensitivity is <1 pg/s for phosphorus and < 10 pg/s for sulfur.

Electrolytic Conductivity Detectors (Hall/ELCDs) An electrolytic conductivity detector (ELCD) is used as a selective detector for sulfur, halogens, and nitrogen-containing compounds. The electrolytic conductivity detector (ELCD) is remarkably suited for pesticide residue analysis because of its selectivity. The selectivity is determined by the chemical properties of the electrolyte, the products formed in the detector reactor, as well as the scrubber used. The reaction furnace is operated under conditions that may include a reductive response (by the hydrogen reaction gas), oxidative response (by oxygen or air reaction gas), or pyrolytic (inert reaction gas) response. It is capable of eliminating much of the matrix response, allowing unimpeded identification of the desired analyte. While the detection mechanism of the ELCD is relatively simple, ELCDs are not regarded as particularly user friendly. The sensitivity of the ELCD depends on the amount of the element investigated, and, hence, it should be independent of the structure of the compound. A direct proportionality between the response and the content of the element in the molecule is found for compounds decomposed under oxidizing conditions and for nitrogen compounds decomposed under catalytic reducing conditions.

5.2.2.3 GC-MS

Many GC instruments are coupled with a mass spectrometer, which is a very good combination of sophisticated analysis. The GC separates the compounds from each other, while the mass spectrometer helps to identify them based on their fragmentation pattern. To obtain the mass spectrum, gaseous molecules or entities desorbed from the condensed phases are subjected to ionization. The resultant ions are accelerated in an electric field and separated according to their mass-to-charge (m/z) ratio. Mass spectrometers are universal detectors for total or single-ion monitoring, with detection limit ranging 25 fg to 100 pg.

5.2.2.4 Rationale of Sample Derivatization

Derivatization methods generally include four types depending on the reaction applied. They include silylation, acylation, alkylation, and esterification. Chemical derivatization of the samples, prior to analysis, is generally done to:

- Increase the volatility and lessen the polarity of the compounds.
- Improve separation and minimize tailing.
- Reduce thermal degradation of samples.
- Increase detector response.

5.2.2.5 Factors Influencing Chromatographic Separation

Vapor Pressure At a given temperature, every compounds (including solids) have a nonzero vapor pressure. The boiling point of a compound is often related to its polarity. The lower the boiling point, the higher the vapor pressure of the compound. The shorter retention time is usually because the compound spends more time in the gas phase. This is one of the main reasons in using low- boiling solvents (i.e., diethyl ether, dichloromethane) to dissolve the samples.

Sample Polarity If the polarity of the stationary phase and compound is similar, the retention time increases due to stronger interaction with the stationary phase. Polar compounds exhibit long retention times on polar stationary phases and shorter retention times on nonpolar columns. Chiral stationary phases that are based on amino acid derivatives, cyclodextrins, and chiral silanes are capable of separating enantiomers because one enantiomer interacts slightly stronger than the other one with the stationary phase, often due to steric effects or other very specific interactions.

Column Temperature An excessively high column temperature results in very short retention time but also in a very poor separation because all components are in the gas phase. However, in order for the separation to occur, the components need to be able to interact with the stationary phase. The retention time decreases, if the compound does not interact with the stationary phase. At the same time, the quality of the separation deteriorates, because the differences in retention times are not as pronounced. The best separations are usually observed for temperature gradients, because the differences in polarity and boiling points are executed.

Carrier Gas Flow Rate Flow rate of the carrier gas is a crucial factor in controlling separation. A high flow rate reduces retention times, but a poor separation would be observed as well. Like above, the components have very little time to interact with the stationary phase and are just being pushed through the column.

Column Length A longer column generally improves the separation. The reason is that the retention time increases proportionally to the column length and a significant peak broadening will be observed as well because of increased longitudinal diffusion inside the column. The peak broadening is inversely proportional to the rate of flow. Broadening is also observed due to the finite rate of mass transfer between the phases and also because the molecules take different paths through the column, as a result of random movement of gas molecules in different directions.

Sample Injection Volume Ideally, the sample peaks in a chromatogram display a symmetric shape (Gaussian curve). Most detectors are relatively sensitive and require only small amount of sample, in order to produce a measureable signal. If too much of the sample is injected, the peaks show a significant tailing, resulting in poor separation. Under standard conditions, only 1–2% of the compound injected into the injection port passes through the column because most GC instruments are operated in split mode, so as to prevent overloading the column and detector. The splitless mode is generally used, only in extremely low sample concentration in terms of the analyte.

5.2.3 Conventional Chromatographic Methods

Generally, the conventional methods of chromatographic principles include thin-layer chromatography (TLC) and paper chromatography. Separation procedures and principle being the same in both conventional methods, the two methods differ only in the stationary phase used. The TLC technique employs a thin layer of silica gel or alumina as the stationary phase, whereas in paper chromatography, the saturated volatile components of the solvent phase, within the porous paper (e.g., Whatman No. 1), serve as the stationary phase.

5.2.3.1 Thin-Layer Chromatography (TLC)

TLC is a type of planar chromatography, employed as a semiquantitative method of analysis. It is routinely used by researchers to identify the most of the nonvolatile compounds in a mixture, owing to its advantage as a cheaper chromatographic technique without sophisticated instrumentation. High-performance thin-layer chromatography/HPTLC (Sect. 5.2.4) is a sophisticated and more precise quantitative version of TLC.

In thin-layer chromatography, the separation depends on the relative affinity of compounds toward stationary and the mobile phases (Vogel et al. 1989). The compounds under the influence of the mobile phase (driven by a capillary action) travel over the surface of the stationary phase. The compounds with higher affinity to stationary phase travel slowly, while the others with higher affinity to mobile phase solvents travel faster enabling separation of components in the mixture. Once separation occurs, the individual components are visualized as spots at a respective level of travel on the plate (R_f values) and are identified by means of suitable detection techniques.

R_f value is the ratio of distance traveled by the sample (solute) to that of the solvent (mobile phase), a value ranging between zero and one.

TLC system components consist of the TLC plate (stationary phase), the development solvent (mobile phase), and the TLC chamber.

(i) *TLC plates (stationary phase)*: The TLC plate consists of a thin plastic sheet covered with a thin layer of silica gel or alumina. These are stable and chemically inert plates, where a very thin layer of stationary phase is applied on its whole surface layer. The stationary phase on the plates is of uniform thickness and is in a fine particle size.

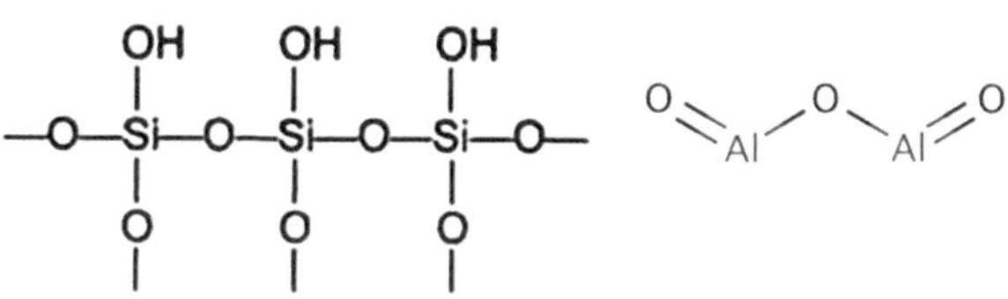

Silica gel **Alumina**

Silica gel consists of a three-dimensional network of thousands of alternating silicon and oxygen bonds, with O-H groups on the outside surface, making it highly polar and is capable of hydrogen bonding. Silica gel is simply very finely ground pure sand. Some commonly used stationary phase materials are given in Table 5.3.

(ii) *Mobile phase*: This comprises of a solvent or solvent mixture. The mobile phase used should be particulate-free and of the highest purity for proper development of TLC spots. The solvents recommended are normally chemically inert with the sample and also with that of the stationary phase.

(iii) *TLC chamber*: This is used for the development of TLC plate. The chamber maintains a uniform environment inside for proper development of spots. It also prevents the evaporation of solvents and keeps the process dust free.

Table 5.3 Stationary phase materials used in TLC plates

Compound	Composition
Silica gel G	Silica gel + $CaSO_4$ (G: Gypsum)
Silica gel H	Silica gel without binder
Silica gel GF	Silica gel + $CaSO_4$ + fluorescent indicator
Magnesol	Magnesium silicate
Al_2O_3 G	Al_2O_3 + $CaSO_4$
Alumina	Al_2O_3 without binder
Fuller's earth	Hydrous magnesium alumina
Powdered cellulose	Cellulose without binder
Sephadex	Cross-linked dextran
Polyamide powder	Polyamide
Kieselguhr G	Diatomaceous earth + binder

TLC consists of three steps – spotting, development, and visualization. The sample to be analyzed is dissolved in a suitable volatile solvent, so as to produce a very dilute (about 1%) solution. Sample spotting is done using a micro pipet/capillary tube to one end of a TLC plate. The spotting solvent evaporates quickly and leaving behind a small spot of the sample material. Chromatogram is developed by placing the spotted TLC plate, inside a chamber containing a shallow pool of pre-saturated mobile phase. The mobile phase travels up the plate by capillary action that can be carried out in linear ascending, linear descending, linear horizontal, circular, and anticircular modes, with each development method having its own advantage and disadvantages. Visualization of the separated samples can be done by exposing the plates to iodine vapor or with the use of fluorescent material impregnated stationary phase plates, which glows under ultraviolet (UV) light, or by uniform spraying of suitable dyes/stains. Visualization of colored compounds is simple, as the spots can be directly observed after development.

5.2.3.2 Paper Chromatography

Chromatography has become a top-ranking technique for the separation and quantification of components of organic liquid or gas phase mixtures. Several innovations of the technique have contributed to speed of analysis and lowering of detection limits; but it is remarkably surprising that all the variants ranging from simplest paper chromatography to thin-layer chromatography and column chromatography. Paper chromatography provides a simple and effective means for separation of components of mixtures. In paper chromatography separations, the component is distributed between the saturated solvent held between the pores of filter paper that acts as the stationary phase and a suitable mobile-phase solvent which travels along the filter paper. Paper essentially consists of cellulose fibers which are polymers having – OH functional groups sticking out of the polymer chains. These groups lead to retention and separation of surface-absorbed molecules.

Even though paper chromatography has evolved over the years and has found widespread applications in separation of molecules of different polarities, there are a few disadvantages compared to other conventional methods.

- Paper chromatography can take several hours to develop the plate, whereas development in thin-layer chromatography can be completed in much shorter time (about half an hour or so).
- Corrosive spray reagents can char the filter paper and can even deteriorate the sample spots. Coated plates used in thin-layer chromatography can withstand corrosive spray reagents to a greater extent.
- Thin-layer chromatography plates can be heated if required for spot development. Paper chromatography sheets can be heated only to a limited range of temperature.

Paper chromatography has not seen much automation over the years, but thin-layer chromatography technique has been automated, which includes automated gel plater, autosampler, and constant volume sample dispenser.

5.2.4 High-Performance Thin-Layer Chromatography (HPTLC)

High-performance thin-layer chromatography (HPTLC) is an automated and sophisticated form of TLC with better and advanced separation efficiency and superior detection limits (Reich and Schibli 2007). The process steps of HPTLC are identical to classical TLC. The main difference between them is in the characteristics of the separation plate. HPTLC plates are based on optimized silica gel 60 with a significantly smaller particle size (5–6 μm) than used for classical TLC (10–12 μm). This allows a higher packing density and a smoother surface. Hence, sample diffusion is reduced, resulting in compact bands or spots. Furthermore, the smaller particle size and thinner layer significantly increase detection sensitivity and analysis speed. The advantages of HPTLC over conventional methods are outlined below.

- Low analysis time as well as lower maintenance and analysis cost. Does not require HPLC grade/UPLC grade solvents.
- A great range of stationary phases with unique selectivity are available, which require no regeneration and are disposable.
- The mobile-phase solvent selection is not restricted to low UV transparency or ultrahigh purity. Corrosive and UV-absorbing mobile phases may also be employed.
- Pretreatment of solvents like filtration, degassing, etc. and sample cleanup are not necessary.
- Densitometric evaluation of the chromatogram to optimize quantification of sample can be achieved under different conditions without repeating experiment, since all sample fractions are retained on the HPTLC plate.

- High sensitivity limits, typically at nanogram (ng) to pictogram (pg) levels. Several samples can be chromatographed simultaneously and high sample throughput. Visual detection is also possible, since it is an open system.
- The samples and standards are analyzed under identical experimental conditions on a single run and hence, high accuracy and precision of quantification.
- The HPTLC concept includes validated methods for qualitative and quantitative analysis and fulfills all quality requirements for use in fully regulated environments.

5.3 Electrophoresis

Electrophoresis is the process of migration of charged molecules through solutions in an applied electric field. Arne Wilhelm Kaurin Tiselius (1902–1971) was a Swedish biochemist who won the Nobel Prize in Chemistry (1948) for his contribution on electrophoresis and adsorption analysis. Simple electrophoresis is the separation of molecules in an electric field at a constant current and pH. Other types of electrophoretic principles include the *isoelectric focusing*, where the separation occurs in a pH gradient, and *isotachophoresis*, where an electric field varies according to the conductivities of the components. These can be used in analytical or preparative scale. Electrophoresis is often classified according to a solid supporting medium or matrix through which the charged molecules move in an electrophoretic system.

Electrophoresis systems can be broadly classified into solution electrophoresis systems and zonal electrophoresis systems. Solution electrophoresis systems employ aqueous buffers in the absence of a solid support medium. Solution electrophoresis systems can suffer from sample mixing due to diffusion of the charged molecules, with resultant loss of resolution during sample application, separation, and removal steps. The problem can be overcome by stabilizing the aqueous solutions in the electrophoresis cell. For example, soluble-gradient electrophoresis systems use varying densities of a nonionic solute (e.g., sucrose or glycerol) to minimize diffusional mixing of the materials being separated during electrophoresis. Even with these refinements, solution electrophoresis systems have only limited application, usually when preparative-scale electrophoretic separation is required. In zonal electrophoresis systems, the aqueous ionic solution is carried in a solid support, and samples are applied as spots or bands of material. Paper electrophoresis, cellulose acetate strip and cellulose nitrate strip, and gel electrophoresis are all examples of zonal electrophoresis systems. Such systems are typically employed for analytical, rather than preparative-scale separations.

All types of electrophoresis are governed by the single set of general principles illustrated by the following equation:

$$\text{Mobility of a molecule} = \frac{(\text{Applied voltage}) \times (\text{Net charge on the molecule})}{\text{Friction of the molecule}}$$

Specific forms of electrophoresis commonly used include:

(i) Paper electrophoresis
(ii) Capillary electrophoresis
(iii) Gel electrophoresis

(i) *Paper electrophoresis*: Paper electrophoresis is commonly used for analysis and resolution of small molecules.

Sample Application

(a) Dry Application Procedure: In dry application procedure, a sample of solution in distilled water or a volatile buffer is applied on the origin line; and after the solvent containing the samples is evaporated, the paper is dampened with electrophoresis buffer – by spraying /dipping/blotting the ends of the paper. Dry application procedure has the advantage of allowing small initial sample spots and better resolution of similarly mobile compounds.
(b) Wet Application Procedure: In wet application procedure, samples dissolved as concentrated solution in distilled water are applied to the paper pre-dampened with electrophoresis buffer. Wet application procedure is simpler to perform, but usually yields larger spots and poor resolution because of sample diffusion.

Disadvantages

- The technique is not suitable for macromolecules (e.g., proteins) because the adsorption and surface tension associated with paper electrophoresis usually alter or denature the macromolecules, causing poor resolution.
- Application of electric field and resultant resistance to current flow in the buffered paper generates heat, which in turn leads to more resistance to current flow – which will distort the migration of the molecule.

(ii) *Capillary electrophoresis (CE)*: As the name suggests, the material to be analyzed and the electrophoresis medium (a conducting liquid, usually aqueous) are placed in a long, fine bore capillary tube, typically 50–100-cm-long and 25–100 μm inner diameter. The technique has been referred to commonly as capillary electrophoresis (CE), likewise the synonyms, viz., capillary zone electrophoresis (CZE), high-performance capillary electrophoresis (HPCE), and free solution capillary electrophoresis (FSCE), are also in use.

Sample Application

A very small sample (in nanoliter range) is placed at one end of the capillary tube and subjected to electrophoresis under fields up to 20–30 kv. Sample application is carried in two ways:

(a) High-voltage injection: With the high-voltage switched off, the buffer reservoir at the positive electrode is replaced by a reservoir containing the sample, and a plug of sample (e.g., 5–30 μm^3 of a 1 mg crude solution) is introduced into the capillary by briefly applying high voltage. The sample reservoir is then removed, the buffer reservoir replaced, voltage again applied, and the separation is then commenced.
(b) Pressure injection: The capillary is removed from the anodic buffer reservoir and inserted through an airtight seal into the sample solution. A second tube provides pressure to the sample solution, which forces the sample into the capillary tube. The capillary is finally removed and replaced in the anodic buffer, and a voltage is applied to initiate electrophoresis.

Advantages

- Extremely high resolution and speed.
- Reduce the problem of heating effects. (Due to small diameter of the capillary tube, there is large surface to volume ratio, which gives enhanced heat dissipation).
- High sensitivity for the analysis of extremely small samples. (But obviously not useful as a preparative method).
- Useful for the separation of DNA molecules that differ in size by as little as a single nucleotide.
- Capillary electrophoresis (CE) is the basis of separation of polynucleotides in some newer designs of DNA sequences.
- CE can also be adapted to the separation of uncharged molecules, by including charged micelles of a detergent (such as SDS) in the aqueous electrophoretic medium. If a mixture of solute molecules that partition between the aqueous medium and the hydrophobic interior of the micelles is introduced into such a system, they can be separated by electrophoresis.
- CE is used to separate a wide spectrum of biological molecules including amino acids, peptides, proteins, nucleic acids, DNA fragments (synthetic oligonucleotides), drugs, or even metal ions.

Capillary electrophoresis is a highly adaptable method, and the range of its applications and optimal methodology are still being explored.

(iii) *Gel electrophoresis*: In gel electrophoresis, molecules are separated in aqueous buffers supported within a polymeric gel matrix. A variety of matrices including starch, agarose, acrylamide, and cellulose acetate can be used for electrophoresis. However, the high-resolution capacity of polyacrylamide and agarose gel electrophoresis makes them the method of choice for most applications.

Gel electrophoresis systems have several advantages:

- They can accommodate larger samples than most paper electrophoresis systems and so can be used for preparative-scale electrophoresis of macromolecules.
- The character of gel matrix can be altered at will to fit a particular application. This is possible because the gel enhances the friction that governs the electrophoretic mobility.
- Low concentration of material and a low degree of cross-linking of monomers in the polymerized gel systems allow them to be used largely as a stabilizing or anticonvection device with relatively low frictional resistance to the migration of macromolecules.
- Alternatively, higher concentrations of matrix material or a higher degree of cross- linking of monomers is used to generate greater friction, which results in molecular sieving. Molecular sieving is a situation in which viscosity and pore size largely define electrophoretic mobility and migration of solutes. As a result, the migration of macromolecules in the system will be substantially determined by molecular weight.
- Agarose gels (a polygalactose polymer) have proven quite successful, particularly when applied to very large macromolecules such as nucleic acids, lipoproteins, and others.
- Polyacrylamide gels are among the most useful and most versatile in gel electrophoresis separations, because they readily resolve a wide array of proteins and nucleic acids.

5.3.1 Polyacrylamide Gel Electrophoresis [PAGE]

Polyacrylamide gels are formed by the polymerization of acrylamide (monomer) and N, N^1-methylenebisacrylamide (cross-linker). Gel polymerization is usually initiated with ammonium per sulfate (APS) and TEMED (N, N, N^1, N^1-tetramethylethylenediamine) – a free radical scavenger. Acrylamide polymerization is an example of "Free radical catalysis" (Fig. 5.8).

- Bisacrylamide (cross-linker) is used as a cross-linking agent and consists of two acrylamide molecules linked by a methylene group.
- The acrylamide monomer and cross-linker are stable by themselves or mixed in solution; but polymerizes readily in the presence of a free radical generating system.

Chemistry of Polymerization The TEMED catalyzed decomposition of persulfate ions give rise to free radicals (i.e., molecules with unpaired electrons). If this free

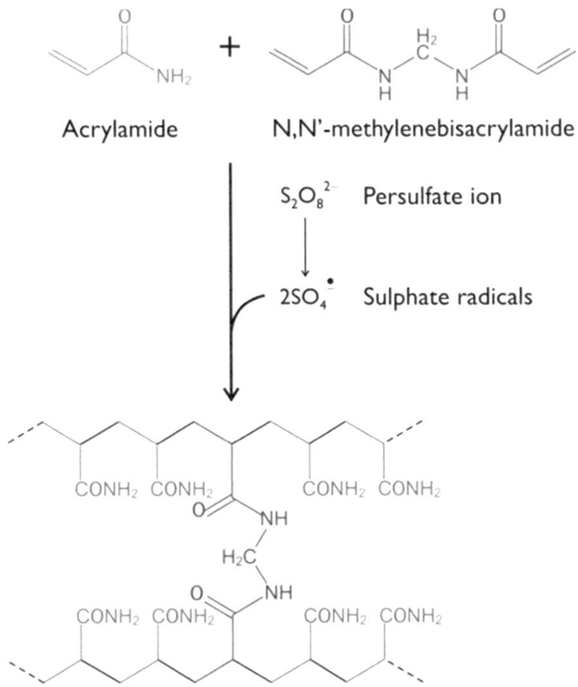

Fig. 5.8 Polymerization of acrylamide and N, N^{1}-methylenebisacrylamide

radical is represented as R∗ (where the ∗ represents an unpaired electron) and M is an acrylamide molecule, then the polymerization is represented as follows:

$$R * + M \rightarrow RM *$$
$$RM * + M \rightarrow RMM *$$
$$RMM * + M \rightarrow RMMM * \quad \rightarrow\rightarrow\rightarrow \text{etc.}$$

Thus, long chains of acrylamide are built up, being cross-linked by the production of the occasional bis-acryl amide molecule into the growing chain. The concentration of the polyacrylamide solution can be adjusted over a wide range according to the size of the molecules being separated.

- Oxygen is an inhibitor of polymerization, since it removes the free radical, and therefore all gel solutions are degassed under vacuum.
- Another optional is "photopolymerization," in which APS is replaced by riboflavin that will generate free radicals when irradiated with UV light.

The different types of polyacrylamide gel electrophoresis (PAGE) include native PAGE, SDS-PAGE, and DGGE.

5.3.1.1 Native-PAGE (Simple/Non-denaturation PAGE)

Native PAGE is the simplest type of polyacrylamide gel electrophoresis in which electrophoresis is performed under non-denaturation conditions, i.e., in the absence of SDS. Proteins remain intact and un-denatured within the gel matrix and retain their physicochemical as well as biochemical properties.

5.3.1.2 SDS-Polyacrylamide Gel Electrophoresis

Sodium dodecyl sulfate-polyacrylamide gel electrophoresis (SDS-PAGE) is used to determine the number and size of protein chains or subunit chains in a protein preparation. SDS (sodium dodecyl sulfate) is an anionic detergent.

$$\text{Structure}: \ CH_3-(CH_2)_{10}-CH_2-O-SO_3-Na^+$$

Most commonly the strongly anionic detergent is used in combination with a reducing agent such as thiol (e.g., β-mercaptoethanol). The thiol reduces all disulfide bonds (–s–s–) present within and/or between peptide units in a protein. The SDS binds to all regions of the proteins and disrupts most noncovalent intermolecular and intramolecular protein interactions. These two components (SDS and β-mercaptoethanol) bring about the total denaturation of the proteins in the sample yielding unfolded, highly anionic (negatively charged) peptide chains. The amount of SDS bound is independent of the sequence of the polypeptide, but proportional to the molecular weight of the polypeptide (Fig. 5.9).

The SDS-coated polypeptides (carrying approximately one SDS molecule per two amino acids) create a situation in which the charge-to-mass ratio of all of the proteins in the sample is approximately the same. At this point, the intrinsic charge on the individual polypeptide chains (i.e., in the absence of SDS) is insignificant as compared with the negative charge imposed on them by the presence of SDS. The friction experienced by the population of molecules as they migrate through the polysaccharide matrix is now the major factor influencing differences in their mobility. In addition, the friction experienced by the molecules during the separation is governed by the pore size of the polyacrylamide matrix; larger polypeptides will experience greater friction when passing through a gel of defined pore size (will

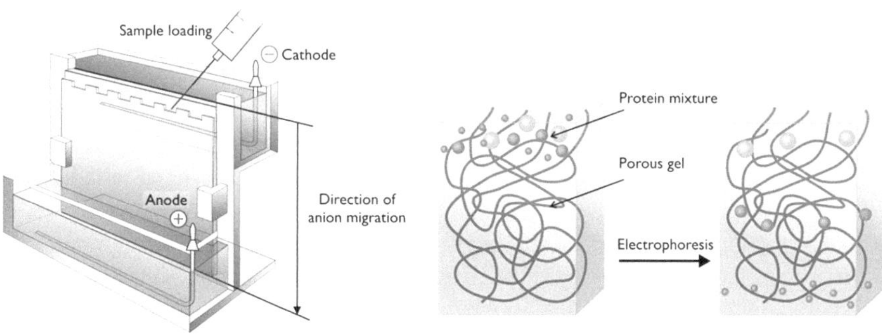

Fig. 5.9 Schematic representation of a polymerized acrylamide gel

migrate more slowly) than smaller polypeptides (which will migrate more rapidly). In summary, SDS-PAGE allows the separation on the basis of size or molecular weight of the proteins.

It is possible to estimate the molecular weight of the polypeptide chain, by the use of markers of known molecular weight. The relative mobility of each of the anionic polypeptide chains is a function of the logarithm of its molecular weight. If a set of polypeptides of known molecular weight markers are included with the sample during the electrophoretic separation, their relative mobilities can be determined and plotted as log molecular weight versus relative mobility. The standard curve produced from this analysis of the polypeptides of known molecular weights can then be used, along with the relative mobilities of the unknown polypeptides in the sample, to estimate their molecular weight.

Buffer Systems Used SDS-PAGE employs two buffer/polyacrylamide gel compositions in a single slab (i.e., discontinuous buffer system). These are referred to as the "stacking gel" and the "running (or resolving) gel." Protein samples are first introduced into wells cast within the stacking gel after they are mixed with a viscous sample buffer (30% glycerol) containing SDS and thiols. The sample and stacking gel contains Tris-Cl, with pH 6.8. The upper and lower reservoir contain Tris-glycine buffer having a pH 8.3, and the resolving gel contains Tris-Cl of pH 8.8. The chloride ions present in the sample and stacking gel form the leading edge of the moving boundary, while glycine molecules are present in the trailing edge. Between the leading and trailing edges of the moving boundary is a zone of lower conductivity and steeper voltage gradient that sweeps the polypeptide from the sample and deposits them on the surface of the resolving gel. Therefore, the higher pH of the resolving gel favors the ionization of glycine, and the resultant glycine ions migrate through stacked polypeptide and travel through the resolving gel immediately behind the chloride ions.

Unrestricted from the moving boundary, the SDS-polypeptide complexes move through the resolving gel in a zone of uniform voltage and pH and are separated according to size by sieving.

5.3.1.3 DGGE: Denaturing Gradient Gel Electrophoresis

The acrylamide concentration can be continuously changed in polymerization gel, using a gradient gel maker. Gradient gels can be used to determine the exact molecular weight of proteins in their native state. The acrylamide concentration and the degree of cross-linking are highly fixed in small pore area in the tight gel matrix, where the protein molecules are trapped. The speed of migration of the individual protein molecules depends on their charge; hence, electrophoresis must be carried out for a long period so that the molecule with the lowest net charge also reaches its end point.

There are various ways of preparing gels with linear or exponential pore gradients based on the principles of polymerization. During casting, the dilute solution is continuously mixed with the concentrated solution so that the concentration in the casting mold increases from top to bottom.

The main advantage of DGGE is that the migrating protein zone is retarded more than the trailing edge, thus resulting in sharpening of the protein bands. As the gradient in pore size increases, proteins with a wide range of molecular weights can be fractionated in a single run. Also proteins with close-molecular-weight values are more likely to be separated with better resolution, in gradient gels than in a linear gel.

5.3.2 Agarose Gel Electrophoresis

Agarose is one of the components of agar that is a mixture of polysaccharides isolated from certain seaweeds (Refer Sect. 5.7.4). Agarose is a linear polysaccharide made up of the basic repeating unit agarobiose that comprises alternating units of galactose (β 1–4) and 3, 6-anhydrogalactose (α 1–4) glycosidic bonds (Fig. 5.10).

Agarose gels are formed by suspending the dry agarose powders in an aqueous Tris-acetate EDTA (TAE) buffer or Tris-borate EDTA (TBE) buffer containing EDTA. The solution is then boiled until a clear gelling solution is obtained. The solution is poured and allowed to cool to room temperature to form a rigid gel. The gelling properties are attributed to both inter- and intramolecular hydrogen bonding within and between the long agarose chains. The induction of this polymerization event does not involve any chemical reaction, unlike the polymerization process of polyacrylamide gels. This cross-linking structure gives the agarose gel good anti-convectional properties.

Pore size of the gel is controlled by the initial concentration of agarose; large pore sizes are formed from low concentrations, and smaller pore sizes are formed from higher agarose concentrations. Although essentially free from charge, substitution of the alternating sugar residues with carboxyl, methoxyl, pyruvate, and especially sulfate groups occurs to varying degrees. This substitution can result in electroendosmosis during electrophoresis and ionic interactions between the gel and sample in all tubes – both unwanted effects. Agarose is therefore sold in

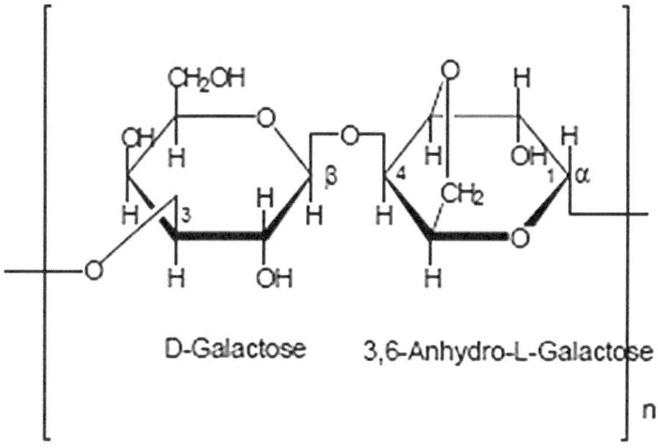

Fig. 5.10 Structure of agarobiose – the repeating unit of agarose (https://www.sigmaaldrich.com/technical-documents/articles/biology/agarose.html)

different purity grades, based on the sulfate concentration. The lower the sulfate content, the higher is its purity.

Agarose gels are used for the electrophoresis of proteins, nucleic acids, and complex carbohydrates. For proteins, the pore sizes of 1% agarose gel are large, relative to the sizes of proteins. Agarose gels are therefore used in techniques such immunoelectrophoresis, or flat-bed isoelectric focusing, where the proteins are required to move unhindered in the gel matrix according to their charge. Such large pore sizes are used to separate much larger molecules like as DNA or RNA.

Further advantage of agarose is the availability of low melting temperature of agarose (62–65 °C). These gels can be reliquefied to 65 °C, and, thus, for example, DNA samples separated in a gel can be returned to solution and recovered.

Owing to poor elasticity of agarose gels and the consequent problems of resolving them from small tubes, the gel rod system (i.e., tube gel electrophoresis) sometimes used for acrylamide is not used. Horizontal slab gels are used routinely for DNA and RNA gels, than vertical systems. Horizontal slab gels are invariably used for isoelectric focusing or immune-electrophoresis in agarose.

5.3.2.1 AGE for Nucleic Acids

The agarose dissolved in aqueous buffer (TAE or TBE) is cooled momentarily and poured into a slab mold fitted at one end with a Teflon or plastic comb. After the solution polymerizes, the comb is removed to create wells into which desired DNA or RNA samples will be applied. The gel is then transferred to an electrophoresis chamber and is completely covered with the same TAE or TBE buffer that was used to cast the gel. Next, the nucleic acid sample is mixed with a viscous buffer (30% glycerol) containing one or more tracking dyes that will be used to monitor the progress of electrophoresis. Bromophenol blue dye will migrate at a same rate as a DNA molecule of about 500 base pairs, while xylene cyanol dye will migrate at the same rate as a DNA molecule of about 4000 base pairs (Fig. 5.11).

The nucleic acid samples are loaded into wells of the gel, along with a sample of DNA fragment of known molecular weight/(number of base pairs) in one of the well, i.e., marker DNA or DNA ladder. These standards are used in the

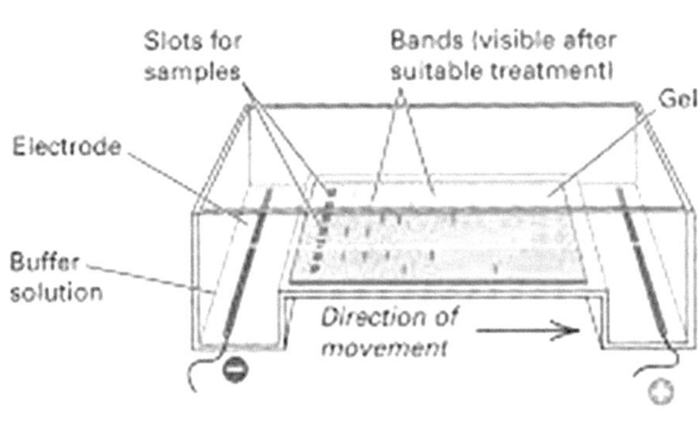

Fig. 5.11 Schematic diagram of agarose gel electrophoresis

electrophoretic separation to aid in the determination of the size/molecular weight of the nucleic acid present in the unknown sample.

Since DNA is negatively charged, the cathode (negative electrode) should be connected to the side of the apparatus nearest to the wells in the gels, while the anode (+ve electrode) is connected to the opposite side of the apparatus. A voltage of 4–6 v/cm is applied to the system (~50–70 mA of current), and electrophoresis continued until the bromophenol dye font reaches the rear end of the gel. Unlike SDS-PAGE, agarose gel electrophoresis does not employ the use of stacking gels. Since the nucleic acids in the sample have a much great fictional component in the gel than they do in the buffer contained in the wells, the nucleic acid focuses very quickly at the buffer gel interface before entering the matrix.

Visualization of the Separated Bands

Ethidium bromide (EtBr) is a fluorescent dye that has the ability to intercalate between the stacked bases of nucleic acid duplexes (i.e., double helix of DNA). After electrophoresis, the gel is transferred to a solution of TBE or TAE buffer containing EtBr (ethidium bromide), which diffuses into the gel and associate with the nucleic acid. Following a short destaining (in a buffer without EtBr), the gel is briefly exposed to ultraviolet light (UV light, 256–300 nm), revealing the nucleic fragments as orange or pink fluorescent bands in the gel. The pattern of fluorescent bands is recorded by photographing the gel in a dark chamber, while the gel is exposed to UV light from below. A filter is included to screen out UV light, so that most of the light that exposes the film is from the fluorescence of the nucleic acid bands that contains tightly bound ethidium bromide.

The method of determining the molecular weight of an unknown nucleic acid sample is exactly the same as that described for the determination of protein molecular weight in SDS-PAGE. The log molecular weights of the nucleic acid samples of known sizes are plotted against their relative mobility to produce a standard curve. From the relative mobility of an unknown nucleic acid fragment on the same gel, the molecular weight and number of base pairs that the fragment contains can readily be determined. (A single base has an average molecular weight of approximately 320 Da, while a base pair has an approx. average molecular weight of 640 Da).

Larger DNAs (up to 10,000 kb in length) can be separated by pulse field gel electrophoresis, in which the direction of electric flux is changed periodically. Agarose gels have a lower resolving power than polyacrylamide gel but have a greater range of separation.

5.3.2.2 AGE for Complex Carbohydrates

Agarose gels are generally used for the electrophoresis of complex proteins, nucleic acids, and other macromolecules as agarose gels are capable of forming gels with large pore sizes. Agarose gels can also be used for the electrophoresis of complex carbohydrates, viz., polyanionic glycans such as the glycosaminoglycans (Volpi and Maccari 2002). A simple and reliable agarose gel electrophoretic method for the identification and preliminary characterization of complex carbohydrates, the glycosaminoglycans (GAGs), is explained below.

Agarose gel electrophoresis of glycosaminoglycans is performed by employing a modified method of Bjornsson (1993). The GAG precipitates are electrophoresed on an agarose gel using 10 mM Tris-acetate buffer having a pH of 8.3. The sulfated glycans, viz., chondroitin-4-sulfate (CS-A), dermatan sulfate (CS-B), chondroitin-6-sulfate (CS-C), and heparin, can be used as standards. Hyaluronic acid is a non-sulfated glycan and may not be visible using toluidine blue dye staining. The gels are electrophoresed on a horizontal gel system at 110 V for 3 h. After electrophoresis, the macromolecules are fixed with 0.1% Cetavlon (CTAB or N-cetyl-N,N,N-trimethylammonium bromide) according to the method of Pavao et al. (1995) and can be stained with 0.02% (w/v) toluidine blue (in 3% acetic acid containing 0.5% (v/v) Triton x 100). The gels are finally destained using 3% acetic acid, and the results are documented as mentioned in Sect. 5.3.2.1. Gel concentrations containing 2.0% agarose gels are most suitable for the characterization of the glycosaminoglycan isolates from marine sources.

Other electrophoretic techniques for the separation of complex carbohydrates are capillary electrophoresis (CE) (Militsopoulou et al. 2003) and FACE (fluorophore-assisted carbohydrate electrophoresis) (Karousou et al. 2005).

5.3.3 Isoelectric Focusing (IEF)

Isoelectric focusing (IEF) is a method used for the separation of proteins according to their isoelectric points (pI), in a stabilized pH gradient (Bjellqvist et al. 1982). It is a type of zone electrophoretic method and also referred to as electrofocusing. Isoelectric point (pI) is defined as the "pH at which a molecule acts as a zwitterion and possesses no overall charge." IEF involves adding an ampholyte solution into immobilized pH gradient (IPG) gels. The method involves casting a layer of polyacrylamide gel containing a mixture of carrier ampholytes, having low-molecular-weight synthetic polyamino and polycarboxylic acids. The carrier ampholytes arrange themselves in an electric field, in the order of increasing pI from the anode to the cathode. Better resolution is achieved by initially establishing a pH gradient along the gel using a mixture of low-M_r carrier ampholytes such as amino acids or small peptides. All proteins with isoelectric point above pH 7.0 will have positive charge and move toward negative electrode. All proteins with isoelectric point below 7.0 will have negative charge and move toward positive electrode. At the region of the gradient where the pH corresponds to the isoelectric point of the protein, the net charge on the protein becomes zero and thus becomes stationary or "focused" at this point. Proteins are therefore separated according to their charge, irrespective of its size or molecular weight. The technique is capable of resolving proteins differing by a single charge. The high resolving power allows the separation of compounds differing even by 0.01 pH units in their pI.

The protein electrophoresis, especially the IEF of the sarcoplasmic proteins, is a well-established technique in the case of species identification for seafood authenticity. In the case of heat-processed fish, the method can be applicable only to those species which possess characteristic patterns of the heat-stable parvalbumins (Josep et al. 1996). Heat-denatured fish muscle proteins may be solubilized by urea or sodium dodecyl sulfate (SDS) and separated by urea-IEF and SDS-PAGE, respectively (Magdalena and Edward 2007).

IEF is traditionally used as first-stage separation for 2D-PAGE, separating proteins by charge prior to second-dimension separation by SDS-PAGE. This additional separation allows resolution of a couple of thousand proteins on a 2D-PAGE gel. Another use for IEF is for fractionation of proteins or peptides prior to mass spectrometry. It can also be used to examine posttranslation modifications of proteins (Mostovenko et al. 2013).

5.4 Nephelometry, Photometry, and Spectrometry

5.4.1 Nephelometry and Turbidimetry (Lawler 2016)

Nephelometry and turbidimetry are methods for estimating the amount of cloudiness, or turbidity, in a solution based on the measurement of transmission and scattering of light. Both nephelometry and turbidimetry are used with Tyndall scattering to quantitatively assay turbid solutions. Turbidity in a liquid is the result of finely divided suspended particles. When a beam of light is allowed to pass through a turbid sample, its intensity is reduced by scattering, and the quantity of light scattered is directly proportional to the concentration and size distribution of the particles. Nephelometry measures the intensity of the scattered light, while turbidimetry measures the intensity of light transmitted through the sample. Nephelometric and turbidimetric measurements are used to determine suspended material in natural waters, in processing streams, and in water supply and waste water management industries. Nephelometry has also been employed for the quantitative analysis of various protein and other antigens in blood and body fluids. Measurement of light scattering has also been employed for the determination of cell size, e.g., in the fluorescence-activated cell sorter (FACS).

5.4.1.1 Instrumentation

The term nephelometry originates from the Greek word "nephele" meaning cloud or mist. The measurement of light scatter is ideally performed in a dilute sample suspension in order to minimize reflection and absorption of light. Nephelometry is the measurement of suspended particles at right angles to the incident beam. This method is mainly used for determination of the low concentration suspensions. The apparatus used for nephelometric measurements differs from that used for turbidimetric measurements in the placement of the detector. In nephelometry, the detector is not aligned with the radiation source and the cell; normally, it is placed perpendicular to the path of the incident radiation. Placing the detector out of the path of the incident radiation eliminates the possibility of measuring its intensity.

Turbidimetry is the measurement of the transmitted light by the suspended particles to the incident beam. This is used for the detection of high concentration suspensions. The measurements are completely analogous to absorption

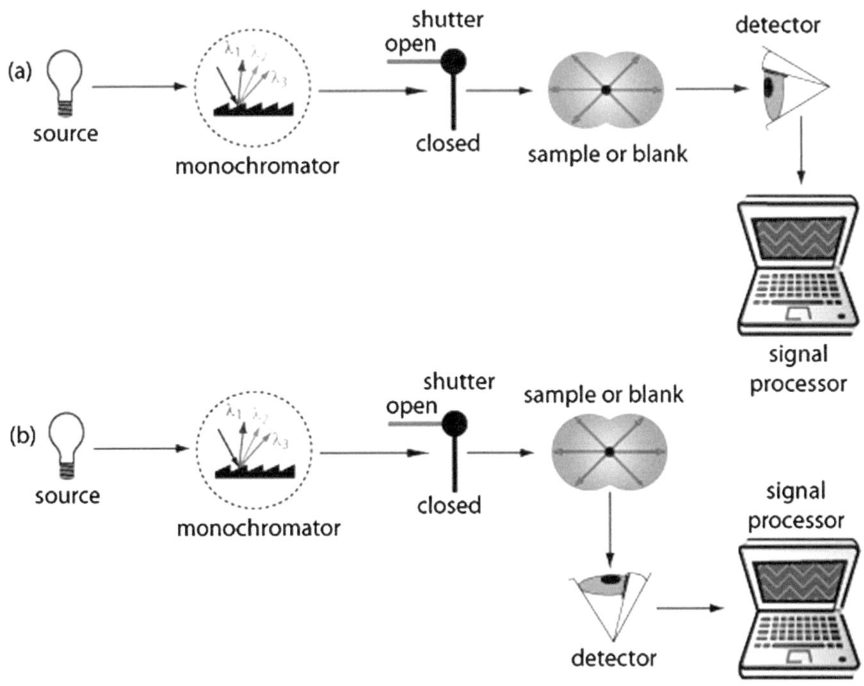

Fig. 5.12 Schematic diagram of (**a**) turbidimetry and (**b**) nephelometry

measurements. The only difference is in the phenomenon that causes the decreased radiative intensity. In the case of absorption measurements, the decreased intensity is directly related to the concentration of the scattering species in the cell at a constant wavelength (Fig. 5.12).

Turbidimetry and nephelometry have found various applications in environmental analysis as well as in the chemical, pharmaceutical, and food industries. Turbidimetry and nephelometry are well-established procedures where filtration processes have to be carried out, monitored, and controlled.

5.4.2 UV-VIS Spectrophotometer

Spectrophotometry is a technique to measure the transmission, reflection, and absorption characteristics of a material as a function of wavelength of light. Spectrophotometry deals with near UV, visible and near IR light region of energy space, where the molecules undergo electronic transitions. The application of spectrophotometers is enormous in different fields of science and technology. It includes semiconductors and laser technology, optics and photonics, printing and forensic examination, as well as in laboratories for the study of materials, chemicals, and biological samples. Based on the principles of the Beer-Lambert law, the technique can be employed in a quantitative way to determine concentrations of an absorbing entities in solution.

$$A = \log_{10}\left(\frac{I}{I_0}\right) = \varepsilon.c.L$$

where A is the measured absorbance, I is the transmitted intensity, I_0 is the intensity of the incident light at a given wavelength, ε is a constant known as the molar extinction coefficient, c is the concentration of the absorbing species, and L is the path length of the cell.

A spectrophotometer, in general, consists of two devices: a spectrometer and a photometer. A spectrometer is a device that produces, typically disperses and measures light. A photometer indicates the photoelectric detector that measures the intensity of light.

- *Spectrometer*: It produces a desired range of wavelength of light. First a collimator (lens) transmits a straight beam of light (photons) that passes through a monochromator (prism) to split it into several component wavelengths (spectrum). Then a wavelength selector (slit) transmits only the desired wavelengths.
- *Photometer*: The desired range of wavelength of light passes through the solution of a sample in the cuvette; a photometer detects the amount of photons that is absorbed and then directs a signal to a galvanometer or a digital display.

5.4.2.1 Instrumentation (Jena et al. 2015)

Spectrophotometry is the quantitative measurement of the light absorbed by a chemical substance on passing a beam of light through the sample. There are two types of spectrophotometer, (i) single-beam spectrophotometer and (ii) double-beam spectrophotometer, which can be used for measuring the transmission in low-absorbing materials.

Single-Beam Spectrophotometer Light from the source lamp is paralleled and then passed through a monochromator, which diffracts the light (Fig. 5.13). The narrow bandwidths of this diffracted spectrum are then passed through a mechanical slit on the output side of the monochromator. Then the intensity of the light transmitted through the sample is measured by detectors and converted to absorbance or transmittance. Generally, halogen lamp is used for VIS-NIR light source, while deuterium lamp is used for UV light source. The most common detectors in the UV-VIS and NIR region are the photomultiplier tubes (PMT) and lead sulfide (PbS) detectors, respectively. The advantages to a single-beam configuration are simpler instrumentation with fewer moving parts and hence less expensive.

Double-Beam Spectrophotometer A double-beam spectrophotometer has two light beams: one of which passes through the sample, while other passes through a reference cell as shown in Fig. 5.13. This allows more reproducible measurements, and multiple-wavelength scanning allows recording of the entire spectrum at once. For transmission measurements in thin-film samples, the spectrophotometer quan-

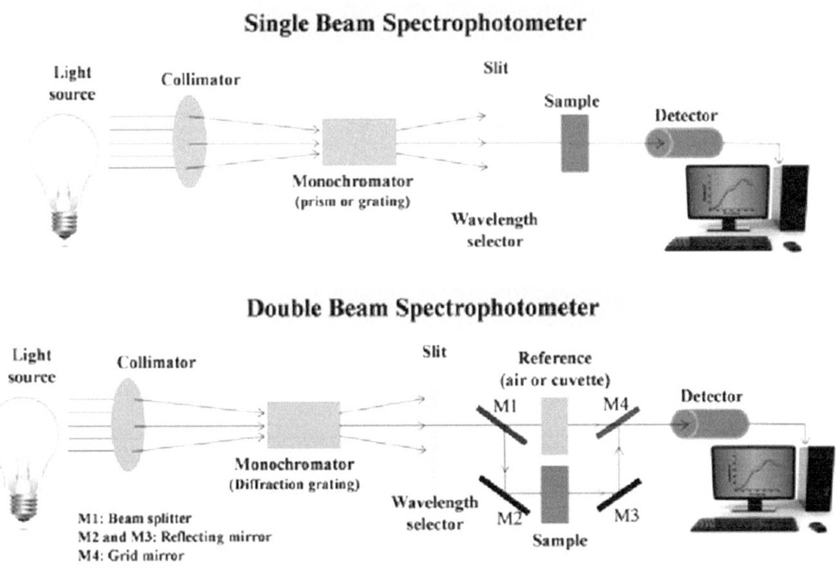

Fig. 5.13 Schematic diagram of a single-beam and double-beam spectrophotometer

titatively compares the fraction of light that passes through a reference (air in our case) and then electronically compares the intensities of the two signals and computes the percentage of transmission of the sample.

5.4.3 Atomic Spectroscopy

Atomic spectroscopy is one of the most reliable instrumental methods for the determination of elements. The technique was introduced by Robert Bunsen and Gustav Kirchhoff in the mid of the nineteenth century, in which they showed that specific elements present in a sample emit characteristic radiation when exposed to a flame. The developments in the instrumentation area led to the widespread application in both the qualitative and quantitative analyses of atomic spectroscopy. Atomic spectroscopy is divided into three types which are absorption, emission, and luminescence spectroscopy. The different branches of atomic absorption spectroscopy are (1) flame photometry or flame atomic emission spectrometry in which the species is examined in the form of atoms, (2) atomic absorption spectrophotometry (AAS), and (3) inductively coupled plasma atomic emission spectrometry (ICP-AES).

The intensity of the light emitted could be described by the Scheibe-Lomakin equation

$$I = K \times C^{n}$$

where I = intensity of emitted light, C = concentration of the element, and K is the constant of proportionality. When n ~1 (at the linear part of the calibration curve):

$$I = K \times C$$

5.4.3.1 Flame Photometry

Flame photometry or flame atomic emission spectrometry (FAES) was established by Robert Bunsen. It is called photometry, even though it is an emission technique. The correlation between the intensity and concentration is linear, and hence the Beer-Lambert law cannot be applied. Flame photometer can be used to determine the concentration of certain metal ions like sodium, potassium, lithium, calcium, cesium, etc. When Na^+ and K^+ are excited, they emit spectra with sharp, bright lines at 589 and 766 nm, respectively, where the amount of light given off by the excited atoms is proportional to the concentration of these metal ions in the solution. The atoms are excited in a flame, and on return to the ground state, a radiation of the characteristic element is emitted, the intensity of which is measured using a flame photometer. The apparatus of FAES contains a nebulizer system, a burner, suitable for different gases like nitrous oxide or methane, a monochromator, and a detector (Fig. 5.14 and Table 5.4).

5.4.3.2 Atomic Absorption Spectrophotometry (Fernandez et al. 2019)

The phenomenon of atomic absorption (AA) was first observed in 1802 by Joseph Von Fraunhofer with the discovery of the Fraunhofer lines in the sun's spectrum. It was not until 1953 that Australian physicist Sir Alan Walsh demonstrated that atomic absorption could be used as a quantitative analytical tool. Atomic absorption analysis employs the measurement of absorption of light by a vaporized ground-state atoms and relating it to its concentration. The incident light beam is attenuated by atomic vapor absorption according to Beer's law.

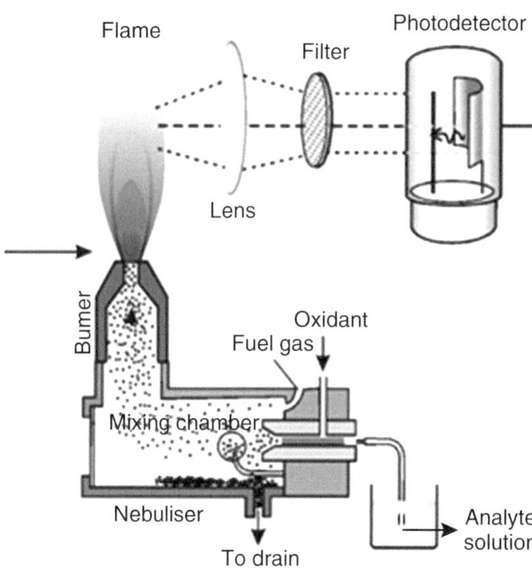

Fig. 5.14 Schematic diagram of a flame photometer (https://www.researchgate.net/figure/Schematic-diagram-showing-flame-photometer_fig6_314502828)

Table 5.4 Emission wavelength and the characteristic color produced by different elements

Element	Emission wavelength (nm)	Color of the flame
Lithium (Li)	670	Red
Sodium (Na)	589	Yellow
Potassium (K)	766	Violet
Barium (Ba)	554	Lime green
Calcium hydroxide	622	Orange

Source: http://www.jenway.com/adminimages/PFP7_Manual(2).pdf

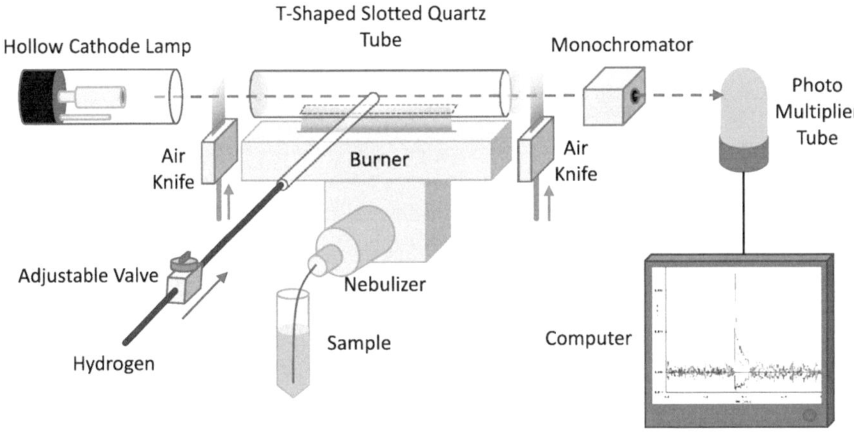

Fig. 5.15 Schematic diagram of an atomic absorption spectrophotometer (https://ars.els-cdn.com/content/image/1-s2.0-S0584854717305451-fx1_lrg.jpg)

The atomic absorption (AA) spectrometer is used to analyze metals at very low concentrations, typically in the parts per million (ppm) or parts per billion (ppb) ranges. A liquid sample containing dissolved material whose concentration is to be measured is aspirated into a thin, wide AA flame or is introduced into a small carbon furnace which is heated to a high temperature. When the ultraviolet or visible light is absorbed by the free atoms, they undergo electronic transitions from the ground state to excited electronic states. At the high temperature of the AA flame, which may be either oxyacetylene or nitrous oxide/acetylene, the sample is ionized, and the concentration of these ions is measured (Fig. 5.15).

5.4.3.3 Inductively Coupled Plasma Atomic Emission Spectrometry (ICP-AES)

ICP-AES is one of the most widely used techniques that can determine concentrations of trace to major elements and can detect most elements in the periodic table. FAAS/FAES instruments require a unique radiation source (lamp) for the approximately 35 elements they can measure. Because the lamp must be changed between each element of interest, FAAS/FAES techniques analyze a single element at a time and are unable to easily analyze metalloids. ICP optical systems, by contrast, can analyze about 60 different elements at the same time with a single source (the plasma). The most common instruments today are inductively coupled plasma – atomic emission spectrometers (ICP-AES) and inductively coupled plasma – mass spectrometers (ICP-MS) (Nordberg et al. 2015).

ICP-AES, also referred to as ICP-optical emission spectrometry (OES), has become the most commonly employed emission spectrometric technique. The argon-based plasma is well-suited with aqueous aerosols and provides very high thermal energy for the drying, dissociation, and atomization of analytes. The temperature of a typical argon ICP is 5500–6500 K, which is highly capable enough to disrupt the molecular bonds and to atomize and ionize most elements, even though with different degrees of efficiency. The relatively high sensitivity of atomic emission spectrometer detectors yields very high sensitivity (ppb) and a very broad dynamic range (for about 70 elements).

The main parts of an ICP include a pneumatic nebulizer for the formation of aerosols and a spray chamber, which acts as a selector for aerosolized droplets. When an aerosolized analyte enters the plasma, they are rapidly desolvated, and a fraction of the atoms are transiently excited. When the excited electrons return to their ground state, they emit electromagnetic radiation characteristic of the change in their energy states. This is then detected by charge-coupled device detectors (photomultiplier tube or PMT) capable of monitoring a large range of wavelengths simultaneously (Fig. 5.16).

Inductively coupled plasma mass spectrometry (ICP-MS) is a technique routinely used to analyze trace levels over a widespread range of inorganic elements. The ICP-MS allows for the detection and quantification of elements with atomic mass ranges 7–250 (lithium to uranium). The typical detection limits are in the parts per billion (ppb) range and even parts per trillion (ppt) in some cases.

5.4.4 FTIR Spectroscopy

FTIR (Fourier transform infrared spectroscopy) is a sensitive technique particularly for identifying organic chemicals as well as some inorganics in a whole range of applications. It is a form of vibrational spectroscopy and a potent tool for identifying different types of chemical bonds in a molecule, thereby producing a characteristic infrared absorption spectrum that acts as a "molecular fingerprint." This technique is useful for analyzing the chemical composition of smaller particles, typically 10–50μm, as well as larger areas on the surface. A Fourier transform infrared (FT-IR) spectrometer is an instrument which acquires broadband near-infrared (NIR) to far-infrared (FIR) spectra. Unlike a dispersive instrument, i.e., a grating monochromator or spectrograph, FTIR spectrometers collect all wavelengths simultaneously. This feature is called the Multiplex or Fellgett's Advantage.

When a material is irradiated with infrared radiation, the absorbed infrared radiation excites the molecules into a higher vibrational state. The wavelength of light

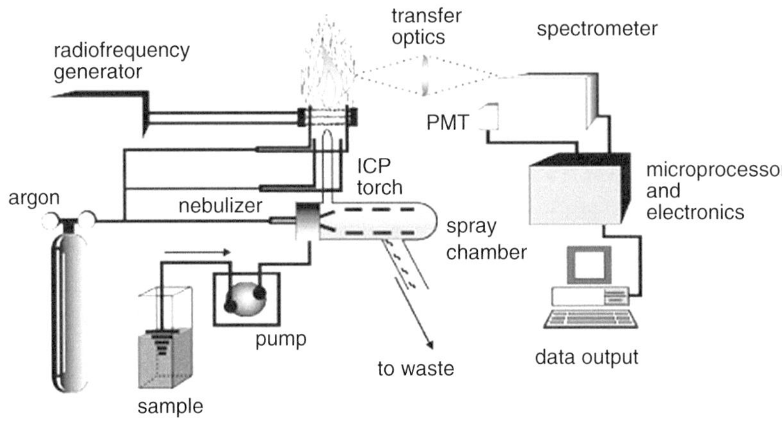

Fig. 5.16 Schematic diagram of an inductively coupled plasma-atomic emission spectrometer (ICP-AES). (Source: http://www.neeri.res.in/division/analytical-inst-division/facilities/inductively-coupled-plasma-%E2%80%93-optical-emission)

absorbed by a particular molecule is a function of the energy difference at the ground and excited vibrational states. The wavelength absorbed by the sample is characteristic of its molecular structure.

5.4.4.1 Instrumentation

The FTIR spectrometer employs an interferometer to modulate the wavelength from a broadband infrared source. The detector measures the intensity of transmitted or reflected light as a function of its wavelength. The signal obtained from the detector is an interferogram, which must be analyzed with a computer using Fourier transforms to obtain a single-beam infrared spectrum. The FTIR spectra are usually presented as plots of intensity versus wavenumber (in cm^{-1}), where a wavenumber is usually the reciprocal of the wavelength. The FTIR instrument sends infrared radiation of about 10,000–100 cm^{-1} through a sample, with some radiation absorbed and some passed through. The absorbed radiation is converted into rotational and/or vibrational energy by the sample molecules. The resulting signal at the detector presents as a spectrum, typically from 4000 to 400 cm^{-1}, representing a molecular fingerprint of the sample. The intensity can be plotted as the percentage of light transmittance or absorbance characteristic of each wavenumber. Each molecule or chemical structure will produce a unique spectral fingerprint, making FTIR analysis a great tool for chemical identification (Fig. 5.17).

In addition to transmission techniques, reflectance techniques are also employed because of the ability to focus and manipulate the incident beam with optics, in the form of ATR (attenuated total reflectance) techniques to introduce and observe the energy. ATR involves using a phenomenon of internal reflectance to propagate the incident energy. The beam is introduced to a crystal at an incident angle which enables internal reflectance "bounces" at the bottom and top of the crystal before it leaves the crystal on the opposite side. The sample is made to make contact with the

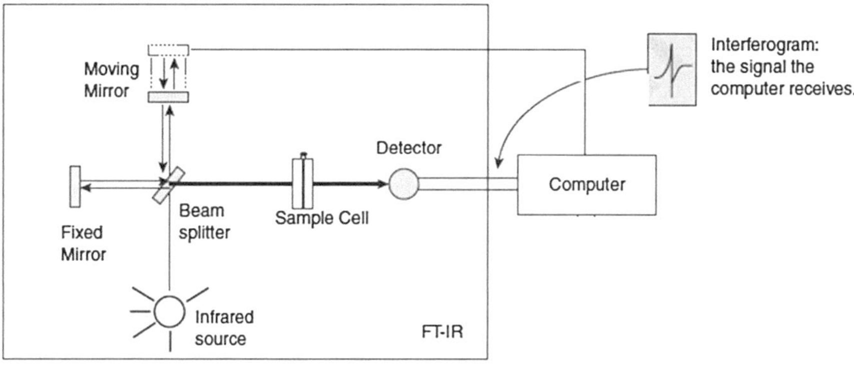

Fig. 5.17 Schematic diagram of typical FTIR spectrophotometer. (Source: https://www.researchgate.net/figure/5-Schematic-diagram-of-typical-FTIRspectrophotometer_fig21_309571420/actions#reference)

crystal at the top such that energy interaction occurs at the crystal and sample interface where the bounce positions are located. Typically, the more bounce positions, the larger the energy transfer (and thus better spectral response); however, single-bounce systems are used when a very small area needs to be analyzed.

5.4.5 NMR Spectroscopy

Nuclear magnetic resonance spectroscopy is a powerful analytical tool employed for the characterization of molecules. Nuclei with an odd mass or odd atomic number have a "nuclear spin," in a similar fashion to the spin of electrons. For example, the ^{1}H and ^{13}C have nuclear spin but not the ^{12}C. The NMR behavior of ^{1}H and ^{13}C nuclei has been exploited by organic chemist since they provide valuable information that can be used to deduce the structure of organic compounds. The spins of nuclei are sufficiently different that NMR experiments can be sensitive for only one particular isotope of one particular element. The net spin of a nucleus is regulated accordingly that if the number of neutrons and the number of protons are both even, then the nucleus has no spin. If the number of neutrons plus the number of protons is odd, then the nucleus has a half-integer spin (i.e., 1/2, 3/2, 5/2), and if the number of neutrons and the number of protons are both odd, then the nucleus has an integer spin (i.e., 1, 2, 3).

Since a nucleus is a charged particle in motion, it will possess a magnetic field. ^{1}H and ^{13}C have nuclear spins of 1/2 and hence behave in a similar fashion to a simple, tiny bar magnet. In the absence of a magnetic field, these are randomly oriented, but when a field is applied, they line up parallel to the applied field, either in a spin-aligned or spin-opposed fashion. In the more highly populated state, the energy spin state in spin-aligned situation is lower. ^{1}H and ^{13}C are not unique in their ability to undergo NMR. All nuclei with the odd number of protons (^{1}H, ^{2}H, ^{14}N, ^{19}F, ^{31}P, etc.) or nuclei with an odd number of neutrons (i.e., ^{13}C) show the magnetic properties

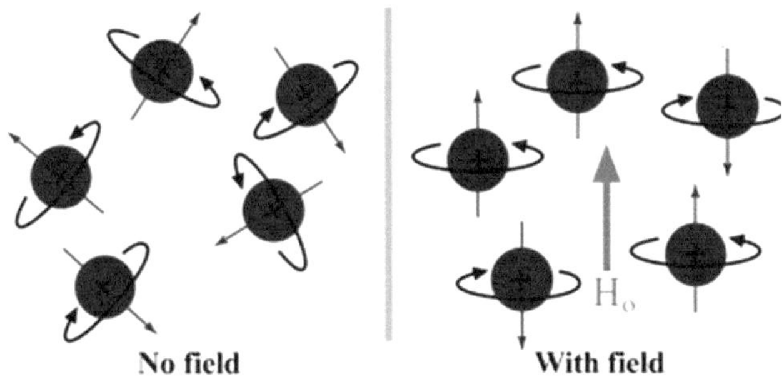

Fig. 5.18 Schematic representation of a nuclear spin in the presence and absence of magnetic field. (Source: http://www.chem.ucalgary.ca/courses/350/Carey5th/Ch13/ch13-nmr-1.html)

required for NMR. Only nuclei with even number of both protons and neutrons (^{12}C and ^{16}O) do not have the required magnetic properties (Fig. 5.18).

5.4.5.1 Instrumentation

The nuclear magnetic resonance (NMR) spectrometer instrument comprises of a superconducting magnet, a sample probe, a radio-frequency (RF) transmitter, a receiver, and a computer controlled and data processing unit. Superconducting magnets can be used to produce very strong magnetic field, on the order of 21 tesla (T). Lower field strengths can also be used, in the range of 4–7 T. A sample (in a small glass tube) is placed in between the poles of a strong magnetic field. A radio frequency generator pulses the sample and excites the nuclei causing a spin-flip. The spin-flip is detected by the detector, and the signal sent to a computer where it is processed (Fig. 5.19).

The position on the plot at which the nuclei absorb is called the chemical shift. Since this has an arbitrary value, a standard reference point must be used. The two most common standards are TMS (tetramethylsilane, ($Si(CH_3)_4$) which has been assigned a chemical shift of zero and $CDCl_3$ (deuterochloroform) which has a chemical shift of 7.26 for ^1H NMR and 77 for ^{13}C NMR.

The scale is normally expressed as parts per million (ppm) which is independent of the spectrometer frequency. The scale is the delta (δ) scale.

$$\delta = \frac{\text{Frequency of signal} - \text{Frequency of standard}}{\text{Spectrometer frequency}} \times 10^6$$

The range at which most NMR absorptions occur is quite narrow. Almost all ^1H absorptions occur downfield within 10 ppm of TMS. For ^{13}C NMR, almost all absorptions occur within 220 ppm downfield of the C atom in TMS.

For the analysis of molecular structure at the atomic level, electron microscopes and X-ray diffraction instruments can also be used, but the advantages of NMR are that sample measurements are nondestructive and there is less sample preparation required.

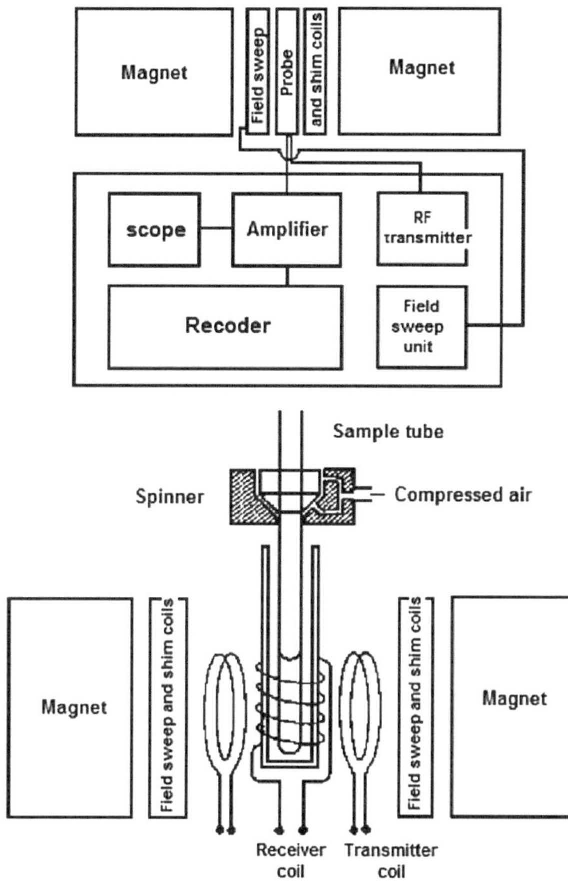

Fig. 5.19 Block diagram of a high-resolution NMR spectrometer and the arrangement of the sample in the probe (cross-coil configuration). (Source: https://www.researchgate.net/figure/Block-diagram-of-a-high-resolution-NMR-spectrometer-and-the-arrangement-of-the-sample-in_fig6_257917183)

5.4.6 Mass Spectrometry

Mass spectrometry is a powerful analytical technique for quantifying trace amounts of known materials, for the identification of unknown compounds, and also for elucidating the structure and chemical properties of different molecules. The process involves the conversion of the sample into its gaseous ions, with or without fragmentation that are then characterized by their mass-to-charge ratios (m/z) and relative abundances. Mass spectrometry (MS) remains unique among the other spectroscopic analyses such as IR, UV, or NMR in that it allows mass discrimination via molecular ionization, thus making this irreversible process a destructive means of analysis. Mass spectrometry performs three main functions – namely, (1) the ionization of molecules, (2) the separation of these molecules based on their mass-to-charge-ratio, and (3) the determination of the respective abundance of each ion produced.

A typical mass spectrometer comprises three parts: ion source, mass analyzer, and detector (Fig. 5.20).

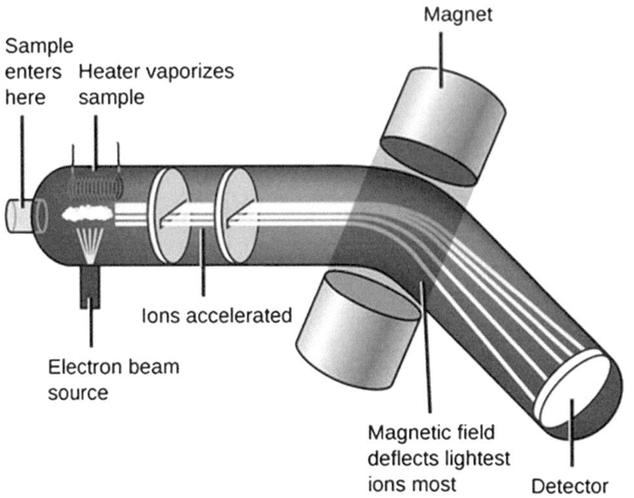

Fig. 5.20 Schematic representation of a mass spectometer (MS). (Source: https://www.bronkhorst.com/int/blog/mass-spectrometry-and-mass-flow-control-a-closer-ion-them/)

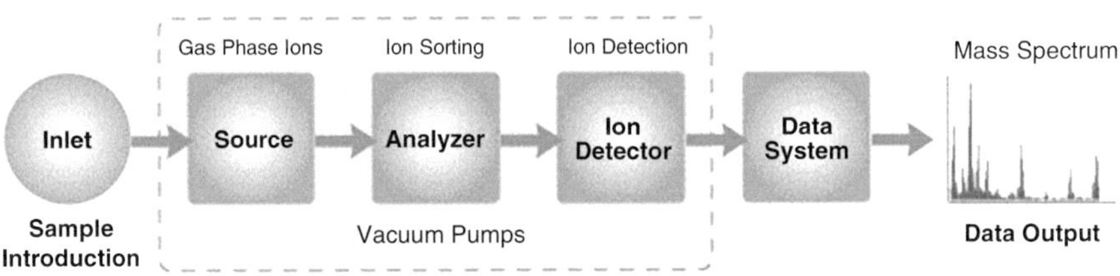

Source: http://www.premierbiosoft.com/tech_notes/mass-spectrometry.html

5.4.6.1 Ion Source

The sample to be analyzed is ionized in the ion source. The different types of sample introduction methods are:

- Gaseous sample introduction which includes electron ionization (EI) and chemical ionization (CI)
- Liquid sample introduction method includes electrospray ionization (ESI), atmospheric pressure chemical ionization (APCI), atmospheric pressure photo ionization (APPI), multimode ionization (MMI), matrix-assisted laser desorption ionization (MALDI), inductively coupled plasma (ICP)

5.4.6.2 Analyzer

After ionization, the ions are then transported by magnetic or electrical fields to the mass analyzer. The mass analyzer separates the ions based on their mass-to-charge ratio (m/z). The mass spectrometer measures the ion signals resulting in a mass spectra, which can provide valuable information about the molecular weight, structure, identity, and quantity of a compound. There are different types of mass analyzers:

- Single quadrupole (SQ)
- Triple quadrupole (QQQ)
- Time-of-flight (TOF)
- Ion trap (IT)
- The electrostatic trap or "Orbitrap"
- Magnetic and electromagnetic analyzers
- Ion cyclotron resonance and Fourier transform mass spectrometry

Many mass spectrometers use two or more mass analyzers for tandem mass spectrometry (MS/MS). Tandem mass spectrometry (MS/MS) is a crucial technique in many applications. MS/MS involves multiple stages of mass analysis.

5.4.6.3 Detector

The final component of the mass spectrometer is the detector. The detector records the current produced or the charge induced when an ion passes by or hits a surface. The different types of detectors include the former, photographic plate and Faraday cup detectors, to most common modern electron multipliers (EM) and electro-optical ion detectors (EOID).

The employability of a mass spectrometer as the detector in gas chromatography or HPLC combines the features of gas or liquid chromatography and mass spectrometry to identify different substances within a sample. In GC/MS or LC/MS, the compounds are chromatographically separated subsequent to their introduction into the ion source and mass spectrometer.

5.5 Microscopy

Microscopes are instruments that produce enlarged images of small objects and samples that cannot be viewed with an unaided eye, thereby allowing the observer an exceedingly close view of minute structures at the scale convenient for examination and analysis. The term originates from the words *micro* = "small" and *scope* = "to look at." The most familiar type of microscope is the optical or light microscope that uses glass lenses to form the image. Generally, optical microscopes can be simple, consisting of a single lens, or compound, consisting of several optical components in line.

Two parameters are especially important in microscopy: magnification and resolution.

- Magnification/magnifying power of a microscope is an expression of the number of times the object being examined appears to be enlarged and is a dimensionless ratio. For instance, the light microscopes magnify up to about 400 times actual size.

- The resolution of a microscope or lens is the smallest distance by which two points can be distinguished as separate entities. The smaller this value, the higher the resolving power of the microscope and hence the better the clarity and details of the image. Resolution is expressed in linear units, usually micrometers (μm).

The hand magnifying glass can magnify about 3–20×. Single-lensed simple microscopes can magnify up to 300× and are capable of revealing bacteria, while compound microscopes can magnify up to 2000×. A simple microscope can resolve below 1 μm, whereas a compound microscope can resolve down to about 0.2 μm.

5.5.1 Light Microscope

Light microscopy or optical microscopy is a general term used for any type of microscopy where light is being transmitted from a source to the objective lens. The light is allowed to pass through a condenser to focus it on the sample to have maximum brightness. After the light has passed through the sample, it goes through the objective lens to magnify the image of the sample and then to the oculars, where the enlarged image is viewed. An image may also be enlarged by many other wave forms, including acoustic (ultra sound), X-ray, or electron beam, and be received by direct or digital imaging or by a combination of these methods.

Simple Microscope

The simple microscope is generally considered to be the first microscope. It was the Dutch scientist, Antonie van Leeuwenhoek (1632–1723), who first built a simple and very effective microscope based on a single lens. With a magnification power of 200× and 300×, it was powerful magnifying power (*M*) of a simple microscope is the ratio of the angle subtended by the image at the eye to the angle subtended by the object seen directly, when both lie at the least distance of distinct vision or the near point.

$$M = 1 + \frac{D}{f}$$

where D is the "least distance of distinct vision" and f is the focal length of the lens.

Today, simple microscopes are not used often because the introduction of a second lens led to the more powerful compound microscope.

Compound Light Microscopes

Compound light microscopes are one of the most familiar of the different types of microscopes. A compound microscope, for instance, is the most common type of microscopes, used basically for research and also referred to as biological microscope. It is called the high-power microscope mainly because its magnification power can range up to 1000 times and some can even go up to 1500 times to 2000 times. The main parts of a compound light microscope are shown in Fig. 5.21.

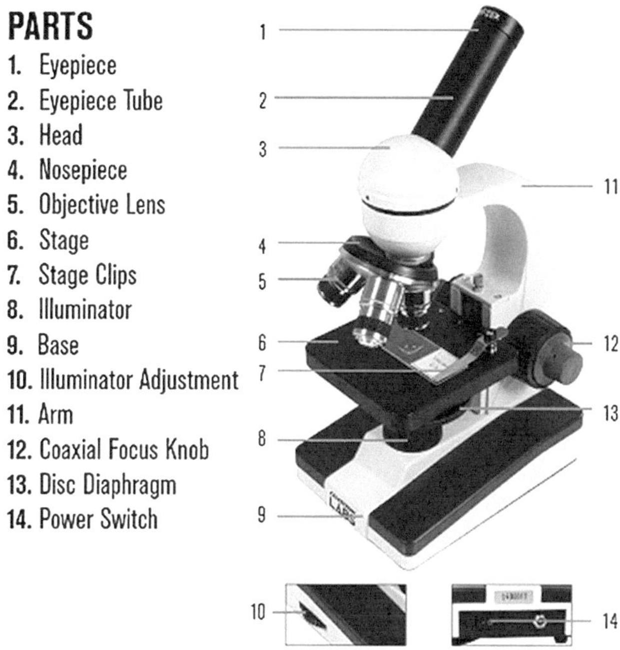

Fig. 5.21 Parts of the compound microscope (https://eco-globe.com/types-of-microscope/)

Magnifying power (M) of a compound microscope will be

$$M = \frac{L}{f_o}\left(\frac{1+D}{f_e}\right)$$

where f_o and f_e are focal length of objective and eye piece, respectively, L is the length a the microscope tube, and D is the least distance of distinct vision.

Stereo Microscope/Dissecting Microscope

The stereo microscope, also called a dissecting microscope, consists of a biconvex lens which is moved up and down by an adjustment screw to bring the object in sharp focus. The object is placed on the platform, and light is focused with the help of a concave mirror fitted below.

A dissecting microscope has two optical paths at slightly different angles allowing the image to be viewed three-dimensionally under the lenses. The stereoscopic microscope is an optical microscope that allows sample observation as a three-dimensional image. This is its main characteristic that distinguishes it from other microscopes where the sample is always observed in two dimensions. Stereo microscopes magnify at low power, typically between 10× and 200×, generally below 100×. Uses for this type of microscope include biological and medical science applications as well as in the electronics industry, such as watch making, plus building, and inspecting circuit boards (Table 5.5).

Table 5.5 The different types of light microscopes and their properties

Microscope type	Description
Brightfield microscope	A typical microscope that uses transmitted light to observe targets at high magnification
Polarizing microscope	A microscope that uses different light transmission characteristics of materials, such as crystalline structures, to produce an image
Phase contrast microscope	A microscope that visualizes minute surface irregularities by using light interference. It is commonly used to observe living cells without staining them
Differential interference contrast microscope	This microscope, similar to the phase contrast, is used to observe minute surface irregularities but at a higher resolution. However, the use of polarized light limits the variety of observable specimen containers
Fluorescence microscope	A biological microscope that observes fluorescence emitted by samples by using special light sources such as mercury lamps. When combined with additional equipment, brightfield microscopes can also perform fluorescence imaging
Total internal reflection fluorescence microscope	A fluorescence microscope that uses an evanescent wave to only illuminate near the surface of a specimen. The region that is viewed is generally very thin compared to conventional microscopes. Observation is possible in molecular units due to reduced background light
Laser microscope (laser scanning confocal microscope)	This microscope uses laser beams for clear observation of thick samples with different focal distances
Multiphoton excitation microscope	The use of multiple excitation lasers reduces damage to cells and allows high-resolution observation of deep areas. This type of microscope is used to observe nerve cells and blood flow in the brain
Structured illumination microscope	A high-resolution microscope with advanced technology to overcome limited resolution found in optical microscopes that is caused by the diffraction of light

Source: https://www.keyence.com/ss/products/microscope/bz-x/study/principle/type.jsp#

5.5.2 Electron Microscope (Hayat 2000)

An electron microscope is a type of high-resolution microscope that utilizes electrons (instead of light) to illuminate a specimen and creates an enlarged image. Electron microscopes have higher resolving power than light microscopes enabling better magnification. The extremely short wavelength and focusing capability of electron beams are accountable to the high resolving power in an electron microscope. Some electron microscopes can magnify specimens up to 2 million times, while the best light microscopes are limited to magnifications of 2000 times. The first electron microscope prototype was built in 1931 by German engineers Ernst Ruska and Max Knoll.

A series of coiled electromagnets focuses a beam of electrons at a specimen placed in a vacuum sample chamber in an electron microscope. The electron microscope uses electrostatic and electromagnetic lenses in forming the image by controlling the electron beam to focus it at a specific plane relative to the specimen in a manner similar to how a light microscope uses glass lenses to focus light on or through a specimen to form an image. Images produced in the electron microscope are in shades of gray, although computerized color may be added in some scopes. The two most common types of electron microscopy are scanning electron microscopy and transmission electron microscopy.

5.5.2.1 Transmission Electron Microscope (TEM)

Transmission electron microscopy (TEM) is the original form of electron microscopy and involves a high-voltage electron beam emitted by an electron gun that is usually fitted with a tungsten filament cathode as the electron source. The electron beam is accelerated by an anode typically at +100 keV (40–400 keV) with respect to the cathode and is focused by electrostatic and electromagnetic lenses that get transmitted through a very thin specimen (usually no more than 100 nm thick). When it emerges from the specimen, the electron beam carries information about the structure of the specimen that is magnified by the objective lens system of the microscope. The electrons further hit a fluorescence screen that forms an image that can be photographically recorded by exposing a photographic film or plate directly to the electron beam, or a high-resolution phosphor may be coupled by means of a fiber-optic light guide to the sensor of a CCD (charge-coupled device) camera. The image detected by the CCD may be displayed on a computer (Fig. 5.22).

Resolution of the TEM is limited primarily by spherical aberration, but a new generation of aberration correctors has been able to partially overcome spherical aberration to increase resolution. Software correction of spherical aberration for the high-resolution TEM (HRTEM) has allowed the production of images with sufficient resolution to even show carbon atoms in diamond separated by only 0.89 Å (89 pm) and atoms in silicon at 0.78 Å (78 pm) at magnifications of 50 million times. The ability to determine the positions of atoms within materials has made the HRTEM an important tool for nanotechnologies research and development.

5.5.2.2 Scanning Electron Microscope (SEM)

In a scanning electron microscope or SEM, a beam of electrons scans the surface of a sample and produces an image by detecting low-energy secondary electrons which are emitted from the surface of the specimen due to excitation by the primary electron beam. The operating environment of a standard scanning electron microscope dictates that specialist preparation techniques are used. Typically, a biological specimen is chemically fixed (using glutaraldehyde) or dehydrated through an acetone or ethanol series and then dried at the critical point. In SEM, the samples are mounted on a stub of metal with adhesive, coated with 40–60 nm of metal such as gold/palladium and then observed in the microscope.

In TEM, the magnitude of resolution is greater than the SEM. The SEM image relies on surface processes rather than transmission, and it is able to image bulk

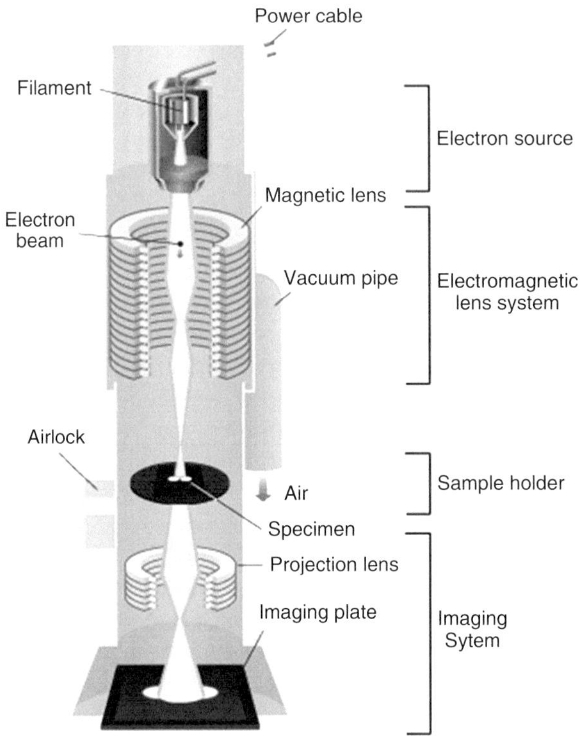

Fig. 5.22 Schematic outline of a transmission electron microscope (TEM). (Source: http://www.hk-phy.org/atomic_world/tem/tem02_e.html)

samples and has a much greater depth of view and so can produce images that are a good representation of the 3D structure of the sample (Fig. 5.23).

A scanning transmission electron microscope or STEM combines the capabilities of both an SEM and a TEM. The electron beam is transmitted across the sample to create an image (TEM), while it also scans a small region on the sample (SEM). The ability to scan the electron beams allows the user to analyze the sample with various techniques such as electron energy loss spectroscopy (EELS) and energy-dispersive X-ray (EDX) spectroscopy which are useful tools to understand the nature of the materials in the sample.

5.6 Refractometer

5.6.1 Refractive Index

Every material with which light interacts has a refractive index (RI, symbolized as η). This constant, dimensionless number describes how fast light moves through a specific medium in relation to how fast it moves through a reference medium

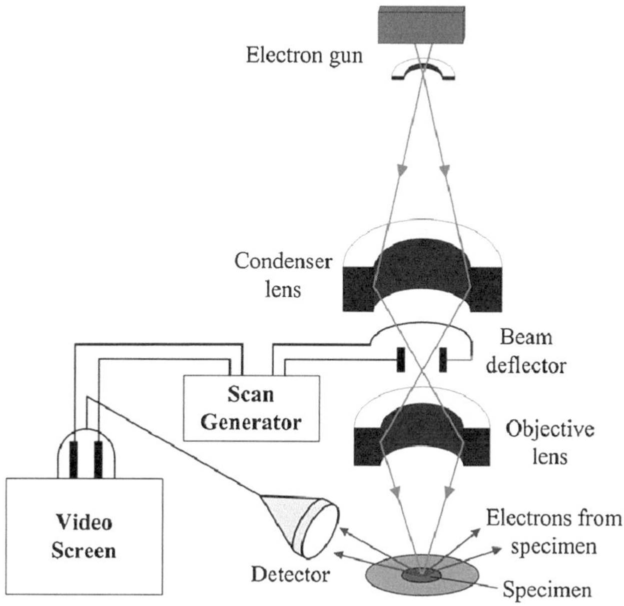

Fig. 5.23 Schematic diagram of SEM working principle (https://www.researchgate.net/figure/Schematic-diagram-of-the-SEM-working-principle-32_fig1_261702149)

(usually vacuum or air). The lower the optical density of the medium, the higher the speed of light in the medium and the lower its refractive index. The refractive index, η is the speed of light in vacuum relative to the speed of light in the considered medium.

$$\eta = \text{Speed of light in vacuum} / \text{Speed of light in medium}$$

For example, water has a refractive index of 1.33 which means that the light travels 1.33 faster in vacuum than in water.

When light enters from a medium with a lower refractive index as, for example, air, into a medium with a higher refractive index as, for example, water, it thus changes its speed (Fig. 5.24).

When light travels through a liquid at an angle, it bends or refracts slowing down and changing direction as it travels through a new medium. This phenomenon can be used to measure the concentration of a liquid solution, as light refracts more when traveling through suspended solids, such as salts or sugars. Using the tool known as a refractometer, the index of refraction can be measured for a liquid and assigned a value on the Brix scale.

Measurement of the refractive index can be used to characterize water, certain oils, and other liquids. Knowledge of the refractive index can be used to control the quality of multicomponent mixtures and check samples for purity (refractive index of fish eye lens could be used as an index of freshness). The refractometer works using the principle of light refraction through liquids. When electromagnetic waves

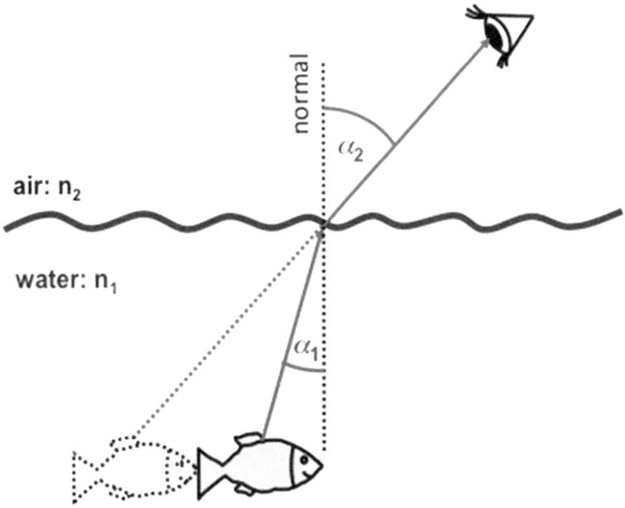

Fig. 5.24 Schematic diagram of refraction of light. (Source: https://wiki.anton-paar.com/fileadmin/wiki/images/basics-of-refractometry/refraction -of-light-2.jpg)

(including visible light) pass through the interface of two different substances, its speed changes and results in a change in the direction of light with an angle called the angle of refraction. This phenomenon is known as "refraction." A refractometer takes the refraction angles on a scale (Brix scale and Plato scale) and correlates them to refractive index (ηD) values that have been established. Using these values, the concentrations of solutions can be determined. Different solutions will have different refractive indexes depending on their concentration and composition. Whenever refractive indices of standard fluids are given, these are prefaced with $\eta 20$ D = a value from 1.3000 to 1.7000, where η indicates the symbol for refractive index, at 20 °C, and D is the wavelength of the light beam (in the D line of the sodium spectrum).

The refractive index, RI (η), is a ratio of the sines of the angles:

$$\eta = \frac{\text{Sine incident ray angle}}{\text{Sine refracted ray angle}}$$

The standard refractive index is typically expressed in the unit ηD. ηD is based on the sodium lamp, which generates a double-wavelength light (588.995 and 589.592 nm). The center of two wavelengths is taken to determine the absolute refraction (589.3 nm). While some applications require other wavelengths, the modern refractometers use the long live LED, instead of sodium lamp. Earlier, the red hydrogen lamp (656.2725 nm, unit nC) or blue hydrogen lamp (486.1327 nm, unit nF) was also used.

5.6.2 Brix Scale and Common Brix %

As light bends while passing through a liquid, the refraction can be measured on the Brix scale. The resulting reading from a refractometer is assigned a value on the Brix scale, making it easy to compare different concentrations in solution. Pure water with no suspended solids has a Brix value of zero, while water with sugars, minerals, or other solids in solution refracts light to produce a higher Brix value.

The Brix scale is calibrated to the number of grams of cane sugar contained in 100 mL of water. Therefore, the Brix % reading equals actual sugar concentration. Another unit, Plato, is also used as a unit of sugar; it is commonly used in the beer industry. Plato represents the sugar in the rye. Therefore, a refractometer is also known as sugar meter.

5.6.3 Factors Influencing Refractive Index

Wavelength For almost all materials, the refractive index differs for different wavelengths, in a pattern characteristic to each material. This is known as the dispersion relation. It is therefore necessary to ensure the precision of the wavelength with an interference filter. Most refractometric measurements were earlier done at 589 (or 589.3) nm, the wavelength of the sodium D-line. (The name "D-line" refers to the symbol given to the spectral line of sodium by Joseph von Fraunhofer (1787–1826), the German physicist credited with the discovery and mapping of over 570 spectral lines.)

Temperature Temperature is another major influence on refractive index and must therefore be precisely controlled and measured during the entire procedure.

Specific Gravity Another factor that affects the refractive index (nD) is the specific gravity of the solution, also known as the mass percentage.

5.6.4 Types of Refractometers

Refractometers are sold as either analog or digital handheld devices for use in the field by scientists, farmers, and students. Both analog and digital refractometers work on the principle that light moves more slowly in more concentrated solutions. When the light slows down, it also changes direction and changes the reading produced by the refractometer.

5.6.4.1 Analog Refractometers

An analog refractometer requires no external power source to function. To obtain a Brix reading for a solution, a few drops are placed on top of a prism, which is then covered by a lid. The refractometer is then brought up to the observer's eye and held perpendicular to a light source. Light is refracted through the liquid, and a Brix reading is taken where the light and dark areas meet on the scale (Fig. 5.26).

5.6.4.2 Digital Refractometers

Digital refractometers produce their own beam of light from an LED, instead of using daylight or other sources of light. Once the liquid is placed inside the receptacle, the light is beamed through the sample, and the software assigns the resulting refraction a numeric value based on the Brix scale. The final value is then displayed on a digital screen.

The modern digital refractometer has innovated the traditional prism refractometers, through the measurement of total reflection method. A refractometer consists of a light source, filtered to a single wavelength, which is directed toward the prism-sample interface by a converging lens. This creates a range of incidence angles, some of which (those less than the critical angle) will be completely reflected. A charge-coupled device (CCD) sensor precisely measures the intensity of the reflected light and determines the exact angle $\alpha_{critical}$ at which light begins to be completely reflected. Because this angle is dependent on the ratio of the refractive index of the prism to that of the sample, the refractive index of the sample can be determined using the known refractive index of the prism. The prism in the refractometer has a greater refractive index than the solution, and the measurements are read at the point where the prism and solution meet. At a low concentration solution, the refractive index of the prism is greater than that of the sample, creating a large refraction angle with low reading and vice versa.

In digital refractometers, the light travels from a prism with a high refractive index (normally glass or artificial sapphire) into the sample. If the angle of incidence exceeds a certain value, the light is reflected at the prism/sample boundary ("total internal reflection"). The reflected light is detected by a CCD sensor. The lower the refractive index of the sample being measured, the smaller the critical angle and the bigger the illuminated surface of the sensor. The refractive index of the sample can thus be calculated by the refractometer, using the ratio of the length of the illuminated and the length of the dark region on the CCD (Figs. 5.25 and 5.26).

The advantage of total reflection is that light does not need to pass through the sample as it will not influence the measurement of opaque, highly viscous, or dark samples. On the contrary, scattered light reduces the accuracy of the reading and is one of the most frequent sources of error when performing refractive index measurements.

5.7 Viscometers

Viscometers measure the viscosity and flow properties of fluids. Viscosity arises from the internal friction of a fluid and is defined as a liquid's resistance to flow or shear stress.

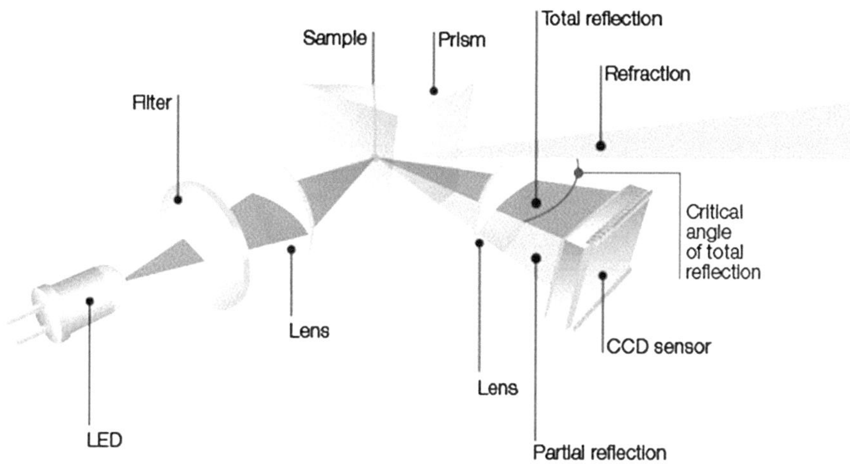

Fig. 5.25 Schematic diagram of a total reflection – digital refractometer. (Source: https://wiki.anton-paar.com/fileadmin/wiki/images/basics-of-refractometry/schematic-setup-abbemat-refractometer.jpg)

$$\text{Viscosity} = \frac{\text{Shear stress}}{\text{Velocity gradient}}$$

or

$$\text{Viscosity, } \eta = \frac{F}{A} \div \frac{dv_x}{dz}$$

Viscometers have wide and varied applications in materials science, chemical industries, food, beverages, and personal care products and pharmaceuticals. Viscosity is a very useful indirect measure of material properties including molecular weight and density, both of which affect flow behavior. Viscometers can therefore be employed to monitor batch consistency and quality control of products.

Viscosity measurement is important when considering process conditions for materials that need to be pumped or piped. Viscosity also affects dipping and coating performance and also the flow properties of liquids. They are also used to determine the activity of enzymes such as cellulase, protease, amylase, and pectinase.

Viscosity measurement can be done in two basic methods, in which the resistance to flow is measured. Viscosity measurement is done by either an object, such as a sphere or a rotor blade, moves through a stationary material (sample), or the sample material flows through or past a stationary object. Six different types of viscometers are generally used in viscometric measurements.

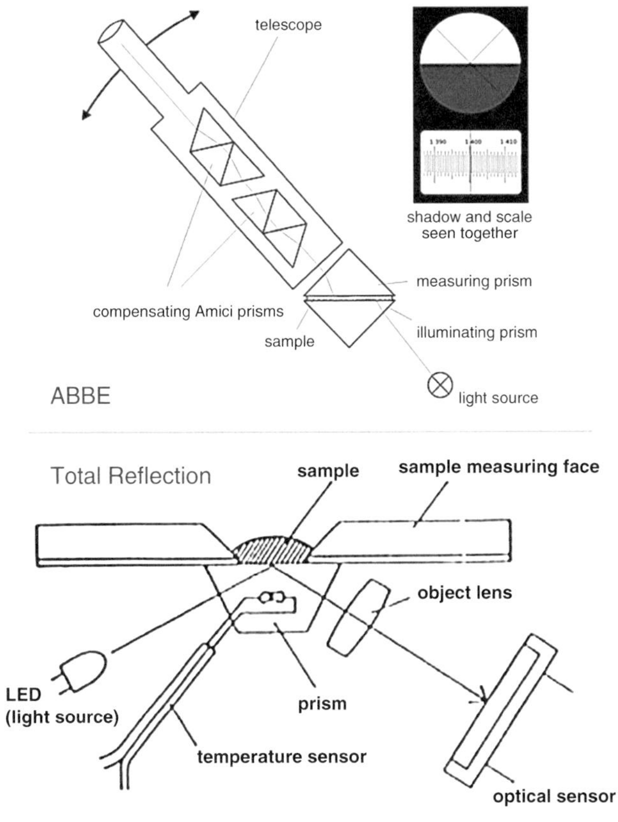

Fig. 5.26 Comparison of the light path in an optical abbe view refractometer and total reflection refractometer. (Source: http://www.acttr.com/en/en-report/en-report-technology/328-en-tech-principle-refractometer-refractive-index.html)

5.7.1 Orifice Viscometers

Orifice viscometers are easy to use manually, by simply dipping into the fluid being analyzed. An orifice viscometer usually consists of a cup with a hole, through which the fluid flows. Viscosity is determined by the time taken by the cup to empty and is measured in cup seconds. Examples include Zahn cups, Ford cups, etc.

5.7.2 Capillary Viscometers

They are also known as U-tube viscometers or glass viscometers. Viscosity is measured by timing how long it takes for a transparent or translucent fluid to flow between two points of a capillary tube. Capillary viscometers include the Ostwald, Ubbelohde variations, Cannon-Fenske capillary, and Houillon capillary viscometers. They are simple to use, consisting of a U-shaped glass tube with two bulbs (one higher and one

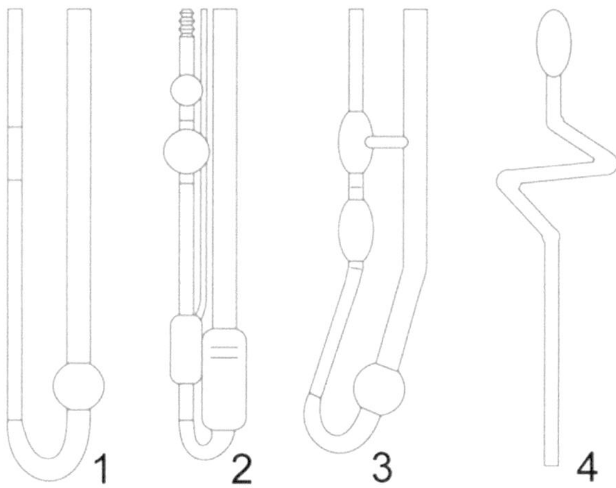

Fig. 5.27 Different types of capillary viscometers. *1*. Ostwald capillary (named after Wilhelm Ostwald), *2*. Ubbelohde capillary (named after Leo Ubbelohde), *3*. Cannon-Fenske capillary (a modified version of Ostwald capillary used for opaque liquids. This reverse-flow type was designed by Dr. Michael R. Cannon and Merrell Robert Fenske), *4*. Houillon capillary. (Source: https://wiki.anton-paar.com/en/how-to-measure-viscosity/)

lower). Fluid passes from the higher bulb to the lower bulb through a capillary, and viscosity is measured by the time taken by the fluid to pass through the tube. Capillary viscometers can be manual or automatic. Automatic instruments use infrared optical sensors for transparent samples or thermal sensors for opaque samples.

Ostwald viscometer is based on Poiseuille's law, where:

$$\text{Viscosity}, \eta = \frac{\pi t p r^4}{8 v l}$$

t = flow in seconds through capillary, p = hydrostatic pressure in dyne/cm^2, r = radius of capillary in cm, v = volume of liquid in ml, l = length of the capillary in cm, and η = viscosity coefficient in poise. (The SI unit of viscosity is pascal.second (Pa.s), i.e., kg.m^{-1}.s^{-1} and the cgs unit of viscosity is poise, i.e., g.cm^{-1}.s^{-1}) (Fig. 5.27).

5.7.3 Falling Piston Viscometers

Falling piston viscometers were invented by Austin Norcross; hence, they are also known as Norcross viscometers. Falling piston viscometers are easy to use and simple to maintain and offer long product life. Falling piston viscometers measure the resistance to the piston falling through the material. They function by drawing the fluid being measured into the piston cylinder while the piston is raised; and the time taken by the piston to fall (time-of-fall seconds) due to the resistance of the fluid is used as a measure of viscosity.

Apart from capillary viscometers driven by gravity, there are also pressurized devices in use. The three different ways of applying pressure include by means of using a weight, a motor (electric), or a gas (inert gas, e.g., nitrogen or pressurized air) pressure (Mezger 2011).

5.7.4 Rotational Viscometers

Rotational viscometers push the upper limit of the potential measuring range further than gravity-based devices. They use a motor drive, which is significantly stronger than the earth's gravitational force. Therefore, they are suited for measuring more highly viscous substances. The resulting quantity is dynamic viscosity, sometimes also referred to as shear viscosity.

Simple rotational viscometers, also known as Brookfield-type viscometers, use a torsion spring to measure the torque required to rotate a spindle in the material. Changing rotor speed and size allow for the measurement of different ranges of viscosity. Rotational viscometers use the Stabinger measuring principle. A light-weight magnetic rotor floats in a liquid-filled tube, which rotates at constant speed. The sample fluid's viscous forces drive the rotor, leading to an equilibrium rotor speed, which is recorded as a measure of the fluid's viscosity.

Different types of rotational instruments (by the way the torque measurement) include the spring instruments and servo motor instrument. In a spring viscometer, a spiral spring connected to both the motor and the rotating bob registers the motor torque as a measure of viscosity. In a servo motor viscometer, the motor current is proportional to the motor torque, which represents the viscosity of the sample.

Different spindle geometries such as cylindrical, coaxial cylinder, parallel plate, and cone plate are used in rheometers to create different shear forces. Rheometers (Sect. 5.8) are more complex instruments than viscometers and are used to measure flow rate-dependent viscosity in non-Newtonian fluids (Macosko 1994).

5.7.5 Falling Ball/Rolling Ball Viscometers

A falling ball/rolling ball viscometer does not measure a liquid's flow time, but the rolling or falling time of a ball. Gravity acts as the driving force. A ball rolls through a closed, sample-filled capillary that is inclined at a defined angle. Inductive sensors determine the ball's rolling time between two marks. In a similar way, falling ball or sphere viscometers measure the time it takes for a ball or sphere to fall under gravity through a sample-filled tube inclined at an angle.

5.7.6 Vibrational Viscometers

Vibrational viscometer uses a powered vibrating rod to measure viscosity. Different fluids are more or less resistant to vibrations depending on viscosity. Therefore, by measuring the reduction in the vibration, or by measuring rate at which the vibration of the viscometer lowers, is used as an index to measure viscosity. Vibrational viscometers are very popular and offer high sensitivity.

5.8 Rheometer

Rheology is the study of flow of matter, in the case of liquids, and also solids in conditions in which they flow, rather than deform elastically. The definition of rheology provided by the Society of Rheology (USA) describes it as the science of the deformation and flow of matter. The term rheology was coined in 1920s, by Professor E C Bingham at Lafayette College in Indiana, inspired by a Greek quotation, "panta rei," meaning "everything flows." In practice, rheology includes the disciplines of elasticity and (Newtonian) fluid mechanics to materials whose mechanical behavior cannot be explained with classical theories. In simple terms, rheology is the study of stress-deformation relationships.

The relationship between stress and deformation is a property of the material

$$\text{Viscosity} = \frac{\text{Stress}}{\text{Shear rate}}$$

$$\text{Modulus} = \frac{\text{Stress}}{\text{Strain}}$$

Viscometers are directly related to fluid measurement analysis, whereas the current rheometers are almost as diverse as the substances and forces they measure. It applies to a variety of substances such as mud, sludges, suspensions, polymers, food products, body fluids, and other biological materials. A rheometer works simply by relating a material property from how hard it's being pushed to how far it moves:

- By commanding torque (stress) and measuring angular displacement (strain)
- By commanding angular displacement (strain) and measuring torque (stress)

Thus, the rheometer is an instrument that measures both viscosity and viscoelasticity of fluids, semisolids, and solids.

In a rheometer, the stress is calculated from the torque. The formula for stress is

$$\text{Stress}, \sigma = M \times K\sigma$$

where σ = stress (Pa or Dyne/cm^2), M = torque in N.m or gm.cm, and $K\sigma$ = stress constant. The stress constant, $K\sigma$, is a geometry dependent factor.

The formula for strain is

$$\text{Strain}, \gamma = K\gamma \times \theta$$

where γ = strain, $K\gamma$ = strain constant, and θ = angular motor deflection (radians). The strain constant, $K\gamma$, is a geometry dependent factor.

5.8.1 Rotational Rheometers

Current rheometers are almost as diverse as the substances and forces that they measure. The main category of rheometers is the rotational or shear rheometers, which rely on rotational motion to achieve a shearing effect. In order to achieve the shearing effect in a rheometer, one must apply a force to the sample through rotation or other movement. The first method was introduced in 1888 by Maurice Couette, and the second method was introduced in 1912 by George Searle. Using the two methods of measurement, rotational rheometers can be classified into four different categories.

- A simple dynamic shear rheometer or DSR can be used to measure the characteristics of a variety of materials under varying conditions of temperature, stress, and strain.
- Rotational cylinder rheometers have two cylinders, one inside the other where the inside rotates at a known speed for calculating shear stresses on a sample. It is sometimes referred to as the "Couetter" rheometer.
- A pipe or capillary rheometer runs liquid through a tube with set dimensions and flow rates to calculate the shear rate on the tube.
- Parallel plate or cone and plate rheometers use rotating plates or a shallow cone and plate with liquid between them to measure the shearing forces on the sample.

Rotational rheometers and viscometers share the same operating principle, but the former have far greater functionality. The accuracy and range over which shear stress can be applied facilitate oscillatory testing and the level of control over the normal force applied during rotational testing. Samples are loaded between two plates or other similar geometries such as cone and plate or alternatively a cup and bob system. The application of a torque to the top plate exerts a rotational shear stress on the material, and the resulting strain or strain rate (shear rate) is quantitatively measured.

Rotational rheometers are used for a broad range of sample types from pastes and gels to the most weakly structured liquids. Applied shear can be precisely controlled into the very low shear stress region making these instruments suitable for stability studies and the measurement of yield stress. However, rotational rheometers are optimized for operation across many decades of torque rather than for the

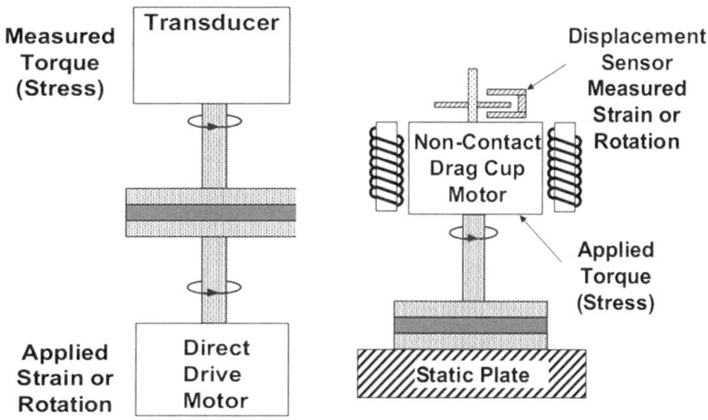

Fig. 5.28 Schematic diagram of a strain-controlled and stress-controlled rotational rheometers. (Source: https://polymerinnovationblog.com/rheology-thermosets-part-2-rheometers/)

precise differentiation of viscosity in low viscosity, weakly structured fluids. In addition, rotational rheometers face mechanical limitations in the high shear region, at shear rates, in excess of 1000 s^{-1} (Fig. 5.28).

Different rheometers use any one of these methods in order to complete their task of measuring samples. Rheological measurements provide information about storage stability, processability, or flow properties of liquid and semisolid formulations, which are essential for quality control and further product development.

5.8.2 Eccentric Viscoelastic Rheometry

Professor Heather A. Cirka and Professor Kristen L. Billiar patented (publication #US20140102180 A1, 2014 and 2016) a new rheometric technique for biological samples. This new technique called eccentric viscoelastic rheometry allows for more accurate measurements of subtle viscoelastic properties in smaller, uniform, or anisotropic samples. Quantifying the viscoelastic characterizations of soft tissue biopsies and fibrous protein gels helps in understanding the complex systems at work in tissue development, wound healing, disease progression, and cell mediation of engineered tissues (Cirka et al. 2012). Additionally, eccentric viscoelastic rheometry can help in analyzing the viscoelastic behavior of newly formed tissues and protein gels that are often unstable and cannot be measured by standard rotational rheometry.

5.9 Texture Analyzer

Texture analyzers, or texturometers, are instruments used to apply scientific methods to the measurement and analysis of product texture. Food texture, through touch, is very important factor for the customer, in addition to its taste and smell.

The equipment enables to evaluate and control the factors affecting quality, processing, handling, and shelf-life, enabling the producer to understand consumer habits and acceptance criteria.

The principle of a texture measurement system involves physical deformation of a test sample in a controlled manner and measurement of its response. The characteristics of the force response are as a result of the sample's mechanical properties, which correlate to specific sensory texture attributes. A texture analyzer applies this principle by performing the procedure automatically and interpreting the results by a software.

The International Organization for Standardization (ISO) defines food texture (in the mouth) as "all the mechanical, geometrical, surface and body attributes of a product perceptible by means of kinesthetic and somesthetic receptors, and (where appropriate) visual and auditory receptors from the first bite to final swallowing" (ISO 5492:2008). According to the terminology of ISO 5492:2008, mechanical attributes are related to the reaction of the product to stress, *viz.*, hardness, cohesiveness, adhesiveness, viscosity, and elasticity; whereas geometrical attributes are related to the shape, size, and arrangement of the particles within a product. Surface attributes involve the sensations produced in the mouth by the moisture and/or fat in and near the surface of the products, and body attributes are those related to the sensations produced in the mouth by the moisture and/or fat in the substance of the products and the way in which these constituents are released.

Szczesniak (2002) defines food texture as "the sensory and functional manifestation of the structural, mechanical and surface properties of food detected through the senses of vision, hearing, touch and kinesthetics."

5.9.1 Instrumentation

Texture profile analysis (TPA) employs a double compression test for determining the textural properties of foods. Cosmetics and pharmaceutical products are also sensory dependent and benefit greatly from a texture analysis capability. During a TPA test, samples are compressed twice using a texture analyzer to provide insight into mimicking a chewing/biting action of mouth. For this reason, the TPA test was often called the "two bite test." The texture of any food is multifaceted and tied to consumers' sensory expectations.

The attractiveness of TPA as an analytical method is that it can quantify multiple textural parameters in an experimental run. A texture analyzer has the capacity to replace human sensory evaluation by providing a numerical value which serves as a quality assurance standard by which products can be evaluated (Fig. 5.29).

Many instrumental measurements are correlated with sensory descriptors and consumer preferences and can be used as predictors of consumer acceptability (Abbott 1999). For this reason, instrumental texture analysis is a well-established analytical technique for evaluating mechanical and physical characteristics of both raw ingredients and finished products in the food industry (Luca et al. 2012). The different parameters of texture and their evaluation using TPA are discussed in Sect. 5.2.

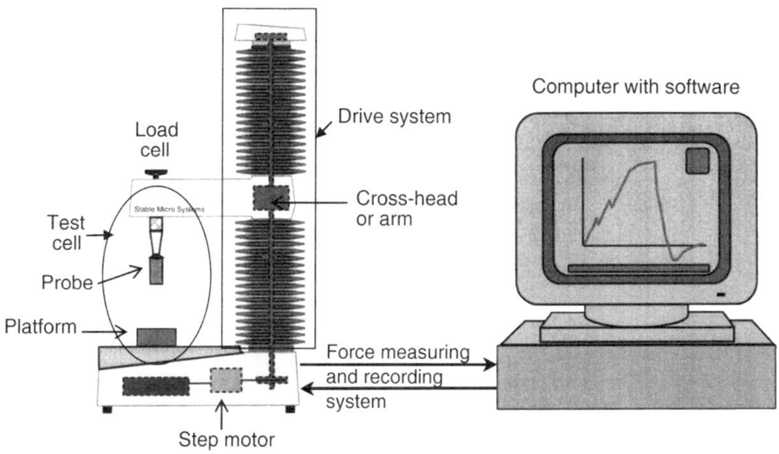

Fig. 5.29 Schematic drawing of a universal testing machine. (Reprinted by permission from Letaief 2007) (Luca et al. 2012; http://www.ajevonline.org/content/63/1/11.figures-only)

5.10 Differential Scanning Calorimeter

The differential scanning calorimeter (DSC) is a vital tool for thermal analysis. Differential scanning calorimetry (DSC) monitors heat effects associated with phase transitions and chemical reactions as a function of temperature (Charu et al. 2017). The DSC technique was initially developed by Watson, E.S., and O'Neill, M. J., in the 1960s.

5.10.1 DSC Instrumentation

DSC instrument measures temperature and heat flows associated with thermal transitions in a material, and therefore the difference in heat flow to the sample and a reference at the same temperature are recorded as a function of temperature. An inert material, such as alumina or an empty aluminum pan, is used as reference. Differential scanning calorimetric (DSC) thermograms are obtained by increasing the temperature of both the sample and reference at a constant rate, using an automatic thermal analyzer system. Generally, temperature calibration can be performed using indium calibration reference standard (transition point: 156.60 °C). Samples are heated in standard aluminum pans and heated from 30 to 500 °C at a heating rate of 10–20 °C/min under a constant purging of a suitable gas, e.g., dry nitrogen at 10–20 mL/min. An empty aluminum pan, sealed in the same way as the sample, is used as a reference. Pans of Al, Cu, Au, Pt, alumina, and graphite are also used based on the sample nature, chosen to avoid reactions with samples and with regard to the temperature range of the measurement. Analytic gas atmospheres of nitrogen, air, oxygen, argon, vacuum, and controlled mixed gases are also employed for the analyses.

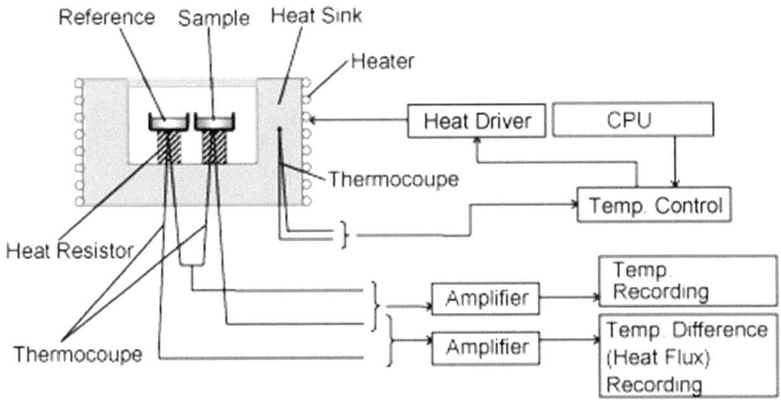

Fig. 5.30 Schematic representation of a Heat flux DSC. (Source: https://www.hitachihightech.com/global/products/science/tech/ana/thermal/descriptions/dsc.html)

DSC analysis is used to measure different thermal parameters of a sample. These include the melting temperature, heat of fusion, latent heat of melting, reaction energy and temperature, glass transition temperature, crystalline phase transition temperature and energy, precipitation energy and temperature, denaturation temperatures, oxidation induction times, and specific heat or heat capacity. The information is used to understand the amorphous and crystalline behavior, polymorph and eutectic transitions, curing and degree of cure, and many other material properties used to design, manufacture, and fabrication of products. DSC instrument is commercially available in two variants: heat flux type and power compensation type. Heat flux DSC utilizes a single furnace. Heat flow into both the sample and reference material is via an electrically heated constant thermoelectric disk and hence is proportional to the difference in output of the two thermocouple junctions. In a power compensation DSC, the temperature difference between the sample and the reference are compensated by varying the heat required to keep both pan at the same temperature. The energy difference is plotted as a function of sample temperature (Fig. 5.30).

5.11 CHNS-O Elemental Analyzer

CHNS-O elemental analyzer is employed for the simultaneous determination of the percentage (%) composition of carbon, hydrogen, nitrogen, sulfur, and oxygen, contained in an organic or inorganic material in a solid or liquid state. They are capable of handling an extensive variety of sample types, including solids, liquids, volatile and viscous samples, polymers, and chemicals, in the fields of pharmaceuticals, environmental monitoring, food composition, and energy analysis (Michael 2008). The elementary analyzer works on the principle of catalytic tube combustion in an oxygenated CO_2 atmosphere at high temperatures. Initially, the combustion gases are freed from foreign gases (for instance, volatile halogen). Helium (He) serves as flushing and carrier gas. The desired measuring components are separated from each other with the help of a specific adsorption column and determined in sequence with a thermal conductivity detector (TCD).

Combustion elemental analyzers are manufactured in a variety of configurations suitable for specific applications, and the choice will depend on the elements of interest, the sample type and size, and the concentration of the analyte. The analyzers are often constructed in modular form such that they can be employed in a number of different configurations to determine, for example, CHN, CHNS, CNS, or N, depending on the application.

5.11.1 Instrumentation

The main components of a CHNS-O elemental analyzer include (Fig. 5.31):

- Autosampler
- Adsorption filters
- Left and right furnaces (temperature up to ca. 1000 °C)
- Left reactor (combustion chamber for CHNS determination)
- Right reactor (pyrolysis chamber for O determination)
- Gas chromatographic column
- Thermal conductivity detector (TCD)
- Data analysis software

5.11.2 Principle of an Elemental Analyzer

CHNS analysis requires high-temperature combustion in an oxygen-rich environment. The combustion can be carried out under both static conditions by introduction of a set volume of oxygen or by dynamic conditions at a constant flow of oxygen for a set period of time. In the "static" type, a preset volume of oxygen is added to the combustion tube before the sample is introduced. In the "dynamic" type, a constant flow of oxygen is set for a specific period of time, at the same time as the sample is introduced. Often, catalysts are also added to the combustion tube in order to aid conversion.

Procedure The sample weighted (in mg) in a tin capsule is dropped into the left reactor for a complete combustion. The combustion products CO_2, SO_2, and NO_2 are carried by a constant flow of carrier gas (helium) that passes through a glass column packed with an oxidation catalyst of tungsten trioxide (WO_3) and a copper reducer, which are kept at 1000 °C. At this temperature, the nitrogen oxide is reduced to N_2. The N_2, CO_2, and SO_2 are then transported by the helium into the GC column for

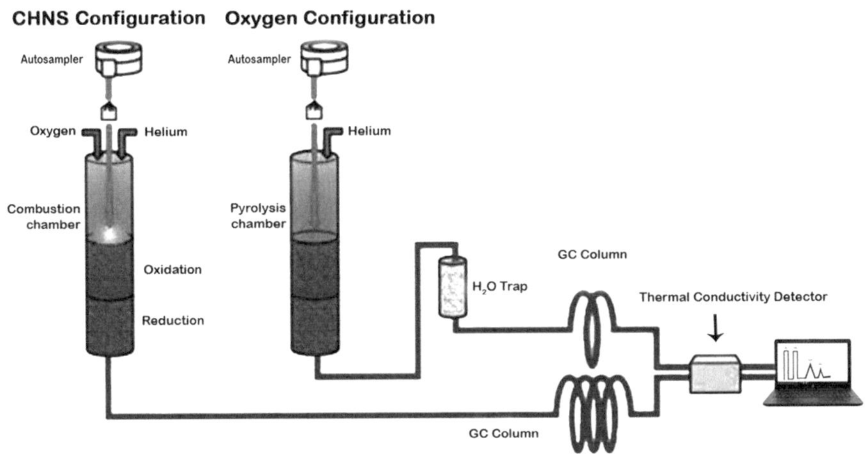

Fig. 5.31 Schematic representation of a CHNS-O analyzer. (Source: https://www.thermofisher.com/blog/materials/characterization-of-polymers-and-plastics-using-an-elemental-anaylser/)

separation, and the combustion mixture will flow and be detected by thermal conductivity detector (TCD – set at ca. 290 °C), in the sequence of N_2, CO_2, H_2O, and SO_2. The interference of phosphorus can be controlled by the addition of vanadium pentoxide (V_2O_5) into the tin capsule, as an oxidizer. The sample weighted (in mg) in a silver capsule is dropped into the right reactor where it undergoes instant pyrolysis, the gas mixture will cross the adsorption filter and flow into the GC column, and the eluted gases are detected by thermal conductivity detector (TCD) to determine oxygen content.

5.12 X-Ray Diffraction

X-ray diffraction (XRD) or X-ray powder diffraction (XRPD) analysis is a unique method in determination of crystallinity of a compound. The technique was pioneered by Max von Laue in 1912, who reported that crystalline substances act as a diffraction grating for X-ray wavelengths similar to the atomic-scale plane spacing in a crystal lattice. XRD analysis depends on the fundamental principles of electron beam and X-ray interactions with solid materials. XRD is the only laboratory technique that reveals structural information, such as chemical composition, crystal structure, crystallite size, strain, preferred orientation, and layer thickness. Therefore, XRD is an indispensable method for material characterization and quality control. XRD is used to analyze a wide range of materials, from solids to thin films and nanomaterials. XRD analysis is based on constructive interference of monochromatic X-rays and a crystalline sample. The X-rays are generated by a cathode ray tube, filtered to produce monochromatic radiation, collimated to concentrate, and directed toward the sample. The interaction of the incident rays with the sample produces constructive interference (and a diffracted ray) when conditions satisfy Bragg's law (Klug and Alexander 1974; Cullity 1978).

5.12.1 Bragg's Law

Bragg's law is the general relationship between wavelength of the incident X-rays, angle of incidence, and spacing between the crystal lattice planes of atoms. It is expressed as n $\lambda = 2d \sin \theta$, where n (an integer) is the "order" of reflection, λ is the wavelength of the incident X-rays on the crystallize sample, d is the interplanar spacing between rows of atoms in the crystal, and θ is the angle of incidence with respect to these planes (Fig. 5.32).

When this equation is satisfied, X-rays scattered by the atoms in the plane of a periodic structure are in phase, and diffraction occurs in the direction defined by the angle θ. The law relates the wavelength of electromagnetic radiation to the diffraction angle and the crystal lattice spacing in a sample. In the simplest instance, an X-ray diffraction experiment consists of a set of diffracted intensities and the angles at which they are observed. The two primary types of XRD analysis include single-crystal XRD and X-ray powder diffraction. Single-crystal X-ray diffraction is used most commonly for determination of unit cell dimensions and the position of atoms within a crystal lattice.

5.12.2 General Principle and Instrumentation

X-rays are electromagnetic radiation with wavelengths between about 0.02 and 100 A (1A = 10^{-10} meters). The wavelength of X-rays is on an atomic level and is much smaller than visible light (3000–8000 A). Since X-rays have a smaller wavelength than visible light, they have higher energy and are more penetrative. Its

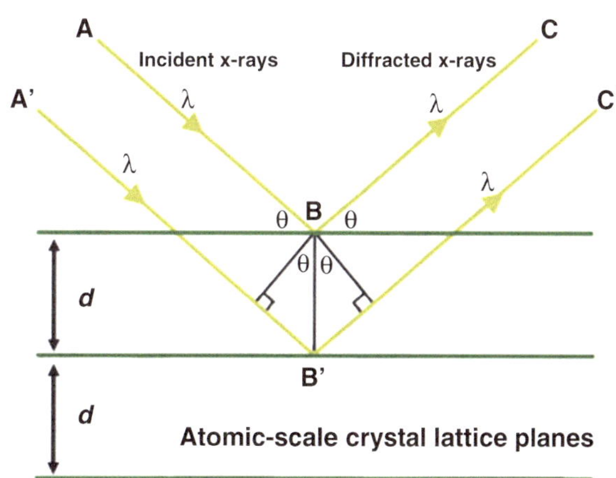

Fig. 5.32 Bragg's law of reflection. The diffracted X-rays exhibit constructive interference when the distance between paths ABC and A′B′C′ differs by an integer number of wavelengths (λ). (https://serc.carleton.edu/research_education/geochemsheets/BraggsLaw.html)

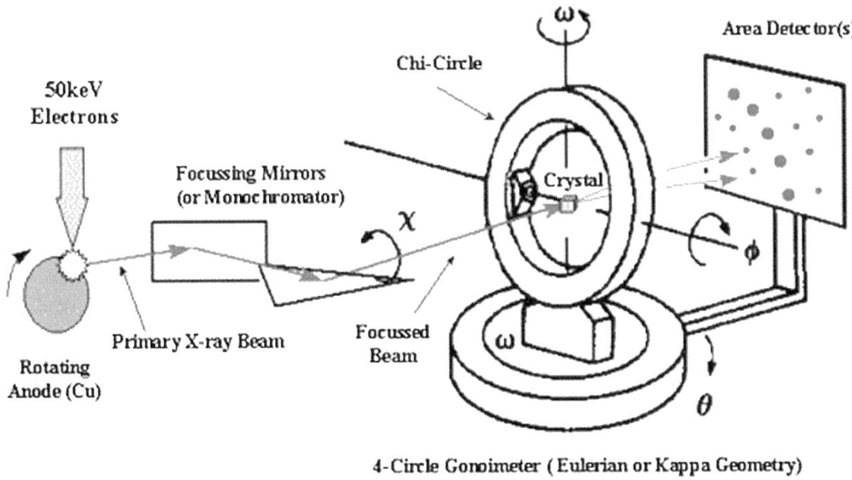

Fig. 5.33 Schematic representation of an X-ray detector (http://pruffle.mit.edu/atomiccontrol/education/xray/xray_diff.php)

ability to penetrate matter irrespective of its density and hence X-rays is useful in exploring structures of atoms (Fig. 5.33).

X-rays are produced in a device called an X-ray tube. The X-ray tube consists of an evacuated chamber with a tungsten filament at one end, called the cathode, and a metal target at the other end, called an anode. When electrical current is put through the filament, excited electrons from the tungsten are emitted. If a high potential difference is placed between the cathode (positive end) and the anode (the negative end), the emitted electrons move at high velocity from the filament to the anode target. The electrons strike the crystal atoms at the target, dislodging inner shell electrons of the target atom. As a result, electrons on the outer shell emit jump down to fill the void in the inner shell. Because the inner shells have lower energy than the outer shells, when an outer shell electrons jump to inner shells, they emit radiation in the form of high-energy X-rays. The characteristic X-ray diffraction pattern generated in a typical XRD analysis provides a unique "fingerprint" of the crystals present in the sample. When properly interpreted, by comparison with standard reference patterns and measurements, this fingerprint allows identification of the crystalline form.

5.12.3 Powder XRD

Reorienting the crystal to every plane and measuring the diffraction angle *theta* to determine the d-spacing of each plane is time-consuming. Thus, the powder method is often used to address the process. In this method, the crystal of study is ground into fine powder which contains pure homogenized granules of crystals with random orientations. The crystal powder randomly oriented on the surface has different atomic planes that lie parallel to the surface. Therefore, when an incident X-ray beam scatters from 0 to 90°, all the diffracted angles correspond to a different atomic spacing. A goniometer is used to rotate both the X-ray tube and the detector through the 90° angle. The instrument used to make these measurements is the X-ray powder diffractometer (Bish and Post 1989).

X-ray powder diffraction is widely used in material science, environmental science, geology, and engineering to rapidly identify unknown crystalline substances (Eby 2004; Michael et al. 1991). Additional uses include detailed characterization of crystalline samples, determination of unit cell dimensions, and quantitative determination of modal amounts of minerals in a sample. X-ray powder diffraction can also be applied to the identification of fine-grained minerals (Putnis 1992). The technique can also be used to distinguish between different clays and sediments that are optically similar (Moore and Reynolds 1997).

Other techniques that operate on the principles of X-ray interactions include X-ray spectroscopy (through energy-dispersive X-ray spectroscopy (EDX)) and wavelength-dispersive spectroscopy (WDS) typically performed using a SEM or EPMA and X-ray fluorescence (XRF) analyses.

5.13 Enzyme-Linked Immunosorbent Assay (ELISA)

Fish species identification has been mostly performed by genetic (PCR-based) and immunological techniques that are suitable for routine analysis of a large number of samples for species authentication (Asensio 2007; Carrera et al. 2014), in addition to the detection of spoilage microorganisms and pathogens. Among immunological methods, the ELISA technique has been the most widely used assay for regulatory purposes in detecting fish species adulteration, because of its simplicity, specificity, and sensitivity (Asensio et al. 2008). ELISA stands for enzyme-linked immunosorbent assay, also referred to as enzyme immunoassay (EIA). ELISA techniques are less expensive and simpler to develop than DNA-based methods.

Even though genetic methods are the most sensitive and specific methods for species identification, the techniques require expensive laboratory equipment and a certain degree of expertise, in addition to their limitations in field tests in a market or landing center. As an alternative, immunological assays can be used as field assays and can relatively reduce the test time and expense. ELISA may prove to be useful even in heat-sterilized products, although they are ineffective at differentiating unrelated species and require the development of antibodies directed against the specific protein of interest.

The technique enables the quantification of a desired marker in a biological sample. The marker can be an antibody, a hormone, a peptide, or a protein. Like other types of immunoassays, an ELISA relies on antibodies to detect a target antigen using highly specific antibody-antigen interactions. ELISA is typically performed in a multi-well plate (96- or 384-wells), thereby enabling to perform multiple samples simultaneously. The microtiter plate provides the solid surface to immobilize the antigen. Immobilization of the analytes facilitates separation of the antigen from the rest of the components in the sample. The antigen is then complexed to a detection antibody conjugated with a molecule amenable for detection such as an enzyme or a fluorophore.

5.13.1 Types of ELISA

ELISA can be performed in different formats with each technique having its own merits and demerits.

5.13.1.1 Direct ELISA

Direct ELISA is considered to be the simplest type of ELISA. This was the ELISA originally developed by Perlmann and Engvall. The surface of the plate is coated with the sample antigen, which is bound by an enzyme-tagged antibody. Incubation is followed by washing, which removes the unbound antibodies from the medium. An appropriate substrate is then added to the medium, producing a signal directly proportional to the amount of antigen in the sample. This correlation can be used to extrapolate the concentration of antigen in an unknown sample from a standard curve. Direct ELISA is appropriate for determining the amount of high-molecular-weight antigens. Only a fewer steps are required making it considerably faster than other types of ELISA. Cross-reactivity of the secondary antibody is also eliminated. However, in direct ELISA, the direct labeling of primary antibodies is time-consuming and expensive and may adversely affect the immune reactivity of the antibody with the targeted antigen (Fig. 5.34).

5.13.1.2 Indirect ELISA

Indirect ELISA is a two-step binding process involving the use of a primary antibody and a labeled secondary antibody. Initially, the primary antibody is incubated with the antigen-coated wells followed by a secondary antibody that recognizes the primary antibody. This secondary antibody is often a polyclonal anti-species antibody. A wide variety of labeled secondary antibodies are prevailing. A substrate is then added to the well to produce a signal amplification. This method is commonly utilized to diagnose infection by bacteria, virus, or a parasite and quantify antibodies against this foreign antigen. Indirect ELISA detection is versatile as different visualization markers can be used with the same primary antibody. Since more than one labeled antibody can be bound per antibody target, indirect ELISA is deemed to

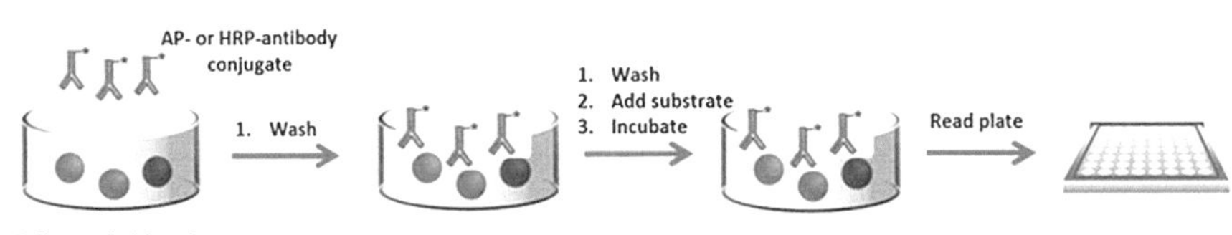

Fig. 5.34 Schematic representation of direct ELISA technique. (Source: http://www.enzolifesciences.com/science-center/technotes/2017/april/what-are-the-differences-between-elisa-assay-types?/)

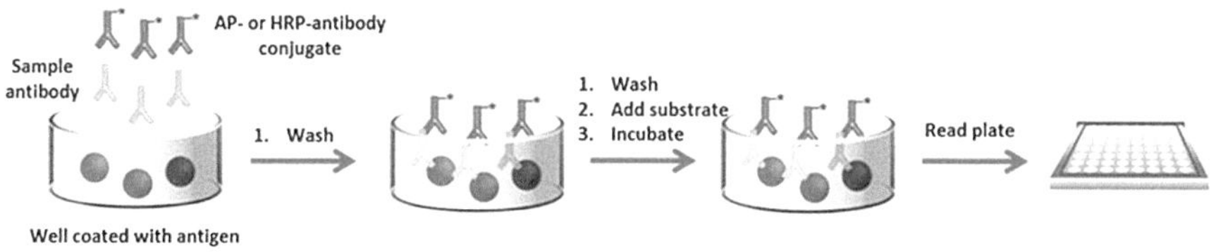

Fig. 5.35 Schematic representation of indirect ELISA technique. (Source: http://www.enzolifesciences.com/science-center/technotes/2017/april/what-are-the-differences-between-elisa-assay-types?/)

be highly sensitive and more flexible than direct ELISA. However, cross-reactivity and a non-specific signal may occur with the secondary antibody (Fig. 5.35).

5.13.1.3 Immunometric/Sandwich ELISA

Immunometric assay, also known as sandwich ELISA, use two antibodies specific to the antigen to capture or "sandwich" antigens in the well for detection. Immunometric assays exhibit a direct correlation between antigen concentration and substrate response. Immunometric assays typically employ a capture antibody coated on the plate to bind the antigen of interest. During a second incubation, the antigen is bound by a second detection antibody that is also specific to the antigen. The detection antibody can either be bound by a secondary antibody-enzyme conjugate or the detection antibody itself is enzyme conjugated. When a chromogenic substrate is added to the assay to develop color, samples with high antigen concentration generate more signal than those with low antigen concentration, producing a signal directly proportional to the amount of antigen in the sample. This correlation can then be used to extrapolate the concentration of antigen in an unknown sample from a standard curve. This type of assay is generally used for low- to high-molecular-weight proteins, such as endothelin-1 (e.g., Endothelin-1 ELISA kit), HSP70 (e.g., AMP'D HSP70 high sensitivity ELISA kit), or survival motor neuron (e.g., SMN ELISA kit). Immunometric ELISAs are highly specific as it relies on a pair of antibodies for capture and detection. They are also considered to be compatible with a variety of complex samples without the prerequisite of sample extraction prior to the analysis (Fig. 5.36).

5.13.1.4 Competitive ELISA

In competitive enzyme immunoassays, the antigen in a sample competes for limited antibody binding sites with antigen conjugated to a reporter enzyme. This produces an inverse relationship between antigen concentration and substrate turnover. Competitive ELISAs typically use a single antibody to a low-molecular-weight antigen, generally less than 10,000 Daltons. During incubation, samples with high antigen content result in unlabeled antigen being bound in greater amounts than conjugated antigen. When chromogenic substrate is added to the assay to develop color, samples with a high antigen concentration generate a lower signal than those containing low antigen concentration, yielding the inverse correlation between antigen concentration in the sample and color development

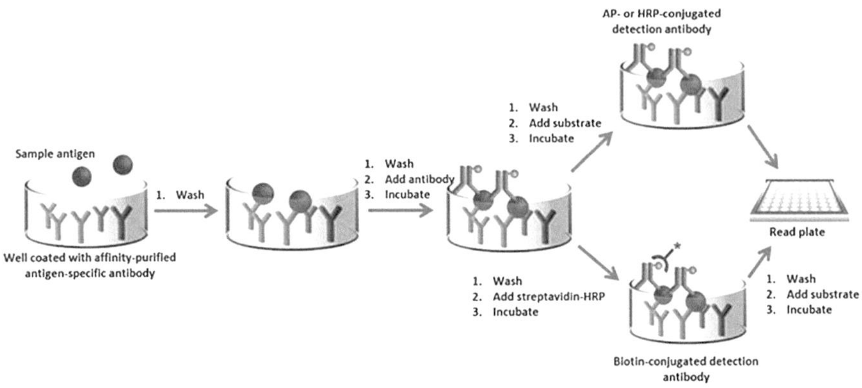

Fig. 5.36 Schematic representation of sandwich ELISA technique. (Source: http://www.enzolifesciences.com/science-center/technotes/2017/april/what-are-the-differences-between-elisa-assay-types?/)

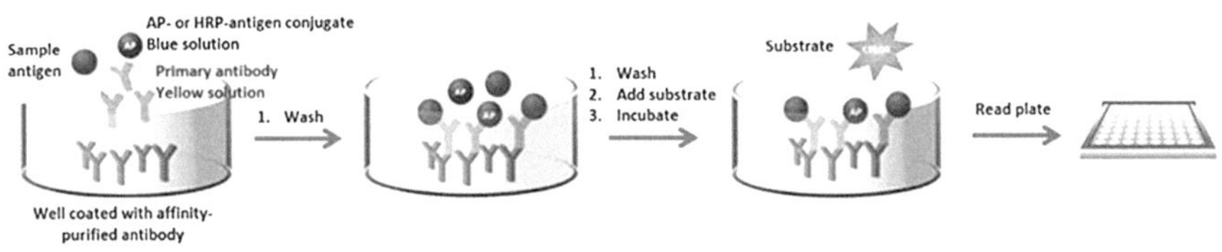

Fig. 5.37 Schematic representation of competitive ELISA technique. (Source: http://www.enzolifesciences.com/science-center/technotes/2017/april/what-are-the-differences-between-elisa-assay-types?/)

in the assay. This relationship can further be used to extrapolate antigen concentration in a test sample from the standard curve. This type of reaction is one of the few methods possible for low-molecular-weight antigens with a limited number of epitopes or antibody-binding sites, such as small molecules (e.g., Direct cAMP ELISA kit), peptides (e.g., Oxytocin ELISA kit), and steroids (e.g., Corticosterone ELISA kit) (Fig. 5.37).

5.14 Polymerase Chain Reaction (PCR)

The authentication of fish and seafood species is an important issue within the seafood industry. Increase in international trade and global consumption of several fish and seafood species has resulted in intentional product mislabeling, and hence, there is a growing need for rapid, reliable, and reproducible tests species authentication. The effects of commercial fish/seafood species substitution are far-reaching enhancing economic fraud, health hazards, and illegal trade of protected species. A variety of DNA-based detection techniques have been developed, including Multiplex PCR, FINS, PCR-RFLP, PCR-RAPD, PCR-AFLP, and PCR-SSCP, which are all based on polymorphisms in the genetic codes of different species (Rosalee and Morrissey 2008).

Even though the analytical diagnosis of fish and seafood by isoelectric focusing (urea – IEF), capillary electrophoresis (CE), SDS-PAGE, high-performance liquid chromatography (HPLC), and immunoassay systems (Moretti et al. 2003) are generally reliable for use with fresh or frozen tissue, analysis of the heat processed or dried food products makes the abovementioned methods impractical (Akasaki et al. 2006) due to changes in the biochemical properties and structural integrity of proteins. Enzyme-linked immunosorbent assay (ELISA) has successfully been used for the identification of several heat-sterilized fish products (Carrera et al. 1997). Yet, immunoassays can be ineffective at differentiating closely related species and require the development of species-specific antibodies (Bartlett and Davidson 1992; Woolfe and Primrose 2004). The DNA-based methods have a number of advantages over protein-based methods, including increased specificity, sensitivity, and reliable performance with highly processed samples. Although DNA molecules can degrade during processing, they are more thermostable than proteins (Rosalee and Morrissey 2008). The DNA barcode (*coxI* gene) was successfully employed in the identification of fish species among 98% of probed marine species and 93% of freshwater species (Ward et al. 2009). Fish adulteration analysis using small portions of fresh or processed fish by using few universal combination primers is also proven to be successful in fish authentication (Steinke and Hanner 2011).

Some DNA-based methods include the use of PCR along with restriction fragment length polymorphism (RFLP), amplified fragment length polymorphism (AFLP), forensically informative nucleotide sequencing (FINS), or single-stranded conformational polymorphism (SSCP).

5.14.1 Conventional PCR Technique

The polymerase chain reaction (PCR) is an artificial method of replicating DNA under laboratory conditions. Polymerase chain reaction (PCR) is a revolutionary method developed by Kary Mullis. The Nobel Prize in Chemistry (1993) was shared by Kary B. Mullis for the invention of the polymerase chain reaction (PCR) technique and also by Michael Smith for his contributions to the study of oligonucleotide-based site-directed mutagenesis. The PCR technique is used to amplify large quantities of a specific sequence of DNA from an initial minute sample. It amplifies a specific DNA (target) sequence lying between known positions (flanks) on a double-stranded (ds) DNA molecule. Each reaction cycle doubles the amount of DNA – a standard PCR sequence of 30 cycles creates over 1 billion copies. The polymerase chain reaction can be used to amplify both double- and single-stranded DNAs.

5.14.2 PCR Amplification Mix

PCR amplification mix typically contains:

- Sample dsDNA with a target sequence.
- Thermostable DNA polymerase: The most commonly used DNA polymerase is Taq DNA polymerase (produced by *Thermus aquaticus*, a thermophilic bacterium) because of high-temperature stability. Pfu DNA polymerase (from *Pyrococcus furiosus*) is also used widely because of its higher fidelity (accuracy of adding complimentary nucleotide).
- Two oligonucleotide primers (the forward primer and the reverse primer): Primers are synthetic DNA strands of about 18–25 nucleotides complementary to 3′ end of the template strand. DNA polymerase starts synthesizing new DNA from the 3′ end of the primer.
- Deoxynucleotide triphosphates (dNTPs of adenine, guanine, thymine, cytosine).
- Reaction buffer, magnesium chloride ($MgCl_2$), and additives: Reaction buffer provides an optimal chemical environment (pH and salt condition) for the polymerase. Mg^{2+} ions in the buffer act as cofactor for DNA polymerase enzyme. Additives such as BSA, DMSO, or formamide are optional and may be used to improve the amplification efficiency and specificity.

5.14.3 Stages of PCR

PCR occurs in a thermal cycler and uses variations in temperature to control the replication process via three steps (Wilson and Walker 2010).

(i) *Heat Denaturation* – DNA sample is heated to separate it into two single strands (~95 °C for 1 min).
(ii) *Primer Annealing* – DNA primers attach to the 3′ ends of the target sequence (~55 °C for 1 min).
(iii) *Strand Elongation* – A heat-tolerant DNA polymerase (Taq) binds to the primer and copies the strand (~72 °C for 2 min).

The different types of PCR generally in use include real-time PCR, quantitative real-time PCR (Q-RT PCR), nested PCR, multiplex PCR, reverse transcriptase PCR (RT-PCR), touchdown PCR, arbitrarily primed PCR, inverse PCR, allele-specific PCR, asymmetric PCR, hot start PCR, core sample PCR, degenerate PCR, in situ PCR, miniprimer PCR, assembly PCR, colony PCR, methylation-specific PCR, variable number of tandem repeats (VNTR) PCR, etc. (Fig. 5.38).

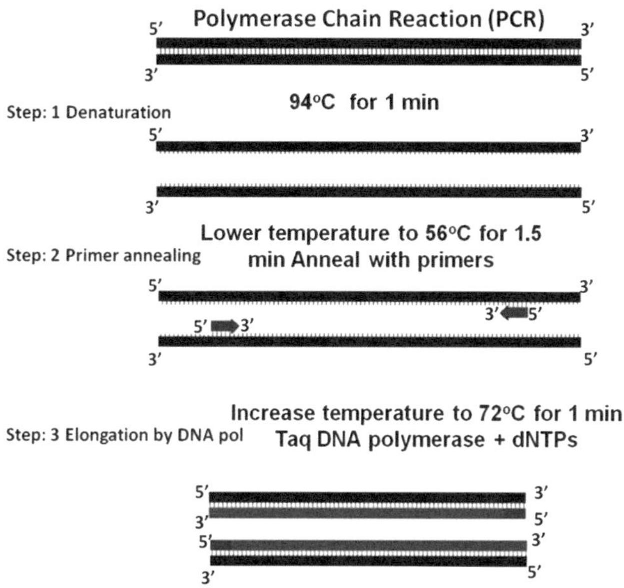

Fig. 5.38 Schematic representation of polymerase chain reaction (PCR)

5.15 Electrical Conductivity Meter

Electrical conductivity/EC of water is its ability to conduct an electric current. Salts or other chemicals that dissolve in water can break down into positively and negatively charged ions. Major positively charged ions that affect the conductivity of water are sodium, potassium, calcium, and magnesium. The major negatively charged ions are chloride, sulfate, carbonate, and bicarbonate. Nitrates and phosphates are minor contributors to conductivity, but they are very important biologically. These free ions in the water conduct electricity, so the water electrical conductivity depends on the concentration of ions. Salinity and total dissolved solids (TDS) are used to calculate the EC of water, which helps to indicate the water's purity. The more pure the water, the lower the conductivity. For example, distilled water is almost an insulator, but salt-water is a very efficient electrical conductor.

The EC of water has a critical influence on aquatic life. Every kind of organism has a typical range that it can tolerate. In a commercial sense, testing the conductivity of water may be specifically useful for the fish/sea food processing industry. Salts and other substances also affect the quality of water for irrigation or drinking, so electrical conductivity meters can be used when testing the output of water treatment plants.

An electrical conductivity meter (EC meter) is an instrument used to measure the electrical conductivity of a solution. It has multiple applications in research and engineering, with common usage in food processing industry, aquaculture, aquaponics, hydroponics, and freshwater irrigation systems to monitor the amount of organics, dissolved salts, or impurities in the water. Water conductivity test equipment can be used to build up an understanding of the typical range of conductivity for a particular water body. This can be used as a baseline for regular monitoring of the water body, and, if there is a significant change in conductivity, this could be an indicator of a discharge or other pollutants entering the water. A change in water electrical conductivity can indicate groundwater seepage or a sewage leak.

5.15.1 Instrumentation

The laboratory conductivity meter employs a potentiometric method having four cylindrical platinum electrodes that are arranged concentrically. An alternating current is applied to the outer pair of the electrodes, and the potential between the inner pair is measured. Conductivity can be determined using the distance between the electrodes and their surface area using Ohm's law, but generally, for accuracy, the calibration is employed using electrolytes with good conductivity.

Industrial conductivity probes often employ an inductive method, which has the advantage that the fluid does not wet the electrical parts of the sensor. Here, two inductively coupled coils are used. One is the driving coil producing a magnetic field, and it is supplied with accurately known voltage. The other forms a secondary coil of a transformer. The liquid passing through a channel in the sensor forms one turn in the secondary winding of the transformer. The induced current is the output of the sensor (Romero et al. 2012) (Fig. 5.39).

Another way is to use four-electrode conductivity sensors that are made from corrosion-resistant materials. Benefit of four-electrode conductivity sensor compared to inductive sensor is scaling compensation and ability to measure low (below 100 μS/cm) conductivities (a feature especially important when measuring near 100% hydrofluoric acid).

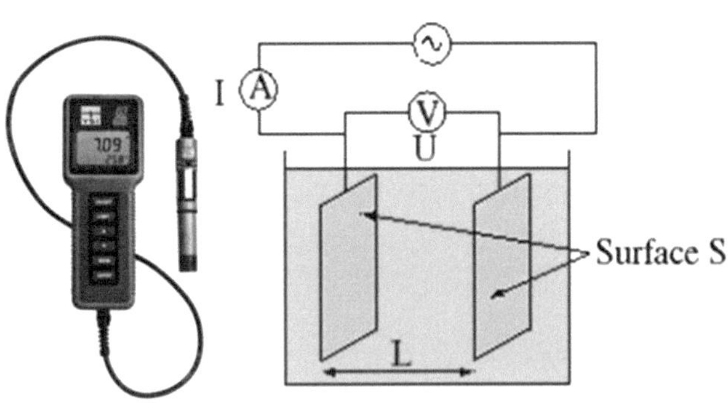

Fig. 5.39 Electrical conductivity meter (EC meter) (https://www.researchgate.net/figure/electrical-conductivity-meter-EC-meter_fig1_280739539)

5.15.2 Factors Affecting Electrical Conductivity

Unlike metals, the conductivity of common electrolytes typically increases with increasing temperature. Since the conductivity of a solution is highly temperature dependent, it is therefore important to either use a temperature compensated instrument or calibrate the instrument at the same temperature as the solution being tested. Over a limited temperature range, the way temperature affects the conductivity of a solution can be modeled linearly, using the following formula:

$$\sigma_{\mathrm{T}} = \sigma_{T_{\mathrm{cal}}} \left[1 + \alpha \left(T - T_{\mathrm{cal}}\right)\right]$$

where T is the temperature of the sample, T_{cal} is the calibration temperature, σ_{T} is the electrical conductivity at the temperature T, σ_{Tcal} is the electrical conductivity at the calibration temperature T_{cal}, and α is the temperature compensation slope of the solution.

5.16 Geiger-Muller Counter

A Geiger counter (Geiger-Muller tube) is a device used for the recognition and measurement of alpha, beta, and gamma radiation. Basically, a Geiger-Muller counter is a type of gaseous ionization detector. The Geiger tube (or GM tube) is the sensing element of a Geiger counter instrument that can detect a single particle of ionizing radiation and typically produce an audible signal upon detection. Hans Geiger and Rutherford, E. invented the device in 1908 (Rutherford and Geiger 1908), and Walther Muller collaborated with Geiger in developing it further in 1928.

5.16.1 Principle of a GM Counter

Ionizing radiations such as an alpha, beta, or gamma particle ionize some of the gas molecules in the tube. An electron is knocked out from these ionized atoms, and the remaining atom is positively charged. The high voltage in the tube produces an electric field inside the tube, where the electrons that were knocked out of the atom are attracted to the positive electrode and the positively charged ions are attracted to the negative electrode. The ion pairs subsequently gain sufficient energy to ionize further gas molecules through collisions during their passage and create a cascade (avalanche) of charged particles. This produces a pulse of current in the connecting wires which passes from the negative electrode to the positive electrode, and the generated pulse is measured by the GM counter. After the pulse is counted, the charged ions become neutralized, and the Geiger counter is ready to record another pulse. The pulse is read by the user through a visual readout in the form of a traditional analog meter or an electronic LCD (liquid crystal display) digital readout. These meters are available in different units, including milliroentgens per hour

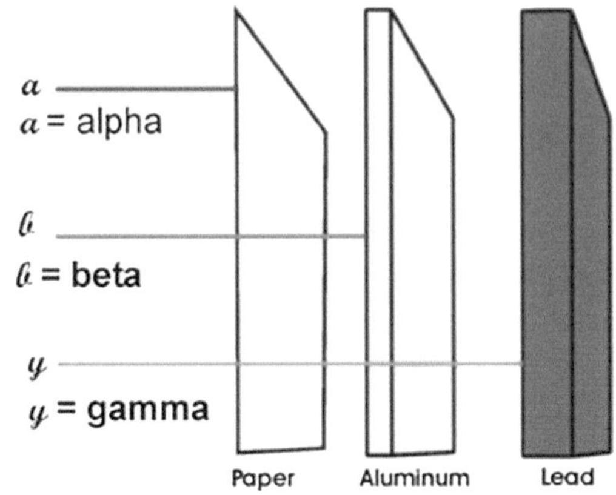

Fig. 5.40 Penetration power of radiation. (Source: https://www.imagesco.com/geiger/about-geiger-counters.html)

(mR/h) (popular in the United States and Israel) and microsieverts per hour (μSv/h) (popular in Canada and other countries) (Fig. 5.40).

5.16.2 Geiger-Muller Tube

A Geiger-Muller tube typically consists of a gas tube filled with a low-pressure (~0.1 atm) inert gas such as neon, argon, or helium and in some cases a penning mixture (named after Frans Michel Penning, it is a mixture of gases used in electric lighting or displaying fixtures) and an organic vapor or a halogen gas. The usual form of GM tube is an "end-window tube." This type is named so, because the tube has a window at one end through which ionizing radiation can easily penetrate. The other end normally has the electrical connectors. There are two types of end-window tubes: the glass-mantle type and the mica window type. The glass window type does not detect alpha radiation, since it is unable to penetrate the glass, but is usually cheaper and detects beta radiation and X-rays. The mica window type detects alpha radiation but is more fragile. Most tubes will detect gamma radiation and usually beta radiation above about 2.5 MeV (Fig. 5.41).

The walls of the tube are entirely metal or nonmetal with their inside surface coated with a conductor to form the cathode, while the anode is a wire passing through the center of this tube. The electrodes within the GM tube have a potential difference of several hundred volts, but without current flowing. In order for the Geiger counter tube to restore itself quickly to its original state after radiation has entered, a gas is added to the tube.

Geiger-Muller tubes will not normally detect neutrons since these do not ionize the gas. However, neutron-sensitive tubes can be produced which either have the inside of the tube coated with boron or contain boron trifluoride or helium-3 gas. The neutrons interact with the boron nuclei, producing alpha particles or with the helium-3 nuclei producing hydrogen and tritium ions and electrons. These charged particles then trigger the normal avalanche/cascade process.

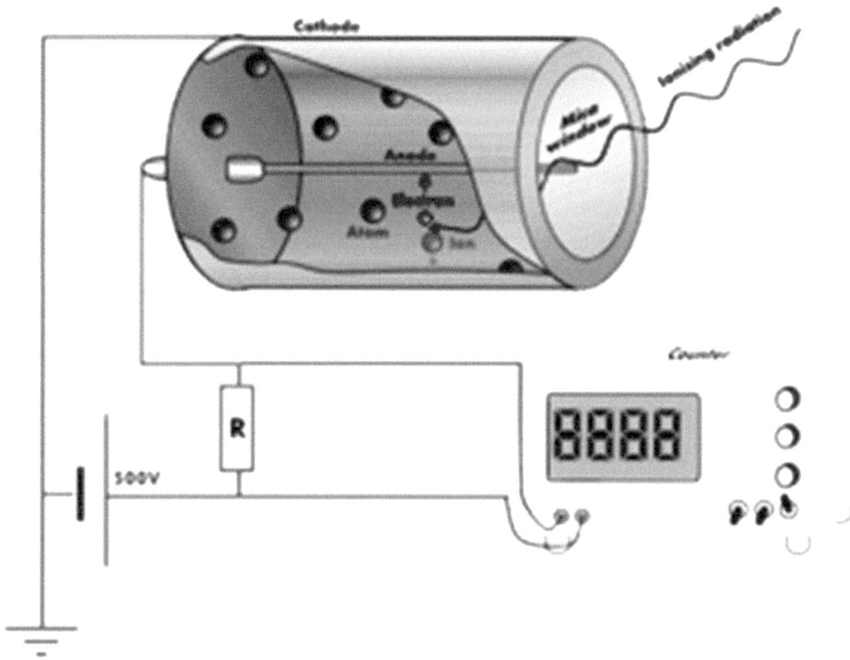

Fig. 5.41 Schematic representation of a GM tube

Although most tubes will detect gamma radiation, standard tubes are relatively inefficient, as most gamma photons will pass through the low-density gas without interacting. Using the heavier noble gases krypton or neon for the fill effects, a small improvement is achieved. Most dedicated gamma detectors use dense cathodes of lead or stainless steel in windowless tubes. The dense cathode then interacts with the gamma flux, producing high-energy electrons, which are then detected.

5.16.2.1 Halogen GM Tube

The halogen GM tube was invented by Sidney H. Liebson in 1947. The discharge mechanism takes advantage of a metastable state of the inert gas atom to more readily ionize a halogen molecule, enabling the tube to operate at much lower voltages, typically 400–600 volts instead of 900–1200 volts. This type of GM tube is therefore by far the most common form now. It has a longer life than tubes quenched with organic compounds, because the halogen ions can recombine, while the organic vapor is gradually destroyed by the discharge process (giving the latter a life of around 10^8 events) (Liebson 1947).

5.16.3 Geiger-Muller Detectors

Most detectors include an audio amplifier that produces an audible signal on discharge. The number of pulses per second measures the intensity of the radiation field. Some Geiger counters display an exposure rate (e.g., mR/h), but this does not relate easily to a dose rate as the instrument does not discriminate between radiation of different energies. The efficiency of a detector is given by the ratio of the (number of particles of radiation detected)/(number of particles of radiation emitted) i.e.,

$$\varepsilon = \frac{\text{Number of particles of radiation detected}}{\text{Number of particles radication emitted}}$$

5.16.3.1 Gamma Scintillation Detectors

Gamma particles are best detected with crystal scintillation detectors. The two main types of crystals used are sodium iodide (NaI) and germanium (Ge). To understand how the gamma detectors work, it is important to understand how the gamma particle interacts with matter. Although the gamma particle is produced in the nucleus, when it travels through matter, it mainly interacts with electrons orbiting the nucleus. Two different types of interaction with the electrons can occur: photo-absorption and Compton scattering.

Photo-Absorption In photo-absorption, the gamma is absorbed by the electron. The gamma particle (photon) enters from the left with a distinct momentum and energy, and the electron is at rest. After the interaction, the gamma particle has been "absorbed" by the electron which travels off to the right. Since energy and momentum are conserved in the interaction, the electron gains the energy and momentum of the gamma photon.

Compton Scattering In Compton scattering, the gamma scatters off the electron. The gamma particle (photon) enters from the left, and the electron is at rest. In this case, the gamma is not absorbed but scatters off the electron. The electron has gained some energy, and the gamma photon has lost some. The scattered gamma photon can interact with other electrons in the material. When a photon approaches an electron, the reaction is unpredictable. There is a certain probability of photo-absorption or a probability of Compton scattering and also a probability that no interaction takes place at all. The angle that the photon scatters is also probabilistic. Using the principles of quantum mechanics, one can calculate the probabilities for each of these possibilities.

In the case of radioactive decay, the probability of each process depends on the energy of the gamma. For photo-absorption, the probability decreases rapidly with the energy of the gamma. For higher energies, the probability for Compton scattering is much larger than for photo-absorption.

NaI Multichannel Analyzer (MCA)

The MCA system is used to detect only gamma and X-ray radiation. However, it detects the radiation well, and the MCA can also determine the energy of gamma and X-ray particles. The MCA system consists of three main parts: the detector itself, the amplifier/power supply, and a computer. The detector has two parts: a scintillation crystal (sodium iodide) and a photomultiplier tube. When a gamma particle interacts with the detector, a number of different outcomes can result. The gamma can be photo-absorbed by an electron in the sodium iodide (NaI) crystal, the gamma can Compton scatter off an electron in the crystal, or the gamma can scatter off an electron outside the crystal and then enter the NaI crystal. Each of these possibilities gives a particular feature to the spectrum.

High-Resolution Germanium Detectors

If an isotope emits more than one gamma, then each gamma will produce a photopeak, a Compton region, backscattering bump, and characteristic X-rays. Although these patterns will overlap with multiple gamma production, the use of high-resolution germanium detectors, enables the detection of very clear and narrow photopeaks, without Compton region or backscattering peaks.

5.16.3.2 Liquid Scintillation Detector

The liquid scintillation detector is the detector often used by biologists. The detector is a liquid, or a fluorescence emitting substance, "fluor," that detects mainly (or only) beta particles, when the sample is placed in the liquid. The detector contains a substance that fluoresces when a charged particle is absorbed by it. The fluorescence that is emitted is in the visible light region. A photomultiplier tube placed around the liquid detects this emitted visible light. Basically, the detector works as follows: When a beta particle is emitted, it enters the fluid and excites the "fluor" as it slows down. The fluor then de-excites and emits photons. The photons are detected by a photomultiplier tube, which sends a current pulse to an amplifier. The amplified current pulse is then passed via a multichannel analyzer similar to the gamma spectrometer. The continuous energy spectrum is recorded by the detector.

Most liquid scintillation detectors do not print out or display the whole spectrum as is done with the gamma detectors. For liquid scintillation detectors, the user usually sets a counting window: the initial and ending channels. The detector prints out and/or displays the sum of all the counts between the initial and final channels. These channels are often referred to as the lower and upper channels. In order to set an appropriate window, the user must know the energy range of the beta that is emitted and the energy/(channel number) for the detector.

The energy of the emitted electron in beta decay is not monoenergetic; the electron shares its energy with the neutrino. Hence, the beta spectrum does not have a distinct peak, but rather a continuous spectrum. Thus, the liquid scintillation detector is not useful in identifying the isotopes in a sample, but rather in counting the amount of a single known isotope in the sample. The detector is mainly used in radioisotope tracing experiments. The main features of a liquid scintillation detector are that the:

(i) The liquid scintillation detector is very efficient since the sample is placed in the detecting material. Biological samples are often in liquid form or can be placed in liquids.
(ii) Hydrogen and carbon are common elements in biological samples and can be tagged using $_3$H (tritium) and $_{14}$C. These two isotopes respective of hydrogen and carbon are beta emitters.

5.16.4 Factors Affecting GM Operation

5.16.4.1 Geiger Plateau

The Geiger plateau is the voltage range in which the Geiger-Muller counteroperates. If a GM tube is exposed to a steady radiation source and the applied voltage increased from zero, at first, the count rate increases rapidly; at a certain voltage, the rate of increase flattens out (only changing a few percent for every 100 V increase).

The slope of the curve in the plateau region is a measure of the quality of the G-M tube. For a good G-M tube, the plateau region should rise at a rate less than 10% per 100 V. That is, for a change of 100 V, (counting rate)/ (average counting rate) should be less than 0.1. An excellent tube could have the plateau slope as low as 3% per 100 V.

The plateau has a slight incline caused by increased sensitivity to low-energy radiation, due to the increased voltage on the device. Normally, when a particle enters the tube and ionizes one of the gas atoms, complete ionization of the gas occurs. Once a low-energy particle enters the counter, it is possible that the kinetic energy in addition to the potential energy of the voltage is insufficient for the additional ionization to occur, and thus, the ion recombines. At higher voltages, the threshold for the minimum radiation level drops; thus, the counter's sensitivity rises. The counting rate for a given radiation source varies slightly as the applied voltage is varied; for standardization of the response of the instrument, a regulated voltage is used to maintain stable counting characteristics.

Depending on the characteristics of the specific tube (manufacturer, size, gas type, etc.), the exact voltage range may vary. In this plateau region, the potential difference in the counter is strong enough to ionize all the gas inside the tube, upon triggering by the incoming ionizing radiation (alpha, beta, or gamma radiation). Below the plateau, the voltage is not high enough to cause complete discharge; a limited avalanche is the result, and the tube acts as a proportional counter, where the output pulse size depends on the initial ionization created by the radiation. Higher voltages give a pulse size independent of the initial ionization energy. If the applied voltage is too high, a continuous glow discharge is formed, and the tube cannot detect radiation.

5.16.4.2 Quenching Time and Dead Time

The ideal GM tube should produce a single pulse on entry of a single ionizing particle. It must not give any spurious pulses and must recover quickly to the passive state. Unfortunately for these requirements, when positive argon ions reach the cathode and become neutral argon atoms again by obtaining electrons from it, the atoms can acquire their electrons in enhanced energy levels. These atoms then return to their ground state by emitting photons which can in turn produce further ionization and hence cause spurious secondary pulse discharges. If it is not quenched, the ionization could even escalate, causing the so-called current avalanche which if prolonged could damage the tube, therefore making some form of quenching of the ionization essential. The disadvantage of quenching is that for a short time after a discharge pulse has occurred (the so-called dead time, which is typically a few microseconds), the tube is rendered insensitive and is thus temporarily unable to detect the arrival of any new ionizing particle. This effectively causes a loss of counts at sufficiently high count rates.

External quenching uses control electronics to temporarily remove the high voltage between the electrodes. Self-quenching or internal-quenching tubes stop the discharge without external assistance, by means of the addition of a small amount of a polyatomic organic vapor such as butane or ethanol, or alternatively a halogen such as bromine or chlorine. If a poor diatomic gas quencher was introduced to the tube, the positive argon ions, during their motion toward the cathode, would have multiple collisions with the quencher gas molecules and transfer their charge and some energy to them, thereby producing neutral argon atoms. The quencher gas ions would reach the cathode instead of argon gas, further gain electrons in excited states which would decay by photon emission, thereby producing spurious tube discharge. However, effective quencher molecules, when excited, do not lose their energy by photon emission but by dissociation into neutral quencher atoms producing no spurious output pulses.

5.16.5 Advantages and Disadvantages of a Geiger Counter

Advantages of using a Geiger counter are that they are relatively inexpensive, durable, and easily portable. They can detect all types of radiation including alpha, beta, and gamma radiation.

The disadvantages of using a Geiger counter are that they cannot differentiate which type of radiation is being detected. Also they cannot be used to determine the exact energy of the radiation as they have a very low detection efficiency.

5.17 Biosensors

A biosensor is an analytical device which converts a biological response into an electrical signal. The term "biosensor" was coined by Cammann in 1977 (Parikha 2016), and its definition was introduced by IUPAC (Thevenot et al. 1999). Biosensors are defined by IUPAC as a self-contained integrated receptor-transducer devices, which are able to provide selective quantitative or semiquantitative analytical information using a biological recognition element (biochemical receptor), which is retained in direct spatial contact with a transduction element (Thevenot et al. 2001; Parikha 2016). The first "true" biosensor was developed by Leland C. Clark, Jr, in 1956 for oxygen detection. He is known as the "father of biosensors," and his invention of the oxygen electrode bears his name, "the Clark electrode" (Heineman and Jensen 2006).

Biosensors are becoming the most extensively studied next-generation analysis discipline because of the ease, low cost, high sensitivity, and speed, making it an ultrasensitive detection technique in various fields. The most common biosensors in use are the pregnancy detection kits and glucose detectors. However, new biosensors for a wide range of applications including food analysis, disease diagnosis, pathogen testing, and drug detection have also been produced for routine use. Biosensors are devices which combine a biological component to detect an analyte and a physicochemical component to produce a signal which is measurable. Conventionally, biosensors consist of a biological recognition element, generally called the bioreceptor, the transducer component, and the electronic system. Biosensors can be classified in terms of the bioreceptor or transducer type used. Bioreceptors allow the binding of analytes of interest to produce a signal measurable by the transducer. Biosensors can be divided into different categories depending on how the signals are translated by the transducers: (1) amperometric biosensors (measure the electrical current), (2) potentiometric biosensors (measure electrical voltage), (3) conductometric biosensors (measure electrical conductance), (4) optical biosensors (measure the absorbed or emitted light), (5) calorimetric biosensors (measure the change of enthalpy), and (6) piezoelectric biosensors (change of resonance frequency of a piezoelectric crystal). There are two broad biosensor categories based on the biorecognition principle: (i) catalytic biosensor, typical of enzyme biosensors and (ii) affinity biosensor, typical of DNA and antibodies. Depending on the bioreceptor type used, biosensors can also be broadly classified into four classes: nucleic acid/DNA, enzymes, antibody-antigen, and cells.

The demonstration of an amperometric enzyme electrode for the detection of glucose by Leland C. Clark in 1962 was followed by the discovery of the first potentiometric biosensor to detect urea in 1969 by Guilbault and Montalvo, Jr (Nikhil et al. 2016). Enzyme biosensors have been devised on immobilization methods, i.e., adsorption of enzymes by van der Waals forces, ionic bonding, or covalent bonding. The commonly used enzymes for this purpose are oxidoreductases, polyphenol oxidases, peroxidases, and aminooxidases (Venugopal 2002; Wang 2008; Akyilmaz et al. 2010). The analyte of interest can be an inhibitor or a substrate of these processes. Rechnitz (1978) developed the first tissue-based sensor for the determination of amino acid, arginine. The first microbe-based or cell-based sensor was actualized by Divies in 1975 (Parikha 2016) and is currently being employed in the detection of a wide variety of pathogens (Torun et al. 2012). The DNA biosensors were devised by Wang (1998), based on the ability of single-strand nucleic acid molecule to recognize and bind to its complementary strand in a sample, whereas *immunosensors* are biosensors concerned solely with the monitoring of antibody-antigen interactions. Phages have been adopted in a variety of applications, e.g., in ecology, medicine (phage therapy in antibiotic-resistant cases), food industry, drug delivery systems (as a preservative), diagnostics (biosensors), and phage display technology (Wittebole et al. 2014). The Nobel Prize for Chemistry was awarded to George P. Smith and Sir Gregory P. Winter "for the phage display of peptides and antibodies." The phage display method was developed by Smith in 1985 and utilized by Winter et al. for production of disease-treating antibodies (Marta et al. 2019) (Fig. 5.42 and Table 5.6).

Biosensors are promising analytical tools that have immense applications in food industry, clinical diagnostics, environmental monitoring, etc., enabling measurement of precise and reliable results, freed from cumbersome analytical instrumentation.

One of the important parameters for fish commercialization and consumer acceptability is its freshness and quality. The traditional method for fish freshness evaluation is sensory analysis (Sect. 5.2). Instrumental methods such as texture analysis and Torry meters are also widely employed as objective alternatives. Also the methods most widely employed for quality testing and detection of fish adulteration are discussed in Sect. 5.5, such as the physicochemical methods, spectroscopic analysis, DNA-based analyses, and MS-based analyses combined with chemometric tools offering validated gradings. Each of these methods has its own advantages

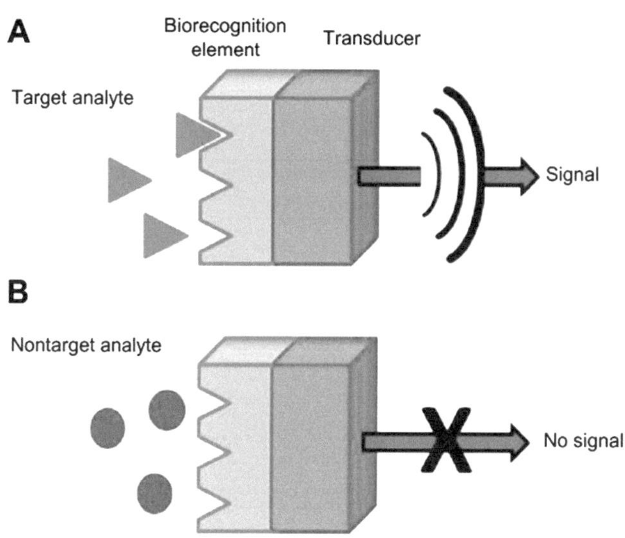

Fig. 5.42 Schematic diagram of a biosensor response in the presence and absence of a target analyte. (Source: https://www.researchgate.net/figure/Schematic-diagram-of-biosensor-with-target-analyte-A-and-nontarget-analyte-B_fig2_281601468)

Table 5.6 Different types of bioreceptors, properties, and application

Bioreceptor	Rationale/properties	Application	References
Antibody bioreceptors	An antigen-specific antibody fits its unique antigen in a highly specific manner	Immunosensors are a powerful analytical tool possessing the ability to recognize molecular structures by binding specifically to chemicals, biomolecules, and even microorganisms	Bhunia et al. (2004) and Vijayalakshmi et al. (2010)
Enzyme bioreceptors	Enzymes have specific binding capability and catalytic activity. The chosen enzyme with a suitable substrate provides sufficient electron transfer to the working electrode	Enzymatic biosensor technology is used in food processing, quality control, and online processes	Liliana et al. (2009) and Ansari and Husain (2012)
Nucleic acid bioreceptors (Genosensors)	Complementary base pairing of nucleotides. Nucleic acid as biorecognition element is simple, rapid, and inexpensive. Nucleic acid recognition layers can be readily synthesized and regenerated	Nucleic acid ligands known as aptamers are valuable for biomedical, environmental, food testing and quality control, drug development, pathogen detection, etc.	Minunni (2003), Tavakoli and Tang (2017), and Senturk et al. (2018)
Cellular bioreceptors	Cellular structures/cell-based bioreceptors with biorecognition of whole cell/ microorganism or a receptor bound to certain species	Pathogenic and spoilage bacteria and fungi are categorized in the cell bioreceptor's group for toxicity detection	Banerjee and Bhunia (2009) and Naik et al. (2017)
Biomimetic receptors	An artificial receptor is fabricated and designed to mimic a bioreceptor (an antigen-antibody, enzyme-substrate, cell receptor) and is often termed as a biomimetic receptor. For example, peptide nucleic acid (PNA), lectin based, molecular imprinting	Biomimetic polymer sensors are used for visual and spectroscopic detection of bacteria	Mecke et al. (2006) and Hussain et al. (2013)
Bacteriophage bioreceptors	Phages specifically recognize the bacterial receptors through its tail spike proteins. Since the recognition is highly specific, it can be used for the typing of bacteria	Specific pathogen detection technologies	Amit et al. (2013) and Marta et al. (2019)

and disadvantages, but none of them can solely be universally proposed for defining and measuring fish freshness, adulteration, and validation, and hence a multi-parametric system is generally adopted in order to assess fish quality.

References

Abbott, J. A. (1999). Quality measurement of fruits and vegetables. *Postharvest Biology and Technology, 15*, 207–225.

Akasaki, T., Yanagimoto, T., Yamakami, K., Tomonaga, H., & Sato, S. (2006). Species identification and PCR-RFLP analysis of cytochrome b gene in cod fish (order Gadiformes) products. *Journal of Food Science, 71*(3), C190–C195.

Akyilmaz, E., Yorganci, E., & Asav, E. (2010). Do copper ions activate tyrosinase enzyme? A biosensor model for the solution. *Bioelectrochemistry, 78*, 155–160.

Amit, S., Somayyeh, P., & Stephane, E. (2013). Recent advances in bacteriophage based biosensors for food-borne pathogen detection. *Sensors, 13*, 1763–1786. https://doi.org/10.3390/s130201763.

Ansari, S. A., & Husain, Q. (2012). Potential applications of enzymes immobilized on/in nano materials: A review. *Biotechnology Advances, 30*, 512–523.

AOAC. (2000). *Official methods of analysis* (17th ed.). Gaithersburg: AOAC International. Also valid are: a second revision of this edition (2003); the 16th edition (1995) and the 15th edition (1990). This last was published in Arlington, Virginia, USA, by AOAC International.

AOAC International. (2007). *Official methods of analysis* (AOAC Method 920.39C for Cereal Fat; AOAC Method 960.39 for Meat Fat) 18th ed., 2005; Current through revision 2, 2007 (Online). , Gaithersburg: AOAC International.

Asensio, G. L. (2007). PCR-based methods for fish and fishery products authentication. *Trends in Food Science and Technology, 18*, 558–566.

Asensio, L., Samaniego, L., Miguel, A. P., González, I., García, T., & Martín, R. (2008). Detection of grouper mislabelling in the fish market by an immunostick colorimetric ELISA assay. *Food and Agricultural Immunology, 19*(2), 141–147. https://doi.org/10.1080/09540100802100202.

Banerjee, P., & Bhunia, A. K. (2009). Mammalian cell-based biosensors for pathogens and toxins. *Trends in Biotechnology, 27*, 179–188.

Bartlett, S. E., & Davidson, W. S. (1992). FINS (forensically informative nucleotide sequencing): a procedure for identifying the animal origin of biological specimens. *BioTechniques, 12*(3), 408–411.

Bhunia, A. K., Geng, T., Lathrop, A., Valadez, A., & Morgan, M. T. (2004). Optical immunosensors for detection of Listeria monocytogenes and salmonella enteritidis from food. *Monitoring Food Safety, Agriculture, and Plant Health, 5271*, 1–6.

Bish, D. L., & Post, J. E. (Eds.). (1989). *Modern powder diffraction* (Reviews in mineralology, 20). Washington, DC: Mineralogical Society of America.

Bjellqvist, B., Kristina, E., Giorgio, R. P., Elisabetta, G., Angelika, G., Reiner, W., & Wilhelm, P. (1982). Isoelectric focusing in immobilized pH gradients: principle, methodology and some applications. *Journal of Biochemical and Biophysical Methods, 6*(4), 317–339. 0165-022X. https://doi.org/10.1016/0165-022X(82)90013-6.

Bjornsson, S. (1993). Size-dependant separation of proteoglycans by electrophoresis in gels of pure agarose. *Analytical Biochemistry, 210*, 292–298.

Carrera, E., Garcia, T., Cespedes, A., Gonzalez, I., Sanz, B., Hernandez, P., & Martin, R. (1997). Immuno stick colorimetric ELISA assay for the identification of smoked salmon (*Salmosalar*), trout (*Oncorhynchusmykiss*) and bream (*Bramaraii*). *Journal of the Science of Food and Agriculture, 74*, 547–550.

Carrera, E., Terni, M., Montero, A., Garcıa, T., Gonzalez, I., & Martın, R. (2014). ELISA-based detection of mislabeled albacore (Thunnus alalunga) fresh and frozen fish fillets. *Food and Agricultural Immunology, 25*(4), 569–577.

Charu, D., Ishan, P., Himanshu, P., Pramod, W. R., Avinash, C. P., Shanti, B. M., & Sandip, P. (2017). Chapter 9, Electrospun nanofibrous scaffold as a potential carrier of antimicrobial therapeutics for diabetic wound healing and tissue regeneration. In A. M. Grumezescu (Ed.), *Nano- and microscale drug delivery systems – Design and fabrication*. Amsterdam: Elsevier.

Cirka, H. A., Koehler, S. A., Farr, W. W., & Billiar, K. L. (2012). Eccentric rheometry for viscoelastic characterization of small, soft, anisotropic, and irregularly shaped biopolymer gels and tissue biopsies. *Annals of Biomedical Engineering, 40*(8), 1654–1665. https://doi.org/10.1007/s10439-012-0532-5.

Corradini, D. (2011). *Handbook of HPLC* (2nd ed.). Boca Raton: CRC Press.

Cullity, B. D. (1978). *Elements of X-ray diffraction* (2nd ed.). Reading: Addison-Wesley.

David, B. M., & Wayne, C. E. (2014). Fat analysis, Chapter 8. In S. S. Nielsen (Ed.), *Food analysis laboratory manual* (Food science texts series) (4th ed.). Boston: Springer.

Eby, G. N. (2004). *Principles of environmental geochemistry* (pp. 212–214). Pacific Grove: Brooks/Cole-Thomson Learning.

Eric, S., Julia, A. D., & Reta, N. (2008). Chapter 8 – Gas chromatography and gas chromatography – Mass spectrometry. In *Fire debris analysis* (pp. 235–293). London: Elsevier. isbn:978-0-12-663971-1 *(For GCdiagram)*.

Fernandez, B., Lobo, L., & Pereiro, R. (2019). Atomic absorption spectrometry – Fundamentals, instrumentation and capabilities. In *Encyclopedia of analytical science* (Reference module in chemistry, molecular sciences and chemical engineering) (3rd ed., pp. 137–143). Amsterdam: Elsevier.

Hayat, M. (2000). *Principles and techniques of electron microscopy, biological applications* (4th ed.). Cambridge: Cambridge University Press.

Heineman, W. R., & Jensen, W. B. (2006). Leland C. Clark Jr. (1918–2005). *Biosensors & Bioelectronics, 21*, 1403–1404.

http://rtilab.com/techniques/ftir-analysis/

https://www.mee-inc.com/hamm/fourier-transform-infrared-spectroscopy-ftir/

Hussain, M., Wackerlig, J., & Lieberzeit, P. A. (2013). Biomimetic strategies for sensing biological species. *Biosensors, 3*, 89–107.

Jena, S., Tokas, R., Thakur, S., & Sahoo, N. K. (2015). Characterization of optical thin films by spectrophotometry and atomic force microscopy. *SMC Bulletin, 6*(1), 1–9.

Jones, D. B. (1941). *Factors for converting percentages of nitrogen in foods and feeds into percentages of proteins* (Circular No. 183). Washington, DC: United States Department of Agriculture.

Josep, S. E., Ingrid, M. Y., Alessandra, B., & Pier, R. (1996). Fish species identification by isoelectric focusing of parvalbumins in immobilized pH gradients. *Electrophoresis, 17*, 1380–1385. https://doi.org/10.1002/elps.1150170817.

Karousou, E. G., Viola, M., Genasetti, A., Vigetti, D., De Luca, G., Karamanos, N. K., & Passi, A. (2005). Application of polyacrylamide gel electrophoresis of fluorophore-labeled saccharides for analysis of hyaluronan and chondroitin sulfate in human and animal tissues and cell cultures. *Biomedical Chromatography, 19*(761–765), 21.

Kjeldahl, J. (1883). Neue Methodezur Bestimmung des Stickstoffs in organischen Körpern [New method for the determination of nitrogen in organic substances]. *Zeitschriftfüranalytische Chemie, 22*(1), 366–383.

Klug, H. P., & Alexander, L. E. (1974). *X-ray diffraction procedures for polycrystalline and amorphous materials* (2nd ed.). New York: Wiley.

Lawler, D. M. (2016). Turbidity, turbidimetry, and nephelometry. In *Encyclopedia of analytical science* (3rd ed., pp. 152–163). Kent: Elsevier.

Lehotay, S., & Hajslova, J. (2002). Application of gas chromatography in food analysis. *Trends in Analytical Chemistry, 21*(9–10), 686–697.

Liebson, S. H. (1947). The discharge mechanism of self-quenching Geiger–Mueller counters. *Physical Review, 72*(7), 602–608.

Liliana, S. C., Ana María, Z. A., & Alfredo, A. A. (2009). Use of enzymatic biosensors as quality indices: A synopsis of present and future trends in the food industry. *Chilean Journal of Agricultural Research, 69*(2), 270–282.

Luca, R., René, S., Susana, R. S., Chantal, M., Vincenzo, G., & Frédérique, J. (2012). Instrumental texture analysis parameters as markers of table-grape and winegrape quality: A review. *American Journal of Enology and Viticulture, 63*(1), 11–28.

Macosko, C. W. (1994). *Rheology – Principles, measurements and applications*. New York: Wiley-VCH.
Magdalena, M., & Edward, P. (2007). Species identification of meat by electrophoretic methods. *Acta Scientiarum Polonorum, Technologia Alimentaria, 6*(1), 5–16.
Marta, J., Iva, M., Joanna, N., & Katarzyna, S. (2019). Recent applications of bacteriophage-based electrodes: A mini-review. *Electrochemistry Communications, 99*, 11–19.
Mecke, A., Dittrich, C., & Meier, W. (2006). Biomimetic membranes designed from amphiphilic block copolymers. *Soft Matter, 2*, 751–759.
Mezger, T. (2011). *The Rheology Handbook* (3rd Rev. ed., , pp. 324–334). Hanover: Vincentz Network.
Michael, T. (2008). AMC technical briefs in Analytical Methods Committee AMCTB, 29, issn:1757-5958.
Michael, L. S., Gene, A. S., Neil, C. S., & Gregg, A. N. (1991). New evidence on the hydrothermal system in Long Valley caldera, California, from wells, fluid sampling, electrical geophysics, and age determinations of hot-spring deposits. *Journal of Volcanology and Geothermal Research, 48*(3–4), 229–263.
Militsopoulou, M., Lamari, F., & Karamanos, N. K. (2003). Capillary electrophoresis: A tool for studying interactions of glycans/proteoglycans with growth factors. *Journal of Pharmaceutical and Biomedical Analysis, 32*, 823–828.
Minunni, M. (2003). Biosensors based on nucleic acid interaction. *Spectroscopy, 17*, 613–625.
Moore, D. M., & Reynolds, R. C., Jr. (1997). *X-ray diffraction and the identification and analysis of clay minerals* (2nd ed.). New York: Oxford University Press.
Moretti, V. M., Turchini, G. M., Bellagamba, F., & Caprino, F. (2003). Traceability issues in fishery and aquaculture products. *Veterinary Research Communications, 27*(1), 497–505.
Mostovenko, E., Hassan, C., Rattke, J., Deelder, A. M., van Veelen, P. A., & Palmblad, M. (2013). Comparison of peptide and protein fractionation methods in proteomics. *EuPA Open Proteomics, 1*, 30–37.
Naik, K. M., Srinivas, D., Sasi, B., & Basha, S. K. J. (2017). Biosensors in food processing – A review. *International Journal of Pure & Applied Bioscience, 5*(4), 1219–1227. https://doi.org/10.18782/2320-7051.5546.
Nikhil, B., Pawan, J., Nello, F., & Pedro, E. (2016). Introduction to biosensors. *Essays in Biochemistry, 60*, 1–8. https://doi.org/10.1042/EBC20150001.
Nordberg, G. F., Fowler, B. A., & Nordberg, M. (2015). *Handbook on the toxicology of metals* (4th ed.). Amsterdam: Elsevier.
Parikha, M. (2016). Biosensors and their applications – A review. *Journal of Oral Biology and Craniofacial Research, 6*, 153–159.
Pavao, M. S. G., Mourao, P. A. S., Mulloy, B., & Tollefsen, D. M. (1995). A unique dermatan sulfate like glycosaminoglycans from ascidian: Its structure and the effect of its unusual sulfation pattern on anticoagulant activity. *The Journal of Biological Chemistry, 270*, 31027–31036.
Putnis, A. (1992). *Introduction to mineral sciences* (Chapter 3, pp. 41–80). Cambridge: Cambridge University Press.
Rechnitz, G. A. (1978). Biochemical electrodes uses tissues slice. *Chemical and Engineering News, 56*, 16–21.
Reich, E., & Schibli, A. (2007). *High-performance thin-layer chromatography for the analysis of medicinal plants* (Illustrated ed.). New York: Thieme. isbn:978-3-13-141601-8.
Romero, G., Díaz, J. R., Sabater, J. M., & Perez, C. (2012). Evaluation of commercial probes for on-line electrical conductivity measurements during goat gland milking process. *Sensors, 12*(4), 4493–4513. https://doi.org/10.3390/s120404493.
Rosalee, S. R., & Morrissey, T. M. (2008). DNA-based methods for the identification of commercial fish and seafood species. *Comprehensive Reviews in Food Science and Food Safety, 7*, 280–295.
Rutherford, E., & Geiger, H. (1908). An electrical method of counting the number of α particles from radioactive substances. *Proceedings of the Royal Society (London), Series A, 81*(546), 141–161.

Senturk, E., Aktop, S., Sanlibaba, P., & Tezel, B. U. (2018). Biosensors: A novel approach to detect food-borne pathogens. *Applied Microbiology, 4*, 151. https://doi.org/10.4172/2471-9315.1000151.

Steinke, D., & Hanner, R. (2011). The FISH-BOL collaborators' protocol. *Mitochondrial DNA, 22*, 10–14.

Szczesniak, A. S. (2002). Texture is a sensory property. *Food Quality and Preference, 13*, 215–225.

Tavakoli, J., & Tang, Y. (2017). Hydrogel based sensors for biomedical applications: An updated review. *Polymers, 9*(8), 364. https://doi.org/10.3390/polym9080364.

Thevenot, D. R., Toth, K., Durst, R. A., & Wilson, G. S. (1999). Electrochemical biosensors: Recommended definitions and classification. *Pure and Applied Chemistry, 71*, 2333–2348.

Thevenot, D. R., Toth, K., Durst, R. A., & Wilson, G. S. (2001). Electrochemical biosensors: Recommended definitions and classification. *Biosensors & Bioelectronics, 16*, 121–131.

Torun, O., Boyaci, I., Temur, E., & Tamer, U. (2012). Comparison of sensing strategies in SPR biosensor for rapid and sensitive enumeration of bacteria. *Biosensors & Bioelectronics, 37*, 53–60.

Venugopal, V. (2002). Biosensors in fish production and quality control. *Biosensors & Bioelectronics, 17*, 147–157.

Vijayalakshmi, V., Khalil, A., Olga, K., Kamila, O., & Catherine, A. (2010). An overview of food borne pathogen detection: In the perspective of biosensors. *Biotechnology Advances, 28*, 232–254.

Vogel, A. I., Tatchell, A. R., Furnis, B. S., Hannaford, A. J., & Smith, P. W. G. (1989). *Vogel's textbook of practical organic chemistry* (5th ed.). Harlow: Longman. isbn:978-0-582-46236-6.

Volpi, N., & Maccari, F. (2002). Detection of submicrogram quantities of glycosaminoglycans on agarose gels by sequential staining with toluidine blue and stains-all. *Electrophoresis, 23*, 4060–4066.

Wang, J. (1998). DNA biosensors based on peptide nucleic acid (PNA) recognition layers. A review. *Biosensors and Bioelectronics, 13*, 757–762. 11.

Wang, J. (2008). Electrochemical glucose biosensors. *Chemical Reviews, 108*, 814–825.

Wang, L., & Weller, C. L. (2006). Recent advances in extraction of nutraceuticals from plants. *Trends in Food Science & Technology, 17*(6), 300–312.

Ward, R. D., Hanner, R., & Hebert, P. D. N. (2009). The campaign to DNA barcode all fishes, FISH-BOL. *Journal of Fish Biology, 74*, 329–356.

Wilson, K., & Walker, J. (2010). *Principles and techniques of biochemistry and molecular biology* (7th ed.). Cambridge: Cambridge University Press.

Wittebole, X., De Roock, S., & Opal, S. M. (2014). A historical overview of bacteriophage therapy as an alternative to antibiotics for the treatment of bacterial pathogens. *Virulence, 5*, 226–235. https://doi.org/10.4161/viru.25991.

Woolfe, M., & Primrose, S. (2004). Food forensics: Using DNA technology to combat misdescription and fraud. *Trends in Biotechnology, 22*(5), 222–226.

Waste Management in Seafood Industry 6

Waste management is a huge problem in all the realms of the world. All industries contribute to the mass of biodegradable and non-biodegradable waste, and seafood industry is no exception in this matter. Till the last decade, seafood waste found their way directly into the sewers and nearby water bodies. But the waste production in those days was not in the same scale as of today. Modern seafood industries had scaled up their production and produce much larger amounts of waste, most of which is not degradable at the rate of deposition. Hence, we must look for better technologies of waste management.

The solid waste from the industry including shells, stomach parts, and bycatches is now in huge demand by many industries. The liquid waste is always a headache with no takers. The industry management must find their own measures for getting rid of it. Most of them depend on reverse osmosis, aerobic, and anaerobic filters for purifying the wastewater.

6.1 Seafood Waste Disposal

Fish waste is often classified as a certified waste with high disposal cost because of high organic content (Knuckey et al. 2004). Most of the seafood processing units have effluent treatment plants (ETP) for discharging liquid waste effluents into the backwaters. In Cochin, India, alone, seafood processing units (87.5%) are depending on private contract for disposing solid waste. Solid waste is utilized for biogas production by 12.5% units. Many institutions are giving awareness programs about by-product valorization possibilities. The high nutritional quality and the possibility of by-product valorization from seafood waste have resulted in a share of the waste generated to be utilized for developing by-products. Shrimp shell waste gets a better price in the market (US dollar 1.098–1.32 per box) than finfish and cephalopod (0.77–0.88). They are mainly used for by-product development, and the major portion for by-product is contributed by cephalopod (19.4%) followed by shrimp shell (15.23%) and finfish (2.2%) waste (Sasidharan and Mathew 2011).

S. Mathew et al., *Fish and Fishery Products Analysis*,
https://doi.org/10.1007/978-981-32-9574-2_6

An effective waste management strategy is comprised of three parts. It includes waste minimization, waste characterization, and waste utilization. Best management practices for handling wastes by plants focus mainly on installation of efficient machinery to increase product yield, reduction of contact of solid waste with water, standardization of water consumption, and early separation/removal of solid waste. To minimize the risk of environmental hazards, the solid and liquid waste generated should undergo treatment.

The size and the seasonal nature are the two major drawbacks regarding fishery processing waste elimination. Fishery wastes differ from municipal wastes with some economic advantages for the fisheries. They have more nitrogen and protein than municipal wastes and hence more potential utilization possibilities. In comparison to other wastes, they fetch more market value. Another advantage over municipal wastes is that wastes from a given fishery would be homogeneous. The need for expensive grinding is relatively minimized due to the small particulate size of fishery solid waste such as fish heads, shrimp shells, etc. Along with the advantages, one major disadvantage of seafood solid waste is that the technology and equipment developed thus far are geared to huge daily volumes of waste. As the volume goes down, the cost per unit for processing waste rapidly increases.

6.1.1 Utilization of Solid Seafood Wastes

Bait – The solid seafood waste is the cheapest bait for both sport and commercial fishermen. This is a convenient method of disposal since most seafood processors are in areas where commercial fisheries and sport fisheries thrive. But the use of seafood waste as bait is limited because only some commercial fisheries like lobster, crab, and long lining use this bait, whereas trawl and seine fisheries do not use it (Green and Mattick 1977).

Pet Foods – Canned pet foods are produced by processing tuna and salmon wastes by many of the larger seafood industries. There are established canning companies that process underutilized fish and fishery wastes into pet food in larger fishing ports.

Fish Meal – This is an established industry for utilizing seafood waste. However, small- or medium-sized seafood processor finds it expensive to process waste into fish meal at a profit. Because of high cost of capitalization, smaller units usually cannot compete with the present market prices for other protein meals.

Other Meals – Crustacean meals can be made from crab and shrimp processing wastes. The protein content in these wastes is moderate, about 25–35%. They contain a fair amount of chitin and $CaCO_3$.

Chitin/Chitosan – Chitin is a polysaccharide made from the monomer N-acetyl-D-glucosamine, and chitosan is its deacetylated derivative. Both chitin and chitosan have many potential uses. A substantial investment in chemical processing equipment and technical personnel is needed for establishing chitin/chitosan plants. A large steady supply of crustacean waste is also required to support such establishments.

Fish Silage (Liquified Fish) – Fish silage is a liquefied fish resulting from the autolysis of fish waste held under acid conditions. This method offers the most promising aspect for seafood processing waste utilization. The process only requires grinding fish, adding formic acid to it, and mixing. The capital investment is very low. It can be carried on in a single drum or a large factory. Silage production has worked well in Europe for fish processing waste. The final product is digested fish which shows good solubility properties. Most of the fish silage is being fed in its liquid form to pigs and cattle. If needed, it can be dried for long-term storage and shipping.

Aerobic Composting – Aerobic composting is a relatively inexpensive method. It is suitable for varying volumes and also seasonal variation in seafood production. Composting to a stabilized state has been achieved within 1 or 2 weeks. The final product is stable, odorless, and safe from a health aspect. Most fishery wastes have little or no available carbohydrates for microbial energy sources. Therefore, fish oils will act as the metabolizable energy sources. For better microbial action, it is necessary to combine carbohydrate wastes with fishery wastes. To reduce odors and achieving maximum composting efficiency, frequent aeration is necessary. Fishery composts will fetch better price because of high nitrogen content.

Protein Hydrolysates – This method is explored to recover peptones and use them in media for the growth of microorganisms. The recovered fish peptones supported microbial growth. Because of its potential use, markets exist for use in both the laboratory and commercial culture of microorganisms.

Food Grade Materials – Deboning recovers 55–65% of the flesh and filleting recovers only 40–42%. Filleting leaves the edible meat on the frames as waste. Methods have been developed to precipitate proteins from fishery effluent wastes. It is achieved by use of alum sulfate, aluminum hydroxide, and lignosulfonic acid derivatives by treatment with mineral acid until isoelectric point, pH 4.4–4.7, is achieved. Protein precipitation by flocculation with chitosan has also been tried.

Other Marketable Materials from Fishery – Fish oils recovered from seafood wastes contain an abundance of long-chain, polyunsaturated fatty acids (PUFA). In the fish meal industry, they are routinely recovered from fish of high oil content. They find use as protective coatings.

6.2 Wastewater Characterization

Wastewater in a seafood industry is the liquid wastes collected during processing and conveyed to a treatment plant. BOD and suspended solids (nonfilterable residue) are the two most important parameters used to define the characteristics of wastewater. A suspended solid concentration of 240 mg/l of BOD is equivalent to 0.20 lb. of BOD in 120 gal. Reduction of suspended solids and BOD in primary sedimentation is approximately 50% and 35%, respectively (Table 6.1). Approximately 70% of the suspended solids are volatile, defined as those lost upon ignition at 550° C (Viessman and Hammer 1999a).

Total solids (residue on evaporation) include organic matter as dissolved salts; the concentration of the latter is dependent to a considerable extent on the hardness

Table 6.1 Change in wastewater parameters after treatments

Parameters	Before sedimentation	After sedimentation	Biologically treated
Total solids	800	680	530
Total volatile solids	440	340	220
Suspended solids	240	120	30
Volatile suspended solids	180	100	20
BOD	200	130	30
Ammonia nitrogen	22	22	24
Total nitrogen	35	30	26
Soluble phosphorus	4	4	4
Total phosphorus	7	6	5

of the water. The concentration of nitrogen in waste is directly related to the concentration of organic matter (BOD). Approximately 60% of the total nitrogen is in solution as ammonia. If raw wastewater has been retained for a long time in collector sewers, a great percentage of ammonia nitrogen results from deamination of the proteins and urea in wastewater. Seven milligrams per liter of phosphorus is approximately equivalent to a 2 lb phosphorus contribution per capita per year.

The surplus of nitrogen and phosphorus in biologically treated wastewater reveals that wastewater contains nutrients and phosphorus in excess for biological needs. The approximate BOD/nitrogen/phosphorus (BOD/N/P) weight ratio required for biological treatment is 100/5/1. The exact BOD/N/P needed for treatment depends on the process and the biological availability of the nitrogen and phosphorus compounds in the wastewater. A minimum of 100/6/1.5 is commonly related to treatment of unsettled sanitary wastewater, while 100/3/0.7 is generally adequate for wastewater where the nitrogen and phosphorus are in soluble forms. The average domestic wastewater listed in Table 6.1. has a ratio of 200/3/0.7 before sedimentation and 100/23/5 after sedimentation, both of which are in excess of the minimum 100/6/1.5. For biological treatment of industrial wastewater deficient in nutrients, soluble phosphorus can be supplied by adding H_3PO_4 and soluble nitrogen by adding NH_4NO_3.

Biodegradable organic matter in wastewater is generally classified into three categories: carbohydrates, proteins, and fats. Carbohydrates are hydrates of carbon with the empirical formula $C_nH_{2n}O_n$ or $C_n(H_2O)_n$. The simplest carbohydrate unit is known as a monosaccharide, although few monosaccharides occur naturally. Proteins in simple forms are long-chain molecules composed of amino acids connected by peptide bonds and are important in both structural and dynamic aspects of living matter. Twenty-one common amino acids when linked together in long peptide chains from a majority of simple proteins found in nature. A mixture of proteins as a bacterial substrate is an excellent growth medium, since proteins contain all the essential nutrients. On the other hand, pure carbohydrate is unsuitable as a growth medium since they do not contain the nitrogen and phosphorus essential for synthesis.

Lipids together with carbohydrates and proteins, form the bulk of organic matter of living cells. The term refers to a heterogeneous collection of biochemical substances having the mutual property of being soluble to varying degrees in organic solvents (e.g., ether ethanol, hexane, and acetone) while being only sparingly soluble in water. In sanitary engineering, the word fats convey the meaning of lipids. The term grease applies to a variety of organic substances in the lipid category that are extracted from aqueous solution or suspension by trichlorotrifluoroethane.

Not all biodegradable organic matter can be classed into these three simple groupings. Many natural compounds have structures that are combinations of carbohydrates, proteins, and fats, such as lipoproteins and nucleoproteins. Approximately 20–40% of the organic matter wastewater appears to be non-biodegradable. Several organic compounds, although biodegradable in the sense that specific bacteria can break them down, must be considered by sanitary engineers as partially biodegradable because of time limitations in waste treatment processes. For example, lignin, a polymer non-carbohydrate material, is for all practical purpose non-biodegradable. Saturated hydrocarbons are a problem in treatment because of their physical properties and resistance to bacterial actions. Alkylbenzene sulfonate is only sparingly biodegradable in water treatment.

6.3 Wastewater Treatment and Monitoring

An industry has three possibilities for disposal of process wastewaters. They may be treated separately in an industrial waste treatment plant prior to discharge to a watercourse; raw wastewater can be discharged to the municipal treatment plant or pretreated at the industrial site prior to discharge in the municipal sewage system. The most feasible method should be adopted. The nitrogenous compounds in sewage include nitrite, nitrate, ammonia, soluble organic nitrogen, and particulate organic nitrogen. Nitrogenous compounds are reduced by nitrification and denitrification process.

Transforming the complex organic molecules in wastewater sludge into an available carbon source is a prime step in biological treatment. Thermal pyrolysis, alkaline hydrolysis, and acid hydrolysis are commonly applied as the pretreatment method for the waste activated sludge (WAS). These technologies increase the soluble organic matter in the hydrolysate, and these can later be used as the carbon source (Mason et al. 1992; Sawayama et al. 1995).

6.3.1 Hydrolysis

Sludge hydrolysis is carried out in a closed, Pyrex glass, cylindrical reactor (Lin et al. 1998). To ensure sufficient mixing of the sludge, a mechanical stirrer can be used. The mixing speed should be maintained at 250 rpm. This disperses the NaOH and prevents sludge from settling. NaOH concentrations ranging from 10 to 50 m-eq. 1-l (l-40 g-NaOH/100 g-TS (total solid)) should be employed for the sludge concentrations ranging from 0.5 to 2% TS. The whole procedure should be done under anoxic condition. The reaction temperature should be in the range of

25 ± 3 °C for a duration of 24 h. Samples should be taken and filtered through a 0.45 pm glass fiber to obtain the filtrate as hydrolysate. At the end of each batch test, TS, BOD, NO_2-N, NO_3-N, and org-N of the hydrolysate should be determined. Solubilization of carbonous, nitrogenous, and phosphorous compounds by alkaline pretreatment should be calculated.

6.3.2 Biodegradation and Bioremediation

Biodegradation is the breakdown of organic contaminants into smaller compounds by metabolic or enzymatic processes of living organism in the environment. The main organisms responsible for such breakdown are bacteria, yeast, and fungi. Conversion of hazardous contaminants into less hazardous substances is called bioremediation. One such conversion is microbial metabolism, which converts organic compounds into energy and carbon to sustain microbial growth.

Normal functions of living organisms are interfered by organic pollutants. Changing the structure of toxic organic pollutants, which always occurs when an electron is lost or gained, will change their properties and may make them less toxic. Eventually, biodegrading breaks organic pollutants into smaller molecules. It finally ends with carbon dioxide, water, and mineral salts if the process goes uninterrupted.

Six basic requirements must be met for biodegradation to occur: (1) suitable degrading organism, (2) electron donors, (3) electron acceptors, (4) carbon, (5) nutrients, and (6) acceptable environmental conditions. Temperature lower than 5 °C tends to inhibit biodegradation in general. Microbial reactions can be divided into two, aerobic and anaerobic reactions. Aerobic reaction requires the presence of oxygen as an electron acceptor. Anaerobic microbial reaction utilizes electron acceptors rather than oxygen. Presence of enough oxygen initiates aerobic reactions. When oxygen is not sufficient or absent, only anaerobic reaction can occur. Aerobic and anaerobic reactions require different kinds of bacteria, and aerobic reactions are considerably faster than anaerobic reactions.

6.3.2.1 Aerobic Treatment

Microbes use oxygen electron as acceptors to transform organic carbon into carbon dioxide. Electrons are transferred from the contaminant to oxygen. The contaminant is oxidized and the oxygen is reduced, forming water and carbon dioxide. The key requirement is adequate oxygen.

Aerobic Respiration by Trickling (Biological) Filters

Trickling filters are fixed-growth biological beds where wastewater is spread on the surface of media, supporting microbial growths. Although the term filter is the accepted designation for this unit, no physical filtration occurs; contaminants are removed by biological actions. The media placed in a tank under the wastewater distributor can be crushed rock, plastic sheet packing formed into modules, or random plastic packaging of various shapes. Trickling filters are preceded by primary clarifiers to remove settleable solids and are followed by final clarifiers to collect microbiological growths that slough from the media. The primary reasons for the popularity of trickling filters are their simplicity, low operating cost, and production of waste sludge that is easy to process.

Biological process in trickling filtration – The biological slime layers on filter media consist of bacteria, protozoans, and fungi. Frequently, noticeable populations of larger organisms such as sludge worms and rotifers are present. The top surface of a bed exposed to sunlight is coated with algae, and the lower portions of deep filter can support nitrifying bacteria. Microorganisms near the surface of the bed where the food concentration is high are in a rapid growth phase, while the microorganisms near the bottom are in a state of starvation.

Figure 6.1 is a schematic diagram illustrating the general biological process at the surface of the medium. Although classified as aerobic treatment, the microbial film is aerobic to a depth of only 0.1–0.2 mm. The zone adjacent to the medium is anaerobic. As the wastewater flows over the microbial film, the soluble organic matter is metabolized, and the colloidal organic matter is absorbed on the film. Dissolved oxygen taken up in the liquid layer is replenished and reoxygenated from the surrounding air. Continuous passage of air through the bed is essential to prevent

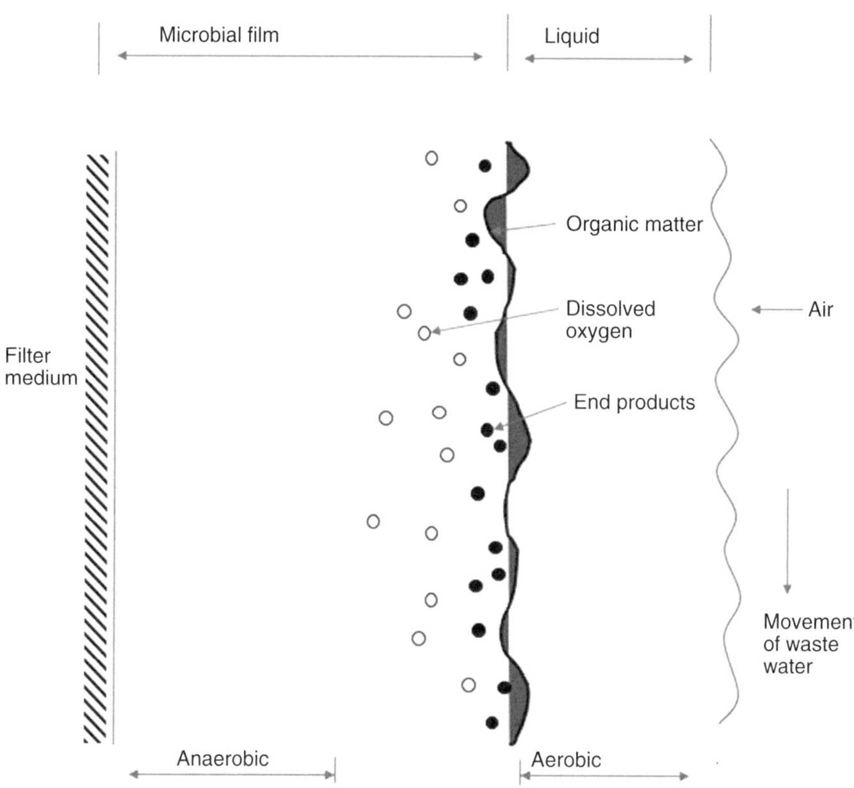

Fig. 6.1 Schematic diagram showing the biological process in a trickling filter

undesirable anaerobic conditions. The actual biological process is more complex than this simplified description and is poorly understood. Physical characteristics, such as media configuration, bed depths associated with offensive odors, bed depth, and hydraulic loading, strongly influence the process. Several forms of manufactured media provide a high specific surface area with a corresponding large percentage of void volume. This permits substantial biological growth without inhibiting passage of air through the bed. A uniform medium also allows loading distribution, and lightweight packing permits the construction of deep beds.

Problems with the biological process that can occur include poor effluent quality and emission with offensive odors, both associated with organic loading and cold weather operation. The anaerobic zone of the film adjacent to the medium surface can produce odorous metabolic end products. If they are not oxidized while moving through the aerobic zone, they can be released and carried out in the easily oxidized microbial film. Cooling of the applied wastewater reduces the biological activity, allowing more organic matter to pass through the bed.

Trickling filter operation and filter media requirements – The major components of a shallow trickling filter are a rotary distributor, underdrain system, and filter medium. The distributor spreads the wastewater at a uniform hydraulic lead per unit area on the surface of the bed. The arms are driven by the reaction of wastewater following out of the center column that supports the arms. The underdrain system, often verified clay blocks with entrance holes to drainage channels, carries away the effluent and permits circulations of air through the bed. The need for free passage of air controls the size of openings in the underdrain. The filter media provide a surface for biological growth and voids for passage of air and water.

Trickling filter secondary systems – A trickling filter secondary system includes a final settling tank to remove biological growths that are washed off the filter media. These sloughed solids are commonly disposed of through a drain line from the bottom of the final clarifier to the head of the plant. This return sludge flow is mixed with the raw wastewater and settled in the primary clarifier.

The raw wastewater in trickling filtration is diluted with recirculated flow so that it is reduced in strength and passes through the filter more than once. A return line from the final clarifier severs a dual function as sludge return and a recirculation line. The combined flow through the filter flow is always sufficient to maintain a minimum hydraulic loading on the media and sufficient flow to turn the distributor arms. The two most common recirculation patterns used in trickling recirculation patterns used in trickling filter system are shown in Fig. 6.2. The recirculation ratio is the ratio of recirculated flow to the quantity of raw wastewater. A common range for recirculation ratio values is 0.5–3.0. Recirculation may be done (1) only during periods of low wastewater flow, (2) at a rate proportional to raw-wastewater flow, (3) at a constant rate at all times, or (4) at two or more constant rates predetermined automatically or by manual control.

Direct recirculation, depicted in the figure, is frequently used in the treatment of stronger wastewaters, where recirculation ratios of 2–3 are desirable. If high rates are used in pattern, the primary clarifier must be sized for the increased flow rate created by the greater volume. In other words, if recirculation flow is routed through

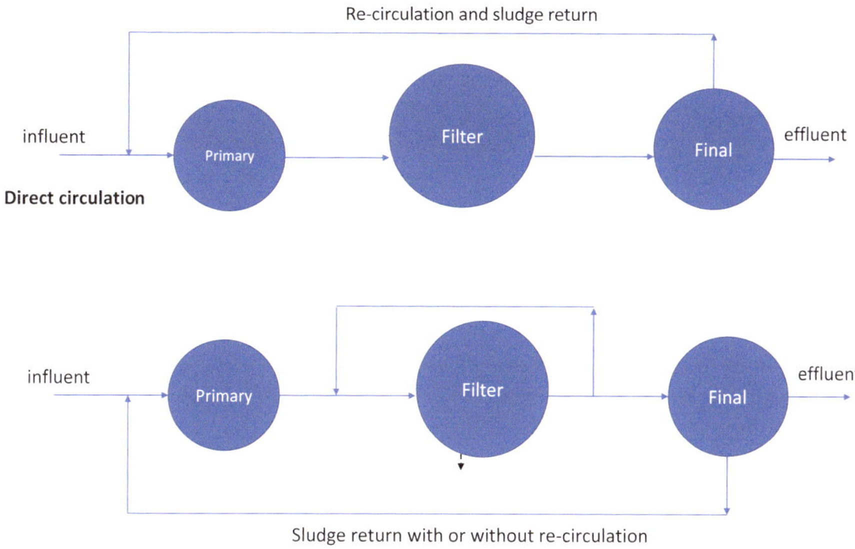

Fig. 6.2 Recirculation system

a clarifier during the peak hourly flows of the raw wastewater, the clarifier must be increased in size to prevent disturbance of the settled solids and resultant loss of removal efficiency in the sedimentation tank. Consider the flow diagram and assume the flow in the sludge return line. Then apply a direct recirculation of 420 gpm around the filter using constant speed pumps. The resultant ratio for the trickling filter is 2.0:0.5 from indirect recirculation.

Deep filters (biological towers) for treating municipal wastewater are generally single-stage units following primary sedimentation, although two stages are installed when processing strong wastewater resulting from industrial discharges. Direct recirculation of this is employed to maintain the desired flow through the tower. Sometimes a portion of the recirculation flow is drawn from the bottom of the final clarifier to develop a microbial floc in the wastewater circulating through the tower. Waste sludge accumulated in the final clarifier can be returned to the plant influent for settling in the primary clarifier.

6.3.2.2 Anaerobic Treatment

Nitrate, sulfate, Mn^{4+}, Fe^{3+}, and CO_2 serve as oxygen substitutes for some microorganisms. Organic carbon is chemically transformed, often to carbon dioxide but sometimes to methane. Electrons are transferred from the contaminant (oxidizing it) to an electron acceptor (reducing it). An adequate supply of electron acceptors and an absence of oxygen are required for this process, and it provides metabolic energy for the microbes.

Anaerobic digestion consists of two distinct stages that occur simultaneously in digesting sludge. The first consists of hydrolysis of the high-molecular-weight organic compounds and conversion to organic acids by acid-forming bacteria. The second stage is gasification of the organic acids to methane and carbon dioxide by the acid-splitting methane-forming bacteria.

Methane bacteria are strict anaerobes and very sensitive to conditions of their environment. It can be adversely affected by excess concentrations of oxidized compounds, volatile acids, soluble salts, and metal cations and show rather extreme substrate specificity. Each species is restricted to the use of only a few compounds, mainly alcohols and organic acids, whereas the normal energy sources, such as carbohydrates and amino acids, are not attacked. An enrichment culture developed on acetic or butyric acid cannot decompose propionic acid. The sensitivity exhibited by methane bacteria in the second stage of anaerobic digestion, coupled with the rugged nature of acid-forming bacteria in the first stage, creates a biological system where the population dynamics are easily upset. Any shift in the environment adverse to the population of methane bacteria causes a buildup of organic acids, which in turn further reduces the metabolism of acid-spilling methane formers.

Single-stage floating-cover digesters – Raw sludge is pumped into the digester through pipes terminating either near the center of the tank or in the gas dome. Pumping sludge into the dome helps to break up the scum layer that forms on the surface. Digested sludge is withdrawn from the tank bottom. The contents are heated in the zone of digesting sludge by pumping them through an external heater and returning the heated slurry through the inlet's lines. The tank contents stratify with a scum layer on top and digested thickened sludge on the bottom. The middle zones consist of a layer of supernatant (water of separation) underlined by the zone of actively digesting sludge. Supernatant is drawn from the digester through any one of a series of pipes extending out the tank wall. Digestion gas from the gas dome is burned as fuel in the external heater.

The weight of the cover is supported by sludge, and the liquid forced up between the tank and the side of the cover provides a gas seal. Gas rises out of the digesting sludge, moves along the ceiling of the cover, and collects in the gas dome. The cover can float on the surface of the sludge between the landing brackets and the height of the overflow pipe.

Three functions of a single-stage floating-cover digester are (1) anaerobic digestion of the volatile solids, (2) gravity thickening, and (3) storage of the digested sludge. A floating-cover feature of the tank provides for a storage volume equal to approximately one-third that of the tank. The unmixed operation of the tank permits gravity thickening of sludge solids and withdrawal of separated supernatant. However, the rate of biological activity is inhibited by the lack of mixing; on the other hand, good mixing would prevent supernatant formation. Therefore, in single-tank operation, the biological process is compromised to allow both digestion and thickening to occur in the same tank.

In the operation of an unmixed digester, raw sludge is pumped to the digester from the bottom of the settling tanks once or twice a day. Supernatant is withdrawn daily and returned to the influent of the treatment plant. It is normally returned by gravity flow to the wet well during periods of low raw-wastewater flow, or in the case of an activated sludge plant, it may be pumped to the head end of the aeration basin. Because of the floating cover, supernatant does not have to be drawn off simultaneously with pumping of raw sludge into digester. Digested sludge is stored in the tank and withdrawn periodically for disposal.

High-rate (completely mixed) digesters – The biological process of anaerobic digestion is significantly improved by complete mixing of the digesting sludge either mechanically or by use of compressed digestion gases. Mechanical mixing is normally accomplished by an impeller suspended from the cover of the digester. The common methods of gas mixing are a) the injection of compressed gas to loft re-circulating sludge from the bottom and split it out on top and b) supplying compressed gas to a number of diffusers mounted in the center at the bottom of the tank.

A completely mixed digester may be either a fixed or a floating-cover tank. Digesting sludge is displaced when raw sludge is pumped into a fixed cover digester. By use of a floating cover, tank volume is available for the storage of digesting sludge, and withdrawals do not have to coincide with introduction of raw sludge.

6.3.3 Filtration/Screening

Filtration is used to separate non-settleable solids from water and wastewater by passing it through a porous medium. The most common system is filtration through a layered bed of granular media underlain by a finer sand (Viessman and Hammer 1999b).

Gravity granular media filtration – Gravity filtration through beds of granular media is the most common method of removing colloidal impurities in water processing and tertiary treatment of wastewater. The mechanisms involved in removing suspended solids consist of interception, straining, flocculation, and sedimentation. Surface straining and interstitial removal result in accumulation of deposits in the upper portion of the filter media. Because of the reduction in the pore area, the velocity of water through the remaining voids increases, shearing off pieces of captured floc and carrying impurities deeper into the filter bed. The effective zone of removal passes deeper and deeper into the filter. Turbulence and the resulting increased particle contact within pores promote flocculation, resulting in trapping of the larger floc particles. Microscopic particulate matter in raw water that has not been chemically treated will pass through the relatively larger pores of a filter bed. Optimum filtration occurs when impurities in the water and coagulant concentration cause in-depth filtration. The impurities neither pass through the bed nor are all strained out on the surface, but a significant amount of flocculated solids are removed throughout the entire depth of the filter.

Traditional filtration – Typical scheme for processing surface supplies to drinking water quality consists of flocculation with a chemical coagulant and sedimentation prior to filtration (Fig. 6.3). Under the force of gravity, water passes downward through the media that collect the floc and particles. When the media become filled or solids break through, a filter bed is cleaned by backwashing where upward flow fluidizes the media and conveys away the impurities that have accumulated in the bed.

Flow control through gravity filters – Constant-level control employs an effluent control valve that automatically adjusts the filter discharge to maintain a preset water level above the bed to the filter box. If the filtration rate is too high, a dropping water level throttles the valve. Conversely, as the rate decreases, resulting from the

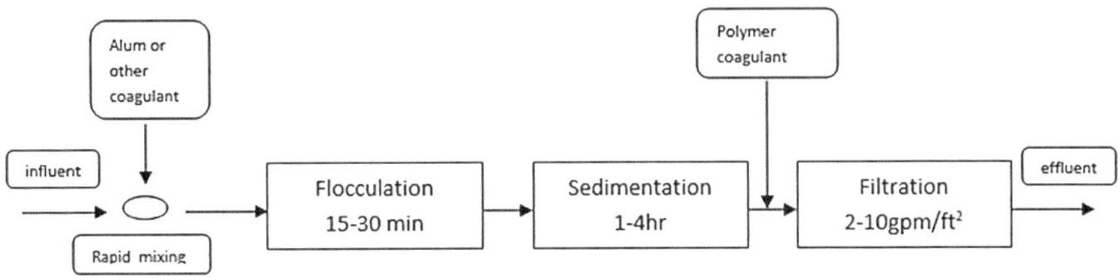

Fig. 6.3 Traditional surface water treatment system

accumulation of impurities in the bed, the tendency of the water level to rise opens the control valve. From uniform operation, the plant inflow must remain unchanged, and the effluent water level in the clear well is held constant because of the hydraulic connection to the water above the bed. In addition to cost and the mechanical complexity, a potential disadvantage for this type of control system is the possibility of momentary surges in filtration rates caused by an increased plant flow or sudden rise in the water level above the operating filters when one unit is taken out of service for backwashing. Sudden filtration rate increases through partially dirty filters can result in accumulated impurities being flushed from the bed in the effluent.

Pressure filters – Pressure filters have the granular media and underdrains contained in a steel tank. Water is pumped through the filter under pressure, and the media are washed by reversing flow through the bed, flushing out the impurities. Filtration rates are comparable to gravity filters; however the maximum head loss can be significantly greater since it is a function of the input pump pressure rather than static water levels.

6.3.4 Miscellaneous

6.3.4.1 Water Coagulation

Coagulation is the process of adding chemicals to surface waters to collect particulate matter and colloids into clusters that can be removed from the solution by subsequent sedimentation and filtration through granular media. In addition to improving aesthetics, the purpose of clarification is to remove substance that interferes with the disinfection of water by chlorine (Viessman and Hammer 1999b).

Two basic mechanisms have been defined in destabilizing a colloidal suspension to produce floc for subsequent solids-separation processes: Coagulation reduces the net electrical repulsive forces at particle surfaces by adding coagulant chemicals, whereas flocculation is agglomeration of the destabilized particles by chemical joining and bridging. The primary purpose of chemical treatment is to agglomerate particulate matter and colloids into floc that can be separated from the water by sedimentation and filtration. In water treatment, coagulation and flocculation are used to destabilize turbidity, color, odor-producing compounds, pathogens, and other contaminants. In wastewater reclamation, coagulation precedes tertiary filtration necessary to clarify a biologically treated effluent for effective chemical disinfection.

Coagulants – The most widely used coagulants for water and wastewater treatment are aluminum and ion salts. The common salt is aluminum sulfate, which is a good coagulant for water containing organic matter. Iron coagulants are effective over a wider pH range and are generally more effective in removing color from water, but they usually cost more. For some waters, cationic polymer is effective as a primary coagulant, but polymers are more commonly applied as coagulant aids. The final choice of coagulants and chemical aids is based on economics.

Aluminum sulfate (filter alum) – Aluminum sulfate is the standard coagulant used in water treatment. The commercial product strength ranges from 15% to 22% as Al_2O_3 with a hydration of about 14 moles of water. A formula used for filter alum is $Al_2\,(CO_4)_3.14.3H_2O$ with a molecular weight of 600. The material is commonly shipped and fed in a dry granular form, although it is available as a powder or liquid alum syrup.

Aluminum sulfate reacts with natural alkalinity in water to form aluminum hydroxide floc. Each mg/l of alum decreases water alkalinity by 0.50 mg/l (as $CaCO_3$ produces 0.44 mg/l of carbon dioxide). Production of carbon dioxide is undesirable since this increases the corrosiveness of water. If water does not contain sufficient alkalinity to react with the alum, lime or soda ash is fed to provide the necessary alkalinity. An advantage of using sodium carbonate (soda ash) is that unlike lime it does not increase water hardness, only corrosiveness of water. The dosage of alum used in water treatment is in the range 5–50 mg/l. The effective pH range for alum coagulation is 5.5–8.0. Alum is preferred in treating relatively high-quality surface waters because it is the only chemical needed for coagulation.

Ferrous sulfate – Commercial ferrous sulfate has strength of 55% $FeSO_4$ and is supplied as green crystal or granulates for dry feeding. Ferrous sulfate reacts with natural alkalinity, but the response is much slower than that between alum and natural alkalinity. Lime is generally added to raise the pH to the point where ferrous ions are predicted as ferric hydroxide by the caustic alkalinity. Treatment using sulfate and lime adds some hardness but no corrosiveness to the water. This process is usually cheaper than alum coagulation, but the dosing operation with two chemicals is more difficult. If excess lime is used, the water may require treatment for stabilization.

Chlorinated copperas treatment is a second method of using ferrous sulfate. In this process chlorine is used to oxidize the ferrous sulfate to ferric sulfate:

$$2\mathrm{FeSO}_{4.}7\mathrm{H}_2\mathrm{O} + 1.5\mathrm{Cl}_2 \rightarrow 2\mathrm{Fe}\left(\mathrm{SO}_4\right)_3 + \mathrm{FeCl}_3 + 21\mathrm{H}_2\mathrm{O}$$

Oxidation is generally performed by adding ferrous sulfate to the discharge from a solution-fed chlorinator. Theoretically, 1.0 lb. of chlorine oxidize 7.8 lb. of copperas. In practice, chlorine feed slightly in excess of the theoretical amount produces good results. Ferric sulfate and ferric chloride react with natural alkalinity or lime, as illustrated by the following reactions with ferric chloride:

$$2\mathrm{FeCl}_3 + 3Ca\left(HCO_3\right)_2 \rightarrow 2Fe\left(OH_3\right) + 3\mathrm{CaCl}_2 + 6CO_2$$
$$2\mathrm{FeCl}_3 + 2Ca\left(OH\right)_2 \rightarrow 2Fe\left(OH_3\right) + 3\mathrm{CaCl}_2$$

Color in water is generally not affected by copperas and lime treatment, whereas chlorinated copperas is effective in the removal of color.

Ferric salts – Ferric salts and ferric chloride are available as coagulant under a variety of trade names. Advantages of the ferric coagulants are that (1) coagulation is possible over a wider pH range, generally pH 4–9 for most waters; (2) the precipitate produced is a heavy quick-settling floc; and (3) they are more effective in the removal of color, taste, and odor compounds. Ferric sulfate is available in crystalline form and may be fed using dry or liquid feeders. Ferric sulfate, although not as aggressive as ferric chloride or chlorinated copperas, must be handled by corrosive-resistant equipment. Ferric chloride is supplied in either crystalline or liquid form. Although ferric chloride can be used in water treatment, its most frequent application is in wastewater treatment (e.g., as a waste sludge-conditioning chemical in combination with lime prior to mechanical dewatering).

Sodium aluminate – Sodium aluminate is an essential alumina dissolved in sodium hydroxide. The principal use of sodium aluminate is as an additional coagulant with aluminum sulfate, generally in the treatment of boiler water. The limited employment of sodium aluminate is dictated by its high cost. Sodium aluminate is alkaline in its reaction, instead of acidic like other coagulants. Reactions of sodium aluminate with aluminum sulfate and carbon dioxide are:

$$6NaA1\text{O}_2 + Al_2\left(\text{SO}_4\right)_{3.}\ 14.3\text{H}_2\text{O} \rightarrow 8Al_2\left(OH_3\right) + 3Na_2\text{SO}_4 + 2.3\text{H}_2\text{O}$$

$$2NaA1O_2 + \text{C}_2\text{O} + 3\text{H}_2\text{O} \rightarrow 2Al\left(OH\right)_3 + Na_2CO_3$$

Polymers – Cationic polymers can be effective for coagulation, without hydrolyzing metals, by producing destabilization through charge neutralization and interparticle bridging. In contrast to metal coagulants, rapid mixing for low-molecular-weight cationic polymer is performed at reduced turbulence relative to metal coagulants to prevent breakup of the more fragile polymer floc. Feed control of polymer coagulants must be precise since effective dosage has a narrow band; overdosing or underdosing results in destabilization of the colloidal particles. Because polymers do not affect pH, they are advantageous for coagulating low-alkalinity waters. Another potential advantage of polymers for coagulation is commonly 0.5–1.5 mg/l, which is much less than the dosage of metal coagulants.

Acids and alkalies – Acids and alkalies are added to water to adjust the pH for optimum coagulation. Typical acids used to lower the pH are sulfuric and phosphoric. Alkalies used to raise the pH are lime, sodium hydroxide, and soda ash. Hydrated lime, with about 70% available CaO, is suitable for dry feeding but costs more than quicklime, which is 90% CaO. The latter must be slaked (combined with water) and fed as lime slurry. Soda is 98% sodium carbonate and can be fed dry but is more expensive than lime. Sodium hydroxide is purchased and fed as a concentrated solution.

6.3.4.2 Chemical Disinfection

Chlorine is the common chemical used for disinfection of water and wastewater. One alternative to chlorine is chlorine dioxide, although it is usually applied for taste and odor control. Another alternative is ozone, but because of high cost it is rarely applied solely for disinfection. Other benefits attributed to ozonation are trihalomethane reduction, taste and odor control, and improved coagulation.

Chemistry of chlorination – Chlorine gas is soluble in water (7160 mg/l at 20 °C and 1 atm) and hydrolyzes rapidly to form hypochlorous acid:

$$\mathrm{Cl_2 + H_2O \rightleftarrows HOCl + H^+ + Cl^{-1}}$$

Hydrolysis goes virtually to completion at pH values and concentrations normally experienced in water treatment and waste treatment operation. The dissociation rate from hypochlorous acid to hypochlorite ion is sufficiently rapid so that equilibrium is maintained even though the former is being continuously consumed. If a reducing agent is put into water containing free available chlorine, the unconsumed residual redistributes itself between HOCl and OCl^-.

Chlorine reacts with ammonia in water to form chloramines as follows:

$$\begin{aligned}
&\mathrm{HOCl} + NH_3 \quad \rightarrow \mathrm{H_2O} + NH_2Cl(\text{monochloramine})\\
&\mathrm{HOCl} + NH_2Cl \rightarrow \mathrm{H_2O} + \mathrm{NHCl_2}(\text{dichloramine})\\
&\mathrm{HOCl} + \mathrm{NHCl_2} \rightarrow \mathrm{H_2O} + NCl_3(\text{dichloramine})
\end{aligned}$$

The chloramines formed depend on the pH of the water, the amount of ammonia available, and the temperature. In the pH range 4.5–8.5, monochloramine and dichloramine are formed. At room temperature, monochloramine exists alone above pH 8.5, and dichloramine occurs alone at pH 4.5. Below pH 4.4, trichloramine is produced.

Free available residual chlorine is that residual chlorine existing in water as hydrochloric acid or hypochlorite ion. Combined available residual chlorine is that residual existing in chemical combination with ammonia (chloramines) or organic nitrogen compounds. Chlorine demand is the difference between the amount added to water and the quantity of free and combined available chlorine remaining at the end of a specified contact period.

When chlorine is added to water containing reducing agents and ammonia, residuals develop that yield a curve similar to that in Fig. 6.4. Chlorine reacts first with reducing agents present and develops no measurable residual, as shown by the portion of the curve extending from A to B. The chlorine dosage at B is the amount required to meet the demand excreted by the reducing agents (those common to water and wastewater include nitrites, ferrous ions, and hydrogen sulfide).

The addition to chlorine in excess of that required up to point B results in the formation of chloramines. Monochloramines and dichloramines are usually considered together because there is little control over which will be formed. The quantities of each are determined primarily by pH. Chloramines thus established show an available chlorine residual and are effective as disinfectants. When all the ammonia has been reacted with, free available chlorine residual increases, the previously produced chloramines are oxidized. These results in the creation of oxidized nitrogen compounds, such as nitrous, as see on the curve between C and D. Once most of the chloramines are oxidized, additional chlorine applied to the water creates an equal residual, as indicated by the rising at point D. Point D is generally referred as the

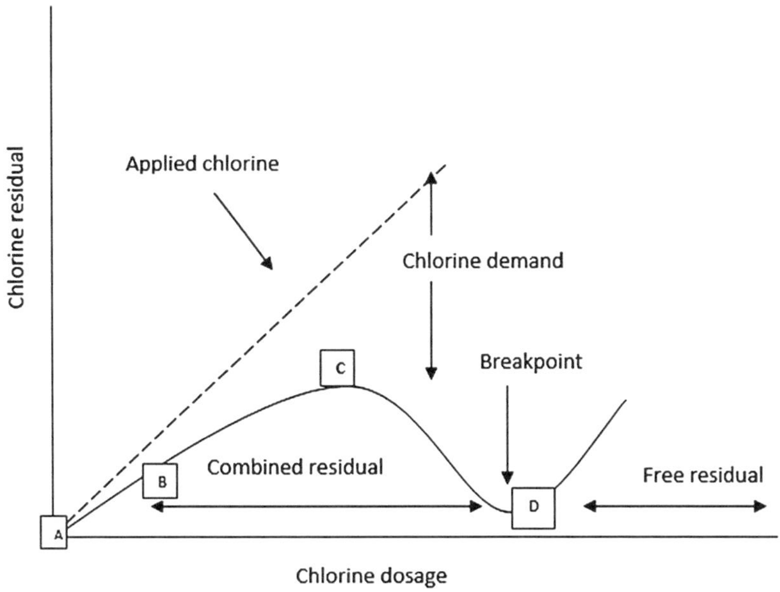

Fig. 6.4 Chlorine residual curve for breakpoint

breakpoint; beyond it, all added residual is free available chlorine; some resistant chloramines can still be present beyond D, but their relative importance is small.

6.3.4.3 Parameters to Be Analyzed After Wastewater Treatment

The major controlling variables of wastewater quality after treatment are pH, alkalinity and acidity, and total dissolved solids (TDS). This section discusses the significant water quality parameters to be checked after subjecting the wastewater to anaerobic or aerobic treatments.

Ammonia and other nitrogenous materials in water tend to be oxidized by aerobic bacteria to nitrite and nitrate. Hence, all organic compounds containing nitrogen is a potential nitrate source. Seafood wastewater will be rich sources of nitrogenous compounds. Due to this reason, wastewater from seafood plants should be analyzed for the level of nitrite, nitrate, and ammonia. Hydrogen sulfide (H_2S) is a colorless gas with the characteristically disagreeable odor of rotten eggs and is formed when sulfide ion (S^{2-}) reacts reversibly with water to form HS^- and H_2S by the reverse reaction. Since seafood is a potential source of essential amino acids including sulfur-containing cysteine and methionine, determination of sulfide content after treatment is recommended.

Acidity from sources other than dissolved CO_2 is often an indicator of pollution. Alkalinity is often a good indicator of total DIC (bicarbonates and carbonate anions) present. Hardness above 100 mg/L can cause significant scale deposits to form in boilers. The softer the water, the greater the tendency to dissolve metals from pipes of water distribution systems. An ideal goal for total hardness is about

70–90 mg/L. Municipal treatment sometimes allows up to 1150 mg/L of total hardness in order to reduce chemical costs and sludge production from precipitation of calcium and magnesium carbonates.

BOD is an indicator of the water body to become depleted in oxygen. It will possibly become anaerobic because of biodegradation. Chemical oxygen (COD) is the amount of oxygen consumed when all the organic matter in a given volume of water is chemically oxidized to CO_2 and H_2O by a strong chemical oxidant. COD is used as a measure of general pollution.

Total solids are the solids remaining after evaporating the water from an unfiltered sample. It includes two subclasses of solids: (1) Total suspended solids in water are organic and mineral particulate matter that do not pass through a 0.45 μm filter. They include silt, clay, metal oxides, algae, bacteria, and fungi. TSS is removed by flocculation and filtering. It contributes to turbidity and limits light penetration for photosynthesis and visibility in water. (2) Total dissolved solids are substances that will pass through a 0.45 μm filter. If the water passed through the filter is evaporated, the TDS will remain behind as a solid residue. TDS may include humic acids, tannin, dissolved minerals and salts, and pyrogens. TDS and salinity both indicate dissolved salts.

References

Green, J. H., & Mattick, J. F. (1977). Possible methods for the utilization or disposal of fishery solid wastes. *Journal of Food Quality, 1*, 229–251.

Knuckey, I., Clive, S., Surapaneni, A., & Ashcroft, W. (2004). Utilization of seafood processing waste – Challenges and opportunities. In 3rd Australian New Zealand soil conference 5–9 December, University of Sydney, Australia.

Lin, J., Ma, Y., & Huang, C. (1998). Alkaline hydrolysis of the sludge generated from a high-strength, nitrogenous-wastewater biological-treatment process. *Bioresource Technology, 65*, 35–42.

Mason, C. A., Haner, A., & Hamer, G. (1992). Aerobic thermophilic waste sludge treatment. *Water Science and Technology, 25*(1), 113–118.

Sasidharan, A., & Mathew, S. (2011). Current status of fish waste management in Kerala, India. *Waste Management, 31*, 1897–1900. https://doi.org/10.1016/j.wasman.2011.05.002. https://www.researchgate.net/publication/321576834

Sawayama, S., Inoue, S., Yagishita, T., Ogi, T., & Yokoyama, S. Y. (1995). Thermochemical liquidization and anaerobic treatment of dewatered sewage sludge. *Journal of Fermentation Bioengineering, 79*(3), 300–302.

Viessman, W., Jr., & Hammer, M. J. (1999a). *Water supply and pollution control* (6th ed., pp. 384–406). Menlo Park: Addison Wesley Longman.

Viessman, W., Jr., & Hammer, M. J. (1999b). *Water supply and pollution control* (6th ed., pp. 432–457). Menlo Park: Addison Wesley Longman.

Bioactive Compounds from Marine Sources

7

Over the past several decades, scientists have given increasing attention to marine organisms and the secondary metabolites they produce. Investigations on marine natural products have focused on areas like marine nutraceuticals, biopolymers, biofilms, CNS-membrane active toxins and ion channel effectors, antifouling, anticancer and antiviral agents, tumor promoters, anti-inflammatory agents, as well as metabolites that control microfilament-mediated processes. The marine natural products field is encouraging a multidisciplinary approach between biologists, chemists, and pharmacologists. It is also important to note that the number of compounds reported annually is increasing steadily, indicating that marine organisms will continue to be sufficient sources of natural bioactive substances.

7.1 Marine Biopolymers and Derivatives

Fish processing generates solid wastes (skin, scales, and bones) that can be as high as 80% of the original raw material and are excellent sources of high value products. The utilization of fish wastes helps to eliminate harmful environmental aspects of pollution and improve quality in fish processing. An important waste reduction strategy for the industry is the recovery of marketable by-products from fish wastes. The two most important marine biopolymers collagen as well as the polysaccharide polymer, chitin, and their derivatives gelatin and chitosan, respectively, are a promising entity in this regard.

7.1.1 Collagen and Gelatin

Collagen is the most abundant and ubiquitous protein in the body of almost all metazoan organisms, as a major component of extracellular matrix (Balian and Bowes 1977) and renowned as the most abundant protein in mammals, making up from 25% to 35% of the whole-body protein. Collagen in the form of elongated

S. Mathew et al., *Fish and Fishery Products Analysis*,
https://doi.org/10.1007/978-981-32-9574-2_7

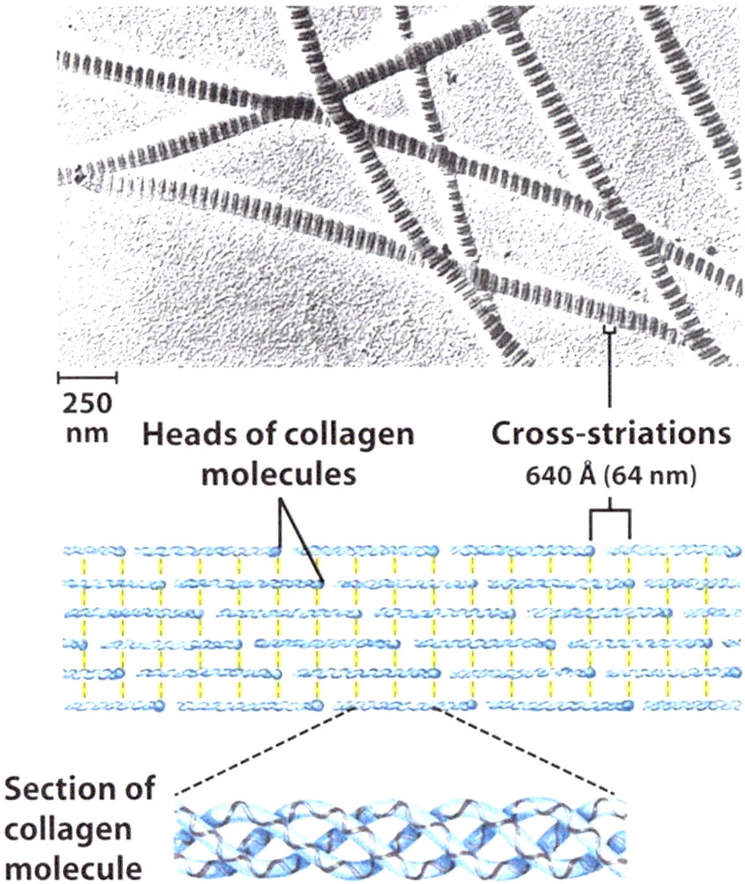

Fig. 7.1 Collagen triple helical structure. (Source: Lehninger Principles of Biochemistry, Sixth edition, 2013. W. H. Freeman and Company)

fibrils is mostly found in fibrous tissues such as tendons, ligaments, and the skin. It is also abundant in corneas, bones, blood vessels, gut, vertebral discs, dentine in teeth, and muscle tissue as a major component of the endomysium. Collagen constitutes 1 to 2% of muscle tissue and accounts for 6% of the weight of strong, tendinous muscles. The fibroblast is the most common cell that creates collagen (Fig. 7.1).

The name collagen comes from the Greek word "kolla" denoting glue and the suffix "gen" meaning producing. Most characteristic feature of this family of proteins is the capacity to form supramolecular aggregates in extracellular spaces (Piez 1984). Collagen molecule comprises of three long left-handed α-helix polypeptide chains, with each polypeptide intertwined around each other towards the right, to form a structure called as the collagen triple helix. In 1954 Dr. G. N. Ramachandran a biophysicist, along with Gopinath Kartha, successfully deciphered the triple helical structure of collagen – the most abundant protein in higher vertebrates. Each polypeptide in the collagen triple helix is characterized by the presence of a glycine-X-Y sequence, in which X and Y are usually amino acids, proline, and hydroxyproline involved in interchain hydrogen bonding and thus conferring significantly towards the stability of the polypeptide (Piez 1984). The collagens that form fibrils are types I, II, III, V, and XL, while types IV, VIII, and X form network-like structures (Gathercole and Keller 1991) (Table 7.1).

7.1.1.1 Fish Collagen

Marine collagen has gained immense attention in the recent past as an appropriate alternative to mammalian collagen. The marine collagen is extracted from various marine sources and their by-products including processing, viz., fish skin, scales, bones, etc. Collagen finds wide application in food, pharmaceutical, and biomedical industries. The type and functional properties of these collagens vary with their molecular structure, extraction process, and source. Marine collagen is usually extracted using mild acids such as acetic acid or enzymes, at low temperature, to prevent denaturation and changes in molecular properties (Raman and Gopakumar 2018).

A number of studies have focused on the extraction of collagen from different species of fish including Antarctic ice fish (Rigby 1968), carp (Kimura et al. 1991), plaice (Montero et al. 1995), hake (*Merluccius hubbsi*) (Fernandez-Diaz et al. 2001), Baltic cod (*Gadus morhua*) (Zelechowska et al. 2010), pufferfish (Nagai and Suzuki 2002), megrim (Montero and Gomez-Guillen 2000), sea bass (Nagai 2004), bone and scales of the subtropical fish black drum (*Pogonias cromis*), sheepshead seabream (*Archosargus probatocephalus*) (Ogawa et al. 2004), Nile perch (*Lates niloticus*) (Muyonga et al. 2004), bigeye snapper (*Priacanthus tayenus*) (Kittiphattanabawon et al. 2005), brown stripe red snapper (*Lutjanus vitta*) (Jongjareonrak et al. 2005), skin of brown-backed toadfish (*Lagocephalus gloveri*) (Senaratne et al. 2006), catfish (*Ictalurus punctatus*) (Liu et al. 2007a), grass carp (*Ctenopharyngodon idella*) (Zhang et al. 2007), deep-sea redfish (*Sebastes mentella*) (Wang et al. 2007), skate (*Raja kenojei*) (Hwang et al. 2007), walleye pollock (*Theragra chalcogramma*) (Yan et al. 2008), and pipefish (*Syngnathus schlegeli*) (Khan et al. 2009), as well as flatfish (Heu et al. 2010), surf smelt (*Hypomesus pretiosus japonicus*) (Nagai et al. 2010), and unicorn leatherjacket (*Aluterus monoceros*) (Ahmad et al. 2010). Sankar et al. (2008) prepared fish collagen sheet from the scales of *Lates calcarifer*. Muthukumar et al. (2014) reported the use of fish scales to prepare porous collagen sponges (CS) incorporated with mupirocin and extracts of *Macrotyloma uniflorum* conferring antimicrobial activity. Chen et al. (2010), Linnartz et al. (2013) reported on the utilization of fish scale-derived scaffold for corneal regeneration.

Collagen has also been studied from a number of marine invertebrates including sea urchins (Barbaglio et al. 2014), sea cucumbers (Arief et al. 2013), and marine sponge (Pozzolini et al. 2018). Marine sources of collagen also have lots of advantages over land animals, such as being environmentally friendly, high quantity of collagen, having biological toxins that are almost negligible, better absorption due to low molecular weight, minimal inflammatory response, less religious and ethical

Table 7.1 Different types of collagen

Collagen family	Type	Tissue distribution	Application
Fibril-forming	I	Skin, bone, tendon (non-cartilage), dermis, cornea, ligament	Membranes for guided tissue regeneration
	II	Cartilage, vitreous humor, nucleus pulposus, lung, cornea, skin, bone	Cartilage repair, arthritis treatment
	III	Extensible connective tissue (skin, lung, vascular system, viz., artery), skin, blood vessel wall, and reticular fibers of tissues (lungs, liver, spleen)	Hemostats and tissue sealants
	V	Co-distributed with type I, especially in the cornea	Corneal treatment
	XI	Along with type II, vitreous body and cartilage	mAbs development for osteoarthritis
Basement membrane	IV	Basement membrane	Attachment enhancer of cell culture membrane(mouse neuroblastoma) and diabetic nephropathy indicator
Microfibrillar	VI	Muscle, dermis, placenta, lungs, intervertebral disc, cartilage	Hemostat
Anchoring	VII	Dermal epidermal junction, skin, cervix, oral mucous	Treatment of dystrophic epidermolysis bullosa (DEB)
FACIT	IX	Along with type II in the cornea, cartilage, and vitreous body	
	XII	Tendon, perichondrium, and ligaments	Regulator in early stages fibrillogenesis
	IX	Along with type I in vessel walls, placenta, liver, dermis, and lungs	
	XIX	Many tissues, human rhabdomyosarcoma	Antiangiogenic and antitumoral properties
	XX	Corneal epithelium, sternal cartilage, embryonic skin, tendon	
	XXI	Many tissues, blood vessel wall	Contribute to matrix assembly of vascular network during blood vessel formation
Transmembrane	XIII	NM junction, skin, hair follicle, intestine, lungs	Involved in inflammation and vasculogenesis, regulate bone mass
	XVII	Epithelia, skin hemidesmosomes	Teeth formation
Multiplexins	XV	Associated with collagens close to basement membranes, kidney, smooth muscle cells, pancreas	
	XVI	Many tissues including keratinocytes and fibroblasts	Drug target and biomarker
	XVIII	Close structural homologue of XV, liver, lungs	Retinal structure, closure of neural tube

Source: Raman and Gopakumar (2018)

constraints, metabolically compatible, and few regulatory and quality control problems (Rahman 2019).

7.1.1.2 Fish Gelatin

Fish gelatin is a protein product produced by partial hydrolysis of collagen-rich fish skin (or) scale material. The gelatin molecule is made up of amino acids joined together by amide linkages in a long molecular chain. Gelatin is a versatile hydrocolloid and is widely applied in food, pharmaceutical, cosmetic, medical, and photographic products. The most important properties of gelatin are thermoreversible gel formation, texturing, thickening, high water-binding capacity, emulsion formation and stabilization, foam formation, protective colloidal function, and adhesion/cohesion. Gelatin is widely used in the food industry such as an ingredient to improve the elasticity, the consistency, and the stability of foods (See et al. 2010).

7.1.1.3 Bioactive Roles of Collagen

Generally, collagen has a wide range of application in cosmetics, biomedical, pharmaceutical, leather, and food industries. It is also regarded as one of the most useful biomaterials, mainly due to its biocompatibility, biodegradability, low toxicity, and immunogenicity, compared to other natural polymers (Bareil et al. 2010). Also, its structure and interaction with host tissue make it an efficient matrix to produce diverse forms of biomaterials and therefore used extensively in various biomedical fields such as scaffold in tissue engineering applications. Although synthetic polymers have been widely used as a biodegradable material in tissue engineering, bio-derived materials such as collagen are being increasingly applied in regenerative medicine.

Sahithi et al. (2013) reported that collagen becomes a very useful carrier for the delivery of various kinds of drugs and agents like growth factors. Collagen is capable of being prepared into cross-linked compacted solids or into latticelike gels. Resorbable forms of collagen have been used to dress oral wounds, for closure of graft and extraction sites, and to promote healing. Collagen-based membranes also have been used in periodontal and implant therapy as barriers to prevent epithelial migration and allow cells with regenerative capacity to repopulate the defective area (Khan et al. 2011). Kruger et al. (2013) reported on the use of collagen scaffolds in bone sialoprotein-mediated bone regeneration. Chattopadhyay and Raines (2014) review the structure and molecular interactions of collagen in vivo on the use of natural collagen in sponges, injectables, films and membranes, dressings, skin grafts, and development of synthetic collagen mimetic peptides as pylons to anchor cytoactive agents in wound beds.

Medical collagen supplements, which are available as collagen creams, collagen injections, and collagen capsules, replenish the level of collagen in the body and reduce the signs of aging. The human skin has difficulty in absorbing animal collagen so that collagen treatment had to be supplemented with a hypodermic injection treatment, which typically is expensive and carries the risk of infection. Animal (terrestrial vertebrates) collagen also has the potential to spread bovine spongiform encephalopathy [BSE] better known as mad cow disease and is another threat to humans. As a protein extracted from fish, it is mainly used as a surgical suture, as biomimetic scaffold, and as a cosmetic aid to decrease the appearance of wrinkles since collagen makes up to 70% of protein in the skin and helps to keep it elastic.

7.1.1.4 Uses of Gelatin (Ledward 2000)

Applications in Food and Confectionery

The food industry is still one of the major consumers of gelatins. Gelatin desserts, all types of jellies, are examples of food products that take advantage of the thermoreversible gel formation and the "melt-in-the-mouth" texture of gelatin (Gomez-Guillen et al. 2011). Gelatin is also essential in some dairy products and pastries to provide the quality of these storable products as required by consumers. Milk is primarily a complex oil-in-water emulsion. The addition of gelatin to dairy products improves the emulsifying capacity as gelatin molecules associate to the surface of the fat droplets and thereby reduce the surface tension towards the aqueous phase. Syneresis is the phenomenon of liquid being exuded from a gel, and this is usually undesirable since a product will become less appealing to the consumers. Gelatin can be added to dairy products to bind whey and in this way hamper secretion of aqueous whey from, e.g., yogurts, curds, and cream cheese. By adding gelatin to foamed milk-based desserts like yogurt, curds, ice creams, and mousses, gelatin depresses the surface tension of water, enabling formation of foam by mechanical whipping or injection of gas. Ice cream gelatin will also influence the size and distribution of ice crystals formed and thereby influence the texture and mouthfeel of the final ice cream. The meat processing industry also applies gelatin to their products for several reasons, but the ability to bind water and meat juices and to secure good texture and taste is very important. Gelatin is also widely used in low-fat (fat replacer), low-carb (binding agent), and low-calorie (fat replacer and binding agent) food products.

Gelatin desserts made from various gelatins may provide variety in texture and gel-melting behavior, offering new product development opportunities. Furthermore, the lower melting temperature in gel desserts made from fish gelatins may accelerate flavor release (Regenstein and Zhou 2007).

The confectionery industry uses gelatin not only for its thermoreversible gelling properties but also for foam formation and stabilization, binding, emulsification, and controlling sugar crystallization. Gelatins with unique properties are developed especially for the confectionary industry. Examples of products containing gelatin are fruit gummies, mallows, meringues, caramels, bar products, and sugar-coated candies.

Pharmaceutical Applications

Gelatin is an important and versatile excipient for pharmaceutical and medical applications. The widespread utilization of gelatin in pharmaceuticals and medical devices is due to several excellent properties such that it is established as a pharmaceutical excipient, tolerated in food (nontoxic, non-allergenic), GRAS status, excellent biocompatibility, high quality (purity), low immunological activity, and controllable physical parameters.

Gelatin is utilized in plasma substitutes used in emergency medicine and surgery; in vitamin coatings, pastilles, and tablets; in the production of globules, paste dressings, and sponges; and in the formulation of new vaccines. About 90% of all gelatin of pharmaceutical grade is applied in capsule production – soft and hard gelatin capsules.

Nutritional and Health Properties

Gelatin is a high-quality source of protein and free of cholesterol and sugar and contains practically no fat. Gelatin is easily digested and completely broken down by the human organism. Gelatin is applied in food products to enrich the protein content and to reduce the amount of carbohydrates, the salt concentration, and the amount of fat in low-fat products and as a carrier for vitamins. Some international studies suggest that gelatin may have a preventive and regenerative effect on the skeleton and locomotor system – especially bones, cartilage, tendons, and ligaments. It is also suggested that gelatin may help fortify hair if taken regularly and strengthen the connective tissue, thus ensuring firm skin, shiny hair, and strong fingernails.

Cosmetic Applications

The healthcare market is one of the fastest growing markets in the world. Collagen and gelatin (from bovine, porcine, and fish sources) play important roles in skin and hair products as functional ingredients. Gelatin hydrolysates are, for example, added to skin creams to improve the water-binding capacity, to reduce transepidermal water loss, and to improve skin feel.

The classical food and cosmetic and pharmaceutical properties of gelatin are based mainly on its gel-forming and viscoelastic properties. Recently, and especially in the food industry, an increasing number of new applications have been found for gelatin in products such as emulsifiers, foaming agents, colloid stabilizers, fining agents, biodegradable packaging materials, and microencapsulating agents, in line with the growing trend to replace synthetic agents with more natural ones (Gomez-Guillen et al. 2011).

Gelatin-Based Biomaterial Scaffolds

The development of antioxidant and/or antimicrobial fish gelatin films has drawn a lot of attention from researchers, mainly by introducing natural components into the polymer matrix. The addition of a variety of natural extracts has also been carried out to investigate their effect on the antioxidant activity of films. Antimicrobial fish skin gelatin films are a promising entity in retaining antibacterial activity and inhibiting bacterial contaminants. Zinc oxide nanoparticles on the antimicrobial activity of films based on tilapia skin and fish protein isolate provide safety and extend the shelf life of packaged foods. Silver-copper nanoparticles had the same positive effect on the antimicrobial properties of fish skin gelatin films and could improve their mechanical and thermal properties. Chitosan is another agent that has been added to fish gelatin-based films due to its antimicrobial action against a variety of fungi and Gram-positive and Gram-negative bacteria. Chitosan-gelatin film could be used as active packaging to preserve fish under refrigerated storage.

7.1.2 Chitin and Chitosan

Chitin and chitosan are known natural polysaccharide-based biodegradable polymers, which are extracted particularly from the exoskeletons of marine invertebrates. Chitin also exists widely in cell walls of some microorganisms such as fungi, molds, and yeasts. Chitin and its most important derivative, chitosan, have a number of useful physical and chemical properties, including high strength, biodegradability, and nontoxicity. Chitosan exists naturally only in a few species of fungi. The structure and properties of chitin are detailed in Sect. 1.1.

Chitosans are deacetylated forms of chitin, and their polymers differ in their degree of acetylation. Chitin, as well as chitosan, has high molecular weight, and both are viscous and highly hydrophobic and are insoluble in water. Many scientific researchers have been focused on beneficial effects of chitosan on human health; thus it has been considered also as functional fiber.

Chitin and chitosan and their derivatives exhibit diverse bioactivities, including antioxidant, antihypertensive, anti-inflammatory, anticoagulant, antitumor and anticancer, antimicrobial, hypocholesterolemic, and antidiabetic effects. By modulating and improving physiological functions, chitin and chitosan and their derivatives may provide novel therapeutic applications for the prevention or treatment of chronic diseases. Antioxidant activity of chitin and chitosan and their derivatives can be attributed to in vitro and in vivo free radical-scavenging activities and is one of the most crucial biomodulatory effects. Antioxidant effect of chitin and chitosan and their derivatives may also be used as functional ingredients in food formulations to promote consumer health and to improve the shelf life of food products (Ngo and Kim 2014).

Chitin has excellent biocompatibility, nontoxicity, and wound healing properties, so it has been widely applied in medical and healthcare fields for applications such as release capsules for drugs, man-made kidney membranes, anticoagulants, and immunity accelerants. Because of its high oxygen permeability, chitosan is used as a material for contact and intraocular lenses. Chitosan has also been found to expedite blood clotting. The fact that chitin compounds are biodegradable makes them particularly appropriate for use in drug delivery systems, as they can act as carriers that release drugs slowly – a property extremely valuable in cancer chemotherapy. Chitin has been evidenced as a potent innate immune stimulator of macrophages and other innate immune cells, and thus chitin is able to suppress allergen-induced type II allergic responses. Chitin and chitin derivatives have nonspecific antiviral and antitumor activities, and it was shown to have an effect on innate and adaptive immune responses including the ability to recruit and activate innate immune cells. Fibers, fabrics, sutures, and nonwoven fabrics from chitin are now available. For example, chitin nonwoven fabric has excellent characteristic as an artificial skin because of its good adhesion to the human body surface and its value in stimulating

Table 7.2 The bioactive and other roles of chitin and chitosan

Chitin/chitosan	Applications
Beads	Use in drug delivery systems
Microspheres	Enzyme immobilization substrate
Nanoparticles	As potent gene delivery vehicles
	Encapsulation of sensitive drugs
Coatings	Used in surface modification and as textile finishes
Fibers	Used as medical textiles and surgical sutures
Nanofibers	Useful for guided bone regeneration
Nonwoven bioactive fibers	Acts as a scaffold for nerve tissue regeneration and aid in the wound healing processes
Films	Application in wound care formulations, as dialysis membrane and antitumor agents
	Used as a semipermeable film for wound dressing
Powder	Used as an adsorbent for pharmaceutical and medical devices, surgical glove powder, and enzyme immobilization
Sponge	Mucosal hemostatic dressing, wound dressing, drug delivery, enzyme entrapment, artificial skin
Shaped objects	Employability in orthopedics and manufacture of contact lenses
Solutions	Used in cosmetics, as bacteriostatic agent, hemostatic agent, anticoagulants, antitumor agent, gene delivery systems, and spermicidal action
Gels	Used as drug delivery vehicle, as implants, coating, and in tissue engineering
	Used in wound dressing for wet treatment
Tablets	Acts as a compressed diluent, disintegrating agent, and excipient
Capsules	Targeted drug delivery vehicle

Source: Younes and Rinaudo (2015)

new skin formation. It promotes no immune antigen-antibody reaction and substantially accelerates healing and reduces pain. A chitosan compound is also being investigated as an inhibitor of the AIDS virus (Vo and Kim 2010). The bioactive roles of chitin and chitosan from marine sources are given in Table 7.2.

7.2 Proteoglycans and Glycosaminoglycans

7.2.1 Proteoglycans

Proteoglycans (PGs) are a family of complex macromolecules distinguished by the covalent attachment of one or more glycosaminoglycan (GAG) chain(s) (N- and /or O-linked oligosaccharides) with matrix proteins (core proteins). They are associated with each other and also with other major structural components of the matrix, viz., collagen and elastin, and play important roles in determining the structural organization of the matrix and in fundamental biological processes such as cell growth and development (Bulow and Hobert 2006). Although originally named and categorized on the basis of the GAG substituent, they are increasingly being viewed as products of gene families that encode the different core proteins. The classification into core proteins that carry KS (keratan sulfate), CS (chondroitin sulfate), DS (dermatan sulfate), or HS (heparin sulfate) is not always stringent because there are a considerable number of proteins that bear more than one kind of GAG side chains.

Proteoglycan core proteins are indispensable for the correct development of almost all metazoan organisms. The core proteins of the known proteoglycans vary in size from 11,000 to about 220,000 daltons (Doege et al. 1987). The number of glycosaminoglycan chains attached to the core varies from 1 to about 100. The GAG chains consist of unbranched, repeated disaccharides that are modified by sulfation and epimerization during synthesis in the Golgi apparatus.

Despite their great diversity, proteoglycans fall under three broad categories: cell surface proteoglycans, small proteoglycans of intracellular granules, and extracellular matrix proteoglycans. The proteoglycans at the cell surface are membrane proteins carrying heparan sulfate and/or chondroitin sulfate chains. A prototype of the small proteoglycans of intracellular granules is serglycin, the yolk sac tumor chondroitin sulfate proteoglycan. Extracellular matrix proteoglycans can further be separated into three groups: the basement membrane proteoglycans, the hyalectans-proteoglycans interacting with hyaluronan and lectins, and the small leucine-rich proteoglycans.

In addition to their structural functions, proteoglycans and GAGs play different roles in diverse processes such as enzyme regulation and cellular adhesion, growth, migration, or differentiation (Turnbull 2001). These processes are mediated by their binding capacities and focal sequestering of a number of biologically active proteins (David and Bernfield 1998). They have a number of important biologic roles like anticoagulant activity, anti-atherosclerotic, anti-inflammatory, and also as fat-clearing agent. Proteoglycans act also as tissue organizers, influence cell growth and the maturation of specialized tissues, play a role as biological filters and modulate growth factor activities, regulate collagen fibrillogenesis and skin tensile strength, affect tumor cell growth and invasion, and influence corneal transparency and neurite outgrowth (Arias and Fernandez 2008).

7.2.2 Glycosaminoglycans

Structurally the simplest and best known of the heteropolysaccharides are the glycosaminoglycans (GAGs). They are multifunctional polysaccharides composed of repeating disaccharide units. Variations are imposed on the disaccharide repeats by post-synthetic modifications in the form of sulfation and epimerization. These modifications often dictate the protein recognition properties and biological activities of GAGs. The disaccharide repeating units of the mucopolysaccharides are similar in structure but varied in composition. The general classification and properties of glycosaminoglycans are discussed in Sect. 1.1.6.3.

Based on the disaccharide composition, on the linkage type, and on the presence of sulfate groups, GAGs have been divided into six main groups: (1) hyaluronic acid

(HA), (2) chondroitin sulfate (CS), (3) dermatan sulfate (DS), (4) heparin, (5) heparan sulfate (HS), and (6) keratan sulfate (KS). Most sulfated GAG chains are covalently linked to core proteins to form proteoglycans. Hyaluronan shows no modifications and is not attached to a protein backbone (Spicer et al. 2002). The amount of carbohydrate in a proteoglycan is usually much greater than found in a glycoprotein and may comprise of up to 95% of its weight. The GAGs present in the proteoglycans are highly negatively charged because of the presence of carboxyl or sulfate groups on many of the sugar residues and hence bind polycations and cations such as sodium or potassium. Modifications of sugar residues create enormous molecular diversity, and GAGs have been suggested to be the most information-dense biological molecules (Turnbull et al. 2001).

GAGs such as hyaluronic acid, heparin, and chondroitin sulfate have several applications in the medical, veterinary, pharmaceutical, and cosmetic field because of their outstanding properties. Even though most interest among scientific community centers on the structures and functions of mammalian glycosaminoglycans, investigations on the occurrence of GAGs in tissues of aquatic vertebrates and invertebrates have shown that GAGs are widely distributed in the animal kingdom and play fundamental biological roles in animal development.

7.2.2.1 Glycosaminoglycans in Aquatic Vertebrates

Glycosaminoglycans in terrestrial vertebrates including mammalian sources have been extensively studied and characterized, whereas studies on glycosaminoglycans in aquatic/marine vertebrates are relatively scanty. The use of powdered shark cartilage for wound healing and as antiangiogenic, anti-arthritic, and antitumor agent from ancient times has been attributed to the presence of glycosaminoglycans. Sugahara et al. (1996) performed the structural analysis of chondroitin ABC lyase susceptible unsaturated hexasaccharides isolated from shark cartilage chondroitin sulfate D (CS-D). Summers et al. (2003) studied on the glycosaminoglycan and proteoglycan components in the fibrocartilaginous tendon of an elasmobranch fish (*Rhinoptera bonasus*). Nandini et al. (2005) isolated chondroitin/dermatan sulfate (CS/DS) hybrid chains from shark skin with an aim to develop therapeutic agents.

Uchisawa et al. (2001) studied on the binding of calcium ions with the chondroitin sulfate glycosaminoglycan chains of salmon nasal cartilage. Pfeiler et al. (2002) analyzed the structure and function of hyaluronan in developing fish larvae (leptocephali). Sakai et al. (2003) purified and characterized dermatan sulfate from the skin of the eel, *Anguilla japonica*. Bernhardt and Schachner (2000) demonstrated the role of chondroitin sulfates in the formation of the segmental motor nerves in **zebrafish embryos**. Bink **et al.** (2003) **reported on the importance of heparan sulfate sulfotransferases during muscle development in zebrafish.** Souza et al. (2007) showed that chondroitin sulfate and keratan sulfate are the major glycosaminoglycans present in the adult zebrafish *Danio rerio* (Chordata-Cyprinidae). Laparra et al. (2008) studied on the role of glycosaminoglycans isolated from cooked haddock in nonheme iron uptake.

7.2.2.2 Glycosaminoglycans in Aquatic Invertebrates

Studies show that invertebrate species are a rich source of sulfated polysaccharides with novel structures. Glycosaminoglycans have been extensively studied and characterized in ascidians, a specialized group of primitive invertebrate chordates. The external supportive tissue called the tunic in ascidians contains large amounts of sulfated polysaccharides, which is different from all previously described mammalian glycosaminoglycans. The preponderant polysaccharide in the ascidian tunic is a high-molecular-weight sulfated α-L-galactan. Dermatan sulfate like glycosaminoglycans have also been isolated and studied in other tunicates like *Ascidia nigra*, *Styela plicata*, and *Halocynthia pyriformis* (Pavao et al. 1998).

Heparin is present in most invertebrates like lobster, prawn, freshwater mussel, clam, shrimp, mangrove crab, sand dollar, etc. Kitagawa et al. (1997) isolated a novel chondroitin sulfate (CS–K) with a unique pentasaccharide sequence GlcA(3-sulfate)(β1-3)GalNAc(4- sulfate)(β1-4)(Fuc-α-3)GlcA(β 1–3)GalNAc(4-sulfate) from king crab cartilage and studied its differential susceptibility to chondroitinases and hyaluronidase. The occurrence of glycosaminoglycans in the body of the shrimp *Penaeus brasiliensis* was shown by Nader and Dietrich (1989).

The body wall of various sea cucumber species contains heterogeneous fucose-branched chondroitin sulfate, heparin, and heparan sulfate. The antithrombotic activity of a fucosylated chondroitin sulfate extracted from the body wall of a sea cucumber was assessed by Pacheco et al. (2000) using a stasis thrombosis model in rats and found that intravenous administration of the polysaccharide reduced thrombosis in a dose-dependent manner compared to the antithrombotic action of standard mammalian glycosaminoglycans, mainly heparin and dermatan sulfate.

A heparin-like substance able to bind antithrombin III (ATIII) is present in the marine clams *Anomalocardia brasiliana* and *Tivela mactroides*. Heparin having an AT-binding region similar to that of mammalian heparin was also purified from *Mercenaria mercenaria* (Ulrich and Boon 2001). Santos et al. (2002) characterized heparin present in epithelial layers of different tissues of the mollusk *Anomalocardia brasiliana* and gave a correlation of heparin and histamine concentration. Heparin with high anticoagulant activity possessing high content of antithrombin III binding site was also isolated and characterized from the clam *Tapes philippinarum* by Cesaretti et al. (2004). Heparin with anticoagulant activities varying from 150 to 320 IU/mg by the USP assay has also been isolated and characterized in the bivalves. Nader et al. (1984) purified CSs with different structures from a gastropod (*Pomacea* sp.) and two bivalves (*Tagelus gibbus* and *Anomalocardia brasiliana*) with the former comprising of a large amount of non-sulfated disaccharide.

Cephalopods including cuttlefish, squid, and octopus are important marine resources in greater world demand due to their rich taste. Hjerpe et al. (1983) examined acidic polysaccharides in head cartilage of the squid *Illex illecebrosus coidentii*. Karamanos et al. (1992) isolated and characterized the glycosaminoglycans in the skin and cornea of squid with peculiar sulfation pattern. Vynios and Tsiganos (1990) assessed the proteoglycans in the cranial cartilage of *Illex illecebrosus coidentii* using immunochemical methods. Kinoshita et al. (1997) isolated novel chondroitin sulfate E glycosaminoglycans from the head cartilage of different squids.

Falshaw et al. (2000) compared the skin and head cartilage glycosaminoglycans of Gould's arrow squid (*Nototodarus gouldi*). Shetty et al. (2009) isolated and characterized a novel chondroitin sulfate from squid liver integument.

7.3 Antioxidant Pigments and Polyphenols

7.3.1 Antioxidant Pigments

Marine natural pigments are a group of chemically heterogeneous molecules that occur across several taxonomical groups. The vast variety of basic pigments formed by the primary producers (photosynthetic bacteria and algae) transfer through the great oceanic food chain, and the accompanying metabolic processes and conversions eventually end up within the tissues of fishes and other aquatic organisms, giving them their characteristic colorations. Rather than the simple display of colors and patterns enabling visual stimulation, cryptic coloration, and behavioral or communication purposes, these wide ranges of pigments exhibit several biological properties and constitute evolutionary adaptations ranging from the signal for fertilization to the very maintenance of fish health as powerful antioxidants.

Marine natural pigments have considerable medicinal potential and promising applications in human health. The major categories of pigments by chemical structure occur either as tetrapyrolles (chlorophyll and heme pigments), carotenoids, indole derivatives (quinones and violacein), alkaloids (prodiginines and tambjamines), polyenes, macrolides, peptides, or terpenoids. Many of these pigments have a variety of biological activities, including antitumor, antibacterial, antioxidant, and anti-inflammatory activity as depicted in Table 7.3.

Carotenoids Carotenoids are widely present in marine environments and responsible for the color of many important marine species of fish, shellfish, and seaweeds. These yellow-to-orange-red pigments, called *carotenoids*, of which there are over 700 different chemical formulations, are typically divided into two subgroups, the *carotenes* (hydrocarbon carotenoids) and the *xanthophylls* (oxygenated carotenoids). Marine photosynthetic organisms such as micro-/macroalgae synthesize β-carotene de novo from isoprenyl diphosphate via phytoene and lycopene and finally alter it to produce other derivatives. Marine animals do not synthesize carotenoids de novo, and those found in the animal bodies are either the result of direct accumulation of carotenoids from food or are partly modified through metabolic reactions (Miyashita 2014). Commonly found carotenoids from aquatic animals are astaxanthin, zeaxanthin, tunaxanthin, lutein, etc. Marine red and brown algae produce β-carotene, zeaxanthin, fucoxanthin, and fucoxanthinol. Astaxanthin is the main carotenoid pigment found in aquatic animals, while fucoxanthin is abundant in aquatic plants. These abundant representative marine carotenoids astaxanthin and fucoxanthin have been reported to be effective antioxidants and have higher biological activity than those of terrestrial origin (Hussein et al. 2006; Hosokawa et al. 2009).

Table 7.3 Natural pigments isolated from marine bacteria

Pigment	Bacterial source	Bioactivity	Color
Astaxanthin	*Agrobacterium aurantiacum*	Antioxidation	Red
Cycloprodigiosin	*Pseudoalteromonas denitrificans*	Anticancer; immunosuppressant; antimalarial	Red
Heptyl prodigiosin	*α-Proteobacteria*	Antiplasmodial	Red
Indigoidine	*Streptomyces chromofuscus* ATCC 49982	Dye-industrial applications	Blue
Melanins	*Vibrio cholerae*, *Shewanella colwelliana*, *Cellulophaga tyrosinoxydans*	Protection from UV irradiation	Black
Phenazine derivatives	*Bacillus* sp.	Cytotoxic	Brownish yellow
Pyocyanin and pyorubrin (Phenazine derivatives)	*Pseudomonas aeruginosa*	Antibacterial	Blue-green
Scytonemin	Cyanobacteria	Protection from UV irradiation	Yellow-green
		Anti-inflammatory, antiproliferative	
Tambjamines	*Pseudoalteromonas tunicata*	Antibiotic, anticancer	Yellow
Tryptanthrin	*Cytophaga*	Antibiotic	Deep yellow
Violacein	*Pseudoalteromonas tunicate*	Antibiotic; antiprotozoan; anticancer	Violet

Source: Soliev et al. (2011)

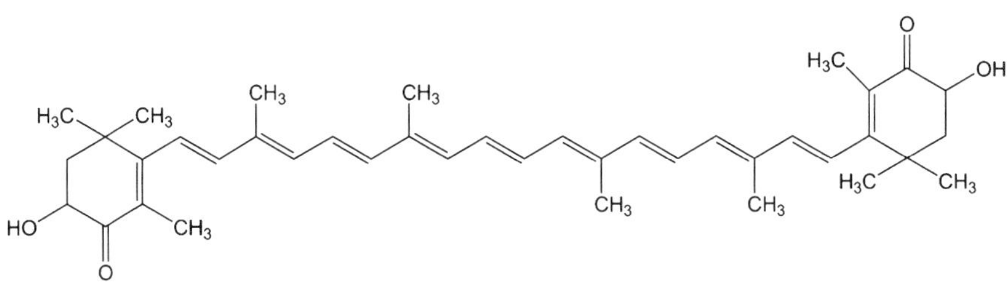

Fig. 7.2 Chemical structure of astaxanthin. (Source: http://www.molvis.org/molvis/v19/1656/mv-v19-1656-f1.html)

7.3.1.1 Astaxanthin

Astaxanthin is the "King of Carotenoids." It is the red pigment that gives salmon, shrimp, and flamingos their pink-red color. The chemical structure of astaxanthin is shown below (Fig. 7.2).

Astaxanthin is a red pigment from the xanthophyll oxygen-carrying group of carotenoids. Currently, the primary natural source for astaxanthin is the microalgae *Haematococcus pluvialis*. Commercially more than 40 g of astaxanthin can be obtained from 1 kg of dry biomass. Also Phaffia yeast *Phaffia rhodozyma* (Fontana et al. 1997) exhibits 100% free, nonesterified astaxanthin, which is readily absorbable and need not be hydrolyzed in the digestive tract of the fish. The pigments are transferred across the food web into higher life forms, e.g., fish (salmon, trout) and birds (flamingo, red ibis), who store the pigments in their skin and fatty tissue proving them characteristic coloration. Many aquatic species including crustaceans utilize these algae and plankton as food and store the pigment in their shell, resulting in their pink-red exterior. Astaxanthin does not undergo bleaching; thus these animals retain their pink-red coloration. *Euphausia pacifica* (Pacific krill), *Euphausia superba* (Antarctic krill), *Pandalus borealis* (Arctic shrimp), snow crab, *Chionectes opilio*, etc. are colored by astaxanthin and are utilized commercially as astaxanthin source.

Research suggests that astaxanthin functions as an intracellular oxygen reserve and fishes and invertebrates fed on a diet containing optimal levels of astaxanthin that show greater survivability rates when subjected to poor oxygen environments as it is a potent antioxidant, comparable to vitamin E. It may also offer protection against UV light. Cultured salmon reared using an astaxanthin-free diet remains white and has less consumer acceptability. The US Food and Drug Administration (FDA) has approved astaxanthin as a food coloring (or color additive) for specific uses in animal and fish foods. The European Commission considers it as food dye, and it is given the E number E161j. Natural astaxanthin is considered and generally recognized as safe (GRAS) by the FDA; but as a food coloring in the United States, it is restricted for use in animal food. Antioxidant activity of astaxanthin has been reported to be several times higher than that of other carotenoids and tocopherols (Olaizola 2008).

7.3.1.2 Fucoxanthin

As an important pigment found in aquatic plants, fucoxanthin is the most abundant carotenoid in nature contributing more than 10% of the estimated total production in nature (Matsuno 2001). Fucoxanthins have a unique structure that includes an unusual allenic bond and 5, 6 monoepoxide. Yan et al. (1999) demonstrated that fucoxanthins possess strong DPPH (2, 2-diphenyl-1-picrylhydrazyl) radical scavenging activity. In addition to antioxidant activity, fucoxanthin from edible seaweeds exhibited anti-obesity effect (Maeda et al. 2005) and antidiabetic effect (Maeda et al. 2007), complemented by DHA synthesis in mouse liver (Tsukui et al. 2007) (Fig. 7.3).

7.3.2 Antioxidant Polyphenols

Marine algae are a rich source of polyphenols. Phlorotannins are the largest group of polyphenols in marine brown algae (Zou et al. 2008). A series of polyphenols including catechins (e.g., gallocatechins, epicatechins, and catechin gallate), flavonols, and flavonoid glycosides have also been identified in several red and brown algae. Algal polyphenols exhibit their bioactive potential as potent free radical

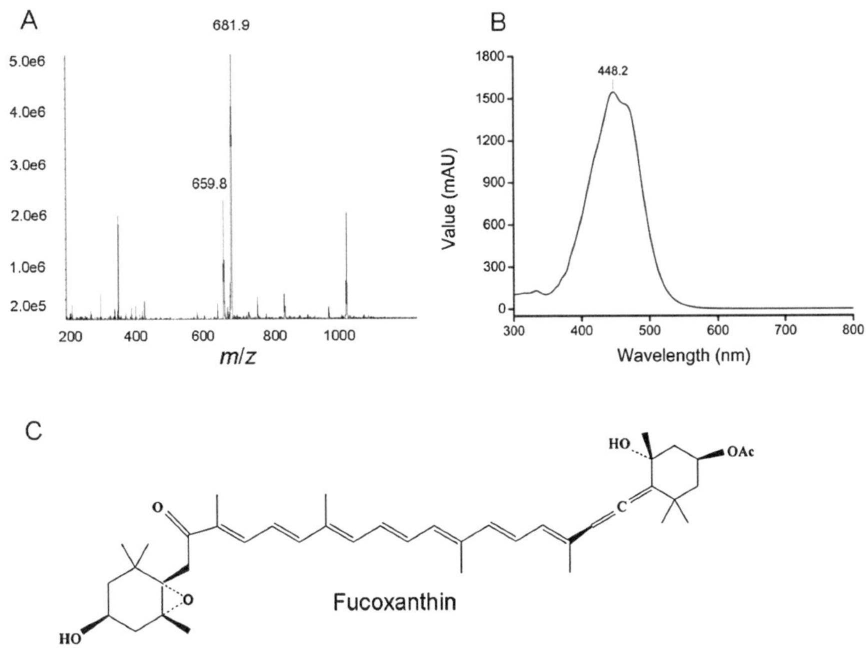

Fig. 7.3 Fucoxanthin from *O. aurita* (**a**) mass fraction, (**b**) UV-visible spectrum, (**c**) chemical structure of all-trans-fucoxanthin. (Source: Xia et al. 2013)

scavengers and primary chain-breaking antioxidants (Wang et al. 2009). Reducing power has been suggested as an important indicator of the overall antioxidant property of phenolic antioxidants.

Algal polyphenols are generally extracted by solvent extraction methods using an aqueous mixture of methanol, ethanol, and water, whereas the enzyme-assisted aqueous extraction methods have several advantages over conventional solvent extraction methods. The enzymes used are carbohydrases and proteases, with proteases having enhanced recovery of polyphenols. Major limitation of enzyme extraction methods is that the digestion of algal cell wall materials by enzymes results in extraction of polysaccharides, proteins, or peptides which may or may not be effective antioxidants as that of algal polyphenols. Highly efficient and environment-friendly techniques such as supercritical fluid extraction (SFE) using CO_2 as extraction fluid, subcritical water extraction (SWE), accelerated solvent extraction (ASE), and pressurized fluid extraction (PFE) are recently used for the extraction of phlorotannins and other natural antioxidants.

7.3.2.1 Phlorotannins

Phlorotannins are a group of polyphloroglucinol phenolics and are mainly stored within cell organelles called physodes (cellular vacuoles). Chemically, they are polymers of phloroglucinols (1, 3, 5-trihydroxybenzene). As secondary metabolites in marine algae, the occurrence of phlorotannins varies among different taxonomic

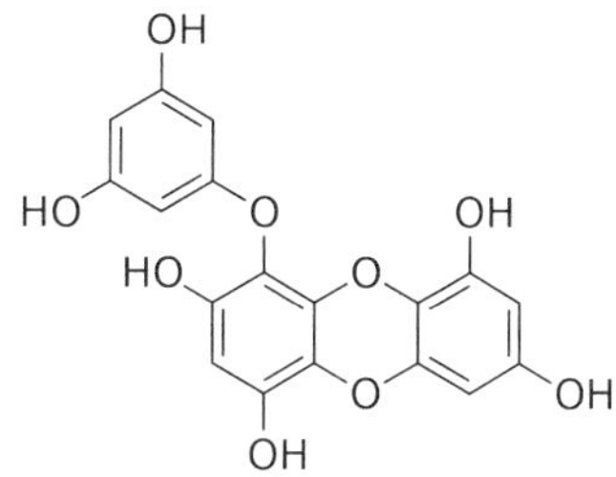

Fig. 7.4 Eckol, a phlorotannin found in several species of the genus *Ecklonia* (Brown algae). (Source: http://www.wikiwand.com/en/Phlorotannin)

groups, and concentrations varying up to 20% of algal dry mass are reported (Targett and Arnold 1998), mainly in the species *Fucus vesiculosus*, *F serratus*, *Ascophyllum nodosum*, *Hizikia fusiformis*, etc. The DPPH radical scavenging activity as a function of antioxidant activity in algal phlorotannins was explained by Zubia et al. (2007) and Wang et al. (2009). Antidiabetic and antioxidant effects of methanol extract from the brown algae, *Ecklonia stolonifera*, were reported by Iwa (2008). The active components of the methanolic extracts of the brown algae *E. stolonifera* were found to be phlorotannins as evidenced by LC/MS, ^{1}H NMR, and ^{13}C NMR analysis (Fig. 7.4).

Based on the structure and chemistry, phlorotannins can be systematically classified into six major subclasses: phlorethols, fucols, fuhalols, fucophlorethols, isofuhalos, and eckols (Parys et al. 2007). Several lines of evidence also indicated that oligomeric and polymeric phlorotannins are more potent antioxidants than the monomer and phloroglucinol. High-molecular-weight phlorotannins from *Sargassum kjellmanianum* were reported to be far more effective than their low-molecular-weight counterparts in preventing rancidity in fish oils (Yan et al. 1996).

7.4 Seaweed Polysaccharides

As an abundant and renewable marine biomass, seaweeds are a part of the human diet in many Asian countries and also traditionally been used as medicines since ancient times (Pereira 2011). The earliest records of herbal medical seaweeds appeared in Chinese literature about 2000 years ago (Chengkui and Junfu 1984). Several species of seaweed have the capacity to produce a diverse array of secondary metabolites, which exhibit important and vital ecological roles as defense and/or signal compounds (Amsler 2008) and are also of biotechnological interest (Abdul Khalil et al. 2018). Seaweeds are also a rich source of bioactive polysaccharides. The remarkable marine polysaccharides found in red and brown algae, viz., alginate, agar, and carrageenan, provide numerous ingredients to the food, pharmaceutical, cosmetic, textiles, paper, and biotechnology industries as stabilizers, thickeners, emulsifier, and fillers (Makkar et al. 2016). Other seaweed hydrocolloids, besides those common seaweed hydrocolloids (alginate, agar, and carrageenan), are fucoidan, laminarin, mannitol, funoran, ulvan, etc.

Polysaccharides are the most abundant and the most complex organic molecules in the ocean. In contrast to land polysaccharides, many marine polysaccharides are highly sulfated. In particular, the cell walls of macroalgae harbor a high diversity of sulfated polysaccharides (carrageenans, agarans, fucoidans, ulvans, etc.). These sulfated polysaccharides, biosynthesized by macro-algal primary producers, represent an important food source for heterotrophic organisms (Helbert 2017). Generally, there are numerous commercial applications of marine polysaccharides as food, beverages, and supplements. In addition, these molecules have many biological functions including antimicrobial, anticoagulant, antiproliferative, antithrombotic, and anti-inflammatory activity. These marine-derived secondary metabolites also have many human health benefits which enable them to be applied as nutraceuticals. In the biomedical field, the polysaccharides from algae can be used in controlled drug delivery, wound management, and regenerative medicine (de Jesus Raposo et al. 2015).

The abundance of sulfate ions in the marine environment and its ubiquity has resulted in a wide distribution of sulfated molecules and more especially sulfated polysaccharides in marine organisms. In contrast to land polysaccharides, a huge number marine polysaccharides are decorated with sulfate ester groups (Helbert 2017). The occurrence of sulfated polysaccharides in macroalgae, marine angiosperms, and seagrasses has been proposed to be an adaptation to marine environments (Aquino et al. 2005), suggesting that the sulfation of polysaccharides is a marker of marine origins. The most studied sulfated marine polysaccharides are those extracted from the cell walls of macroalgae due to the industrial interest in their hydrocolloid properties. Seaweeds are classified into four principal group based on their pigments, namely, red, brown, blue, and blue-green. Only the red (Rhodophyceae) and brown (Phaeophyceae) seaweeds are important sources of hydrocolloids (Glicksman 1987). The algal sources, as well as the constituent bioactive polysaccharide molecules, are discussed in the following sections.

7.4.1 Agar, Carrageenan, and Alginate

The term "hydrocolloid" is derived from the Greek word *hydro* and *kolla* which means "water" and "glue." Chemically, hydrocolloids are substances with affinity to bind water; in other words they are macromolecular hydrophilic substances and probably water soluble. However, not all of them are water soluble and form colloidal solutions, as some hydrocolloids are only able to swell in water and can be dispersed only by means of shear forces (Abdul Khalil et al. 2018). Agarans and carrageenans are the most exploited polysaccharides from red algae (Usov 2011).

Agar

Agar is commercially extracted from red seaweed (Rhodophyceae), specifically *Gracilaria* sp., *Gelidium* sp., and *Pterocladia*/*Gelidiella* sp. (Usov 2011). Agar consists of a mixture of two polysaccharides, namely, agarose and agaropectin, with agarose making up about 70% of the mixture. Agarose is responsible for gelling, while agaropectin is responsible for thickening properties. Agar is the oldest utilized hydrocolloids from seaweed discovered by Minoya Tarazaemon in 1658 in Japan,

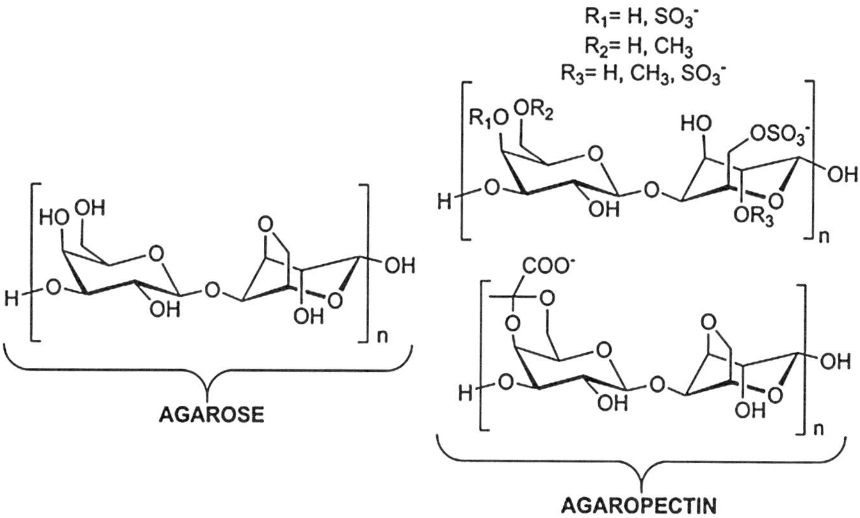

Fig. 7.5 Structure of agarose and agaropectin (https://www.sciencedirect.com/science/article/pii/S0165237016307203)

where it is called kanten, literally means "frozen sky," but can also referred to by many names such as "grass jelly," "seaweed jelly," and "vegetable gelatin" (Pereira 2011) (Fig. 7.5).

Chemically, agar contains hydrophilic galactans consisting of galactopyranose units with alternating α-1,3 and β-1,4 linkages, whereby the α-linked galactopyranose is made up of L-galactopyranose 6-sulfate.

Agar is the first hydrocolloid with European registration number of E406 approved as GRAS (generally recognized as safe) by the Food and Drug Administration (FDA) which employed it as an additive in the food industry (Abdul Khalil et al. 2018). About 80% of agar produced globally is used for food applications, while remaining 10–20% is used in pharmaceutical and biotechnology industries (Pereira 2011). Low-quality agar is used basically in food stuff (frozen foods, bakery icings, meringues, dessert gels, candies, etc.), while some may extend to industry applications for paper sizing, coating, adhesives, textile printing/dyeing, casting, and impression. About one-sixth of agar in the United States is normally used for microbial culture medium because they are not easy to metabolize and non-digestible and have good gel firmness, elasticity, clarity, and stability (Wu 1990). This kind of agars, usually referred to as medium-quality agar, can also be used to obtain monoclonal antibodies, interferons, steroids, and alkaloids, plus act as bulking agents, laxatives, suppositories, capsules, tablets, and anticoagulants in medicinal/pharmaceutical fields (Pereira 2011). The most highly purified agar, usually obtained from fraction of agar called agarose (refer to Sect. 5.3.2), is used for separation in molecular biology: electrophoresis, immune diffusion, and gel chromatography (Rhein-Knudsen et al. 2015).

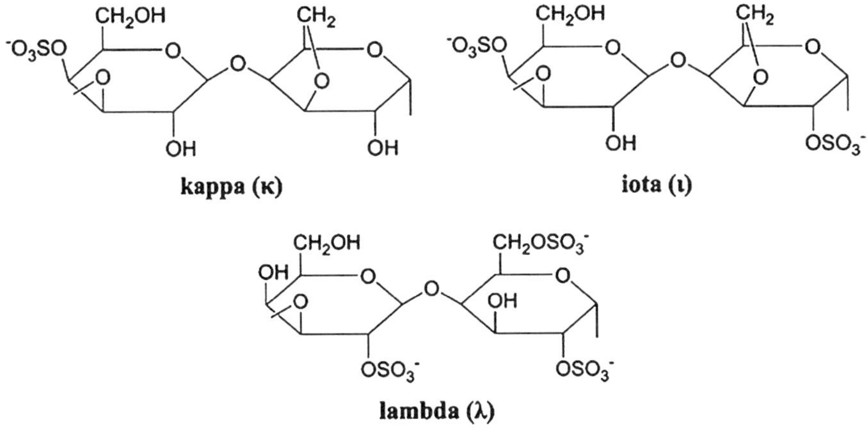

Fig. 7.6 Structure of *iota-*, *kappa-*, and *lambda*-carrageenan. (https://www.researchgate.net/figure/Structure-of-kappa-iota-and-lambda-carrageenans_fig2_305773172)

Carrageenan

Carrageenan, first named as the "carrageenin," was discovered by British pharmacist Stanford in year 1862, who extracted it from Irish moss (*Chondrus crispus*). The name was later changed to carrageenan so as to comply with the "-an" suffix for the names of polysaccharides (Pereira 2011). Carrageenan is a vegetarian and vegan alternative to gelatin in some applications or may be used to replace gelatin in confectionery.

Carrageenans are a class of sulfated galactans extracted from different species of red seaweed (Rhodophyta). The main seaweed species used in industrial scale for carrageenan production are *Kappaphycus alvarezii*, *Eucheuma denticulatum*, *Chondrus crispus*, and *Gigartina* sp. (McHugh 2003), and also *Furcellaria lumbricalis* as a source of furcellaran (Danish Agar) is also classified together with carrageenan (Fig. 7.6).

Carrageenans are hydrophilic linear hydrocolloids, consisting mainly of potassium, sodium, magnesium, and calcium salts of sulfated esters of D-galactose and 3,6-anhydro-D-galactose (Laurienzo 2010). Traditionally, these polysaccharides are split into six basic forms: *iota-*, *kappa-*, *lambda-*, *mu-*, *nu-*, and *theta*-carrageenan. This nomenclature is relevant both for their chemical classification and commercial production. *Iota-*, *kappa-*, and *lambda*-carrageenan are the three most prevalent and valuable forms, according to the number and position of sulfate groups in the repeating units. *Kappa*-carrageenan has only one sulfate ester per disaccharide (and tends to form a strong and rigid gel), while *iota-* and *lambda*-carrageenan have two and three sulfates per dimer, respectively. *Iota*-carrageenan gives soft elastic thermoreversible gels, whereas *lambda*-carrageenan is non-gelling.

Carrageenans present a variety of physicochemical properties, which are the basis for their applications in the food, cosmetic, and pharmaceutical industries. Carrageenan has been used as a gelling and thickening agent in foods and in traditional medical remedies for many hundreds of years. Today it is the leading seaweed-derived food hydrocolloid and is used widely for its textural functionality, particularly in dairy products, desserts, cooked, processed meats, and infant formulations. Carrageenan is an important production used in the cosmetic industry because of its physical and functional ability and antioxidant activity. Products such as skin lotions, toothpaste binders, and shaving foams are available from carrageenan. Other key applications are in pet food and in air fresher gels. Another application of carrageenan in food industry is its use as a protective coating on fresh cut packaged food whereby carrageenan acts as gas barrier, changing the cut surfaces of the fruit and reducing respiration which consequently slows down the discolorations and maintains texture throughout the shelf life (Lin et al. 2018).

Carrageenan has been widely used for drug delivery applications in different isoforms including nanoparticles, micro stabilizers, gelling agent, beads, microspheres, and microcapsules. Carrageenan-based hydrogels and carrageenan-graphene oxide composites are widely employed for bone tissue engineering (Santo et al. 2009). One of the most important uses of carrageenan is its employability as biosensors (Sect. 5.17). k-carrageenan gel immobilized with superoxide dismutase enzyme is used as a membrane receptor in biosensor for the determination of superoxide radicals.

The use of food-grade carrageenan is allowed in the United States by the FDA in organic and nonorganic foods, including juices, chocolate milk, and organic infant formula, and considered as safe when used in amounts necessary as emulsifiers, stabilizers, or thickeners. However, the use of carrageenan in infant formula, organic or otherwise, is prohibited in the European Union for precautionary reasons but is permitted in other food.

Alginate

Alginate is distinguished from other hydrocolloids (agar and carrageenan) because it is isolated from brown seaweed (Phaeophyceae) specifically from outer layer of brown algae cell wall, as inner layer makes up mostly of cellulose. Alginate was first described by the British chemist E.C.C. Stanford in 1881 and exists as the most abundant polysaccharide in the brown algae comprising of up to 40% of the dry matter. These alginate molecules provide both flexibility and strength to the plants (Pereira 2011). Besides brown seaweed, several bacteria also produce alginate exocellularly, viz., *Azotobacter vinelandii* and *Pseudomonas* sp. (Remminghorst and Rehm 2006). At present, all commercial alginates stem from algal sources. Brown seaweed species like *Laminaria hyperborean*, *Macrocystis pyrifera*, *Laminaria digitata*, and *Ascophyllum nodosum* are mainly used for commercial alginate production. Species like *Sargassum* spp., *Laminaria japonica*, *Ecklonia maxima*, and *Lessonia nigrescens* are also used (Abdul Khalil et al. 2018) (Fig. 7.7).

Alginates are linear polymers build up by two monomeric uronic acids, β-D-mannuronic acid (M) and α-L-guluronic acid (G). The ratio of these two uronic acids is usually 2:1, but it may vary depending on species, age of the plants, and the type of tissue that the alginates are extracted. The two uronic acids are arranged in an asymmetrical pattern of varying proportions of MM, MG, and GG blocks, depending on seaweed source, extraction technique, and harvest time (Abdul Khalil

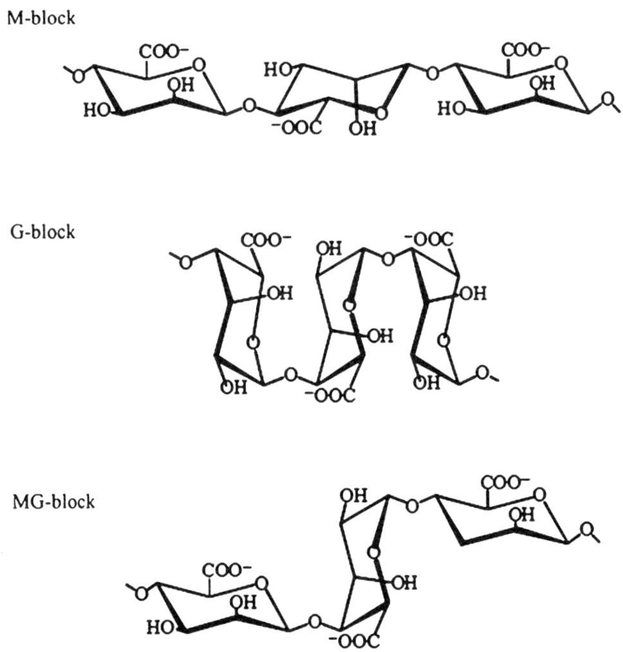

Fig. 7.7 Chemical structures of mannuronate (M)-block, guluronate (G)-block, and MG-block of alginate. (Source: https://www.sciencedirect.com/science/article/pii/S1043452616300365)

et al. 2018). The M/G ratio and block structure have major impact on the physicochemical properties of alginate. By raising the guluronic acid (G) content or molecular weight, more stronger and brittle alginate gels may be achieved. Inversely, more flexible gels may form due to the high amount of alginate-M blocks (Fertah et al. 2017).

Alginate is widely used in industry because of its ability to retain water and its gelling, viscosifying, and stabilizing properties. Alginate is an excellent stabilizing and thickening agent in food such as jelly, ice cream, desserts, etc. due to its properties to chelate metal ions to form highly viscous solutions (Cardozo et al. 2007). The food additive E numbers for the various alginates are as follows: E400, alginic acid; E401, sodium alginate; E402, potassium alginate; E403, ammonium alginate; E404, calcium alginate; and E405, propane-1,2-diol alginate.

Alginates from macroalgae have been most used in several biomedical applications including controlled drug release (Lucinda-Silva et al. 2010), microencapsulation (Hori et al. 2009), tissue engineering and scaffold in regeneration of almost all human tissues (Lee and Mooney 2012), or even preparation of molds in dentistry. Hydrophilic, pleasant odor and taste, nonstaining, ease of mixing, and effective use in the presence of saliva make it suitable for application in dental impression materials (Madhavan and Abirami 2015). A new form of medical textiles (wound dressings) incorporated with alginate is used in wound management products that are nontoxic. These products with good mechanical properties may also be used to incorporate medication (Sun and Tan 2013). Recently, alginate oligosaccharides, low molecular polymer fragments obtained by enzymatic depolymerization, and acid hydrolysis have received much attention and are widely reported due to their unique pharmaceutical properties such as antioxidant, antifungal, anti-inflammatory, and antibacterial activities (Szekalska et al. 2016).

7.4.2 Fucoidan, Laminarin, and Ulvan

Fucoidan

Brown algae also contain highly complex, sulfated matrix polysaccharides called fucoidans. Sulfated fucans, usually referred to as fucans or fucoidans, were first isolated by Kylin in 1913 (Chevolot et al. 2001). Fucoidans represent a class of sulfated polysaccharides composed mainly of a-L-fucose found in the cell walls of brown algae and also in some marine invertebrates (i.e., worms, sea urchins, and sea cucumber). The sulfated fucans are the most abundant marine-sulfated polysaccharides, as the brown algae are the predominant algae in the near-shore environment. The structure of fucoidans depends on their biological origin and differs among algal species. Fucoidans are coproducts of the alginate industry, which make them attractive as bioactive compounds, and several potential applications have been examined (Pomin and Mourão 2008). Since its first description, fucan has been explored for biological activities in different mammalian systems. Their medicinal properties include antitumor, immunomodulation, antiviral, antifungal, hepatoprotective, antiprotozoal, anti-inflammatory, antioxidant, and anticoagulant activity. Apart from these, fucans are used also in drug targeting and regenerative studies. The role of fucan polysaccharides as nutraceuticals and their biomedical applications are depicted in Tables 7.4 and 7.5, respectively.

Fucoidans have found new applications in biomedicine, as nanoparticles of fucoidan and chitosan (Yoon et al. 2014), for the synthesis of biohybrid glycopolymers used to treat dermal burns and regeneration of the epithelial tissue (Sezer et al. 2007). The sulfated polysaccharides decrease wound secretions and scar formation and also reduce bacterial inflammation (Motta 1989). Fucoidans effectively inhibit proliferation and colony formation of cancer cells in vitro (Ermakova et al. 2011) and also possess inhibitory activity in tumors growing in vivo (Ye et al. 2008).

Osteoarthritis symptoms were significantly inhibited and reduced by 52% on oral administration of fucoidan-rich seaweed extract (Irhimeh et al. 2007). Orally administered *Undaria pinnatifida* fucoidan also successfully inhibited pain (Park et al. 2010).

Laminarin

Laminarin is a polysaccharide composed of D-glucopyranose with 1, 3-glycosidic linkages with average length of 20–25 disaccharides. They are generally known as β 1, 3-glucans (Patil et al. 2018). They are found in some brown algae and in high concentrations in members of the genus *Laminaria* (kelp) where it composes

Table 7.4 Bioactive properties of marine algal hydrocolloids

Seaweed	Compound	Activity	References
Ulva reticulata	Polysaccharides	Antifouling agent against *Hydroides elegans* settlement	Harder et al. (2004)
Fucus vesiculosus	Fucoidan	Anticoagulant	Bernardi and Springer (1962)
Phyllophora brodiei	S-λ-carrageenan	Anticoagulant	Prajapati et al. (2014)
Eucheuma cottonii	Sulfated galactan	Antioxidant	Costa et al. (2010)
U. pertusa	Ulvans	Antioxidant	Huimin et al. (2005)
		Antihyperlipidemic	Pengzhan et al. (2003)
E. prolifera	Sulfated polysaccharides	Immunomodulator	Kim et al. (2011)
Sargassum vulgare	Alginate	Anticancer	Sousa et al. (2007)
Kappaphycus striatum	Kappa-carrageenan	Anticancer	Yuan et al. (2006)
Acanthophora spicifera, *Hydroclathrus clathratus*	Carrageenan	Antiviral (herpes simplex virus 1)	Hanaa and Gihan (2015)
Chondrus crispus, *Codium fragile*	Crude enzyme extracts	Antiviral (herpes simplex virus 1)	Garima et al. (2015)
Cladosiphon okamuranus	Fucan	Antiviral (anti-dengue activity)	Hidari et al. (2008)
Schizymenia dubyi	Sulfated glucuronogalactan	Antiviral (anti-HIV activity)	Bourgougnon et al. (1996)
Corallina pilulifera	Methanol extract	Anti-photoaging activity	Ryu et al. (2009)

30–50% of the dry weight as well as other species of Laminariales and Fucales (Holdt and Kraan 2011). Laminarin can be chemically sulfated, which increases its biological activity in many cases.

Similar to other marine polysaccharides, laminarin sulfate possesses lipid lowering and anticoagulant properties (Shanmugam and Mody 2000). Dietary supplementation with laminarin reduces hepatotoxicity in rats treated with lipopolysaccharide (LPS), suggesting an anti-inflammatory effect of laminarins (Neyrinck et al. 2007). Laminarin also modifies the gut microflora providing potential benefits to preventing or reducing atherosclerosis and also the potential to shift the gut microbiome to reduce the impact of a high-fat diet on atherosclerosis.

Laminarin works as an antioxidant and free radical scavenger with an EC_{50} (dose for 50% free radical neutralization) of 460 μg/ml (Tsiapali et al. 2001). Laminarin may also affect immune modulatory pathways that are associated with atherosclerotic progression or risk factors for atherosclerosis. The (1,3)-β-glucans have also

Table 7.5 Biomedical applications of marine algal polysaccharides

Seaweed	Compound	Activity	References
B. montaigne, *Gloiopeltis* spp., *P. tenera*, *P. capillacea*	Agaroids	Cell encapsulation	Gong et al. (2007)
		Scaffolds for tissue engineering	
		Wound healing and dressings, revascularization	Bao et al. (2008)
Laminaria spp., *A. nodosum*, *Ecklonia* sp.	Alginates	Drugs carriers	George and Abraham (2006)
M. pyrifera, *Durvillaea* sp., *Lessonia* sp.		Encapsulation, scaffolds for ligaments and tissue engineering, and regeneration	Lee and Mooney (2012)
		Wound healing and dressings	Qin (2008)
		Molding in dentistry	Madhavan and Abirami (2015)
		Scaffold for dental-derived stem cells	Moshaverinia et al. (2012)
Eucheuma cottonii, *E. cottonii*	Carrageenan	Carrageenan – Calcite ($CaCO_3$) biopolymer composites for bone tissue regeneration	Lucas et al. (2016)
		Antibacterial bio-nanocomposite hydrogels of carrageenan/silver nanoparticles (Ag-NPs)	Azizi et al. (2017)
U. pinnatifida, *A. platensis*, *Porphyridium* sp.	Fucoidans from microalgae	Anticancer activity	Yang et al. (2008)
		Nanofiber production	De Morais et al. (2010)
		BioGlue for soft tissue closure after surgery, lubricants for bone joints	Laurienzo (2010)
A. nodosum, *E. bicyclis*, *Fucus* sp., *Laminaria* sp., *U. pinnatifida*, *C. vulgaris*	Laminarin (β-glucans)	Tissue regeneration	Sezer et al. (2007)
		Burn wound dressings	Delatte et al. (2001)
		Wound healing	Kofuji et al. (2010)
Ulva rigida, *Ulva* spp.	Ulvans	Nanofibers	Toskas et al. (2012)
		Tissue engineering	Morelli and Chiellini (2010)
		Wound dressings	Alves et al. (2013)

been used to help healing wounds, by inducing the migration of macrophages to the wound site, and accelerating the healing process as a constituent in composites (Kofuji et al. 2010), and in burn wound dressings, thereby reducing the need for analgesics (Delatte et al. 2001).

Ulvan

Ulvan is a sulfated heteropolysaccharide that is composed of disaccharide units of D-glucuronic acid or L-iduronic acid linked *via* sulfated 1, 4 L-rhamnose residue with traces of D-xylose and D-glucose (Wijesekara and Karunarathna 2017; Patil et al. 2018). They are extracted from the green algae, which belong to the phyla Chlorophyta, family Ulvanacaea, and genuses *Ulva* and *Enteromorpha*. The chemical composition of ulvan varies according to the species it is extracted from, geographical location, and harvest. They are commonly extracted from *Ulva pertusa* and *Ulva lactuca* (Yaich et al. 2017), also called by the common name "sea lettuce." Ulvans are commonly used as food in many parts of Asia and have also been utilized in traditional Chinese medicines (Li et al. 2018). Ulvans are utilized for human consumption due to their rich vitamin, oligo element, mineral, carbohydrate, and dietary fiber content and have been explored for their medicinal properties and efficacy to be used as nutritional supplements.

High sulfate content (Qi and Sun 2015) and molecular weight (Yaich et al. 2017) of ulvans lead to increased antioxidant activity. Ulvans reduce triglycerides (TG), total cholesterol (TC), and total lipid levels by 46, 69, and 30%, respectively, in hypercholesterolemic rats (Borai et al. 2015). They can be chemically modified to attach functional groups to the polysaccharide and improve its antioxidant abilities. Studies show that acetylated and benzoylated ulvans exhibited higher antioxidant effects compared to natural ulvan and modulate liver and kidney function in high-fat diet-induced hypercholesterolemic rats (Rizk et al. 2016).

Ulvan, as a sulfated water-soluble polysaccharide has great potential in food, pharmaceutical, agricultural, and chemical applications. Ulvan membrane is proposed to be a novel candidate in drug delivery and in wound treatment (Alves et al. 2013). Ulvan incorporated in scaffolds is a good bone graft substitute due to its nontoxic nature and improves the osteogenic activity. Biofunctionalized ulvan hydrogels with ALP enzyme serve as mineralization inducers in osteoblastic differentiation (Dash et al. 2014) and cytocompatibility. Scaffolds containing ulvan from *Ulva lactuca* and chitosan extracted from *Loligo forbesi* squid pen were evaluated as suitable bone cement candidate in vitro and in vivo (Barros et al. 2013). The intermolecular charges of ulvan (anionic charges) and chitosan (cationic charges) produce stable supramolecular structures via electrostatic interactions that form stabilized membranes.

7.4.3 Algal Polysaccharide Extraction and Processing

Traditional production of algal hydrocolloids generally involves multistage processing, with a few principal steps such as cleaning/washing, pretreatment, solid/liquid separation (extraction), precipitation and filtration, and drying and milling (refer to Appendix). However, the existing methods require high amount of chemicals, water, and energy as well as control of waste generated throughout the whole process, which are less environmentally friendly and cost-effective. The innovation and design of "green" extraction methods for natural products are currently a popular topic in multidisciplinary areas. Some of the green methods employed to extract bioactive compounds are supercritical fluid extraction (SPE), microwave-assisted extraction (MAE), enzyme- assisted extraction (EAE), ultrasound-assisted extraction (UAE), pressurized solvent extraction (PSE), reactive extrusion, and photo-bleaching processes (Abdul Khalil et al. 2018).

7.5 Antimicrobial Peptides

Antimicrobial peptides (AMPs) are considered to play an important role as host defense molecules in both vertebrates and invertebrates. The living organisms use AMPs as a vital tool for survival, as crucial effector molecules of the innate immune response. The widespread use of antibiotics in recent years has led to the rapid emergence of antibiotic resistance. Naturally occurring cationic antimicrobial peptides (AMPs) may be one of the suitable templates to meet the demand as novel therapeutic agents. AMPs possess broad-spectrum antimicrobial activities against pathogens, including Gram-positive and Gram-negative bacteria, fungi, and protozoa, and they exhibit antiviral and anticancer properties also (Park et al. 2011), whereas the likelihood of pathogens developing AMP resistance is very low. AMPs destroy microorganisms using diverse range of mechanisms.

One of the first explained and also the most famous antimicrobial peptides (AMPs), nisin, is a 34-residue peptide produced by a strain of lactic acid bacteria, *Lactococcus lactis*. It belongs to the lantibiotics class of a highly posttranslationally modified bacteriocin. Nisin contains unusual amino acids (unsaturated and stereo inverted amino acids) as well as thioether bridges. Another ancient and nonetheless famous AMP is melittin, the main peptide component of the honey bee (*Apis mellifera*) venom. Unlike nisin, melittin is a 26 non-modified residue peptide.

7.5.1 Antimicrobial Peptides (AMPs) in Marine Invertebrates

AMPs are a major component of the innate immune defense mechanism in marine invertebrates (Cole 2005). Marine organisms are a rich source of uniquely structured AMPs and have been isolated using biochemical in silico and genetic approaches. Hemolymph is the main source of AMPs in marine invertebrates, even though other tissues may also contain these peptides (Rajanbabu et al. 2015). Prominent among AMPs isolated from crustaceans are penaeidins from the shrimp *Penaeus vannamei*, while in marine mollusks, defensins, myticins, and mytilins isolated from *Mytilus galloprovincialis* and *Mytilus edulis*, respectively, have been studied for their specific roles in innate immune mechanisms. Huang et al. (2006) identified various protein fractions from the hemolymph of *S. serrata* with antimicrobial action against various Gram-positive and Gram-negative bacteria and also isolated a 10.8 kDa anionic protein named scygonadin from the male reproductive tract of *S. serrata* that inhibited growth of *Micrococcus luteus* and *Aeromonas hydrophila*.

AMPs have also been shown to demonstrate in vitro microbicidal properties against various clinical pathogens, including the sexually transmitted infection

(STI)-causing *Treponema pallidum*, *Chlamydia trachomatis*, and HIV. These peptides have been isolated from a wide variety of vertebrate and invertebrate phyla including mammals, reptiles, ascidians, insects, chelicerates, annelids, and mollusks. Some examples of AMPs isolated from marine invertebrates and hemichordates are given in Tables 7.6 and 7.7, respectively.

7.5.2 Antimicrobial Peptides (AMPs) in Marine and Freshwater Fishes (Jorge and Diamond 2014)

AMPs participate in the innate host defense, and fish is a great source of these peptides, as they are frequently exposed to a wide variety of pathogens.

The following AMPs are identified in a variety of fish species.

7.5.2.1 Piscidins

Piscidins display potent antimicrobial activity against a variety of microorganisms. They are widely active against several bacteria including *Streptococcus*, *Pseudomonas*, *Bacillus*, and *Vibrio* species. They also possess antifungal (Cho and Lee 2011), anti-parasitic (Zahran and Noga 2010), and antiviral activity (Wang et al. 2010a). Piscidins are mainly expressed in the gills, skin, intestine, kidney, and spleen as well as ubiquitously detected in chondrocytes, the heart, oocytes, exocrine and endocrine glands, swim bladder, and other tissues in fishes.

The first member of the family identified was a 25-residue peptide isolated and characterized from skin mucous secretions of the winter flounder, *Pseudopleuronectes americanus*, named as pleurocidin (Cole et al. 1997). Further research identified other homologous peptides, including misgurin, moronecidin, epinecidin, dicentracin, etc., in related fish species (Pan et al. 2008), and the similarities of the mature peptide secondary structure (Cole et al. 1997) as well as analysis of their gene structures suggested that they all belong to the same evolutionarily related family, which were then referred to as the piscidins. Primary amino acid sequences confirm that piscidins as a group have little direct sequence identity but are homologous as a group and have an amphipathic α-helical structure (Cole et al. 1997), similar to that of magainins (amphibian AMPs) and cecropins (insect AMPs).

Among other attractive features of piscidins includes their ability to retain antibacterial activity at high salt concentrations (Lauth et al. 2002), thermostability (Sun et al. 2012), and relatively low cytotoxicity against mammalian cells (Kim et al. 2010).

7.5.2.2 Cathelicidins

Cathelicidins are a group of unusual AMPs that share little sequence homology between the mature peptides. The first cathelicidins identified in fish were initially isolated as antimicrobial peptides from the Atlantic hagfish, *Myxine glutinosa* (Uzzell et al. 2003). Fish cathelicidins can be subdivided into two classes – the linear peptides and those with a characteristic disulfide bond. In contrast to mammalian cathelicidins, there is significant sequence homology among members of the

Table 7.6 Antimicrobial peptides (AMPs) in marine invertebrates

Class/order		Producer/organism	Antimicrobial compound	Activity	References
Porifera		Bahamian marine sponge (*Stylissa caribica*)	Stylissamide G	Antifungal	Dahiya et al. (2016)
		Marine sponge (*Siliquariaspongia mirabilis*)	Mirabamides A–D	Antimicrobial	Plaza et al. (2007)
		Marine sponge (*Discodermia calyx*)	Calyxamides A and B	Cytotoxic	Kimura et al. (2012)
		Lithistid sponge (*Aciculites orientalis*)	Aciculitins A–C	Cytotoxic and antifungal	Bewley et al. (1996)
		Sidonops microspinosa	Microspinosamide – cyclic depsipeptide	Anti HIV	Rashid et al. (2001)
		Theonella swinhoei	Cyclolithistide A	Antifungal	Clark et al. (1998)
		Marine sponge (*Ircinia* sp.)	Polydiscamides B, C, D	[a]SNSR agonist	Feng et al. (2008)
		Marine sponge *Jaspis splendens*	Jaspamide	Cytotoxic	Ebada et al. (2009)
		Australian marine sponges (*Phoriospongia* sp. and *Callyspongia bilamellata*)	Phoriospongin A and B	Nematocide	Capon et al. (2002)
Cnidaria	Scyphozoa	Jelly fish (*Aurelia aurita*)	Aurelin	Antimicrobial	Ovchinnikova et al. (2006)
	Anthozoa	Sea anemone	Blood-depressing substance I (BDS-I)	Potassium channel blocker	Liu et al. (2012)
		Sea anemone (*Anemonia sulcata*)	ATX-II	Antibacterial and neurotoxic (sodium channel blocker)	Trapani et al. (2014)
Annelida		*Nereis diversicolor*	Hedistin	Antimicrobial	Tasiemski et al. (2007)
		Perinereis aibuhitensis (Grube, 1878)	Perinerin	Antimicrobial	Pan et al. (2004)
		Arenicola marina	Arenicin	Antimicrobial	Ovchinnikova et al. (2004)

(continued)

Table 7.6 (continued)

Class/order		Producer/organism	Antimicrobial compound	Activity	References
Arthropoda	Crustaceans	Pacific white shrimp (*Litopenaeus vannamei*)	Crustins *Lv*1 to 6	Antibacterial	Bartlett et al. (2002)
			Penaeidins 1 to 3	Antimicrobial	Destoumieux-Garzon et al. (1997)
			2.7 kDa AMPs	Antifungal	Destoumieux-Garzon et al. (2001)
		Litopenaeus stylirostris	7.9 and 8.3 Da AMPs	Antifungal	Destoumieux-Garzon et al. (2001)
			Crustins Lv1 to 3	Antibacterial	Bartlett et al. (2002)
			Penaeidin 4	Antimicrobial	Cuthbertson et al. (2002)
			Ls-Stylicin1	Antifungal	Rolland et al. (2010)
		Black tiger shrimp (*Penaeus monodon*)	Crustin Pm1 and crustin Pm7	Antimicrobial	Krusong et al. (2012)
			Penaeidin-5	Antimicrobial	Hu et al. (2006)
		Green shore crab (*Carcinus maenas*)	6.5 kDa	Antibacterial	Schnapp et al. (1996)
			Cysteine-rich 11.5-kDa antibacterial protein (crustin cm-1)	Antibacterial	Relf et al. (1999)
		Mud crab (*Scylla paramamosain*)	Arasin-like *Sp* and GRP*Sp*	Antimicrobial	Imjongjirak et al. (2011)
		Atlantic blue crab (*Callinectes sapidus*)	Callinectin	Antibacterial	Khoo et al. (1999)
		Spider crab, *Hyas araneus*	Arasin 1	Antimicrobial	Stensvag et al. (2008)
		Indian mud crab, *Scylla serrata*	*SSAP* (*Scylla serrata* antimicrobial protein)	Antibacterial	Yedery and Reddy (2009)
		American lobster, Homarus americanus	Homarin	Antibacterial	Battison et al. (2008)

	Xiphosura	Horseshoe crab (*Tachypleus tridentatus*)	Defensin	Antimicrobial	Saito et al. (1995)
		Horseshoe crab (*Tachypleus tridentatus*)	Tachycitin	Antimicrobial	Kawabata et al. (1996)
		Horseshoe crab (*Tachypleus tridentatus*)	Tachyplesin	Antimicrobial	Nakamura et al. (1988)
Mollusca		Ark shell (*Scapharca broughtonii*)	Defensin	Antimicrobial	Li et al. (2012)
		Mytilus edulis	Defensins A and B, mytilin A and B, mytimicin	Antimicrobial	Charlet et al. (1996)
		Mytilus galloprovincialis	Myticin	Antimicrobial	Mitta et al. (1999a)
			Defensins 1 and 2	Antibacterial	Mitta et al. (1999b)
			Mytilin B, C, D, and G1	Antimicrobial	Mitta et al. (2000)
			Myticin C	Antiviral	Balseiro et al. (2011)
		Scallop (*Argopecten irradians*)	Defensin recombinant protein – AiBD	Antimicrobial	Zhao et al. (2007)
		Pacific oyster (*Crassostrea gigas*)	Defensin	Antimicrobial	Gueguen et al. (2006)
		Sea slug (*Elysia rufescens*)	Kahalalides	Antimicrobial, Cancer therapy	Cheung Randy Chi Fai et al. (2015)
		Sea hare (*Dolabella auricularia*)	Dolabellanin A, C, E, and P	Antimicrobial	Iijima et al. (2003)
Echinodermata		*Green sea urchin* (*Strongylocentrotus droebachiensis*)	Centrocins, strongylocins	Antimicrobial	Li et al. (2008, 2010)

[a]SNSR: Human sensory neuron-specific G-protein-coupled receptor

Table 7.7 Antimicrobial peptides in tunicates (Hemichordata)

Class/order	Producer/organism	AMP	References
Tunicates	*Styela clava*	Styelin, clavanin	Leher et al. (2003)
		Clavaspirin	Lee et al. (2001a)
	Styela plicata	Plictamide	Tincu et al. (2003)
	Halocynthia aurantium	Dicynthaurin	Lee et al. (2001b)
		Halocidin	Jang et al. (2002)
	Halocynthia roretzi	Halocyamines	Azumi et al. (1990)
	Ciona intestinalis	A salt-tolerant AMP	Fedders et al. (2008)

classes (up to 90%) and little homology between the classes. In addition, the recently identified cathelicidins found in cod appear to comprise a third class, as it lacks sequence homology compared with the other two classes (Maier et al. 2008).

The in vitro antibacterial activities of cathelicidins appear to exhibit significant variability with respect to selectivity and species. For example, cod cathelicidin (cod-CATH) is highly active against Gram-negative bacterial species whereas inactive against the Gram-positive species. Cathelicidins also exhibit potent antifungal activity against *C. albicans* (Broekman et al. 2011). In contrast, the hagfish cathelicidins are active against both Gram-positive and Gram-negative bacteria but inactive against *Candida* (Uzzell et al. 2003). Rainbow trout cathelicidins are active against *Y. ruckeri*, while Atlantic salmon cathelicidins are not (Bridle et al. 2011).Cathelicidin expression is observed early in embryonic development. The immunomodulatory activities seen by mammalian cathelicidins may be shared by their fish counterparts, indicating an evolutionarily conserved mechanism of innate immune regulation.

7.5.2.3 β-Defensins

Defensins are a group of cysteine-rich cationic (a net positive charge that can vary from +1 to +5) antimicrobial peptides found in plants, fungi, invertebrates, and vertebrates (Bulet et al. 2004). Defensins exhibit a general conformation made by cysteine-stabilized α-helical and β-sheet folds (Zhu and Gao 2013). Fish defensins were first identified in zebrafish, fugu, and *Tetraodon* by a database-mining approach (Zou et al. 2007), and currently defensins have also been identified in a number of marine and freshwater fish species. In mammals three types of defensins have been identified based on their structure, α-, β-, and θ-defensins (Tang et al. 1999).

The human β-defensin gene has two exons and one intron, fairly typical of most β-defensin genes. Furthermore, mammalian β-defensins are translated as prepeptides, with the mature peptide sequence immediately downstream from the signal sequence (Beckloff and Diamond 2008). In fish, three exons and two introns encode a prepropeptide (including signal peptide, propeptide, and mature peptide) comprising of 60 to 77 amino acids and a mature peptide from 38 to 45 amino acids (Casadei et al. 2009). Among all vertebrate β-defensins, there are six conserved cysteines, located in a single exon in human and birds, while in fish they span two exons (Nam et al. 2010).

7.5.2.4 Hepcidins

Hepcidins were first discovered in humans. Park et al. (2001) isolated a new peptide from human urine and named it hepcidin, based on its site of synthesis (the liver, hep-)

and antibacterial properties in vitro (-cidin). They are cysteine-rich peptides with antimicrobial activity that and have been identified in many other vertebrates including reptiles, amphibians, and fish. Fish hepcidin was first identified and isolated from the hybrid striped bass (Shike et al. 2002). Fish hepcidin genes have undergone duplication and diversification processes that have produced multiple gene copies (Xu et al. 2008), and up to eight copies have been identified (Rodrigues et al. 2006).

Hepcidin isoforms have different expression pattern and kinetics in larval development with gene expression. Hepcidin genes are composed of three exons and two introns encoding a signal peptide, a prodomain and a mature peptide (Rodrigues et al. 2006). The pre-prohepcidin size can range from 81 to 96 amino acids and the mature hepcidin from 19 to 31, with a molecular weight around 2–3 kDa. However, sequence analysis of fish hepcidins has shown the presence of hepcidins that contain 7, 6, or 4 cysteines. Fish hepcidins are active against a wide variety of bacteria, both Gram-positive and Gram-negative bacteria, and can be induced by exposure at a low μM range, similar to other AMP genes (Xu et al. 2008).

In addition to antimicrobial, cytotoxic, and immunomodulatory effect, hepcidin is better known for its key iron regulator activity, in controlling ferroportin, which is able to degrade by internalization, thereby decreasing iron transfer into blood. Although ferroportin internalization by hepcidin has not been proven in fish, there is evidence suggesting that fish hepcidin also controls iron (Rodrigues et al. 2006).

7.5.2.5 Histone-Derived Peptides

Novel amphibian antimicrobial peptides were described by Park et al. (1996) as a new peptide from the Asiatic toad, *Bufo gargarizans*, called buforin I, identical to the *N*-terminal portion of histone 2A. Later, antimicrobial activity of histone fragments from numerous species was demonstrated. Histone-derived AMPs have been identified in a number of fish species, with broad-spectrum activity against both human and fish pathogens (Noga et al. 2011). They are expressed and secreted in fish skin and found in other tissues, including the gills, spleen, and gut. Histone-derived AMP genes are induced under conditions of stress in specific tissues in different fish species (Terova et al. 2011).

Antimicrobial peptides from aquatic animals are considered promising candidates with a broad spectrum of antimicrobial activity. Najafian and Babji (2012) reported that almost all fish antimicrobial peptides have antibacterial or bacteriostatic functions against both Gram-positive and Gram-negative strains. Apart from antimicrobial activity, they also possess antioxidative, cardioprotective (antihypertensive, antiatherosclerotic, and anticoagulant), antitumor, immunomodulatory, analgesic, antidiabetic, appetite suppressing, and neuroprotective activities. Marine peptides have high potential nutraceutical and medicinal values because of their broad spectra of bioactivities.

7.6 Nutraceutical Peptides

The concept of "nutraceutical" was introduced by Stephen DeFelice in 1989, by combining the terms "nutrition" and "pharmaceutical." The term refers to raw foods, fortified foods, or dietary supplements containing biologically active molecules, also known as bioactive molecules (Palthur et al. 2010) that provide health benefits beyond basic nutrition.

Marine peptides are important bioactive natural products that are present in many marine species. These marine peptides have high potential nutraceutical and medicinal values because of their broad spectra of bioactivities. The role of marine peptides as antimicrobial peptides (AMPs) possessing antibacterial, antiviral, antifungal, and antiprotozoal functions is discussed in the previous section (refer to Sect. 7.5). Nutraceuticals or health-promoting products are food-derived components (naturally occurring or enzymatically generated) that, in addition to their nutritional value, exert a physiological effect on the body (Suarez-Jimenez et al. 2012). They are usually claimed to prevent chronic diseases, enhance the immune system, manage stressful conditions, control body weight, regulate the blood glucose level, improve cognitive function, and delay the onset of aging process or increase life expectancy (Cheung et al. 2015). Most marine peptides or their derivatives have high commercial values and have reached the pharmaceutical and nutraceutical markets. A large number of them are already in different phases of the clinical and preclinical trial in the United States and Europe.

Bioactive peptides or protein hydrolysates can be extracted and isolated from the protein of the marine species by various methods in industrial-scale production. Organic solvent extraction method was used traditionally, but as a time-consuming, expensive, and environmentally unfriendly technique, it has been replaced by better extraction techniques like supercritical fluid extraction, pressurized solvent extraction, microwave-assisted extraction, ultrasound-assisted extraction, pulsed electric field-assisted extraction, and enzyme-assisted extraction methods.

Marine nutraceutical peptides are derived from a diverse range of sources that provide a myriad of bioactive molecules. Protein hydrolysates have attracted increasing attention in the past decade mainly due to their nutritional value and improved functionalities arising from the formation of short peptides or free amino acids during hydrolysis. These sources, as well as the constituent bioactive molecules, are discussed in the following sections and summarized in Table 7.8.

7.6.1 Antioxidant Peptides

Antioxidant compounds play an important role as health-protecting factors. They protect the body against reactive oxygen species (ROS), which exert oxidative damage to membrane lipids, protein, and DNA. All organisms have antioxidant mechanisms to counteract oxidation and its deleterious effect. Dietary intake of antioxidants is now becoming popular and considered as an effective way to fuel up our body with antioxidant reserve. As the synthetic antioxidants exhibit some potential health risks, the search for natural antioxidants in foodstuff has increased. Protein hydrolysates formed by the enzymatic digestion of aquatic products and by-products have been proven to be a good source of antioxidant peptides.

Bougatef et al. (2010) indicated that peptides containing histidine, tryptophan, and tyrosine possessed antioxidative activity. The amino acids proline, alanine, and leucine contribute to radical scavenging activity when located in the C-terminal in an amino

Table 7.8 Bioactive peptides from fish protein hydrolysates (FPH)

Hydrolysate source	Enzyme used	Bioactivity	References
Tuna gelatin	Alcalase	Antioxidant, antimicrobial	Gomez-Guillen et al. (2011)
Tuna cooking juice	Protease XXIII	Antioxidant	Jao and Ko (2002)
	Orientase		Hsu et al. (2009)
Bigeye tuna dark muscle	Pepsin	Antioxidant	Je et al. (2008)
Horse mackerel skin	Pepsin	Antioxidant	Sampath Kumar et al. (2011)
Head and viscera of sardine (*Sardinella aurita*)	Crude enzyme – sardine viscera	Antioxidant	Bougatef et al. (2010)
Jumbo squid skin gelatin (*Dosidicus gigas*)	Trypsin	Antioxidant	Mendis et al. (2005)
Silver carp (*Hypophthalmichthys molitrix*)	–	Antioxidant	Dong et al. (2008)
Squid gelatin hydrolysate	Alcalase	ACE-inhibiting peptide	Aleman et al. (2011)
Leather jacket (*Meuchenia* sp.)	Bromelain	Antibacterial	Salampessy et al. (2010)
Snow crab	–	Antibacterial	Doyen et al. (2012)
Oyster (*Crassostrea gigas*)	Alcalase, bromelain	Antibacterial, antifungal	Liu et al. (2008)
	Protease	Antitumor	Wang et al. (2010b)
Yellowfin sole (*Limanda aspera*)	Pepsin	Antioxidant, ACE-inhibiting peptide	Jun et al. (2004) and Jung et al. (2006a)
Cuttlefish (*Sepia officinalis*)	Proteases	ACE-inhibiting peptide	Balti et al. (2010)
Alaska pollock skin (*Theragra chalcogramma*)	Alcalase, pronase E, collagenase	ACE-inhibiting peptide	Byun and Kim (2001)
		Antioxidant	Kim et al. (2001)
Sardinella aurita	Serine proteases	ACE-inhibiting peptide	Bougatef et al. (2008)
Tuna frame fish protein hydrolysate	–	ACE-inhibiting peptide	Lee et al. (2010)
Catfish	Protamex	ACE-inhibiting peptide	Theodore and Kristinsson (2007)
Tilapia	–	ACE-inhibiting peptide	Raghavan and Kristinsson (2009)
Hoki (*Johnius belengerii*)	Pepsin	Ca-binding activity	Jung and Kim (2007)
	–	Antioxidant	Kim et al. (2007)
Alaska pollock (*Theragra chalcogramma*) backbone peptide	Pepsin	Affinity to Ca on hydroxyapatite	Jung et al. (2006b)
Goby muscle protein hydrolysate	Protease	Anticoagulant	Nasri et al. (2012)

(continued)

Table 7.8 (continued)

Hydrolysate source	Enzyme used	Bioactivity	References
Mytilus edulis	–	Anticoagulant oligopeptide	Jung and Kim (2009)
Pacific hake (*Merluccius productus*)	–	Cryoprotectant in frozen cod fillet mince	Cheung et al. (2009)
		Antioxidant and ACE inhibiting peptide	Samaranayaka et al. (2010)
Gelatin hydrolysate	–	Inhibit ice crystal growth in ice cream mix	Damodaran (2007)
Marine salmon skin	–	Antidiabetic	Zhu et al. (2010a)
Collagen peptides	–	Antidiabetic	Zhu et al. (2010b)
Disk abalone (*Haliotis discus*)	–	Wound healing	Wickramaarachchi et al. (2013)

acid sequence. Antioxidant peptide derived from different marine animals exhibits varying potencies to free radical scavenging and various antioxidant mechanisms.

7.6.2 Antitumor/Cytotoxic Peptides

More than 100 species of marine fish are traditionally used in medicines to treat tumors. A tumor is a solid or fluid-filled lesion that may or not be formed by an abnormal growth of neoplastic cells which become enlarged in size. It can be benign, premalignant, and malignant or represent a lesion without any cancerous potential. The antitumor/cytotoxic marine peptides can be classified into four types: (i) linear depsipeptides (e.g., dolastatin 10); (ii) cyclic depsipeptides (e.g., kahalalide F); (iii) linear peptides (e.g., pardaxin); and (iv) marine protein hydrolysates. Protein hydrolysates from marine fish are a potential source of bioactive peptides with antitumor activity. The digested hydrolysates of peptides from tuna dark muscle exhibited antiproliferative activity on human breast cancer cell line MCF-7 (Hsu et al. 2011). Peptide hydrolysates from oyster extract expressed in vivo antitumor activity in BALB/c mice (Wang et al. 2010b). Tunicate extract (Jumeri and Kim 2011) and squid gelatin hydrolysates (Aleman et al. (2011)) also demonstrated in vitro cytotoxic effect on several human cancer cell lines.

7.6.3 Antihypertensive Peptides

Angiotensin-I converting enzyme (ACE) is a dipeptidyl carboxypeptidase (EC. 3.4.15.1) important for the regulation of blood pressure and normal heart function. ACE inhibitors are used to treat congestive heart failure and high blood pressure (hypertension). Antihypertensive peptides or angiotensin-I converting enzyme (ACE) inhibitory peptides have been isolated from fish protein hydrolysates. Two peptides, with potent antihypertensive and antioxidant activities, were produced by hydrolysis of skate (*Okamejei kenojei*) gelatin using Alcalase and protease (Ngo et al. 2014). ACE-inhibiting peptides have been isolated from fish protein hydrolysate such as Alaska pollock skin hydrolysate, *Sardinella aurita*, yellowfin sole, salmon muscle, sea bream scales, and tuna frame proteins as well as squid and cuttlefish hydrolysates. ACE-inhibiting peptides from aquatic protein hydrolysates are summarized in Table 7.8.

7.6.4 Cardiovascular Protective Peptides

The cardiovascular protective peptides include bioactive peptides which exhibit activities important to cardiovascular health, including effects on blood pressure (Sect. 7.6.3), oxidative stress (Sect. 7.6.1), coagulation, atherosclerosis, and lipid metabolism. The anticoagulant peptide binds to coagulation factors in the blood clotting intrinsic pathway and inhibits their molecular interaction. The anti-atherosclerotic peptides showed their cardiovascular protection effect by suppressing the inflammatory responses in histamine-stimulated endothelial cells that may be related to early atherosclerosis. The mechanisms involved in the lipid reduction include binding of bile acid and disruption of cholesterol micelles and their adsorption in the digestive system and modification of hepatic and adipocytic enzyme activity and gene expression of lipogenic proteins (Howard and Udenigwe 2013). Fish protein lowers serum cholesterol level by retarding absorption and promoting excretion of cholesterol and bile acid attributed to lowered micellar solubility of cholesterol and enhanced bile acid-binding capacity.

7.6.5 Neuropeptides and Neuroprotective Peptides

Neuropeptides Neuropeptides derived from marine species have potential as natural and safe alternative to opioid drugs for anxiety and stress management. Opioid-like peptides exhibit a positive effect on motivation, emotion, behavior, stress, appetite, and pain management by interacting with specific opioid receptors in the nervous, endocrine, immune, and digestive systems. The gonadotropin-releasing hormone, which increases the secretion of pituitary gonadotropin, has been detected in hagfish (Nozaki 2013), in Japanese anchovy brain (Sukhan et al. 2013), and in the gonads of prosobranch *Patella caerulea* (De Lisa et al. 2013).

Neuroprotective Peptides Marine proteins and peptides suppress the development of neurodegenerative diseases like Parkinson's disease, Alzheimer's disease, and multiple sclerosis. Their neuroprotective action is brought about by the direct interaction of absorbed protein and peptide with a variety of cellular and molecular targets with enzyme/ion channels (Ryu and Kim 2013). The neuroprotective peptide (HTP-1) Gly-Thr-Glu-Asp-Glu-Leu-Asp-Lys from the seahorse *Hippocampus tri-*

maculatus exerted a protective action on PC12 cells and prevented them from the deleterious action of β amyloid plaques (amyloid β42), which is associated with the pathogenesis of Alzheimer's disease (Pangestuti et al. 2013).

7.6.6 Other Bioactive Roles of Marine Peptides

Analgesic Peptides Conopeptides from marine snail venoms have attracted much interest in drug design for chronic neuropathic pain.

Appetite-Suppressing Peptides Low-molecular-weight peptides from fish and shrimp head protein hydrolysates have been found to be effective for stimulating cholecystokinin release in STC-1 cells. These peptides are thus suggested as a promising functional food against obesity via regulation of cholecystokinin release. Marine peptides may be employed as anorexigenic agents and may interact with appetite-suppressing gut hormones including cholecystokinin and glucagon-like peptide 1 to produce anti-obesity effects (Cudennec and Ravallec 2013).

Immunomodulatory Peptides Hydrolysates are usually added to fish feed as immunostimulants in controlling infectious diseases in fish farms. The hydrolysate from Atlantic cod promoted the oxidative burst of Atlantic salmon leucocytes and stimulated the bactericidal power of phagocytes (Gildberg et al. 1996). Bagni et al. (2000) reported on the potential use of protein hydrolysate oral administration in sea bass (*Dicentrarchus labrax*) as an immunostimulant diet enhancing the innate immunity in cultured fish. Yang et al. (2009) reported that chum salmon hydrolysate stimulated lymphocyte proliferation, cytokine secretion, and enhancement of the cytotoxic activity.

Food Preservation According to Gomez-Guillen et al. (2011), antimicrobial peptides can also be used in natural biopreservation to control spoilage and foodborne pathogens. The peptide *CgPep33* from Pacific oysters is used to inhibit the activity against gray mold disease in strawberries. *CgPep33* can therefore be used as a substitute for fungicide to control postharvest diseases and to extend the shelf life of fruits and vegetables (Liu et al. 2007b). Protein hydrolysates also function as cryoprotectants, reduce lipid oxidation, enhance water retention, and suppress dehydration-induced denaturation of myofibrils in foods (Benjakul et al. 2014).

7.7 Marine Bioceramics

Biomineralization is a phenomenon in nature leading to the formation of a variety of solid inorganic structures by living organisms, such as intracellular crystals in prokaryotes; exoskeletons in protozoa, algae, and invertebrates; spicules and lenses; bone, teeth, statoliths, and otoliths; eggshells; plant mineral structures; oyster pearls; and also pathological biomineralization such as gallstones and kidney stones.

Calcification involving the formation of $CaCO_3$ crystals is among the most common normal and pathological biomineralizations (Addadi et al. 1987).

Three crystal phases of anhydrous calcium carbonate are encountered in nature. The most stable phase thermodynamically is calcite followed by aragonite and vaterite. Under ambient conditions $CaCO_3$ crystal commonly exists in the stable form, calcite; aragonite is the high-pressure polymorph, whereas vaterite is thermodynamically unstable (Zhou and Zheng 1998). Vaterite acts as a precursor in the formation of aragonite or calcite and is a very unstable phase of calcium carbonate rarely occurring in nature. The assembly of these biominerals consists of a four-stage process. It starts with the fabrication of a hydrophobic solid organic substrate or scaffolding onto which nucleation of the crystalline phase takes place closely associated with some polyanionic macromolecules. Crystal growth is then controlled by the addition of gel-structuring polyanionic macromolecules, and finally mineralization arrest is accompanied by the secretion of a new inert scaffolding of the same type or the deposition of other hydrophobic inhibitory macromolecules.

Aragonite's crystal lattice differs from that of calcite, resulting in a different crystal shape, an orthorhombic system with acicular crystals. It is formed by biological and physical processes, including precipitation from marine and freshwater environments. Aragonite forms naturally in almost all mollusk shells and as the calcareous endoskeleton of warm- and cold-water corals (Scleractinia). Because the mineral deposition in mollusk shells is strongly biologically controlled, some crystal forms are distinctively different from those of inorganic aragonite (Checa et al. 2015). In some mollusks, the entire shell is aragonite; in others, aragonite forms only discrete parts of a bimineralic shell (aragonite plus calcite). Aragonite also forms in the ocean and in caves as inorganic precipitates called marine cements and speleothems, respectively. The nacreous layer of the aragonite fossil shells of some extinct ammonites forms an iridescent material called ammolite. Ammolite is primarily aragonite with impurities that make it iridescent and valuable as a gemstone. Aragonite is metastable and is thus commonly replaced by calcite in fossils.

The mineral vaterite, also known as μ-$CaCO_3$, is another phase of calcium carbonate that is metastable at ambient conditions typical of earth's surface and decomposes even more readily than aragonite. Biovaterite was first reported to occur in the repair tissues of shells of certain gastropods. The existence of the vaterite is one of the key factors influencing the quality of freshwater cultured pearls (Qiao et al. 2007).

7.7.1 Hydroxyapatite

Among the existing calcium orthophosphates, only certain compounds are useful for biomedical applications. Commercialization of the dental and surgical applications of calcium orthophosphate (mainly, hydroxyapatite) bioceramics occurred in the 1980s. Currently hydroxyapatite (HA) has become a bioceramic of reference in the field of calcium orthophosphates for biomedical applications.

Hydroxyapatite (HAp), $Ca_{10}(PO_4)_6(OH)_2$, is widely employed as a biomaterial, especially as a bone substitute. Presently, calcium orthophosphate bioceramics are available in the form of particulates, blocks, cements, coatings, customized designs for specific applications, and as injectable composites in a polymer carrier. Recent biomedical applications include artificial replacements for the hips, knees, teeth, tendons, and ligaments, as well as repair for periodontal disease, maxillofacial reconstruction, augmentation, and stabilization of the jawbone, spinal fusion, and bone fillers after tumor surgery. Studies demonstrate potential applications of calcium orthophosphate bioceramics as scaffolds, drug delivery systems, as well as carriers of growth factors, bioactive peptides, and/or various types of cells for tissue engineering purposes (Dorozhkin 2010).

Human bone anatomy has the architecture of a nanocomposite material, made of 60–70% mineral component, up to 30% organic components (mostly type I collagen), and approximately 10% water. The need for restoring damaged bone tissues leads to the development of various bone reconstruction and tissue engineering solutions. The current market offers a wide range of calcium phosphate-based biomaterials as substitutes for bone tissue. Most representative materials are hydroxyapatite (HA), beta-tricalcium phosphate (β-TCP), and different combinations of these, generally named biphasic calcium phosphates (BCP) (Maidaniuc et al. 2018). Land resources such as animal bone tissue – generally bovine bone (Brzezińska-Miecznik et al. 2015) or other calcified structures such as eggshells – are used as cost-effective raw materials (Macha et al. 2015) for the preparation of hydroxyapatite.

$CaCO_3$ is a common biomineral, and extensive attention has been paid to its mineralization mechanism and biomimetic synthesis (Zhang and Vecchio 2013). Studies offer extensive information regarding marine resources, such as vertebrates bones, fish bones (Sunil and Jagannatham 2016), and fish scales (Kongsri et al. 2013) and calcified structures of invertebrates – coral, mussel shells (Shariffuddin et al. 2013), oyster shell (Rujitanapanich et al. 2014), cuttlefish bone (Ivankovic et al. 2007), sea urchin (Alvarez-Lloret et al. 2010), and marine algae (Walsh et al. 2010).

These entire natural resources harbor, besides the calcium resources required for calcium phosphate synthesis, many beneficial chemical elements such as Mg, Na, K, Cl, F, Si, and Sr, which are found in human bones in various amounts. This provides a unique advantage to naturally derived biomaterials against conventional calcium phosphates because the synthesis and preparation methods do not need additional procedures for doping these stoichiometric compounds. If not, the main $CaCO_3$ resource available on land – "marble" – would become a suitable alternative. Ceramics of marine origin such as corals, nacres, sponges, and fish bone provide significant amounts of ceramic materials for biomedical applications.

Synthesis of hydroxyapatite using $CaCO_3$ precursors generally involves two methods – direct synthesis and indirect synthesis. Direct synthesis implies precursor treatment with phosphorus-based reagents. Indirect synthesis requires thermal dissociation of $CaCO_3$ to calcium oxide (CaO) prior to phosphate-based reagent treatment. Carbonated HA nanoparticles constitute the inorganic part of enamel, teeth, and bones, and its concentration ranges a few percent in weight (Feki et al. 1991). Therefore, CO_3^{2-}-substituted HA nanoparticles are considered to be a good candidate for bone and dental applications (Landi et al. 2003). Crystalline hydroxyapatite is also used as an adsorbent in chromatography to separate mixtures of proteins or nucleic acids (Andrews-Pfannkoch et al. 2010). The affinity of double-stranded DNA for hydroxyapatite is so high that it can be employed to selectively remove double-stranded DNA from RNA and proteins in cell extracts (Yu et al. 2008). Table 7.9 summarizes the uses and potent bioactive roles of marine-derived hydroxyapatite.

Table 7.9 Marine and aquatic source-derived hydroxyapatite

Organism/source	Use	References
Marine algae		
Rodophycophyta	$CaPO_4$ biomaterials	Felício-Fernandes and Laranjeira (2000)
Phymatolithon calcareum	Composite non-cytotoxic tissue scaffolds	Kusmanto et al. (2008)
Corallina officinalis	$CaPO_4$ bone filler	Walsh et al. (2010)
Rodophycophyta	Hydroxyapatite fucoidan nanocomposites	Jeong et al. (2013)
Rodophycophyta	Carbonate substituted hydroxyapatite	Feki et al. (1991)
Marine invertebrates		
Coral	Phase-pure HA	Murugan and Ramakrishna (2004)
	Highly porous Si-incorporated hydroxyapatite	Kim et al. (2005)
Mollusk shell (*Pomacea lineata*)	Flower like hydroxyapatite	de Paula et al. (2010)
Mussel shells (*Perna canaliculus*)	Nanosized, crystalline HA	Shariffuddin et al. (2013)
Cuttlefish bone (*Sepia officinalis*)	Highly porous carbonate substituted hydroxyapatite	Ivankovic et al. (2010)
Sea urchin spines (*Paracentrotus lividus*)	Nanosized elongated apatite nanocrystals	Alvarez-Lloret et al. (2010)
Fish scales		
Tilapia nilotica	Nanocrystalline hydroxyapatite, cHA	Kongsri et al. (2013)
Oreochromis sp.	Porous hydroxyapatite for tissue engineering	Huang et al. (2011)
Fish bones		
Bones of swordfish (*Xiphias gladius*) and tuna (*Thunnus*)	Carbonate substituted hydroxyapatite	Boutinguiza et al. (2012)
Pseudoplatystoma corruscans, *Paulicea lutkeni*, *Pseudoplatystoma fasciatum*	Nanostructured hydroxyapatite	Coelho et al. (2006)
Thunnus obesus	Crystalline hydroxyapatite, cHA	Venkatesan et al. (2011)
Fish (*Fringescale sardinella*) bones	Hydroxyapatite based sunscreen formulation in cosmetics	Mohd Zul et al. (2018)

7.7.2 Biosilica

Biogenic silica, commonly known as biosilica, is an important bioceramide produced by several aquatic organisms, in addition to terrestrial organisms. Biosilica containing marine organisms includes diatoms, sponges, radiolarians, and choanoflagellates involved in the production of glassy amorphous silica via enzyme-mediated processes (Schroder et al. 2008). In the case of diatoms, unicellular microalgae, the protoplasm is enclosed in a nanoporous microshell, made of hydrogenated amorphous silica, called frustule (Delalat et al. 2015). Most sponges have internal skeletons made of the collagenous, spongin, and/or spicules of calcium carbonate or silicon dioxide (Muller et al. 2007). The choanoflagellates build complex basket-shaped cell coverings called lorica formed from several silica strips cemented together (Leadbeater and Thomsen 2000). Porous hierarchically organized microstructures; the biosilica can be obtained by removal of the organic biological matter from the organism of choice (Fig. 7.8).

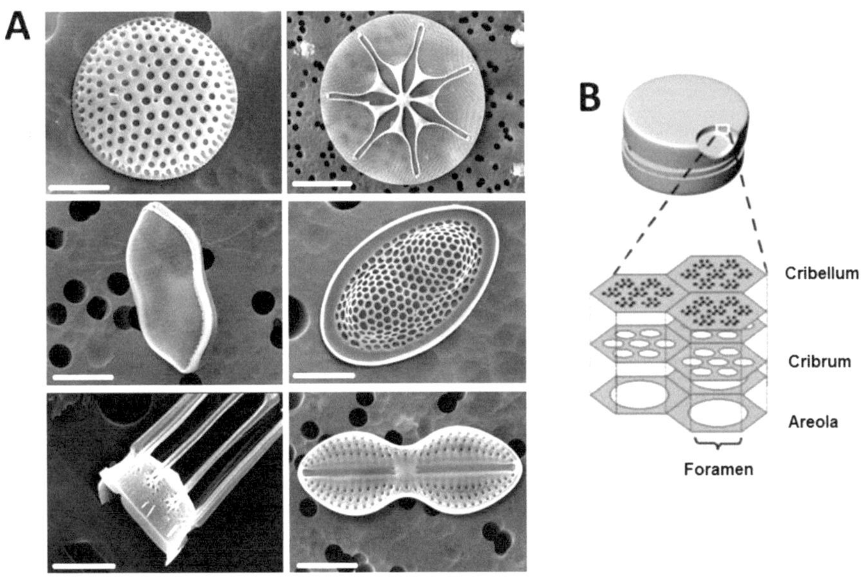

Fig. 7.8 The biosilica architecture. (**a**) Scanning electron microscopy (SEM) images of the cell walls of different diatom species showing (**b**) illustration of a centric diatom frustule with cross-sectional profile of the silica wall typically formed by three overlapping porous layers: cribellum, cribrum, and areola. (Losic et al. 2007)

Natural sources of mesoporous silica have wide applications ranging from biomedical to photonic fields (Panwar and Dutta 2019). They are the most potential naturally available materials for the development of cost-effective drug delivery systems because of their excellent biocompatibility, high surface area, pore size, drug-holding capacity, low cost, and ease of surface modification. Uthappa et al. (2018) review on diatom biosilica as a nature gifted potential drug delivery vehicle. Gale et al. (2009) opined on the use of biosilica frustules as a biosensor platform. The genetically engineered biosilica frustules exhibit an IgG-binding domain of protein G on the biosilica surface, which acts as versatile packs for the targeted drug delivery of poorly water-soluble anticancer drugs. Leone et al. (2017) investigated on the use of diatom frustules as scaffold for bone tissue engineering. Advantage of diatom frustules in possessing the nanostructured surfaces and the possibility to chemically modify the biosilica was utilized as calcium-doped biosilica supports for bone tissue regeneration. Ferrara et al. (2014) studied on the optical properties of nanostructured biosilica in the diatom *Arachnoidiscus* sp. and recommended on their intricate employability in next-generation photonic devices.

Appendix (Table 7.10 and Figs. 7.9, 7.10 and 7.11)

Table 7.10 Properties of agar, carrageenan and alginate

Seaweed hydrocolloids properties	Agar	Carrageenan	Alginate
Marine algae	**Rhodophyceae**	**Rhodophyceae**	**Phaeophyceae**
	Gracilaria sp., *Gelidium* sp., *Pterocladia*/*Gelidiella* sp.	*Kappaphycus alvarezii*, *Eucheuma denticulatum*, *Chondrus crispus*, *Gigartina* sp., *Furcellaria* sp., *Hypnea* sp., *Iridaea* sp.	*Laminaria hyperborean*, *Macrocystis pyrifera*, *Laminaria digitata*, *Ascophyllum nodosum*, *Eclonia* sp., *Lessonia* sp., *Durvillea* sp., *Sargassum* sp.
Appearance	Yellowish powder	Yellowish powder	White to yellowish-brown powder
Water solubility	Boiling water	*Iota*-, *kappa*-, and *lambda*-carrageenan soluble at 80 °C	Na^+, K^+, NH_4^+ alginates are soluble in hot and cold water. Ca_2^+, Mg_2^+ alginates are insoluble in water
pH in aqueous solution	6.8–7	7–10	Acidity (pKa) 1.5–3.5
Molecular weight (kDa)	340–380	100–1000	10–600
Gelling temperature	32–45 °C	30–50 °C	Sodium alginate does not have a boiling point or melting point
Melting temperature	85–95 °C	50–70 °C	The sol/gel transition of alginates is not temperature dependent
Viscosity in water (1.5 w/v at 75 °C)	10–100 centipoise	30–300 cP	Na^+, K^+ alginates: Thickening agent Ca_2^+, Mg_2^+ alginates: Forms gels
Gel strength (1.5 w/v at 20 °C)	700–1000 g/cm^2	100–350 g/cm^2	The gelling property is also dependant on the amount of guluronate present in the alginate

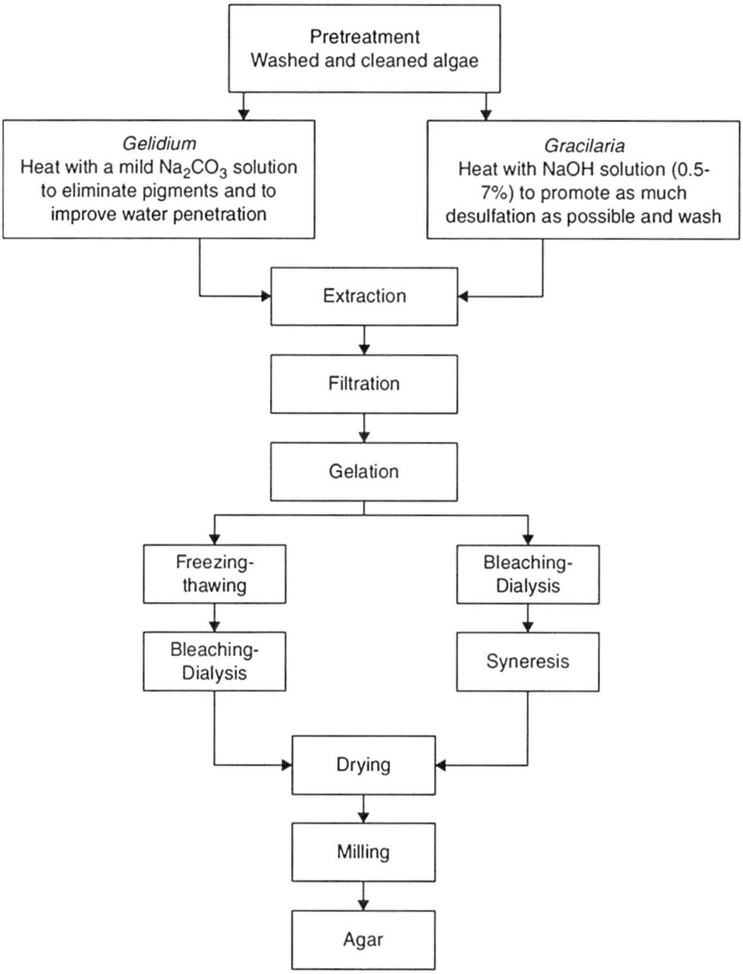

Fig. 7.9 Flow chart showing the procedure for isolation of agar. https://www.sciencedirect.com/science/article/pii/B978085709512150014X

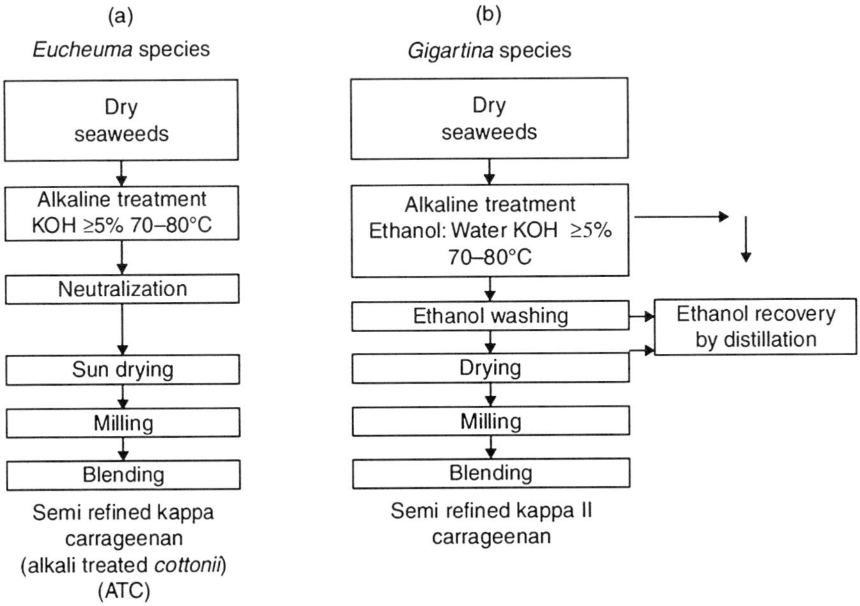

Fig. 7.10 Flow chart showing the procedure for isolation of carrageenan.https://www.sciencedirect.com/science/article/pii/B978085709512150014X

Fig. 7.11 Flow chart showing the procedure for isolation of alginate. https://www.sciencedirect.com/topics/agricultural-and-biological-sciences/alginic-acid

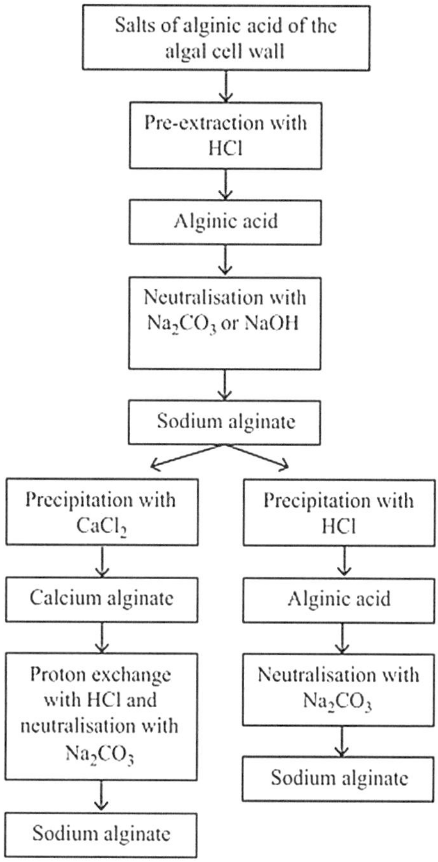

References

Abdul Khalil, H. P. S., Lai, T. K., Tye, Y. Y., Rizal, S., Chong, E. W. N., Yap, S. W., Hamzah, A. A., Nurul Fazita, M. R., & Paridah, M. T. (2018). A review of extractions of seaweed hydrocolloids: Properties and applications. *eXPRESS Polymer Letters, 12*(4), 296–317.

Addadi, L., Moradian, J., Shay, E., Maroudas, N. G., & Weiner, S. (1987). A chemical model for the cooperation of sulfates and carboxylates in calcite crystal nucleation: Relevance to biomineralization. *Proceedings of the National Academy of Sciences, 84*, 2732–2736.

Ahmad, M., Benjakul, S., & Nalinanon, S. (2010). Compositional and physicochemical characteristics of acid solubilized collagen extracted from the skin of unicorn leatherjacket (*Aluterus monoceros*). *Food Hydrocollids, 24*, 588–594.

Aleman, A., Pérez-Santín, E., Bordenave-Juchereau, S., Arnaudin, I., Gómez-Guillén, M. C., & Montero, P. (2011). Squid gelatin hydrolysates with antihypertensive, anticancer and antioxidant activity. *Food Research International, 44*, 1044–1051.

Alvarez-Lloret, P., Rodríguez-Navarro, A. B., Falini, G., Fermani, S., & Ortega-Huertas, M. (2010). Crystallographic control of the hydrothermal conversion of calcitic sea urchin spine (*Paracentrotus lividus*) into apatite. *Crystal Growth & Design, 10*, 5227–5232.

Alves, A., Sousa, R. A., & Reis, R. L. (2013). Processing of degradable ulvan 3D porous structures for biomedical applications. *Journal of Biomedical Materials Research. Part A, 101*, 998–1006. https://doi.org/10.1002/jbm.a.34403.

Amsler, C. D. (2008). *Algal chemical ecology*. Berlin: Springer.

Andrews-Pfannkoch, C., Fadrosh, D. W., Thorpe, J., & Williamson, S. J. (2010). Hydroxyapatite-mediated separation of double-stranded DNA, single-stranded DNA, and RNA genomes from natural viral assemblages. *Applied and Environmental Microbiology, 76*(15), 5039–5045. https://doi.org/10.1128/AEM.00204-10.

Aquino, R. S., Landeira-Fernandez, A. M., Valente, A. P., Andrade, L. R., & Mourão, P. A. (2005). Occurrence of sulfated galactans in marine angiosperms: Evolutionary implications. *Glycobiology, 15*, 11–20. https://doi.org/10.1093/glycob/cwh138.

Arias, J. L., & Fernandez, M. S. (2008). Polysaccharides and proteins in calcium carbonate based biomineralization. *Chemical Reviews, 108*, 4475–4482.

Arief, E. M., Siddiqui, Y. D., Yusoff, A., Suzina, A. H., & Abdullah, S. Y. (2013). Isolation of pepsin- solubilized collagen (PSC) from crude collagen extracted from body wall of sea cucumber (*Bohadschia* spp.). *International Journal of Pharmacy and Pharmaceutical Sciences, 5*, 555–559.

Azizi, S., Mohamad, R., Abdul Rahim, R., Mohammadinejad, R., & Bin Ariff, A. (2017). Hydrogel beads bio-nanocomposite based on kappa-carrageenan and green synthesized silver nanoparticles for biomedical applications. *International Journal of Biological Macromolecules, 104*, 423–431.

Azumi, K., Yokosawa, H., & Ishi, S. (1990). Haolcyamines: Novel antimicrobial tetrapeptide-like substances isolated from the hemocytes of the solitary ascidian *Halocynthia roretzi*. *Biochemistry, 29*, 156–165.

Bagni, M., Archetti, L., Amadori, M., & Marino, G. (2000). Effect of long-term oral administration of an immunostimulant diet on innate immunity in sea bass (*Dicentrarchus labrax*). *Journal of Veterinary Medicine, Series B, 47*, 745–751.

Balian, G., & Bowes, J. H. (1977). The structure and properties of collagen. In A. Ward & A. Courts (Eds.), *The science and technology of gelatin* (pp. 1–27). London: Academic.

Balseiro, P., Falcó, A., Romero, A., Dios, S., Martínez-López, A., Figueras, A., Estepa, A., & Novoa, B. (2011). *Mytilus galloprovincialis* Myticin C: A chemotactic molecule with antiviral activity and immunoregulatory properties. *PLoS One, 6*(8), e23140.

Balti, R., Nedjar-Arroume, N., Bougatef, A., Guillochon, D., & Nasri, M. (2010). Three novel angiotensin I-converting enzyme (ACE) inhibitory peptides from cuttlefish (*Sepia officinalis*) using digestive proteases. *Food Research International, 43*, 1136–1143.

Bao, L., Yang, W., Mao, X., Mou, S., & Tang, S. (2008). Agar/collagen membrane as skin dressing for wounds. *Biomedical Materials, 3*, 1–7.

Barbaglio, A., Benedetto, C. D., Martinello, T., Alongi, V., Fassini, D., Cullora, E., Patruno, M., Bonasoro, F., Barbosa, M. A., Carnevali, M. D. C., & Sugni, M. (2014). Production, characterization and biocompatibility of marine collagen matrices from an alternative and sustainable source: The sea urchin *Paracentrotus lividus*. *Marine Drugs, 12*(9), 4912–4933.

Bareil, R. P., Gauvin, R., & Berthod, F. (2010). Collagen-based biomaterials for tissue engineering applications. *Materials, 3*, 1863–1887.

Barros, A., Alves, A., Nunes, C., Coimbra, M. A., Pires, R. A., & Reis, R. L. (2013). Carboxymethylation of ulvan and chitosan and their use as polymeric components of bone cements. *Acta Biomaterialia, 9*, 9086–9097.

Bartlett, T., Cuthbertson, B. J., Shepard, E., Chapman, R., Gross, P., & Warr, G. (2002). Crustins, homologues of an 11.5-kDa antibacterial peptide, from two species of penaeid shrimp, *Litopenaeus vannamei* and *Litopenaeus setiferus*. *Marine Biotechnology, 4*, 278–293.

Battison, A. L., Summerfield, R., & Patrzykat, A. (2008). Isolation and characterisation of two antimicrobial peptides from haemocytes of the American lobster *Homarus americanus*. *Fish & Shellfish Immunology, 25*(1–2), 181–187.

Beckloff, N., & Diamond, G. (2008). Computational analysis suggests beta-defensins are processed to mature peptides by signal peptidase. *Protein and Peptide Letters, 15*, 536–540.

Benjakul, S., Yarnpakdee, S., Senphan, T., Halldorsdottir, S., & Kristinsson, H. (2014). Fish protein hydrolysates: Production, bioactivities, and applications. In *Antioxidants and functional components in aquatic foods* (pp. 237–282). New York: Wiley.

Bernardi, G., & Springer, G. F. (1962). Properties of highly purified fucan. *The Journal of Biological Chemistry, 237*, 75–80.

Bernhardt, R. R., & Schachner, M. (2000). Chondroitin sulfates affect the formation of the segmental motor nerves in zebrafish embryos. *Developmental Biology, 221*, 206–219.

Bewley, C. A., He, H. Y., Williams, D. H., & Faulkner, D. J. (1996). Aciculitins A-C: Cytotoxic and antifungal cyclic peptides from the lithistid sponge *Aciculites orientalis*. *Journal of the American Chemical Society, 118*(18), 4314–4321.

Bink, R. J., Habuchi, H., Lele, Z., Dolk, E., Joore, J., Rauch, G. J., Geisler, R., Wilson, S. W., Hertog, J., Kimata, K., & Zivkovica, D. (2003). Heparan sulfate 6-*O*-sulfotransferase is essential for muscle development in zebrafish. *Journal of Biological Chemistry, 278*, 31118–31127.

Borai, I. H., Ezz, M. K., Rizk, M. Z., El-Sherbiny, M., Matloub, A. A., Aly, H. F., Farrag, A. E. R., & Fouad, G. I. (2015). Hypolipidemic and anti-atherogenic effect of sulphated polysaccharides from the green alga *Ulva fasciata*. *International Journal of Pharmaceutical Sciences and Research, 31*(1), 1–12.

Bougatef, A., Nedjar-Arroume, N., Ravallec-Ple, R., Leroy, Y., Guillochon, D., Barkia, A., & Nasri, M. (2008). Angiotensin I- convertising enzyme (ACE) inhibitory activities of sardinella (*Sardinella aurita*) by-products protein hydrolysates obtained by treatment with microbial and visceral fish serine proteases. *Food Chemistry, 111*, 350–356.

Bougatef, A., Nedjar-Arroume, N., Manni, L., Ravallec, R., Barkia, A., Guillochon, D., & Nasri, M. (2010). Purification and identification of novel antioxidant peptides from enzymatic hydrolysates of sardinella (*Sardinella aurita*) by-products proteins. *Food Chemistry, 118*, 559–565.

Bourgougnon, N., Lahaye, M., Quemener, B., Chermann, J. C., Rimbert, M., Cormaci, M., Furnari, G., & Komprobst, J. M. (1996). Annual variation in composition and *in vitro* anti-HIV-1 activity of the sulfated glucuronogalactan from *Schizymenia dubyi* (Rhodophyta, Gigartinales). *Journal of Applied Phycology, 8*, 155–161.

Boutinguiza, M., Pou, J., Comesana, R., Lusquinos, F., Carlos, A., & Leon, B. (2012). Biological hydroxyapatite obtained from fish bones. *Materials Science and Engineering C Materials for Biological Applications, 32*, 478–486.

Bridle, A., Nosworthy, E., Polinski, M., & Nowak, B. (2011). Evidence of an antimicrobial-immunomodulatory role of Atlantic salmon cathelicidins during infection with *Yersinia ruckeri*. *PLoS One, 6*, e23417.

Broekman, D. C., Zenz, A., Gudmundsdottir, B. K., Lohner, K., Maier, V. H., & Gudmundsson, G. H. (2011). Functional characterization of codCath, the mature cathelicidin antimicrobial peptide from Atlantic cod (*Gadus morhua*). *Peptides, 32*, 2044–2051.

Brzezińska-Miecznik, J., Haberko, K., Sitarz, M., Bućko, M. M., & Macherzyńska, B. (2015). Hydroxyapatite from animal bones—Extraction and properties. *Ceramics International, 41*(3, Part B), 4841–4846.

Bulet, P., Stocklin, R., & Menin, L. (2004). Anti-microbial peptides: From invertebrates to vertebrates. *Immunological Reviews, 198*, 169–184.

Bulow, H. E., & Hobert, O. (2006). The molecular diversity of glycosaminoglycans shapes animal development. *Annual Review of Cell and Developmental Biology, 22*, 375–407.

Byun, H. G., & Kim, S. K. (2001). Purification and characterization of angiotensin I-converting enzyme (ACE) inhibitory peptides from Alaska pollack (*Theragra chalcogramma*) skin. *Process Biochemistry, 36*, 1155–1162.

Capon, R. J., Ford, J., Lacey, E., Gill, J. H., Heiland, K., & Friedel, T. (2002). Phoriospongin A and B: Two new nematocidal depsipeptides from the Australian marine sponges *Phoriospongia* sp. and *Callyspongia bilamellata*. *Journal of Natural Products, 65*(3), 358–363.

Cardozo, K. H. M., Guaratini, T., Barros, M. P., Falcao, V. R., Tonon, A. P., Lopes, N. P., Campos, S., Torres, M. A., Souza, A. O., Colepicolo, P., & Pinto, E. (2007). Metabolites from algae with economical impact. *Comparative Biochemistry and Physiology Part C, Toxicology and Pharmacology, 146*, 60–78.

Casadei, E., Wang, T., Zou, J., Gonzalez Vecino, J. L., Wadsworth, S., & Secombes, C. J. (2009). Characterization of three novel beta-defensin antimicrobial peptides in rainbow trout (*Oncorhynchus mykiss*). *Molecular Immunology, 46*, 3358–3366.

Cesaretti, M., Luppi, E., Maccari, F., & Volpi, N. (2004). Isolation and characterization of a heparin with high anticoagulant activity from the clam *Tapes phylippinarum*: Evidence for the presence of a high content of antithrombin III binding site. *Glycobiology, 14*, 1275–1284.

Charlet, M., Chernysh, S., Philippe, H., Hetru, C., Hoffmann, J. A., & Bulet, P. (1996). Innate immunity: Isolation of several cysteine-rich antimicrobial peptides from the blood of a mollusc, *Mytilus edulis*. *The Journal of Biological Chemistry, 271*, 21808–21813.

Chattopadhyay, S., & Raines, R. T. (2014). Review collagen-based biomaterials for wound healing. *Biopolymers, 101*, 821–833.

Checa, A. G., Cartwright, J. H., Sánchez-Almazo, I., Andrade, J. P., & Ruiz-Raya, F. (2015). The cuttlefish *Sepia officinalis* (Sepiidae, Cephalopoda) constructs cuttlebone from a liquid-crystal precursor. *Scientific Reports, 5*, 11513.

Chen, C. L., Ritch, R., Ming, S. L., Hui, M. N., Chung, Y. C., & Lung, Y. C. (2010). A new fish scale derived scaffold for corneal regeneration. *European Cells and Materials, 19*, 50–57.

Chengkui, Z., & Junfu, Z. (1984). Chinese seaweeds in herbal medicine. In *Developments in hydrobiology eleventh international seaweed symposium* (Vol. 22, pp. 152–154). Dordrecht: Springer. https://doi.org/10.1007/978-94-009-6560-7_24.

Cheung Randy Chi Fai, Tzi Bun Ng, & Jack Ho Wong. (2015). Marine peptides: Bioactivities and applications. *Marine Drugs, 13*, 4006–4043. https://doi.org/10.3390/md13074006.

Cheung, I. W. Y., Liceaga, A. M., & Li-Chan, E. C. Y. (2009). Pacific hake (*Merluccius productus*) hydrolysates as a cryoprotective agents in frozen Pacific cod fillet mince. *Journal of Food Science, 74*, 588–594.

Chevolot, L., Mulloy, B., Ratiskol, J., Foucault, A., & Colliec-Jouault, S. (2001). A disaccharide repeat unit is the major structure in fucoidans from two species of brown algae. *Carbohydrate Research, 330*, 529–535.

Cho, J., & Lee, D. G. (2011). Oxidative stress by antimicrobial peptide pleurocidin triggers apoptosis in *Candida albicans*. *Biochimie, 93*, 1873–1879.

Clark, D. P., Carroll, J., Naylor, S., & Crews, P. (1998). An antifungal cyclodepsipeptide, cyclolithistide A, from the sponge *Theonella swinhoei*. *The Journal of Organic Chemistry, 63*(24), 8757–8764.

Coelho, T. M., Nogueira, E. S., Steimacher, A., Medina, A. N., Weinand, W. R., Lima, W. M., Baesso, M. L., & Bento, A. C. (2006). Characterization of natural nanostructured hydroxyapatite obtained from the bones of Brazilian river fish. *Journal of Applied Physics, 100*, 094312–094316.

Cole, A. M. (2005). Antimicrobial peptide microbicides targeting HIV. *Protein and Peptide Letters, 12*, 41–47.

Cole, A. M., Weis, P., & Diamond, G. (1997). Isolation and characterization of pleurocidin, an antimicrobial peptide in the skin secretions of winter flounder. *The Journal of Biological Chemistry, 272*, 12008–12013.

Costa, L. S., Fidelis, G. P., Cordeiro, S. L., Oliveira, R. M., Sabry, D. A., Câmara, R. B., Nobre, L. T., Costa, M. S., Almeida-Lima, J., Farias, E. H., Leite, E. L., & Rocha, H. A. (2010). Biological activities of sulfated polysaccharides from tropical seaweeds. *Biomedicine & Pharmacotherapy, 64*, 21–28.

Cudennec, B., & Ravallec, R. (2013). Biological active peptides from marine sources related to gut hormones. *Current Protein & Peptide Science, 14*, 231–234.

Cuthbertson, B. J., Shepard, E., Chapman, R., & Gross, P. (2002). Diversity of penaeidin antimicrobial peptides in two shrimp species. *Immunogenetics, 54*, 442–445.

Dahiya, R., Singh, R., Sharma, A., Chennupati, S., & Maharaj, S. (2016). First total synthesis and biological screening of a proline-rich cyclopeptide from a Caribbean marine sponge. *Marine Drugs, 14*(12), E228. https://doi.org/10.3390/md14120228.

Damodaran, S. (2007). Inhibition of ice crystal growth in ice cream mix by gelatin hydrolysate. *Journal of Agricultural and Food Chemistry, 55*, 10918–10923.

Dash, M., Samal, S. K., Bartoli, C., Morelli, A., Philippe, F. S., Peter, D., & Federica, C. (2014). Biofunctionalization of ulvan scaffolds for bone tissue engineering. *ACS Applied Materials & Interfaces, 6*(5), 3211–3218.

David, G., & Bernfield, M. (1998). The emerging roles of cell surface heparan sulfate proteoglycans. *Matrix Biology, 17*, 461–463.

de Jesus Raposo, M. F., de Morais, A. M. B., & de Morais, R. M. S. C. (2015). Marine polysaccharides from algae with potential biomedical applications. *Marine Drugs, 13*, 2967–3028. https://doi.org/10.3390/md13052967.

De Lisa, E., Carella, F., De Vico, G., & Di Cosmo, A. (2013). The gonadotropin releasing hormone (GNRH)-like molecule in prosobranch *Patella caerulea*: Potential biomarker of endocrine – Disrupting compounds in marine environments. *Zoological Science, 30*, 135–140.

De Morais, M. G., Stillings, C., Dersch, R., Rudisile, M., Pranke, P., Costa, J. A. V., & Wendorff, J. (2010). Preparation of nanofibers containing the microalga *Spirulina* (*Arthrospira*). *Bioresource Technology, 101*, 2872–2876.

Delalat, B., Sheppard, V. C., Ghaemi, S. R., Rao, S., Prestidge, C. A., McPhee, G., & Voelcker, N. (2015). Targeted drug delivery using genetically engineered diatom biosilica. *Nature Communications, 6*(8791), 1–11. https://doi.org/10.1038/ncomms9791.

Delatte, S. J., Evans, J., Hebra, A., Adamson, W., Othersen, H. B., & Tagge, E. P. (2001). Effectiveness of beta-glucan collagen for treatment of partial-thickness burns in children. *Journal of Pediatric Surgery, 36*, 113–118.

Destoumieux-Garzon, D., Bulet, P., Loew, D., VanDorsselaer, A., Rodriguez, J., & Bachere, E. (1997). Penaeidins, a new family of antimicrobial peptides isolated from the shrimp *Penaeus vannamei*. *The Journal of Biological Chemistry, 272*, 28398–28406.

Destoumieux-Garzon, D., Saulnier, D., Garnier, J., Jouffrey, C., Bulet, P., & Bachere, E. (2001). Crustacean immunity: Antifungal peptides are generated from the C terminus of shrimp hemocyanin in response to microbial challenge. *The Journal of Biological Chemistry, 275*, 47070–47077.

Doege, K., Sasaki, M., Horigan, E., Hassell, J. R., & Yamada, Y. (1987). Complete primary structure of the rat cartilage proteoglycans core protein deduced from cDNA clones. *The Journal of Biological Chemistry, 262*, 17757–17767.

Dong, S., Zeng, M., Wang, D., Liu, Z., Zhao, Y., & Yang, H. (2008). Antioxidant and biochemical properties of protein hydrolysates repared from silver carp. (*Hypophthalmichthys molitrix*). *Food Chemistry, 107*, 1485–1493.

Dorozhkin, S. V. (2010). Calcium orthophosphates as bioceramics: State of the art. *Journal of Functional Biomaterials, 1*, 22–107. https://doi.org/10.3390/jfb1010022.

Doyen, A., Saucier, L., Beaulieu, L., Pouliot, Y., & Bazinet, L. (2012). Electroseparation of an antibacterial peptide fraction from snow crab by-products hydrolysate by electrodialysis with ultrafiltration membranes. *Food Chemistry, 132*, 1177–1184.

Ebada, S. S., Wray, V., De Voogd, N. J., Deng, Z., Lin, W., & Proksch, P. (2009). Two new jaspamide derivatives from the marine sponge *Jaspis splendens*. *Marine Drugs, 7*, 435–444.

Ermakova, S., Sokolova, R., Kim, S. M., Um, B. H., Isakov, V., & Zvyagintseva, T. (2011). Fucoidans from brown seaweeds *Sargassum hornery*, *Ecklonia cava*, *Costaria costata*: Structural characteristics and anticancer activity. *Applied Biochemistry and Biotechnology, 164*, 841–850.

Falshaw, R., Hubl, U., Ofman, D., Slim, G. C., Tariq, M. A., Watt, D. K., & Yorke, S. C. (2000). Comparison of the glycosaminoglycans isolated from the skin and head cartilage of Gould's arrow squid (*Nototodarus gouldi*). *Carbohydrate Polymers, 41*, 357–364.

Fedders, H., Michalek, M., Grotzinger, J., & Leippe, M. (2008). An exceptional salt-tolerant antimicrobial peptide derived from a novel gene family of haemocytes of the marine invertebrate *Ciona intestinalis*. *The Biochemical Journal, 416*(1), 65–75.

Feki, H. C., Rey, C., & Vignoles, M. (1991). Carbonate ions in apatites: Infrared investigations in the ly 4 CO3 domain. *Calcified Tissue International, 49*, 269–274.

Felício-Fernandes, G., & Laranjeira Mauro, C. M. (2000). Calcium phosphate biomaterials from marine algae. Hydrothermal synthesis and characterization. *Química Nova, 23*(4), 441–446.

Feng, Y., Carroll, A. R., Pass, D. M., Archbold, J. K., Avery, V. M., & Quinn, R. J. (2008). Polydiscamides B-D from a marine sponge Ircinia sp. as potent human sensory neuron-specific G protein coupled receptor agonists. *Journal of Natural Products, 71*(1), 8–11.

Fernandez-Diaz, M. D., Montero, P., & Gomez-Guillen, M. C. (2001). Gel properties of collagens from skins of cod (*Gadus morhua*) and hake (*Merluccius merluccius*) and their modification by the co-enhancers magnesium sulphate, glycerol and transglutaminase. *Food Chemistry, 74*, 161–167.

Ferrara, M. A., Dardano, P., De Stefano, L., Rea, I., Coppola, G., Rendina, I., Congestri, R., Antonucci, A., De Stefano, M., & De Tommasi, E. (2014). Optical properties of diatom nanostructured biosilica in *Arachnoidiscus* sp.: Micro-optics from mother nature. *PLoS One, 9*(7), e103750.

Fertah, M., Belfkira, A., Dahmane, E. M., Taourirte, M., & Brouillette, F. (2017). Extraction and characterization of sodium alginate from Moroccan *Laminaria digitata* brown seaweed. *Arabian Journal of Chemistry, 10*, S3707–S3714. https://doi.org/10.1016/j.arabjc.2014.05.003.

Fontana, J. D., Chocial, M. B., Baron, M., Guimaraes, M. F., Maraschin, M., Ulhoa, C., Florencio, J. A., & Bonfim, T. M. (1997). Astaxanthinogenesis in the yeast *Phaffia rhodozyma*: Optimization of low-cost culture media and yeast cell-wall lysis. *Applied Biochemistry and Biotechnology, 63-65*, 305–314.

Gale, D. K., Gutu, T., Jiao, J., Chang, C.-H., & Rorrer, G. L. (2009). Photoluminescence detection of biomolecules by antibody-functionalized diatom biosilica. *Advances Functional Materials, 19*(6), 926–933. https://doi.org/10.1002/adfm.200801137.

Garima, K., Anne-Sophie, B., Christel, M., Alan, C., Jeff, H., Gilles, B., Nathalie, B., & Balakrishnan, P. (2015). Enzyme-assisted extraction of bioactive material from *Chondrus crispus* and *Codium fragile* and its effect on *Herpes simplex* virus (HSV-1). *Marine Drugs, 13*, 558–580. https://doi.org/10.3390/md13010558.

Gathercole, L. J., & Keller, A. (1991). Crimp morphology in the fibre-forming collagens. *Matrix, 11*, 214–234.

George, M., & Abraham, T. E. (2006). Polyionic hydrocolloids for the intestinal delivery of protein drugs, alginate and chitosan – A review. *Journal of Controlled Release, 114*, 1–14.

Gildberg, A., Bøgwald, J., Johansen, A., & Stenberg, E. (1996). Isolation of acid peptide fractions from a fish protein hydrolysate with strong stimulatory effect on Atlantic salmon (*Salmo salar*) head kidney leucocytes. *Comparative Biochemistry and Physiology. Part B, Biochemistry & Molecular Biology, 114*, 97–101.

Glicksman, M. (1987). Utilization of seaweed hydrocolloids in the food industry. In M. A. Ragan & C. J. Bird (Eds.), *Twelfth international seaweed symposium* (Vol. 41, pp. 31–47). Dordrecht: Springer. https://doi.org/10.1007/978-94-009-4057-4_3.

Gomez-Guillen, M. C., Gimenez, B., Lopez-Caballero, M. E., & Montero, M. P. (2011). Functional and bioactive properties of collagen and gelatin from alternative sources: A review. *Food Hydrocolloids, 25*, 1813–1827.

Gong, Y., He, L., Li, J., Zhou, Q., Ma, Z., Gao, C., & Shen, J. (2007). Hydrogel-filled polylactide porous scaffolds for cartilage tissue engineering. *Journal of Biomedical Materials Research. Part B, Applied Biomaterials, 82*, 192–204.

Gueguen, Y., Herpin, A., Aumelas, A., Garnier, J., Fievet, J., Escoubas, J. M., Bulet, P., Gonzalez, M., Lelong, C., Favrel, P., & Bachère, E. (2006). Characterization of a defensin from the oyster *Crassostrea gigas*. Recombinant production, folding, solution structure, antimicrobial activities, and gene expression. *The Journal of Biological Chemistry, 281*(1), 313–323.

Hanaa, H. A. G., & Gihan, A. E. (2015). Antiviral activity of sulfated polysaccharides carrageenan from some marine seaweeds. *International Journal of Current Pharmaceutical Review and Research, 7*(1), 34–42.

Harder, T., Dobretsov, S., & Qian, P. Y. (2004). Waterborne polar macromolecules act as algal antifoulants in the seaweed *Ulva reticulata*. *Marine Ecology Progress Series, 274*, 133–141.

Helbert, W. (2017). Marine polysaccharide sulfatases. *Frontiers in Marine Science, 4*(6), 1–10. https://doi.org/10.3389/fmars.2017.00006.

Heu, M. S., Lee, J. H., Kim, H. J., Jee, S. J., Lee, J. S., Jeon, Y., Shahidi, F., & Kim, J. (2010). Characterization of acid- and pepsin-soluble collagens from flatfish skin. *Food Science and Biotechnology, 19*, 27–33.

Hidari, K. I., Takahashi, N., Arihara, M., Nagaoka, M., Morita, K., & Suzuki, T. (2008). Structure and anti-dengue virus activity of sulfated polysaccharide from a marine alga. *Biochemical and Biophysical Research Communications, 376*(1), 91–95.

Hjerpe, A., Engfeldt, B., Tsegenidis, T., Antonopoulos, C. A., Vynios, D. H., & Tsiganos, C. P. (1983). Analysis of the acid polysaccharides from squid cranial cartilage and examination of a novel polysaccharide. *Biochimica et Biophysica Acta, 757*, 85–91.

Holdt, S., & Kraan, S. (2011). Bioactive compounds in seaweed: Functional food applications and legislation. *Journal of Applied Phycology, 23*, 543–597. https://doi.org/10.1007/s10811-010-9632-5.

Hori, Y., Winans, A. M., & Irvine, D. J. (2009). Modular injectable matrices based on alginate solution/microsphere mixtures that gel in situ and co-deliver immunomodulatory factors. *Acta Biomaterialia, 5*, 969–982.

Hosokawa, M., Okada, T., Mikami, N., Konishi, I., & Miyashta, K. (2009). Bio-functions of marine carotenoids. *Food Science and Biotehnology, 18*(1), 1–11.

Howard, A., & Udenigwe, C. C. (2013). Mechanisms and prospects of food protein hydrolysates and peptide-induced hypolipidaemia. *Food & Function, 4*, 40–51.

Hsu, K. C., Lu, G. H., & Jao, C. L. (2009). Antioxidative properties of peptides prepared from tuna cooking juice hydrolysates with orientase (*Bacillus subtilis*). *Food Research International, 42*, 647–652.

Hsu, K. C., Li-Chan, E. C. Y., & Jao, C. L. (2011). Antiproliferative activity of peptides prepared from enzymatic hydrolysates of tuna dark muscle on human breast cancer cell line MCF-7. *Food Chemistry, 126*, 617–622.

Hu, S., Huang, J., Huang, W., Yeh, Y., Chen, M. H., Gong, H. Y., Chiou, T. T., Yang, T. H., Chen, T. T., Lu, J. K., & Wu, J. L. (2006). Structure and functison of antimicrobial peptide penaeidin-5 from the black tiger shrimp *Penaeus monodon*. *Aquaculture, 260*, 61–68.

Huang, W. S., Wang, K. J., Yang, M., Cai, J. J., Li, S. J., & Wang, G. Z. (2006). Purification and part characterization of a novel antibacterial protein scygonadin, isolated from the seminal plasma of mud crab, *Scylla serrata* (Forskal). *Journal of Experimental Marine Biology and Ecology, 339*, 37–42.

Huang, Y. C., Hsiao, P. C., & Chai, H. J. (2011). Hydroxyapatite extracted from fish scale: Effects on MG63 osteoblast-like cells. *Ceramics International, 37*, 1825–1831.

Huimin, Q., Tingting, Z., Quanbin, Z., Zhien, L., Zengqin, Z., & Ronge, X. (2005). Antioxidant activity of different molecular weight sulfated polysaccharides from *Ulva pertusa* Kjellm (Chlorophyta). *Journal of Applied Phycology, 17*, 527–534.

Hussein, G., Sankawa, U., Goto, H., Matsumoto, K., & Watanabe, H. (2006). Astaxanthin, a carotenoid with potential in human health and nutrition. *Journal of Natural Products, 69*(3), 443–449.

Hwang, J. H., Miuta, S., Yokoa, Y., & Yoshinaka, R. (2007). Purification and characterization of molecular species of collagen in the skin of skate (*Raja kenojei*). *Food Chemistry, 100*, 921–925.

Iijima, R., Kisugi, J., & Yamazaki, M. (2003). A novel antimicrobial peptide from the sea hare *Dolabella auricularia*. *Developmental and Comparative Immunology, 27*(4), 305–311.

Imjongjirak, A., Amparyup, P., & Tassanakajon, A. (2011). Two novel antimicrobial peptides, arasin-like Sp and GRPSp, from the mud crab *Scylla paramamosain*, exhibit the activity against some crustacean pathogenic bacteria. *Fish & Shellfish Immunology, 30*(2), 706–712.

Irhimeh, M. R., Fitton, J. H., & Lowenthal, R. M. (2007). Fucoidan ingestion increases the expression of CXCR4 on human CD34C cells. *Experimental Hematology, 35*, 989–994.

Ivankovic, H., Orlic, S., Tkalcec, E., & Gallego Ferrer, G. (2007). Kinetics of hydroxyapatite formation from cuttlefish bones. In J. G. Heinrich & C. Aneziris (Eds.), *Proceedings of 10th ECerS conference* (pp. 942–947). Baden-Baden: Goller Verlag. ISBN:3-87264-022-4.

Ivankovic, H., Tkalcec, E., Orlic, S., Ferrer, G. G., & Schaupererl, Z. (2010). Hydroxyapatite formation from cuttlefish bones: Kinetics. *Journal of Materials Science: Materials in Medicine, 21*, 2711–2722.

Iwa, K. (2008). Antidiabetic and antioxidant effects of polyphenols in brown alga *Ecklonia stolonifera* in genetically diabetic KK-A[y]mice. *Plant Foods for Human Nutrition, 63*, 163–169.

Jang, W., Kim, K., Lee, Y., Nam, M., & Lee, I. (2002). Halocidin: A new antimicrobial peptide from hemocytes of the solitary tunicate, Halocynthia aurantium. *FEBS Letters, 521*, 81–86.

Jao, C. L., & Ko, W. E. N. C. (2002). Utilization of cooking juice of young tuna processed into canned tuna as condiments: Effect of enzymatic hydrolysis and membrane treatment. *Fisheries Science, 70*, 1121–1129.

Je, J. Y., Qian, Z. J., Lee, S. H., Byun, H. G., & Kim, S. K. (2008). Purification and antioxidant properties of bigeye tuna (*Thunnus obesus*) dark muscle peptide on free radical-mediated oxidative systems. *Journal of Medicinal Food, 11*, 629–637.

Jeong, H. S., Venkatesan, J., & Kim, S. K. (2013). Hydroxyapatite fucoidan nanocomposites for bone tissue engineering. *International Journal of Biological Macromolecules, 57*, 138–141.

Jongjareonrak, A., Benjakul, S., Visesanguan, W., Nagai, T., & Tanaka, M. (2005). Isolation and characterization of acid and pepsin-solubilised collagens from the skin of Brownstripe red snapper (*Lutjanus vitta*). *Food Chemistry, 93*, 475–484.

Jorge, A. M., & Diamond, G. (2014). Antimicrobial peptides from fish. *Pharmaceuticals, 7*(3), 265–310. https://doi.org/10.3390/ph7030265.

Jumeri, & Kim, S. M. (2011). Antioxidant and anticancer activities of enzymatic hydrolysates of solitary tunicate (*Styela clava*). *Food Science and Biotechnology, 20*, 1075–1085.

Jun, S. Y., Park, P. J., Jung, W. K., & Kim, S. K. (2004). Purification and characterization of an antioxidative peptide from enzymatic hydrolysate of yellowfin sole (*Limanda aspera*)frame protein. *European Food Research and Technology, 219*, 20–26.

Jung, W. K., & Kim, S. K. (2007). Calcium-binding peptide derived from pepsinolytic hydrolysates of hoki (*Johnius belengerii*) frame. *European Food Research and Technology, 224*, 763–767.

Jung, W. K., & Kim, S. K. (2009). Isolation and characterization of an anticoagulant oligopeptide from blue mussel, Mytilus edulis. *Food Chemistry., 117*, 687–692.

Jung, W. K., Mendis, E., Je, J. Y., Park, P. J., Son, B., Kim, H. C., Choi, Y. K., & Kim, S. K. (2006a). Angiotensin I-converting enzyme inhibitory peptide from yellowfin sole (*Limanda aspera*) frame protein and its antihypertensive effect in spontaneously hypertensive rats. *Food Chemistry, 94*, 26–32.

Jung, W. K., Kaarawita, R., Heo, S. J., Lee, B. J., Kim, S. K., & Jeon, Y. J. (2006b). Recovery of a novel Ca-binding peptide from Alaska pollock (*Theragra chalcogramma*) backbone by pepsinolytic hydrolysis. *Process Biochemistry, 41*, 2097–2100.

Karamanos, N. K., Aletras, A. J., Tsegenids, T., Tsiganos, C. P., & Antonopoulos, C. A. (1992). Isolation, characterization and properties of the oversulphated chondroitin sulphate proteoglycan from squid skin with peculiar glycosaminoglycan sulphation pattern. *European Journal of Biochemistry, 204*, 553–560.

Kawabata, S., Nagayama, R., Hirata, M., Shigenaga, T., Agarwala, K. L., Saito, T., Cho, J., Nakajima, H., Takagi, T., & Iwanaga, S. (1996). Tachycitin, a small granular component in horseshoe crab hemocytes, is an antimicrobial protein with chitin-binding activity. *Journal of Biochemistry, 120*(6), 1253–1260.

Khan, S. B., Qian, Z., Ryu, B. M., & Kim, S. (2009). Isolation and biochemical characterization of collagens from seaweed pipefish, *Syngnathus schlegeli. Biotechnology and Bioprocess Engineering, 14*, 436–442.

Khan, R., Khan, M. H., & Bey, A. (2011). Use of collagen as an implantable material in the reconstructive procedures – An over view. *Biology and Medicine, 3*, 25–32.

Khoo, L., Robinette, D., & Noga, E. (1999). Callinectin, an antibacterial peptide from blue crab, *Callinectes sapidus*, hemocytes. *Marine Biotechnology, 1*, 44–51.

Kim, S. K., Kim, Y., Byun, H. G., Nam, K. S., Joo, D. S., & Shahidi, F. (2001). Isolation and characterization of antioxidative peptide from gelatin hydrolysates of Alaska pollock skin. *Journal of Agricultural and Food Chemistry, 49*, 1984–1989.

Kim, Y. H., Song, H., Riu, D. H., Kim, S. R., Kim, H. J., & Moon, J. H. (2005). Preparation of porous Si-incorporated hydroxyapatite. *Current Applied Physics, 5*, 538–541.

Kim, S. Y., Je, J. Y., & Kim, S. K. (2007). Purification and characterization of antioxidant peptide from hoki (*Johnius belengerii*) frame protein by gastrointestinal digestion. *Journal of Nutritional Biochemistry, 18*, 31–38.

Kim, J. K., Lee, S. A., Shin, S., Lee, J. Y., Jeong, K. W., Nan, Y. H., Park, Y. S., Shin, S. Y., & Kim, Y. (2010). Structural flexibility and the positive charges are the key factors in bacterial cell selectivity and membrane penetration of peptoid-substituted analog of Piscidin 1. *Biochimica et Biophysica Acta, 1798*, 1913–1925.

Kim, J. K., Cho, M. L., Karnjanapratum, S., Shin, I. S., & You, S. G. (2011). *In vitro* and *in vivo* immunomodulatory activity of sulfated polysaccharides from *Enteromorpha prolifera*. *International Journal of Biological Macromolecules, 49*(5), 1051–1058.

Kimura, S., Miyauchi, Y., & Uchida, N. (1991). Scale and bone type I collagens of carp (*Cyprinus carpio*). *Comparative Biochemistry and Physiology, 99B*, 473–476.

Kimura, M., Wakimoto, T., Egami, Y., Tan, K. C., Ise, Y., & Abe, I. (2012). Calyxamides A and B, cytotoxic cyclic peptides from the marine sponge *Discodermia calyx*. *Journal of Natural Products, 75*(2), 290–294.

Kinoshita, A., Yamada, S., Haslam, S. M., Morris, H. R., Dell, A., & Sugahara, K. (1997). Novel tetrasaccharides isolated from squid cartilage chondroitin sulfate E contain unusual sulfated disaccharide units GlcA(3- O- sulfate)β1-3GalNAc(6 – O – sulfate) or GlcA (3 –O – sulfate) β 1-3GalNAc(4,6 – O- disulfate). *The Journal of Biological Chemistry, 272*, 19656–19665.

Kitagawa, H., Tanaka, Y., Yamada, S., Seno, N., Haslam, S. M., Morris, H. R., Dell, A., & Sugahara, K. (1997). A novel pentasaccharide sequence GlcA(3 – sulfate)(β1 -3)GalNAc(4- sulfate)(β1 -4)(Fucα –3) – GlcA(β 1-3)GalNAc(4 – sulfate) in the oligosaccharides isolated from King Crab cartilage chondroitin sulfate K and its differential susceptibility to chondroitinases and hyaluronidase. *Biochemistry, 36*, 3998–4008.

Kittiphattanabawon, P., Benjakul, S., Visesanguan, W., Nagai, T., & Tanaka, M. (2005). Characterization of acid-soluble collagen from skin and bone of bigeye snapper (*Priacanthus tayenus*). *Food Chemistry, 89*, 363–372.

Kofuji, K., Huang, Y., Tsubaki, K., Kokido, F., Nishikawa, K., Isobe, T., & Murata, Y. (2010). Preparation and evaluation of a novel wound dressing sheet comprised of beta-glucan-chitosan complex. *Reactive and Functional Polymers, 70*, 784–789.

Kongsri, S., Janparadit, K., Buapa, K., Techawongstien, S., & Chanthai, S. (2013). Nanocrystalline hydroxyapatite from fish scale waste: Preparation, characterization and application for selenium adsorption in aqueous solution. *Chemical Engineering Journal, 215–216*, 522–532.

Kruger, T. E., Miller, A. H., & Wang, J. (2013). Collagen scaffolds in bone sialoprotein-mediated bone regeneration. *Scientific World Journal, 2013*, 1–6. https://doi.org/10.1155/2013/812718.

Krusong, K., Poolpipat, P., Supungul, P., & Tassanakajon, A. (2012). A comparative study of antimicrobial properties of crustinPm1 and crustinPm7 from the black tiger shrimp *Penaeus monodon*. *Developmental and Comparative Immunology, 36*(1), 208–215.

Kusmanto, F., Walker, G., Gan, Q., Walsh, P., Bushanan, F., Dickson, G., McCaigue Maggs, C., & Dring, M. (2008). Development of composite tissue scaffolds containing naturally sourced microporous hydroxyapatite. *Chemical Engineering Journal, 139*, 398–407.

Landi, E., Celotti, G., Logroscino, G., & Tampieri, A. (2003). Carbonated hydroxyapatite as bone substitute. *Journal of the European Ceramic Society, 23*, 2931–2937.

Laparra, J. M., Tako, E., Glahn, R. P., & Miller, D. D. (2008). Isolated glycosaminoglycans from cooked haddock enhance nonheme iron uptake by Caco-2 cells. *Journal of Agricultural and Food Chemistry, 56*, 10346–10351.

Laurienzo, P. (2010). Marine polysaccharides in pharmaceutical applications: An overview. *Marine Drugs, 8*, 2435–2465. https://doi.org/10.3390/md8092435.

Lauth, X., Shike, H., Burns, J. C., Westerman, M. E., Ostland, V. E., Carlberg, J. M., van Olst, J. C., Nizet, V., Taylor, S. W., Shimizu, C., & Bulet, P. (2002). Discovery and characterization of two isoforms of moronecidin, a novel antimicrobial peptide from hybrid striped bass. *The Journal of Biological Chemistry, 277*, 5030–5039.

Leadbeater, B. S., & Thomsen, H. (2000). *Order choanoflagellida. An illustrated guide to the protozoa* (Vol. 451, 2nd ed., pp. 14–38). Lawrence: Society of Protozoologists.
Ledward, C. A. (2000). Chapter 4: Gelatin. In G. O. Phillips & P. A. Williams (Eds.), *Handbook of hydrocolloids*. Boca Raton: CRC Press.
Lee, K. Y., & Mooney, D. J. (2012). Alginate: Properties and biomedical applications. *Progress in Polymer Science, 7*, 106–126.
Lee, I., Zhao, C., Nguyen, T., Menzel, L., Waring, A., Sherman, M., & Lehrer, R. (2001a). Clavaspirin, an antimicrobial and hemolytic peptide from *Styela clava*. *Journal of Peptide Research, 58*, 445–456.
Lee, I. H., Lee, Y. S., Kim, C., Chung-Ryul, K., Hong, T., Menzel, L., Lee, B. M., Pohl, J., Sherman, M. A., Waring, A., & Lehrer, R. (2001b). Dicynthaurin: An antimicrobial peptide from hemocytes of the solitary tunicate, Halocynthia aurantium. *Biochimica et Biophysica Acta, 1527*, 141–148.
Lee, S. H., Qian, Z. J., & Kim, S. K. (2010). A novel angiotensin I-converting enzyme inhibitory peptide from tuna frame protein hydrolysate and its hypertensive effect in spontaneously hypertensive rats. *Food Chemistry, 118*, 96–102.
Leone, G., Vona, D., Lo, P. M., Urbano, L., Cicco, S., Gristina, R., Palumbo, F., Ragni, R., & Farinola, G. M. (2017). Ca2+–*in vivo* doped biosilica from living *Thalassiosira weissflogii* diatoms: Investigation on Saos-2 biocompatibility. *MRS Advances, 2*, 1047–1058. https://doi.org/10.1557/adv.2017.49.
Li, C., Haug, T., Styrvold, O. B., Jorgensen, T. O., & Stensvag, K. (2008). Strongylocins, novel antimicrobial peptides from the green sea urchin, *Strongylocentrotus droebachiensis*. *Developmental and Comparative Immunology, 32*(12), 1430–1440.
Li, C., Haug, T., Moe, M. K., Styrvold, O. B., & Stensvag, K. (2010). Centrocins: Isolation and characterization of novel dimeric antimicrobial peptides from the green sea urchin, *Strongylocentrotus droebachiensis*. *Developmental and Comparative Immunology, 34*(9), 959–968.
Li, M., Zhu, L., Zhou, C. Y., Sun, S., Fan, Y. J., & Zhuang, Z. M. (2012). Molecular characterization and expression of a novel big defensin (Sb-BDef1) from ark shell, Scapharca broughtonii. *Fish and Shellfish Immunology, 33*, 1167–1173.
Li, W., Jiang, N., Li, B., Wan, M., Chang, X., Liu, H., Zhang, L., Yin, S., Qi, H., & Liu, S. (2018). Antioxidant activity of purified ulvan in hyperlipidemic mice. *International Journal of Biological Macromolecules, 113*, 971–975. https://doi.org/10.1016/j.ijbiomac.2018.02.104.
Lin, M. G., Lasekan, O., Saari, N., & Khairunniza-Bejo, S. (2018). Effect of chitosan and carrageenan-based edible coatings on post-harvested longan (*Dimocarpus longan*) fruits. *CyTA – Journal of Food, 16*(1), 490–497. https://doi.org/10.1080/19476337.2017.1414078.
Linnartz, H., Huibertus, T. V. S., Lin, C. C., Karim, H., Mass, S., Lai, H., Van den Berg, T., Salvatori, D., Luyten, G. P. M., & Jager, M. J. (2013). A fish scale – Derived collagen matrix as artificial cornea in rats: Properties and potential. *Nanotechnology and Regenerative Medicine, 54*, 3224–3233.
Liu, H. Y., Li, D., & Guo, S. D. (2007a). Studies on collagen from the skin of channel catfish (*Ictalurus punctatus*). *Food Chemistry, 101*, 621–625.
Liu, Z., Zeng, M., Dong, S., Xu, J., Song, H., & Zhao, Y. (2007b). Effect of antifungal peptide from oyster enzymatic hydrolysates for control of gray mold (*Botrytis cinerea*) on harvested strawberries. *Postharvest Biology and Technology, 46*, 95–98.
Liu, Z., Dong, S., Xu, J., Zeng, M., Song, H., & Zhao, Y. (2008). Production of cysteine rich antimicrobial peptide by digestion of oyster (*Crassostrea gigas*) with Alcalase and bromelin. *Food Control, 19*, 231–235.
Liu, P., Jo, S., & Bean, B. P. (2012). Modulation of neuronal sodium channels by the sea anemone peptide BDS-I. *Journal of Neurophysiology, 107*(11), 3155–3167.
Losic, D., Pillar, R. J., Dilger, T., Mitchell, J. G., & Voelcker, N. H. (2007). Atomic force microscopy (AFM) characterization of the porous silica nanostructure of two centric diatoms. *Journal of Porous Materials, 14*, 61–69. (for Biosilica SEM image permission required).

Lucas, F. B. N., Bianca, C. M., Lourivaldo, S. P., Delia, R. T., & Ana, P. R. (2016). Formation of carrageenan-$CaCO_3$ bioactive membranes. *Materials Science and Engineering C, 58*, 1–6.

Lucinda-Silva, R. M., Salgado, H. R. N., & Evangelista, R. C. (2010). Alginate–chitosan systems: *In vitro* controlled release of triamcinolone and *in vivo* gastrointestinal transit. *Carbohydrate Polymers, 81*, 260–268.

Macha, I. J., Ozyegin, L., Oktar, F. N., & Ben-Nissan, B. (2015). Conversion of ostrich eggshells (*Struthio camelus*) to calcium phosphates. *Journal of the Australian Ceramic Society, 51*(1), 125–133.

Madhavan, S., & Abirami. (2015). A review on hydrocolloids-agar and alginate. *Journal of Pharmaceutical Sciences and Research, 7*, 704–707.

Maeda, H., Hosokawa, M., Sashima, T., Funayama, K., & Miyashita, K. (2005). Fucoxanthin from edible seaweed, *Undaria pinnatifida*, shows antiobesity effect through UCPI expression in white adipose tissues. *Biochemical and Biophysical Research Communications, 332*(2), 392–397.

Maeda, H., Hosokawa, M., Sashima, T., & Miyashita, K. (2007). Dietary combination of fucoxanthin and fish oil attenuates the weight gain of white adipose tissue and decrease blood glucose in obese/diabetic KK-A^y mice. *Journal of Agricultural and Food Chemistry, 55*(19), 7701–7706.

Maidaniuc, A., Dascălu, C.-A., Miculescu, M., Voicu, Ș. I., & Ciocoiu, R.-C. (2018). Chapter 6: Biomimetic calcium phosphates derived from marine and land bioresource. In *Hydroxyapatite – Advances in composite nanomaterials, biomedical applications and its technological facets*. Rijeka: Intech Open. https://doi.org/10.5772/intechopen.71489.

Maier, V. H., Dorn, K. V., Gudmundsdottir, B. K., & Gudmundsson, G. H. (2008). Characterisation of cathelicidin gene family members in divergent fish species. *Molecular Immunology, 45*, 3723–3730.

Makkar, H. P. S., Tran, G., Heuze, V., Giger-Reverdin, S., Lessire, M., Lebas, F., & Ankers, P. (2016). Seaweeds for livestock diets: A review. *Animal Feed Science and Technology, 212*, 1–17.

Matsuno, T. (2001). Aquatic animal carotenoids. *Fisheries Sciences, 67*(5), 771–783.

McHugh, D. J. (2003). *A guide to seaweed industry*. Rome: FAO Fisheries and Aquaculture Department.

Mendis, E., Rajapakse, N., Byun, H. G., & Kim, S. K. (2005). Investigation of jumbo squid (*Dosidicus gigas*) skin gelatin peptides for their in vitro antioxidant effects. *Life Sciences, 77*, 2166–2178.

Mitta, G., Hubert, F., Noel, T., & Roch, P. (1999a). Myticin, a novel cysteine-rich antimicrobial peptide isolated from haemocytes and plasma of the mussel *Mytilus galloprovincialis*. *European Journal of Biochemistry, 265*, 71–78.

Mitta, G., Vandenbulcke, F., Hubert, F., & Roch, P. (1999b). Mussel defensins are synthesized and processed in granulocytes then released into the plasma after bacterial challenge. *Journal of Cell Science, 112*, 4233–4242.

Mitta, G., Vandenbulcke, F., Hubert, F., Salzet, M., & Roch, P. (2000). Involvement of mytilins in mussel antimicrobial defense. *The Journal of Biological Chemistry, 275*, 12954–12962.

Miyashita, K. (2014). Marine antioxidants: Polyphenols and carotenoids from algae. In H. G. Kristinsson (Ed.), *Antioxidants and functional components in aquatic foods* (pp. 219–235). Chichester: Wiley.

Montero, P., & Gomez-Guillen, M. C. (2000). Extracting conditions for megrim (*Lepidorhombus boscii*) skin collagen affect functional properties of the resulting gelatin. *Journal of Food Science, 65*, 434–438.

Montero, P., Alvarez, C., Marti, M. A., & Borderias, A. J. (1995). Plaice skin collagen extraction and functional properties. *Journal of Food Science, 60*, 1–3.

Morelli, A., & Chiellini, F. (2010). Ulvan as a new type of biomaterial from renewable resources: Functionalization and hydrogel preparation. *Macromolecular Chemistry and Physics, 211*, 821–832. https://doi.org/10.1002/macp.200900562.

Moshaverinia, A., Chen, C., Akiyama, K., Ansari, S., Xu, X., Chee, W. W., Schricker, S. R., & Shi, S. (2012). Alginate hydrogel as a promising scaffold for dental-derived stem cells: An *in vitro* study. *Journal of Materials Science: Materials in Medicine, 23*, 3041–3051.

Motta, G. J. (1989). Calcium alginate topical wound dressings, A new dimension in the cost-effective treatment for exudating dermal wounds and pressure sores. *Ostomy Wound Management, 25*, 52–56.

Muller, W. E. G., Li, J., Schröder, H. C., Qiao, L., & Wang, X. (2007). The unique skeleton of siliceous sponges (Porifera; Hexactinellida and Demospongiae) that evolved first from the Urmetazoa during the Proterozoic: A review. *Biogeosciences, 4*(2), 219–232. https://doi.org/10.5194/bg-4-219-2007.

Murugan, R., & Ramakrishna, S. (2004). Crystallographic study of hydroxyapatite bioceramics derived from various sources. *Crystal Growth and Design, 5*, 111–112.

Muthukumar, T., Prabu, P., Ghosh, K., & Sastry, T. P. (2014). Fish scale collagen sponge incorporated with *Macrotyloma uniflorum* plant extract as a possible wound or burn dressing material. *Colloids and Surfaces B: Biointerfaces, 113*, 207–212. https://doi.org/10.1016/j.colsurfb.2013.09.019.

Muyonga, J. H., Cole, C. G. B., & Duodu, K. G. (2004). Characterisation of acid soluble collagen from skins of young and adult Nile perch (*Lates niloticus*). *Food Chemistry, 85*, 81–89.

Nader, H. B., & Dietrich, C. P. (1989). Natural occurrence and possible biological role of heparin. In D. A. Lane & U. Lindahl (Eds.), *Heparin: Chemical and biological properties, clinical applications* (pp. 81–96). London: Edward Arnold Publishers.

Nader, H. B., Ferreira, T. M. P. C., Paiva, J. F., Medeiros, M. G. L., Jeronimo, S. M. B., Paiva, V. M. P., & Dietrich, C. P. (1984). Isolation and structural studies of heparin sulfates and chondroitin sulfates from three species of molluscs. *The Journal of Biological Chemistry, 259*, 1431–1435.

Nagai, T. (2004). Characterization of collagen from Japanese sea bass caudal fin as waste material. *European Food Research and Technology, 218*, 424–427.

Nagai, T., & Suzuki, N. (2002). Collagen of the skin of ocellate puffer fish (*Takifugu rubripes*). *Food Chemistry, 78*, 173–177.

Nagai, T., Suzuki, N., Tanoue, Y., Kai, N., & Nagashima, T. (2010). Characterization of acid-soluble collagen from skins of surf smelt (*Hypomesus pretiosus japonicus* Brevoort). *Food and Nutrition Sciences, 1*, 59–66.

Najafian, L., & Babji, A. S. (2012). A review of fish-derived antioxidant and antimicrobial peptides: Their production, assessment, and applications. *Peptides, 33*, 178–185.

Nakamura, T., Furunaka, H., Miyata, T., Tokunaga, F., Muta, T., & Iwanaga, S. (1988). Tachyplesin, a class of antimicrobial peptide from the hemocytes of the horseshoe crab (*Tachypleus tridentatus*). Isolation and chemical structure. *The Journal of Biological Chemistry, 263*(32), 16709–16713.

Nam, B. H., Moon, J. Y., Kim, Y. O., Kong, H. J., Kim, W. J., Lee, S. J., & Kim, K. K. (2010). Multiple beta-defensin isoforms identified in early developmental stages of the teleost *Paralichthys olivaceus*. *Fish & Shellfish Immunology, 28*, 267–274.

Nandini, C. D., Itoh, N., & Sugahara, K. (2005). Novel 70-kDa chondroitin sulfate/dermatan sulfate hybrid chains with a unique heterogenous sulfation pattern from shark skin, which exhibit neuritogenic activity and binding activities for growth factors and neurotrophic factors. *The Journal of Biological Chemistry, 280*, 4058–4069.

Nasri, R., Amor, I. B., Bougatef, A., Nedjar-Arroume, N., Ghulster, P., Gargouri, J., Chaabouni, M. K., & Nasri, M. (2012). Anticoagulant activities of Goby muscle protein hydrolysates. *Food Chemistry, 133*, 835–841.

Neyrinck, A. M., Mouson, A., & Delzenne, N. M. (2007). Dietary supplementation with laminarin, a fermentable marine beta (1-3) glucan, protects against hepatotoxicity induced by LPS in rat by modulating immune response in the hepatic tissue. *International Immunopharmacology, 7*, 1497–1506. https://doi.org/10.1016/j.intimp.2007.06.011.

Ngo, D. H., & Kim, S. K. (2014). Antioxidant effects of chitin, chitosan, and their derivatives. *Advances in Food and Nutrition Research, 73*, 15–31. https://doi.org/10.1016/B978-0-12-800268-1.00002-0.

Ngo, D. H., Ryu, B., & Kim, S. K. (2014). Active peptides from skate (*Okamejei kenojei*) skin gelatin diminish angiotensin-I converting enzyme activity and intracellular free radical-mediated oxidation. *Food Chemistry, 143*, 246–255.

Noga, E. J., Ullal, A. J., Corrales, J., & Fernandes, J. M. (2011). Application of antimicrobial polypeptide host defenses to aquaculture: Exploitation of downregulation and upregulation responses. *Comparative Biochemistry and Physiology. Part D, Genomics & Proteomics, 6*(1), 44–54.

Nozaki, M. (2013). Hypothalamic-pituitary-gonadal endocrine system in the hagfish. *Frontiers in Endocrinology (Lausanne), 4*, 200.

Ogawa, M., Portier, R. J., Moody, M. W., Bell, J., Schexnayder, M. A., & Losso, J. N. (2004). Biochemical properties of bone and scale collagens isolated from the subtropical fish black drum (*Pogonias cromis*) and sheepshead seabream (*Archosargus probatocephalus*). *Food Chemistry, 88*, 495–501.

Olaizola, M. (2008). The production and health benefits of astaxanthin. In C. Barrow & F. Shahidi (Eds.), *Marine nutraceuticals and functional foods* (pp. 321–343). New York: CRC/Taylor & Francis.

Ovchinnikova, T. V., Aleshina, G. M., Balandin, S. V., Krasnosdembskaya, A. D., Markelov, M. L., Frolova, E. I., Leonova, Y. F., Tagaev, A. A., Krasnodembsky, E. G., & Kokryakov, V. N. (2004). Purification and primary structure of two isoforms of arenicin, a novel antimicrobial peptide from marine polychaeta *Arenicola marina*. *FEBS Letters, 577*(1–2), 209–214.

Ovchinnikova, T. V., Balandin, S. V., Aleshina, G. M., Tagaev, A. A., Leonova, Y. F., Krasnodembsky, E. D., Menshenin, A. V., & Kokryakov, V. N. (2006). Aurelin, a novel antimicrobial peptide from jellyfish Aurelia aurita with structural features of defensins and channel-blocking toxins. *Biochemical and Biophysical Research Communications, 348*(2), 514–523.

Pacheco, R. G., Vicente, C. P., Zancan, P., & Mourao, P. A. S. (2000). Different antithrombotic mechanisms among glycosaminoglycans revealed with a new fucosylated chondroitin sulfate from an echinoderm. *Blood Coagulation & Fibrinolysis, 11*, 563–573.

Palthur, M. P., Sajala Palthur, S. S., & Chitta, S. K. (2010). Nutraceuticals: Concept and regulatory scenario. *International Journal of Pharmacy and Pharmaceutical Sciences, 2*, 14–20.

Pan, W., Liu, X., Ge, F., Han, J., & Zheng, T. (2004). Perinerin, a novel antimicrobial peptide purified from the clamworm *Perinereis aibuhitensis* grube and its partial characterization. *Journal of Biochemistry, 135*(3), 297–304.

Pan, C. Y., Chen, J. Y., Ni, I. H., Wu, J. L., & Kuo, C. M. (2008). Organization and promoter analysis of the grouper (*Epinephelus coioides*) epinecidin-1 gene. *Comparative Biochemistry and Physiology. Part B, Biochemistry & Molecular Biology, 150*, 358–367.

Pangestuti, R., Ryu, B., Himaya, S., & Kim, S. K. (2013). Optimization of hydrolysis conditions, isolation, and identification of neuroprotective peptides derived from seahorse *Hippocampus trimaculatus*. *Amino Acids, 45*, 369–381.

Panwar, V., & Dutta, T. (2019). Diatom biogenic silica as a felicitous platform for biochemical engineering: Expanding frontiers. *ACS Applied Bio Materials, 2*(6), 2295–2316.

Park, C. B., Kim, M. S., & Kim, S. C. (1996). A novel antimicrobial peptide from *Bufo bufo gargarizans*. *Biochemical and Biophysical Research Communications, 218*, 408–413.

Park, C. H., Valore, E. V., Waring, A. J., & Ganz, T. (2001). Hepcidin, a urinary antimicrobial peptide synthesized in the liver. *The Journal of Biological Chemistry, 276*, 7806–7810.

Park, S. B., Chun, K. R., Kim, J. K., Suk, K., Jung, Y. M., & Lee, W. H. (2010). The differential effect of high and low molecular weight fucoidans on the severity of collagen-induced arthritis in mice. *Phytotherapy Research, 24*(9), 1384–1391.

Park, S. C., Park, Y., & Hahm, K. S. (2011). The role of antimicrobial peptides in preventing multidrug-resistant bacterial infections and biofilm formation. *International Journal of Molecular Sciences, 12*(9), 5971–5992. https://doi.org/10.3390/ijms12095971.

Parys, S., Rosenbaum, A., Kehraus, S., Reher, G., Glombitza, K. W., & Konig, G. M. (2007). Evaluation of quantitative methods for the determination of polyphenols in algal extracts. *Journal of Natural Products, 70*(12), 1865–1870.

Patil, N. P., Le, V., Sligar, A. D., Mei, L., Chavarria, D., Yang, E. Y., & Baker, A. B. (2018). Algal polysaccharides as therapeutic agents for atherosclerosis. *Frontiers inCardiovascular Medicine, 5*, 153. https://doi.org/10.3389/fcvm.2018.00153.

Pavao, M. S. G., Aiello, K. R. M., Werneck, C. C., Silva, L. C., Valente, A. P., Mulloy, B., Colwell, M. S., Tollefsen, D. M., & Mourao, P. A. S. (1998). Highly sulfated dermatan sulfate from ascidians. Structure versus anticoagulant activity of these glycosaminoglycans. *The Journal of Biological Chemistry, 273*, 27848–27857.

Pengzhan, Y., Ning, L., Xiguang, L., Gefei, Z., Quanbin, Z., & Pengcheng, L. (2003). Antihyperlipidemic effects of different molecular weight sulfated polysaccharides from *Ulva pertusa* (Chlorophyta). *Pharmacological Research, 48*, 543–549. https://doi.org/10.1016/S1043-6618(03)00215-9.

Pereira, L. (2011). A review of the nutrient composition of selected edible seaweeds. In V. H. Pomin (Ed.), *Seaweed, ecology, nutrient composition and medicinal uses* (pp. 15–47). New York: Nova Science Publishers.

Pfeiler, E., Toyoda, H., Williams, M. D., & Nieman, R. A. (2002). Identification, structural analysis and function of hyaluronan in developing fish larvae (leptocephali). *Comparative Biochemistry and Physiology Part B: Biochemistry and Molecular Biology, 132*, 443–451.

Piez, K. A. (1984). Molecular and aggregate structure of the collagens in extracellular matrix. *Research Journal of Pharmaceutical, Biological and Chemical Sciences, 42*, 36–49.

Plaza, A., Gustchina, E., Baker, H. L., Kelly, M., & Bewley, C. A. (2007). Mirabamides A-D. Depsipeptides from the sponge *Siliquariaspongia mirabilis* that inhibit HIV-1 fusion. *Journal of Natural Products, 70*(11), 1753–1760.

Pomin, V. H., & Mourão, P. A. (2008). Structure, biology, evolution, and medical importance of sulfated fucans and galactans. *Glycobiology, 18*, 1016–1027. https://doi.org/10.1093/glycob/cwn085.

Pozzolini, M., Scarfì, S., Gallus, L., Castellano, M., Vicini, S., Cortese, K., Gagliani, M. C., Bertolino, M., Costa, G., & Giovine, M. (2018). Production, characterization and biocompatibility evaluation of collagen membranes derived from marine sponge *Chondrosia reniformis* Nardo, 1847. *Marine Drugs, 16*, 111. https://doi.org/10.3390/md16040111.

Prajapati, V. D., Mahereriya, P. M., Jani, G. K., & Soalnki, H. K. (2014). Carrageenan: A natural seaweed polysaccharide and its applications. *Carbohydrate Polymers, 105*, 97–112.

Qi, H., & Sun, Y. (2015). Antioxidant activity of high sulfate content derivative of ulvan in hyperlipidemic rats. *International Journal of Biological Macromolecules, 76*, 326–329. https://doi.org/10.1016/j.ijbiomac.2015.03.006.

Qiao, L., Feng, Q., & Li, Z. (2007). Special vaterite found in freshwater lackluster pearls. *Crystal Growth and Design, 7*, 275–279.

Qin, Y. (2008). Alginate fibers, An overview of the production processes and applications in wound management. *Polymer International, 57*, 171–108.

Raghavan, S., & Kristinsson, H. G. (2009). ACE-inhibitory activity of tilapia protein hydrolysates. *Food Chemistry, 117*, 582–588.

Rahman, A. M. (2019). Collagen of extracellular matrix from marine invertebrates and its medical applications. *Marine Drugs, 17*, 118. https://doi.org/10.3390/md17020118.

Rajanbabu, V., Chen, J.-Y., & Wu, J.-L. (2015). Antimicrobial peptides from marine organisms. In S. K. Kim (Ed.), *Springer handbook of marine biotechnology*. Berlin/Heidelberg: Springer.

Raman, M., & Gopakumar, K. (2018). Fish collagen and its applications in food and pharmaceutical industry: A review. *EC Nutrition, 13*(12), 752–767.

Rashid, M. A., Gustafson, K. R., Cartner, L. K., Shigematsu, N., Pannell, L. K., & Boyd, M. R. (2001). Microspinosamide, a new HIV-inhibitory cyclic depsipeptide from the marine sponge *Sidonops microspinosa*. *Journal of Natural Products, 64*, 117–121.

Regenstein, J., & Zhou, P. (2007). Collagen and gelatin from marine by-product. In *Maximising the value of marine by-products* (pp. 279–303). Boca Raton: CRC Press. https://doi.org/10.1533/9781845692087.2.279.

Relf, J., Chisholm, J., Kemp, G., & Smith, V. (1999). Purification and characterization of a cysteine-rich 11.5-kDa antibacterial protein from the granular haemocytes of the shore crab, *Carcinus maenas*. *European Journal of Biochemistry, 264*(2), 350–357.

Remminghorst, U., & Rehm, B. H. A. (2006). Bacterial alginates: From biosynthesis to applications. *Biotechnology Letters, 28*, 1701–1712. https://doi.org/10.1007/s10529-006-9156-x.

Rhein-Knudsen, N., Ale, M. T., & Meyer, A. S. (2015). Seaweed hydrocolloid production: An update on enzyme assisted extraction and modification technologies. *Marine Drugs, 13*, 3340–3359.

Rigby, B. J. (1968). Aminoacid composition and thermal stability of the skin collagen of the Antartic ice fish. *Nature, 219*, 166–167.

Rizk, M. Z., El-sherbiny, M., Borai, I. H., Ezz, M. K., Aly, H. F., Matloub, A. A., Farrag, A. E. R., & Fouad, G. I. (2016). Sulphated polysaccharides (SPS) from the green alga *Ulva fasciata* extract modulates liver and kidney function in high fat diet-induced hypercholesterolemic rats. *International Journal of Pharmacy and Pharmaceutical Sciences, 8*(6), 43–55.

Rodrigues, P. N., Vazquez-Dorado, S., Neves, J. V., & Wilson, J. M. (2006). Dual function of fish hepcidin: Response to experimental iron overload and bacterial infection in sea bass (*Dicentrarchus labrax*). *Developmental and Comparative Immunology, 30*, 1156–1167.

Rolland, J. L., Abdelouahab, M., Dupont, J., Lefevre, F., Bachère, E., & Romestand, B. (2010). Stylicins, a new family of antimicrobial peptides from the Pacific blue shrimp *Litopenaeus stylirostris*. *Molecular Immunology, 47*(6), 1269–1277.

Rujitanapanich, S., Kumpapan, P., & Wanjanoi, P. (2014). Synthesis of hydroxyapatite from oyster shell via precipitation. *Energy Procedia, 56*, 112–117.

Ryu, B., & Kim, S. K. (2013). Potential beneficial effects of marine peptide on human neuron health. *Current Protein & Peptide Science, 14*, 173–176.

Ryu, B., Qian, Z.-J., Kim, M.-M., Nam, K. W., & Kim, S.-K. (2009). Anti-photoaging activity and inhibition of matrix metalloproteinase (MMP) by marine red alga, *Corallina pilulifera* methanol extract. *Radiation Physics and Chemistry, 78*, 98–105.

Sahithi, B., Ansari, S. K., Hameeda, S. K., Sahithya, G., Durga Prasad, M., & Lakshmi, Y. (2013). A review on collagen based drug delivery systems. *Indian Journal of Research in Pharmacy and Biotechnology, 1*, 461–468.

Saito, T., Kawabata, S., Shigenaga, T., Takayenoki, Y., Cho, J., Nakajima, H., Hirata, M., & Iwanaga, S. (1995). A novel big defensin identified in horseshoe crab hemocytes: Isolation, amino acid sequence, and antibacterial activity. *Journal of Biochemistry, 117*(5), 1131–1137.

Sakai, S., Kim, W. S., Lee, I. S., Kim, Y. S., Nakamura, A., Toida, T., & Imanari, T. (2003). Purification and characterization of dermatan sulfate from the skin of the eel, *Anguilla japonica*. *Carbohydrate Research, 338*, 263–269.

Salampessy, J., Philips, M., Seneweera, S., & Kailasapathy, K. (2010). Release of antimicrobial peptides through bromelain hydrolysis of leatherjacket (*Meuchenia* sp.) insoluble proteins. *Food Chemistry, 120*, 556–560.

Samaranayaka, A. G. P., Kitts, D. D., & Li-Chan, E. C. Y. (2010). Antioxidative and angiotensin I-converting enzyme inhibitory potential of a Pacific hake (*Merluccius productus*) fish protein hydrolysate subjected to stimulated gastrointestinal digestion and caco-2 cell permeation. *Journal of Agricultural and Food Chemistry, 58*, 1535–1542.

Sampath Kumar, N. S., Nazeer, R. A., & Jaiganesh, R. (2011). Purification and identification of antioxidant peptides from the skin protein hydrolysate of two marine fishes, horse mackerel (*Megalaspis cordyla*) and croaker (*Otolithes ruber*). *Amino Acids, 42*, 1641–1649.

Sankar, S., Sekar, S., Mohan, R., Rani, S., Sundaraseelan, J., & Sastry, T. P. (2008). Preparation and partial characterization of collagen sheet from fish (*Lates calcarifer*) scales. *International Journal of Biological Macromolecules, 42*, 6–9.

Santo, V. E., Frias, A. M., Carida, M., Cancedda, R., Gomes, M. E., Mano, J. F., & Reis, R. L. (2009). Carrageenan-based hydrogels for the controlled delivery of PDGF-BB in bone tissue engineering applications. *Biomacromolecules, 10*, 1392–1401.

Santos, E. A., Rocha, L. R. M., Pereira, N. M. L., Andrade, G. P. V., Nader, H. B., & Dietrich, C. P. (2002). Mast cells are present in epithelial layers of different tissues of the mollusk

Anomalocardia brasiliana. *In situ* characterization of heparin and a correlation of heparin and histamine concentration. *The Histochemical Journal, 34*, 553–558.
Schnapp, D., Kemp, G., & Smith, V. (1996). Purification and characterization of a proline-rich antibacterial peptide, with sequence similarity to bactenecin-7, from the haemocytes of the shore crab, *Carcinus maenas*. *European Journal of Biochemistry, 240*, 532–539.
Schroder, H. C., Wang, X., Tremel, W., Ushijima, H., & Muller, W. E. (2008). Biofabrication of biosilica-glass by living organisms. *Natural Product Reports, 25*, 455–474.
See, S. F., Hong, P. K., Ng, K. L., Wan Aida, W. M., & Babji, A. S. (2010). Physicochemical properties of gelatins extracted from skins of different freshwater fish species. *International Food Research Journal, 17*, 806–816.
Senaratne, L. S., Park, P. J., & Kim, S. K. (2006). Isolation and characterization of collagen from brown backed toadfish (*Lagocephalus gloveri*) skin. *Bioresource Technology, 97*, 191–197.
Sezer, A. D., Hatipoğlu, F., Cevher, E., Oğurtan, Z., Baş, A. L., & Akbuğa, J. (2007). Chitosan films containing fucoidan as a wound dressing for dermal burn healing, preparation and *in vitro/in vivo* evaluation. *AAPS PharmSciTech, 8*, E94–E101.
Shanmugam, M., & Mody, K. H. (2000). Heparinoid-active sulphated polysaccharides from marine algae as potential blood anticoagulant agents. *Current Science India, 79*, 1672–1683.
Shariffuddin, J. H., Jones, M. I., & Patterson, D. A. (2013). Greener photocatalysts: Hydroxyapatite derived from waste mussel shells for the photocatalytic degradation of a model azo dye wastewater. *Chemical Engineering Research and Design*. https://doi.org/10.1016/j.cherd.2013.04.018.
Shetty, A. K., Kobayashi, T., Mizumoto, S., Narumi, M., Kudo, Y., Yamada, S., & Sugahara, K. (2009). Isolation and characterization of a novel chondroitin sulfate from squid liver integument rich in N- acetylgalactosamine(4,6-disulfate) and glucuronate(3-sulfate) residues. *Carbohydrate Research, 344*, 1526–1532.
Shike, H., Lauth, X., Westerman, M. E., Ostland, V. E., Carlberg, J. M., van Olst, J. C., Shimizu, C., Bulet, P., & Burns, J. C. (2002). Bass hepcidin is a novel antimicrobial peptide induced by bacterial challenge. *European Journal of Biochemistry, 269*, 2232–2237.
Soliev, A. V., Hosokawa, K., & Enomoto, K. (2011). Bioactive pigments from marine bacteria: Applications and physiological roles. *Evidence-Based Complementary and Alternative Medicine, 2011*, ID 670349, 17p. https://doi.org/10.1155/2011/670349.
Sousa, A. P. A., Torres, M. R., Pessoa, C., Moraes, M. O., Rocha-Filho, F. D., Alves, A. P. N. N., & Costa-Lotufo, L. V. (2007). *In vivo* growth-inhibition of sarcoma 180 tumor by alginates from brown seaweed *Sargassum vulgare*. *Carbohydrate Polymers, 69*, 7–13.
Souza, A. R. C., Kozlowski, E. O., Cerqueira, V. R., Castelo-Branco, M. T. L., Costa, M. L., & Pavao, M. S. G. (2007). Chondroitin sulfate and keratan sulfate are the major glycosaminoglycans present in the adult zebrafish *Danio rerio* (Chordata-Cyprinidae). *Glycoconjugate Journal, 24*, 521–530.
Spicer, A. P., Tien, J. L., Joo, A., & Bowling, R. A., Jr. (2002). Investigation of hyaluronan function in the mouse through targeted mutagenesis. *Glycoconjugate Journal, 19*, 341–345.
Stensvag, K., Haug, T., Sperstad, S. V., Rekdal, O., Indrevoll, B., & Styrvold, O. B. (2008). Arasin 1, a proline-arginine-rich antimicrobial peptide isolated from the spider crab, Hyas araneus. *Developmental and Comparative Immunology, 32*(3), 275–285.
Suarez-Jimenez, G. M., Burgos-Hernandez, A., & Ezquerra-Brauer, J. M. (2012). Bioactive peptides and depsipeptides with anticancer potential: Sources from marine animals. *Marine Drugs, 10*, 963–986.
Sugahara, K., Nadanaka, S., Takeda, K., & Kojima, T. (1996). Structural analysis of unsaturated hexasaccharides isolated from shark cartilage chondroitin sulfate D that are substrates for the exolytic action of chondroitin ABC lyase. *European Journal of Biochemistry, 239*, 871–880.
Sukhan, Z. P., Kitano, H., Selvaraj, S., Yoneda, M., Yamaguchi, A., & Matsuyama, M. (2013). Identification and distribution of three gonadotropin-releasing hormone (GNRH) isoforms in the brain of a clupeiform fish, *Engraulis japonicus*. *Zoological Science, 30*, 1081–1091.
Summers, A. P., Koob-Emunds, M. M., Kajiura, S. M., & Koob, T. J. (2003). A novel fibrocartilaginous tendon from an elasmobranch fish (*Rhinoptera bonasus*). *Cell and Tissue Research, 312*, 221–227.

Sun, J. C., & Tan, H. P. (2013). Alginate-based biomaterials for regenerative medicine applications. *Materials, 6*, 1285–1309.

Sun, D., Wu, S., Jing, C., Zhang, N., Liang, D., & Xu, A. (2012). Identification, synthesis and characterization of a novel antimicrobial peptide HKPLP derived from *Hippocampus kuda* Bleeker. *Journal of Antibiotics (Tokyo), 65*, 117–121.

Sunil, B. R., & Jagannatham, M. (2016). Producing hydroxyapatite from fish bones by heat treatment. *Materials Letters, 185*, 411–414.

Szekalska, M., Puciłowska, A., Szymanska, E., Ciosek, P., & Winnicka, K. (2016). Alginate: Current use and future perspectives in pharmaceutical and biomedical applications. *International Journal of Polymer Science, 2016*, 7697031/1–7697031/17. https://doi.org/10.1155/2016/7697031.

Tang, Y. Q., Yuan, J., Osapay, G., Osapay, K., Tran, D., Miller, C. J., Ouellette, A. J., & Selsted, M. E. (1999). A cyclic antimicrobial peptide produced in primate leukocytes by the ligation of two truncated alpha-defensins. *Science, 286*, 498–502.

Targett, N. M., & Arnold, T. M. (1998). Predicting the effects of brown algal phlorotannins on marine herbivores in tropical and temperate oceans. *Journal of Phycology, 34*, 195–205.

Tasiemski, A., Schikorski, D., Le Marrec-Croq, F., Pontoire-Van Camp, C., Boidin-Wichlacz, C., & Sautière, P. E. (2007). Hedistin: A novel antimicrobial peptide containing bromotryptophan constitutively expressed in the NK cells-like of the marine annelid, *Nereis diversicolor. Developmental and Comparative Immunology, 31*(8), 749–762.

Terova, G., Cattaneo, A. G., Preziosa, E., Bernardini, G., & Saroglia, M. (2011). Impact of acute stress on antimicrobial polypeptides mRNA copy number in several tissues of marine sea bass (Dicentrarchus labrax). *BMC Immunology, 12*, 69. https://doi.org/10.1186/1471-2172-12-69.

Theodore, A. E., & Kristinsson, H. G. (2007). Angiotensin converting enzyme inhibition of fish protein hydrolysates prepared from alkaline-aided channel catfish protein isolate. *Journal of the Science of Food and Agriculture, 87*, 2353–2357.

Tincu, J. A., Menzel, L. P., Azimov, R., Sands, J., Hong, T., Waring, A. J., Taylor, S. W., & Lehrer, R. I. (2003). Plicatamide, an antimicrobial octapeptide from *Styela plicata hemocytes. The Journal of Biological Chemistry, 278*(15), 13546–13553.

Toskas, G., Heinemann, S., Heinemann, C., Cherif, C., Hund, R. D., Roussis, V., & Hanke, T. (2012). Ulvan and ulvan/chitosan polyelectrolyte nanofibrous membranes as a potential substrate material for the cultivation of osteoblasts. *Carbohydrate Polymers, 89*(3), 997–1002.

Trapani, M. R., Parisi, M. G., Toubiana, M., Coquet, L., Jouenne, T., Roch, P., & Cammarata, M. (2014). First evidence of antimicrobial activity of neurotoxin-2 from *Anemonia sulcata* (Cnidaria). *Invertebrate Survival Journal, 11*(1), 182–191.

Tsiapali, E., Whaley, S., Kalbfleisch, J., Ensley, H. E., Browder, I. W., & Williams, D. L. (2001). Glucans exhibit weak antioxidant activity, but stimulate macrophage free radical activity. *Free Radical Biology & Medicine, 30*(4), 393–402. https://doi.org/10.1016/S0891-5849(00)00485-8.

Tsukui, T., Konno, K., Hosokawa, M., Maeda, H., Sashima, T., & Miyashitta, K. (2007). Fucoxanthin and fucoxanthinol enhance the amount of docosahexaenoic acid in the liver of KK-A^y mice. *Journal of Agricultural and Food Chemistry, 55*(13), 5025–5029.

Turnbull, J., Powell, A., & Guimond, S. (2001). Heperan sulfate: Decoding a dynamic multifunctional cell regulator. *Trends in Cell Biology, 11*, 75–82.

Uchisawa, H., Okuzaki, B., Ichita, J., & Matsue, H. (2001). Binding between calcium ions and chondroitin sulfate chains of salmon nasal cartilage glycosaminoglycan. *International Congress Series, 1223*, 205–220.

Ulrich, P. N., & Boon, J. K. (2001). The histological localization of heparin in the northern quahog clam, *Mercenaria mercenaria. J. Invert. Pathol., 78*, 155–159.

Usov, A. I. (2011). Polysaccharides of the red algae. *Advances in Carbohydrate Chemistry and Biochemistry, 65*, 115–217.

Uthappa, U. T., Brahmkhatri, V., Sriram, G., Jung, H. Y., Yu, J., Kurkuri, N., Aminabhavi, T. M., Altalhi, T., Neelgund, G. M., & Kurkuri, M. D. (2018). Nature engineered diatom biosilica as drug delivery systems. *Journal of Controlled Release, 281*, 70–83. https://doi.org/10.1016/j.jconrel.2018.05.013.

Uzzell, T., Stolzenberg, E. D., Shinnar, A. E., & Zasloff, M. (2003). Hagfish intestinal antimicrobial peptides are ancient cathelicidins. *Peptides, 24*(11), 1655–1667.

Venkatesan, J., Qian, Z. J., Ryu, M., Thomas, N. V., & Kim, S. K. (2011). A comparative study of thermal calcination and an alkaline hydrolysis method in the isolation of hydroxyapatite from *Thunnus obesus* bone. *Biomedical Materials, 6*(035003), 12.

Vo, T. S., & Kim, S. K. (2010). Potential anti-HIV agents from marine resources: An overview. *Marine Drugs, 8*(12), 2871–2892. https://doi.org/10.3390/md8122871.

Vynios, D. H., & Tsiganos, C. P. (1990). Squid proteoglycans: Isolation and characterization of three populations from cranial cartilage. *Biochimica et Biophysica Acta, 1033*, 139–147.

Walsh, P. J., Walker, G. M., Maggs, C. A., & Buchanan, F. J. (2010). Thermal preparation of highly porous calcium phosphate bone filler derived from marine algae. *Journal of Materials Science: Materials in Medicine, 21*, 2281–2286.

Wang, L., An, X., Xin, Z., Zaho, L., & Hu, Q. (2007). Isolation and characterization of collagen from the skin of deep-sea redfish (*Sebastes mentella*). *Journal of Food Science, 72*, E450–E455.

Wang, T., Jonsdottir, R., & Olafsdottir, G. (2009). Total phenolic compounds, radical scavenging and metal chelation of extracts from Icelandic seaweeds. *Food Chemistry, 116*, 240–248.

Wang, Y. D., Kung, C. W., & Chen, J. Y. (2010a). Antiviral activity by fish antimicrobial peptides of epinecidin-1 and hepcidin 1–5 against nervous necrosis virus in medaka. *Peptides, 31*, 1026–1033.

Wang, Y. K., He, H. L., Wang, G. F., Wu, H., Zhou, B. C., Chen, X. L., & Zhang, Y. Z. (2010b). Oyster (*Crassostrea gigas*) hydrolysates produced on a plant scale have antitumor activity and immunostimulating effects in BALB/c mice. *Marine Drugs, 8*, 255–268.

Wickramaarachchi, W. D., De Zoysa, M., Whang, I., Wan, Q., & Lee, J. (2013). Kazal-type proteinase inhibitor from disk abalone (*Haliotis discus discus*): Molecular characterization and transcriptional response upon immune stimulation. *Fish & Shellfish Immunology, 35*, 1039–1043.

Wijesekara, I., & Karunarathna, W. K. D. S. (2017). Chapter 18 – Usage of seaweed polysaccharides as nutraceuticals. In *Seaweed polysaccharides as nutraceuticals – Isolation, biological and biomedical applications* (pp. 341–348). Saint Louis: Elsevier. https://doi.org/10.1016/B978-0-12-809816-5.00018-9.

Wu, C. (1990). Properties, manufacture, and application of seaweed polysaccharides-agar, carrageenan, and algin. In C. Wu (Ed.), *Training manual on Gracilaria culture and seaweed processing in China* (pp. 2–46). Zhanjiang: FAO Fisheries and Aquaculture Department.

Xia, S., Wang, K., Wan, L., Li, A., Hu, Q., & Zhang, C. (2013). Production, characterization, and antioxidant activity of fucoxanthin from the marine diatom *Odontella aurita*. *Marine Drugs, 11*, 2667–2681. https://doi.org/10.3390/md11072667.

Xu, Q., Cheng, C. H., Hu, P., Ye, H., Chen, Z., Cao, L., Chen, L., Shen, Y., & Chen, L. (2008). Adaptive evolution of hepcidin genes in antarctic notothenioid fishes. *Molecular Biology and Evolution, 25*, 1099–1112.

Yaich, H., Amira, A. B., Abbes, F., Bouaziz, M., Besbes, S., Richel, A., Blecker, C., Attia, H., & Garna, H. (2017). Effect of extraction procedures on structural, thermal and antioxidant properties of ulvan from *Ulva lactuca* collected in Monastir coast. *International Journal of Biological Macromolecules, 105*, 1430–1439. https://doi.org/10.1016/j.ijbiomac.2017.07.141.

Yan, X. J., Li, X. C., Zhou, C. X., & Fan, X. (1996). Preservation of fish oil rancidity by phlorotannins from *Sargassum kjellmanianum*. *Journal of Applied Phycology, 8*, 201–203.

Yan, X., Chuda, Y., Suzuki, M., & Nagata, T. (1999). Fucoxanthin as the major antioxidant in *Hijikia fusiformis*, a common edible sea weed. *Bioscience. Biotechnology and Biochemistry, 63*(3), 605–607.

Yan, M., Li, B., Zhao, X., Ren, G., Zhuang, Y., Hou, H., Zhang, X., Chen, L., & Fan, Y. (2008). Characterization of acid- soluble collagen from the skin of walleye Pollock (*Theragra chalcogramma*). *Food Chemistry, 107*, 1581–1586.

Yang, C., Chung, D., Shin, I. S., Lee, H. Y., Kim, J. C., Lee, Y. J., & You, S. G. (2008). Effects of molecular weight and hydrolysis conditions on anticancer activity of fucoidans from sporophyll of *Undaria pinnatifida*. *International Journal of Biological Macromolecules, 43*, 433–437.

Yang, R., Zhang, Z., Pei, X., Han, X., Wang, J., Wang, L., Long, Z., Shen, X., & Li, Y. (2009). Immunomodulatory effects of marine oligopeptide preparation from chum salmon (*Oncorhynchus keta*) in mice. *Food Chemistry, 113*, 464–470.

Ye, H., Wang, K., Zhou, C., Liu, J., & Zeng, X. (2008). Purification, antitumor and antioxidant activities *in vitro* of polysaccharides from the brown seaweed *Sargassum pallidum*. *Food Chemistry, 111*, 428–432.

Yedery, R. D., & Reddy, K. V. (2009). Purification and characterization of antibacterial proteins from granular hemocytes of Indian mud crab, *Scylla serrate*. *Acta Biochimica Polonica, 56*(1), 71–82.

Yoon, H. Y., Son, S., Lee, S. J., You, D. G., Yhee, J. Y., Park, J. H., Swierczewska, M., Lee, S., Kwon, I. C., Kim, S. H., Kim, K., & Pomper, M. G. (2014). Glycol chitosan nanoparticles as specialized cancer therapeutic vehicles: Sequential delivery of doxorubicin and Bcl-2 siRNA. *Scientific Reports, 4*, 6878. https://doi.org/10.1038/srep06878.

Younes, I., & Rinaudo, M. (2015). Chitin and chitosan preparation from marine sources. Structure, properties and applications. *Marine drugs, 13*, 1133–1174. https://doi.org/10.3390/md13031133.

Yu, S., Geng, J., Zhou, P., Wang, J., Chen, X., & Hu, J. (2008). New hydroxyapatite monolithic column for DNA extraction and its application in the purification of *Bacillus subtilis* crude lysate. *Journal of Chromatography. A, 1183*(1–2), 29–37. https://doi.org/10.1016/j.chroma.2007.11.120. Epub 2007 Dec 26.

Yuan, H., Song, J., Li, X., Li, N., & Dai, J. (2006). Immunomodulation and antitumor activity of κ-carrageenan oligosaccharides. *Cancer Letters, 243*, 228–234.

Zahran, E., & Noga, E. J. (2010). Evidence for synergism of the antimicrobial peptide piscidin 2 with antiparasitic and antioomycete drugs. *Journal of Fish Diseases, 33*, 995–1003.

Zelechowska, E., Sadowska, M., & Turk, M. (2010). Isolation and some properties of collagen from the backbone of Baltic cod (*Gadus morhua*). *Food Hydrocolloids, 24*, 325–329.

Zhang, X., & Vecchio, K. S. (2013). Conversion of natural marine skeletons as scaffolds for bone tissue engineering. *Frontiers of Materials Science, 7*(2), 103–117.

Zhang, Y., Liu, W. T., Li, G. Y., Shi, B., Miao, Y. Q., & Wu, X. H. (2007). Isolation and partial characterization of pepsin- soluble collagen from the skin of grass carp (*Ctenopharyngodon idella*). *Food Chemistry, 103*, 906–912.

Zhao, J., Song, L., Li, C., Ni, D., Wu, L., Zhu, L., Wang, H., & Xu, W. (2007). Molecular cloning, expression of a big defensin gene from bay scallop *Argopecten irradians* and the antimicrobial activity of its recombinant protein. *Molecular Immunology, 44*(4), 360–368.

Zhou, G. T., & Zheng, Y. F. (1998). Synthesis of aragonite-type calcium carbonate by overgrowth technique at atmospheric pressure. *Journal of Materials Science Letters, 17*, 905–908.

Zhu, S., & Gao, B. (2013). Evolutionary origin of beta-defensins. *Developmental and Comparative Immunology, 39*(1–2), 79–84.

Zhu, C. F., Li, G. Z., Peng, H. B., Zhang, F., Chen, Y., & Li, Y. (2010a). Treatment with marine collagen peptides modulates glucose and lipid metabolism in Chinese patients with type 2 diabetes mellitus. *Applied Physiology, Nutrition, and Metabolism, 35*, 797–804.

Zhu, C. F., Peng, H. B., Liu, G. Q., Zhang, F., & Li, Y. (2010b). Beneficial effects of oligopeptides from marine salmon skin in a rat model of type 2 diabetes. *Nutrition, 26*, 1014–1020.

Zou, J., Mercier, C., Koussounadis, A., & Secombes, C. (2007). Discovery of multiple beta-defensin like homologues in teleost fish. *Molecular Immunology, 44*(4), 638–647.

Zou, Y., Qian, Z. J., Li, Y., Kim, M. M., Lee, S. H., & Kim, S. K. (2008). Antioxidant effect of phlorotannins isolated from *Ishige okamurae* in free radical mediated oxidative systems. *Journal of Agricultural and Food Chemistry, 56*(16), 7001–7009.

Zubia, M., Robledo, D., & Freile-Pelegrin, Y. (2007). Antioxidant activities in tropical marine microalgae from the Yucatan Peninsula, Mexico. *Journal of Applied Phycology, 19*, 449–458. https://doi.org/10.1007/s10811-006-9152-5.

MIX
Papier aus verantwortungsvollen Quellen
Paper from responsible sources
FSC® C105338
FSC
www.fsc.org

If you have any concerns about our products,
you can contact us on
ProductSafety@springernature.com

In case Publisher is established outside the EU,
the EU authorized representative is:
Springer Nature Customer Service Center GmbH
Europaplatz 3, 69115 Heidelberg, Germany

Printed by Libri Plureos GmbH
in Hamburg, Germany